AMNIOTE PALEOBIOLOGY

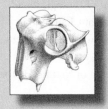
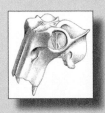
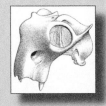
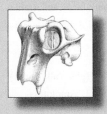

AMNIOTE PALEOBIOLOGY

Perspectives

on the Evolution of

Mammals, Birds, and

Reptiles

A volume honoring

James Allen Hopson

EDITED BY

Matthew T. Carrano,

Timothy J. Gaudin,

Richard W. Blob, *and*

John R. Wible

The University of Chicago Press

Chicago and London

Matthew Carrano is curator of dinosaurs in the Department of Paleobiology at the Smithsonian Institution.

Timothy Gaudin is associate professor in the Department of Biological and Environmental Sciences at the University of Tennessee at Chattanooga.

Richard Blob is assistant professor in the Department of Biological Sciences at Clemson University.

John Wible is curator of mammals at the Carnegie Museum of Natural History.

The University of Chicago Press, Chicago 60637
The University of Chicago Press, Ltd., London
© 2006 by The University of Chicago
All rights reserved. Published 2006
Printed and bound by CPI Group (UK) Ltd,
Croydon, CR0 4YY

15 14 13 12 11 10 09 08 07 06   1 2 3 4 5

ISBN: 978-0-226-09477-9 (cloth)
ISBN: 978-0-226-09478-6 (paper)

Library of Congress Cataloging-in-Publication Data

Amniote paleobiology : perspectives on the evolution of mammals, birds, and reptiles :
    a volume honoring James Allen Hopson / edited by Matthew T. Carrano . . . [et al.].
      p. cm.
    Includes bibliographical references and index.
    ISBN 0-226-09477-4 (cloth : alk. paper)—ISBN 0-226-09478-2 (pbk. : alk. paper)
    1. Hopson, James Allen. 2. Amniotes, Fossil—Evolution. 3. Amniotes—
Evolution.   I. Carrano, Matthew T.

    QE847A462 2006
    566—dc22   2005026549

CONTENTS

1 Introduction

Timothy J. Gaudin
Matthew T. Carrano
Richard W. Blob, *and*
John R. Wible

Jim Hopson and Amniote Paleobiology

Living amniotes—including all extant mammals, birds, crocodilians, lizards, snakes, and turtles—together comprise an extraordinarily varied array of more than 21,000 species (Pough et al., 1999). These familiar animals are found in every major habitat on earth and are characterized by a dazzling range of morphological, ecological, and behavioral adaptations. The fossil record of the Amniota extends back in time over 300 million years (Carroll, 1988; Benton, 1993; Gauthier, 1994) and encompasses additional thousands of extinct genera (Carroll, 1988). In many respects, this impressive fossil assemblage is even more diverse than are the living amniote taxa. Given that more than ninety-nine percent of all species that have ever existed are extinct (Novacek & Wheeler, 1992), it is not surprising that extant amniote species provide only a glimpse of the total diversity of form and function that have existed historically. As a result, paleobiological analyses of the fossil record are crucial to understanding the origins of modern biological diversity and the processes that shape the patterns we observe today.

Two critical questions are at the core of many paleobiological studies: (1) how are species related to one another and (2) what do species relationships reveal about changes in anatomy and function through time? These two perspectives—phylogenetic and functional—lie at the heart of this volume, which highlights the modern contributions of amniote paleobiology to evolutionary studies. At the same time, this volume serves as tribute to Dr. James A. Hopson, in honor of his retirement after more than thirty-five years of teaching and research at the University of Chicago. Throughout his career, Jim has been a strong advocate for both phylogenetic and functional approaches to studies of vertebrate evolution. Most importantly, Jim's work stands as an example of the successful integration of these two perspectives in amniote paleobiology.

The contributions in this volume focus primarily on phylogenetic and functional themes, but they also reflect the broad range of approaches used by modern paleobiologists. Therefore, the chapters are divided into four sections; the first two (entitled "New Fossils and Phylogenies" and "Large-Scale Evolutionary Patterns") are primarily phylogenetic in emphasis, and the second two (entitled "Functional Morphology" and "Ontogeny and Evolution") focus on other biological issues. However, the reader may be struck by the fact that many of these papers do not fit neatly into the categories that we, the editors, have erected. Indeed, to a certain degree such distinctions are difficult to make in any modern paleobiological compendium, given the importance of bringing varied perspectives to the task of elucidating any group's evolutionary history. This difficulty in discretely categorizing the various contributions also occurs in part by design, as the contributors were encouraged to integrate phylogenetic and functional perspectives into their work. Thus, it also testifies to the degree of integration increasingly seen in all aspects of modern paleobiology.

New Fossils and Phylogenies

Our knowledge of the amniote fossil record has increased rapidly over the past several decades. For example, Sereno (1997) noted that the diversity of dinosaur taxa alone has doubled in the past twenty years. The discovery, excavation, preparation, and description of new fossils is the most fundamental component of the science of paleontology. Because these actions represent the collection and integration of primary data, they underlie all other aspects of paleontological study. As long as new fossils remain to be discovered, paleontology will maintain this emphasis on descriptive work.

Early descriptions of fossil discoveries in the nineteenth century (and before) are typically used as benchmarks for the slow birth of paleontology as a field of study. Yet even the earliest paleontologists recognized that morphology does not exist in isolation but serves to illuminate the varied and complex connections between organisms. For example, Baron Georges Cuvier's comments on the first known European dinosaurs included prescient insights into their paleoecological relationships (Taquet, 1994).

With the advent of evolutionary theory and its profound impact on paleontology came an appreciation for the roles fossils could play in understanding the interrelationships of organisms. Description increasingly fed more directly into analyses of phylogenetic relationships for the simple reason that morphology formed the primary basis for such hypotheses until the advent of molecular studies in very recent decades.

New fossils were seen as important markers of transitions between previously known forms and ultimately between distantly related modern groups. Indeed, few new fossils are described even today without accompanying statements or analyses of their phylogenetic placement, and these descriptions typically are written with phylogenetic informativeness in mind. Thus, description and phylogeny have become intertwined.

As is true of many workers in vertebrate paleontology, Jim Hopson's first scientific papers were descriptive (Hopson, 1964a, 1964b), and he continued to offer such studies throughout his later career. It is fitting, therefore, that several of the contributors to this volume describe new fossil taxa or significant new material from previously described taxa. In addition to their particular scientific content, they provide an excellent illustration of the modern state of the oldest of paleontological methods.

Lombard and Bolt (chapter 2) describe the mandibular morphology in *Whatcheeria deltae,* one of the oldest well-preserved tetrapods in North America. The authors then analyze the systematic implications of this morphology for basal tetrapods. As the authors explain, *Whatcheeria* was thought to represent a critical link in our understanding of basal tetrapod relationships, given its proposed phylogenetic position at the base of the anthracosaur radiation (Lombard & Bolt, 1995), which ultimately gave rise to all living amniotes. However, more recent analyses have thrown doubt on its anthracosaur affinities (see references in chapter 2). Nevertheless, Lombard and Bolt's analysis suggests that *Whatcheeria* provides novel insights into mandibular evolution in early tetrapods, and has improved our understanding of this important character complex.

In chapter 3, Munter and Clark describe new remains of small theropod (predatory) dinosaurs from the latest Early Jurassic of Mexico. A formal description is followed by a series of phylogenetic analyses that demonstrate the coelophysoid affinities of these fragmentary remains. The new coelophysoids represent the youngest known fossils of this clade. The group is common throughout Late Triassic and Early Jurassic terrestrial faunas but had disappeared by the Late Jurassic. The poorly known Middle Jurassic terrestrial record has hampered our understanding of their demise, but new fossils such as these ultimately may shed light on this process.

Sidor and Rubidge (chapter 4) discuss the morphology and phylogenetic affinities of a new biarmosuchian, which they named in honor of Jim Hopson. This new taxon is based on a nearly complete skull and broadens our understanding of this relatively poorly known group. The biarmosuchians are particularly important to studies of synapsid evolution, as they represent the oldest and most basal clade of therapsids. The authors use modern numerical cladistic techniques to elucidate the systematic relationships of the new taxon, ultimately supporting the monophyly of Biarmosuchia.

Sues and Jenkins (chapter 5) present a detailed description of the postcranial skeleton of the tritylodontid therapsid *Kayentatherium*. This taxon contributes greatly to our knowledge of the morphology of these cynodonts (specifically by suggesting that tritylodonts are not nearly as mammal-like as previously had been supposed) and simultaneously offers new clues on their relationships. The authors mine this wealth of new data for important phylogenetic implications, some of which bear directly on the origin of mammals themselves. As they point out, the phylogenetic placement of *Kayentatherium* makes it a particularly appropriate subject for a volume honoring Jim Hopson.

It is noteworthy that all these contributions include data matrices and/or some form of systematic analysis. The ability to place fossils into their appropriate systematic context has long been a central focus of paleontology. However, the role of systematics has changed dramatically since Jim Hopson began his career in the 1960s. At that time, investigations of higher-level vertebrate phylogeny were almost exclusively the domain of paleontologists (Ross, 1974; Patterson, 1981). Groups that lacked a significant fossil record typically were portrayed as having unknown (and by implication unknowable) phylogenetic histories (Patterson, 1981). Patterson (1981; p. 195) described the logical sequence whereby paleontology was granted its pivotal role in the reconstruction of phylogenetic or evolutionary relationships: "evolution is a theory about the history of life; evolutionary relationships are historical relationships; fossils are the only concrete historical evidence of life; therefore fossils must be the arbiters of evolutionary relationships."

The ensuing years have witnessed a dramatic revolution both in systematic practice and in the role of paleontology in systematics. Beginning in the late 1960s and 1970s, two important objections were raised to traditional systematic practices. The first objection concerned subjectivity in systematics. The practices of traditional systematists came to be viewed as overly dependent on intuition and authority, with groupings often established on the basis of subjective or cryptic selection and weighting of characteristics (e.g., see criticisms by Patterson, 1980; Wiley et al., 1991). Cladistics or phylogenetic systematics rose to prominence in response to such concerns and enjoyed particular success among systematic paleontologists over the past few decades. Indeed, in even a brief scan of important systematic paleontology journals such as the *Journal of Pale-ontology* or *Journal of Vertebrate Paleontology,* one would be hard pressed to find more than a handful of papers that used traditional ("evo-lutionary" *sensu* Mayr [1980]) systematic methodologies over the past de-cade. In recent years, the incorporation of computer-driven analyses into cladistic methodology has allowed ever more sophisticated investigations of ever larger systematic data sets. This transition in the vertebrate pale-ontological community from traditional "evolutionary" systematics to cladistic methodology to computer-driven analyses was embraced by Jim Hopson and his students and is mirrored in their systematic publications (e.g., see Hopson & Crompton, 1969; Hopson & Barghusen, 1986; Sidor & Hopson 1998).

The contribution by Gaudin and Wible (chapter 6) is another example of the use of such computer-driven cladistic analysis to analyze large data sets. They examine the phylogenetic relationships among living and extinct armadillos based on a data set of 163 craniodental characters sampled across nineteen taxa. The armadillos are the most diverse of the living members of the placental mammalian order Xenarthra. Along with their extinct kin, armadillos have been placed in the clade Cingulata, a group whose systematic relationships remain poorly investigated (Gaudin, 2003). This contribution overturns many of the traditional ideas about armadillo interrelationships and is notable in that the phylogenetic results for living taxa are changed substantially by adding fossil taxa to the analysis.

The results reported in chapter 6 are particularly pertinent in relation to the second primary objection to traditional systematics that began to surface, particularly in the 1970s and 1980s. This second objection con-cerned the role of paleontology in systematics. Influential papers pub-lished by Patterson (1981), Rosen et al. (1981), and Gardiner (1982) pointed out that, because of gaps in the fossil record, phylogeny could rarely if ever be read from the rocks in any straightforward manner. Cladistic methodology did not depend on stratigraphic information, meaning that living taxa could play a role in cladistic analyses equivalent to that of fossil taxa. Furthermore, because fossils typically do not pre-serve information on soft tissue structure or molecular composition, they contain only a fraction of the potentially useful systematic information provided by living forms. Patterson (1981, 218) claimed that "instances of fossils overturning theories of relationship based on Recent organisms are very rare, and may be nonexistent." Patterson (1981), Rosen et al. (1981) and Gardiner (1982) suggested that fossils may play only a minor role in phylogenetic reconstruction. A number of recent papers addressed their challenge of the relevance of paleontology to systematic analysis (e.g., Gauthier et al., 1988; Smith & Littlewood, 1994; O'Leary, 1999), and readers are referred to these works for more detailed arguments advocating the continued importance of fossils in investigations of phylogenetic relationships. We believe the papers throughout this volume provide further corroboration for the continued relevance of vertebrate fossils in advancing our understanding of amniote phylogeny.

Large-Scale Evolutionary Patterns

One of the benefits of cladistics' emphasis on the importance of living taxa in phylogenetic analysis has been a more widespread realization of the centrality of phylogeny to comparative biology and evolutionary biology. Many authors have discussed the importance of using phylogenetic information to better understand the evolution of myriad aspects of organismal biology, from functional morphology and physiology (e.g., see Lauder et al., 1995, and references therein) to behavior and ecology (Brooks & McLennan, 1991). Morphological, functional, ecological, and behavioral characters can be incorporated into cladograms as a means of elucidating evolutionary patterns and thereby enhancing our understanding of evolutionary processes. The importance of using cladograms to further our understanding of the biology of extinct vertebrates has also been recognized by the vertebrate paleontological community (e.g., see Polly & Sidor, 1999).

Such systematics-driven analyses of evolutionary patterns have long figured importantly in the work of Jim Hopson, who used such analyses to improve our understanding of a variety of topics from the evolution of phalangeal formulas in nonmammalian therapsids (Hopson, 1995) to the evolution of the mammalian ear (Allin & Hopson, 1992).

Modern workers have also developed an array of statistical techniques to address macroevolutionary questions regardless of whether they are based explicitly on a particular phylogeny. Trend analyses (e.g., Stanley, 1973; McShea, 1998; Alroy, 2000) complement cladogram-based analyses by investigating the distributional, variational, and probabilistic aspects of large-scale evolutionary trends. A number of our contributions use both systematic and trend-based techniques as a means of understanding evolutionary patterns of morphological change.

Parrish's work in chapter 7 draws paleoecological inferences from specific morphological observations. On the basis of previously published observations of fossil taxa, Parrish quantifies ranges of motion for cervical vertebrae in sauropodomorph dinosaurs (often using computer modeling as a guide). These ranges allow him to hypothesize potential feeding regimes for primitive members of this group and subsequently to place them into a larger phylogenetic and faunal context. This type of analysis is especially meaningful for taxa like most dinosaurs that have no obvious functional or behavioral analogues among the modern fauna.

In chapter 8, Carrano takes an even broader look at dinosaur history, quantifying and analyzing patterns of body size evolution in this group. Using several different analytical techniques, he tracks specimen-based size measures through the entire phylogeny of Dinosauria. This allows size changes to be quantified, revealing that most dinosaur lineages seem to follow Cope's Rule, showing marked size increases over time. A few exceptions are also identified, among both theropods and sauropods, that may prove worthy of future paleobiological study.

Finally, Rougier and Wible (chapter 9) examine the evolution of the mammalian ear region and basicranium. This anatomical region is rich in morphological characters, allowing them to investigate important phylogenetic questions surrounding the early evolution of modern mammals. They are also able to draw on a series of previous works to discuss anatomical changes in the mammalian ear region and basicranium in great detail. Of particular note are instances of correlated changes between ear osteology and soft-tissue structures, simplification in vascular patterns, and independent acquisition of several middle ear characters in different mammalian groups.

Functional Morphology

The study of functional morphology is essentially an investigation into the relationship between organismal form and function. Such research is in evidence long before the advent of evolutionary theory (Russell, 1916; Padian, 1995; Ross, 1999). Since its beginnings, functional morphology has had a long and mutually enlightening relationship with the science of engineering, relying substantially on physics to explain the how and why of function (Vogel, 1998). In fact, early concepts of animal and plant "architecture" drew heavily from concepts of human architecture (Vogel, 1998). However, matches between form and function generally were viewed as static reflections of overarching principles of design and frequently were argued to be both purposeful and unalterable (Padian, 1995; Ross, 1999).

The theory of evolution, and specifically the concept of natural selection, altered the relationship between biological study and physics. If the form of a structure could affect its function, it also could affect the reproductive success of its parent organism (Arnold, 1983; Ross, 1999). Thus, organismal design had the potential to be invested with a selective value. Working alongside studies of developmental biology, sexual selection, and population biology, functional morphology has come to assume a prominent role in the study of evolution, contributing to the broad goal of understanding the origins, evolution, and adaptation of species (Lauder, 1991, 1996).

But why study function in fossil taxa? Links between form and function can be tested effectively in living organisms, and extant taxa are the only source for many kinds of functional data, such as kinematics, muscle activity patterns, and musculoskeletal stress and strain. However, extant species provide only a snapshot of a single cohort of taxa at one evolutionary instant. The study of evolution is the study of biological change through time, and intermediate stages in the transitions toward the character states displayed by living taxa often are documented only in the fossil record. Thus, the interpretation of function in fossil taxa is critical for the examination of such transitions (e.g., Gauthier et al., 1988). Moreover, by studying the functional morphology of extinct species, we can evaluate the full range of possible shapes and behaviors that organisms have exploited through time, gaining insight into constraints on diversity.

The techniques available for investigating functional morphology in organisms are varied, and they reflect both the types of data being collected and the conceptual framework of the questions being addressed.

What was once a largely qualitative, descriptive enterprise (answering the question, "what does it do and how does it do it?") has become increasingly quantitative and experimental (Ashley-Ross & Gillis, 2002). Studies of function can encompass everything from a complete behavior—which includes a great complexity of activities, from physical to neurological—down to a single motion or action. Experimental studies of biomechanics (i.e., the mechanical roles of biological structures) now play a central role in examinations of function among extant species, providing data on kinematics, transmitted forces, skeletal stresses and strains, muscle contractions, and neurological inputs (Biewener, 1992; Vogel, 2003). Such studies form the ultimate basis for our understanding of functional morphology in extinct species (e.g., Witmer, 1995). As a result, experimental data from extant species are applied as an integral component of many analyses of function in fossil taxa (e.g., Crompton & Hiiemae, 1970; Thomason, 1985; Carrano, 1998; Blob, 2001).

A striking pattern that has emerged in functional studies in recent years has been that the dominant structural materials of the amniote body—bone, cartilage, muscles, and tendons—tend to have conservative material properties across most taxa (Currey, 1984; Erickson et al., 2002; for exceptions, see Espinoza, 2000; Blob & LaBarbera, 2001). When coupled with the universality of gravity and other physical laws, two powerful conclusions emerge: (1) amniotes face largely similar problems, structurally speaking, and (2) amniotes are restricted to a few, common avenues of response when dealing with these problems. Thus, structural convergences are quite common among amniotes (indeed, as they are among many other organisms), often extending to a remarkable degree. These convergences suggest that similar selective pressures frequently may act in organisms of very different phylogenetic lineages.

Two main approaches have been applied to study the functional morphology of fossil taxa: *historical* (or *phylogenetic*) and *ahistorical* (or *paradigmatic*) (Lauder, 1995; Weishampel, 1995; Ross, 1999). In the former, phylogenetic principles are applied to evaluate the validity of functional inferences in fossil taxa (Bryant & Russell, 1992; Witmer, 1995; Lauder, 1995). Using the Extant Phylogenetic Bracket method (formalized by Witmer, 1995), for example, fossil taxa are analyzed in the context of at least two extant sister lineages. If osteological correlates of the function can be identified, then examination of fossils for those markers provides a test of functional inferences (Witmer, 1995; but see Lauder, 1995). However, historical analyses of function can also draw conclusions strictly from the most parsimonious distribution of features (Lauder, 1995; Witmer, 1995).

In contrast, ahistorical analyses depend critically on morphological data from the fossils in question (DeMar, 1976; Lauder, 1995; Ross, 1999) along with four central methods: qualitative correlation, quantitative correlation, physical modeling, and mathematical modeling. Qualitative correlation is rooted firmly in descriptive anatomy: conclusions about the function of a structure in an extinct species are based on whether its morphology suggests that it appears to have been good at performing that function (e.g., Molnar, 1977). With quantitative correlation, measured morphometric correlations between shape and function in living taxa are used as a guide to draw inferences about function in fossil species (e.g., Coombs, 1978; Van Valkenburgh, 1985; Carrano, 1998). Under physical modeling, replicates of fossil taxa are constructed and their functional responses to various stimuli are measured to evaluate the effects of shape (e.g., Bennett, 1996; Erickson et al., 1996). Finally, mathematical modeling uses physical principles and experimental data from living animals to derive equations that predict functional capabilities and constraints in fossil taxa (e.g., Alexander, 1989; Thomason, 1985; Blob, 2001). Such "ahistorical" approaches can enlighten evolutionary studies of function when applied within a phylogenetic context.

Two of our contributions focus on questions of functional morphology. They illustrate the breadth of perspectives and techniques available as well as the use of both extinct and extant taxa as focal groups.

Sereno (chapter 10) discusses the morphology of the multituberculate shoulder girdle based on an exceptional specimen from Mongolia that shows well-preserved three-dimensional structures. The specimen provides important data on the evolution of this region in mammals and sheds light on the controversial phylogenetic position of the Multituberculata. This is, perhaps, a best-case scenario for functional inference based on the description of a fossil amniote specimen; here the materials, though lacking soft tissues, nevertheless retain a wealth of morphological information.

In chapter 11, Crompton et al. review mandibular condyle morphology across extant herbivorous mammals, seeking to explain observable variations in functional terms. Their work is grounded in the biomechanics of feeding and a broad examination of living taxa. Using kinematic and morphological data from the bones and teeth, they are able to draw conclusions about the functional implications of mammalian jaw geometry and the changes in morphology in different lineages.

Ontogeny and Evolution

Groundbreaking nineteenth-century studies of vertebrate embryology by workers such as Haeckel and von Baer laid the foundation for a dialogue between researchers in evolution and development (Gould, 1977). Recent decades have witnessed a resurgence of interest in the relationships between development and evolution, and a mutual understanding between workers in these fields has flourished with the widespread use of new analytical and investigative techniques. As finer-scale developmental processes become better understood, so too do their relationships to the evolutionary history of the organisms in which they occur (e.g., Alberch, 1982; Shubin & Marshall, 2000; Donoghue, 2002). The current success of numerous "evo-devo" research programs is but one illustration of the integration of these approaches to understanding organismal diversity.

One of the outgrowths of this renewed relationship between studies in evolution and development has been a widespread appreciation for the importance of ontogenetic and developmental information in studies of paleontology. Ontogeny—with its constituent phenomena such as heterochrony—is now seen as an integral part of organismal evolution and therefore is of specific interest to paleontologists. Although the determination of specific ontogenetic processes can be difficult at the paleontological scale (e.g., Benton & Kirkpatrick, 1989; Jones & Gould, 1999), such efforts have the potential to yield great insights into the evolutionary process (e.g., Nelson, 1978; Larsson, 1998). As a result, paleontologists have taken new interest in studying the ontogenies of extinct species wherever possible, using information from sources as wide ranging as morphometrics (e.g., Zelditch et al., 2003), growth rates (e.g., Curry, 1999) and curves (e.g., Heinrich et al., 1993; Rinehart & Lucas, 2001), and the developmental biology of modern forms (Brochu, 1996; Maisano, 2002).

Three of our contributions draw on ontogenetic approaches to investigate vertebrate paleobiology and evolution. Together, these chapters provide a broad view of some of the varied ways paleobiological studies are currently emphasizing and utilizing ontogenetic data.

O'Keefe (chapter 12) provides a detailed description of the braincase in several basal sauropterygians and plesiosaurs, representing the onset of an important radiation of Mesozoic marine reptiles. By placing these observations into a phylogenetic context, he is able to determine the plesiomorphic condition for Plesiosauria. This condition is much more similar to that of basal diapsids than to that of basal sauropterygians, providing evidence for character reversal. O'Keefe's study brings an ontogenetic perspective to this pattern by applying a heterochronic interpretation.

In his paper (chapter 13), Blob examines patterns of within-taxon scaling (size-related changes in limb bone morphology and proportions) in cynodont therapsid taxa for which juvenile and subadult fossils have been preserved. Data from extant vertebrates have suggested that such patterns are linked to metabolic rate and could provide insight into the metabolic status of fossil taxa. Blob tests this proposal by analyzing data from a critical fossil lineage, allowing examination of the evolution of locomotor ontogeny and endothermy in the ancestors of modern mammals.

Finally, in chapter 14, Grine et al. tackle the leviathan problems surrounding the taxonomy and ontogeny of the dicynodont therapsid *Lystrosaurus*. This widespread genus has housed an ever-proliferating array of species since its original diagnosis. Grine et al. use several morphometric techniques to analyze the large sample of *Lystrosaurus* skull material in the context of size and stratigraphic position. They conclude that growth changes account for much of the variation seen in this taxon but that valid taxonomy differences are detectable within this context. The bewildering list of *Lystrosaurus* species is reduced to a few well-documented forms.

Acknowledgments

As editors, we offer sincere thanks to the many contributors to this volume. They responded to our request for papers enthusiastically, ultimately providing some of their finest work. We are pleased and proud to offer this volume to Jim on their behalf and hope our work as editors can measure up to theirs as authors.

We initially approached many of Jim Hopson's former students, postdocs, and collaborators with the idea of creating a volume to honor Jim's career and work. We received an overwhelmingly positive response and are pleased that so many people were able to provide papers for this volume. Indeed, even those who for various reasons were unable to contribute nonetheless provided unsolicited praise for Jim and his work.

It has been a pleasure to work with the University of Chicago Press on this project. From the beginning, the editors were unanimous in their feeling that the Press would be the most appropriate home for a book honoring the career of Jim Hopson, and we were pleased to find our opinion shared by people at the Press. Nonetheless, this work would have been much more difficult, and its success less assured, without the sincere efforts, enthusiasm, and support offered to us by Christie Henry. The importance of her guidance to these four novice editors should not be underestimated.

Claire Vanderslice has worked as a scientific illustrator with Jim Hopson for more than twenty years, and her skills are evident in many of Jim's publications. Here again, we thought she was the single most appropriate person to work with in designing the cover of this volume. We thank her for producing the wonderful image of *Thrinaxodon* that graces this book's cover. We also thank Claire, Jim Hopson, and Edgar Allin for providing the previously published illustrations that are placed at the head of each section of the book as a tribute to their long and productive collaborations.

Finally, this book could not have proceeded without the generous efforts of numerous outside reviewers, who agreed to treat these manuscripts as they would submissions to any scientific journal. Their expertise and assistance elevated both the content and the organization of this work. We thank numerous anonymous reviewers as well as Gerardo de Iuliis, Zofia Kielan-Jaworowska, Timothy O. Koneval, Michael W. Maisch, Sean Modesto, Michael J. Novacek, Olivier C. Rieppel, and Sergio Vizcaíno.

It has been our honor and privilege to serve as coeditors of this volume honoring the occasion of Jim Hopson's retirement from the University of Chicago. Three of us (M.T.C., T.J.G., R.W.B.) are former graduate students of Jim's, and the other (J.R.W.) is a former post-doc. Jim has shared with each of us his tremendous enthusiasm for paleontology and his interest in our work and our development as scientists. Jim's knowledge of paleobiology across the entire taxonomic spectrum of vertebrates is extraordinary; he has used this knowledge to encourage each of us to pursue our own taxonomic and methodological interests in a rigorous fashion. We have enjoyed his wide-ranging intellectual pursuits and his willingness to share these interests with those around him. We have benefited from his and his family's hospitality on many occasions as well as from their kindness and friendship. It is therefore with love and gratitude that we offer this volume in tribute and profess our hope that he finds it a scientific enterprise worthy of his own high standards of professional excellence.

Literature Cited

Alberch, P. 1982. Developmental constraints in evolutionary processes; pp. 313–332 *in* J. T. Bonner (ed.), *Evolution and Development.* New York: Springer-Verlag.

Alexander, R. M. 1989. Mechanics of fossil vertebrates. *Journal of the Geological Society, London* 146:41–52.

Allin, E. F. and J. A. Hopson. 1992. Evolution of the auditory system in Synapsida ("mammal-like reptiles" and primitive mammals) as seen in the fossil record; pp. 587–614 *in* D. B. Webster, R. R. Fay, and A. N. Popper (eds.), *The Evolutionary Biology of Hearing.* New York: Springer-Verlag.

Alroy, J. 2000. New methods for quantifying macroevolutionary patterns and processes. *Paleobiology* 26:707–733.

Arnold, S. J. 1983. Morphology, performance, and fitness. *American Zoologist* 23:347–361.

Ashley-Ross, M. A. and G. B. Gillis. 2002. A brief history of vertebrate functional morphology. *Integrative and Comparative Biology* 42:183–189.

Bennett, S. C. 1996. Aerodynamics and thermoregulatory function of the dorsal sail of *Edaphosaurus. Paleobiology* 22:496–506.

Benton, M. J. 1993. Reptilia; pp. 681–715 *in* M. J. Benton (ed.), *The Fossil Record 2.* New York: Chapman and Hall Inc.

Benton, M. J. and R. Kirkpatrick. 1989. Heterochrony in a fossil reptile: juveniles of the rhynchosaur *Scaphonyx fischeri* from the late Triassic of Brazil. *Palaeontology* 32:335–353.

Biewener, A. A. 1992. *Biomechanics Structures and Systems: A Practical Approach.* Oxford: IRL Press.

Blob, R. W. 2001. Evolution of hindlimb posture in nonmammalian therapsids: biomechanical tests of paleontological hypotheses. *Paleobiology* 27:14–38.

Blob, R. W. and M. LaBarbera. 2001. Correlates of variation in deer antler stiff-
ness: age, mineral content, intra-antler location, habitat, and phylogeny. *Biological Journal of the Linnean Society* 74:113–120.

Brochu, C. A. 1996. Closure of neurocentral sutures during crocodilian ontogeny: implications for maturity assessment in fossil archosaurs. *Journal of
Vertebrate Paleontology* 16:49–62.

Brooks, D. R. and D. A. McLennan. 1991. *Phylogeny, Ecology and Behavior.*
Chicago: University of Chicago Press.

Bryant, H. N. and A. P. Russell. 1992. The role of phylogenetic analysis in the
inference of unpreserved attributes of extinct taxa. *Philosophical Transactions of the Royal Society of London B* 337:405–418.

Carrano, M. T. 1998. Locomotion in non-avian dinosaurs: integrating data from
hindlimb kinematics, in vivo strains, and bone morphology. *Paleobiology*
24:450–469.

Carroll, R. L. 1988. *Vertebrate Paleontology and Evolution.* New York: W.H.
Freeman and Company.

Coombs, W. P. 1978. Theoretical aspects of cursorial adaptations in dinosaurs.
Quarterly Review of Biology 53:393–418.

Crompton, A. W. and K. Hiiemae. 1970. Molar occlusion and mandibular movements during occlusion in the American opossum, *Didelphis marsupialis* L.
Zoological Journal of the Linnean Society 49:21–47.

Currey, J. D. 1984. *The Mechanical Adaptations of Bones.* Princeton, NJ: Princeton University Press.

Curry, K. A. 1999. Ontogenetic histology of *Apatosaurus* (Dinosauria:
Sauropoda): new insights on growth rates and longevity. *Journal of Vertebrate
Paleontology* 19:654–665.

DeMar, R. E. 1976. Functional morphological models: evolutionary and non-
evolutionary. *Fieldiana (Geology)* 33:339–354.

Donoghue, P. C. J. 2002. Evolution of development of the vertebrate dermal
and oral skeletons: unraveling concepts, regulatory theories, and homologies. *Paleobiology* 28:474–507.

Erickson, G. M., S. D. Van Kirk, J. Su, M. E. Levenston, W. E. Caler, and
D. R. Carter. 1996. Bite-force estimation for *Tyrannosaurus rex* from tooth-
marked bones. *Nature* 382:706–708.

Erickson, G. M., J. Catanese III, and T. M. Keaveny. 2002. Evolution of the biomechanical material properties of the femur. *Anatomical Record* 268:115–124.

Espinoza, N. R. 2000. Scaling of the Appendicular Musculoskeletal System of
Frogs (Order Anura): Effects on Jumping Performance. Ph.D. Dissertation,
University of Chicago.

Gardiner, B. 1982. Tetrapod classification. *Zoological Journal of the Linnean
Society* 74:207–232.

Gaudin, T. J. 2003. Phylogeny of the Xenarthra (Mammalia) *in* R. A. Fariña,
S. F. Vizcaíno, and G. Storch (eds.), Morphological studies in fossil and extant Xenarthra (Mammalia). *Senckenbergiana Biologica* 83: 27–40.

Gauthier, J. A. 1994. The diversification of the amniotes; pp. 129–159 *in* D. R. Prothero and R. M. Schoch (eds.), *Major Features of Vertebrate Evolution. Short Courses in Paleontology, Number 7.* Knoxville, TN: The Paleontological Society, The University of Tennessee Press.

Gauthier, J. A., A. G. Kluge, and T. Rowe 1988. Amniote phylogeny and the importance of fossils. *Cladistics* 4:105–209.

Gould, S. J. 1977. *Ontogeny and Phylogeny.* Cambridge, MA: Harvard University Press.

Heinrich, R. E., C. B. Ruff, and D. B. Weishampel. 1993. Femoral ontogeny and locomotor biomechanics of *Dryosaurus lettowvorbecki* (Dinosauria, Iguanodontia). *Zoological Journal of the Linnean Society* 108:179–196.

Hopson, J. A. 1964a. *Pseudodontornis* and other large marine birds from the Miocene of South Carolina. *Yale Peabody Museum Postilla* 83:1–19.

———. 1964b. The braincase of the advanced mammal-like reptile *Bienotherium. Yale Peabody Museum Postilla* 87:1–30.

———. 1995. Patterns of evolution in the manus and pes of non-mammalian therapsids. *Journal of Vertebrate Paleontology* 15(3):615–639.

Hopson, J. A. and H. R. Barghusen. 1986. An analysis of therapsid relationships; pp. 83–106 *in* N. Hotton, P. D. MacLean, J. J. Roth and E. C. Roth (eds.), *The Ecology and Biology of Mammal-like Reptiles.* Washington, DC: Smithsonian Institution Press.

Hopson, J. A. and A. W. Crompton. 1969. Origin of mammals; pp. 15–72 *in* T. Dobzhansky, M. K. Hecht, and W. C. Steere (eds.), *Evolutionary Biology,* Volume 3. New York: Appleton-Century-Crofts.

Jones, D. S. and S. J. Gould. 1999. Direct measurement of age in fossil *Gryphaea:* the solution to a classic problem in heterochrony. *Paleobiology* 25: 158–187.

Larsson, H. C. E. 1998. A new method for comparing ontogenetic and phylogenetic data and its application to the evolution of the crocodilian secondary palate. *Neues Jahrbuch für Geologie und Paläontologie Abhandlungen* 210:345–368.

Lauder, G. V. 1991. Biomechanics and evolution: integrating physical and historical biology in the study of complex systems; pp. 1–19 *in* J. M. V. Rayner and R. J. Wooton (eds.), *Biomechanics in Evolution.* Cambridge: Cambridge University Press.

———. 1995. On the inference of function from structure; pp. 1–18 *in* J. J. Thomason (ed.), *Functional Morphology in Vertebrate Paleontology.* Cambridge: Cambridge University Press.

———. 1996. The argument from design; pp. 55–91 *in* M. R. Rose and G. V. Lauder (eds.), *Adaptation.* San Diego: Academic Press.

Lauder, G. V., R. B. Huey, R. K. Monson, and R. J. Jensen. 1995. Systematics and the study of organismal form and function. *Bioscience* 45:696–704.

Lombard, R. E. and J. R. Bolt. 1995. A new primitive tetrapod, *Whatcheeria deltae,* from the Lower Carboniferous of Iowa. *Palaeontology* 38:471–494.

Maisano, J. A. 2002. Terminal fusions of skeletal elements as indicators of maturity in squamates. *Journal of Vertebrate Paleontology* 22:268–275.

Mayr, E. 1980. Biological classification: toward a synthesis of opposing methodologies. *Science* 214:510–516.

McShea, D. W. 1998. Possible largest-scale trends in organismal evolution: eight "live hypotheses." *Annual Review of Ecology and Systematics* 29: 293–318.

Molnar, R. E. 1977. Analogies in the evolution of combat and display structures in ornithopods and ungulates. *Evolutionary Theory* 3:165–190.

Nelson, G. 1978. Ontogeny, phylogeny, paleontology, and the biogenetic law. *Systematic Zoology* 27:324–345.

Novacek, M. J. and Q. D. Wheeler. 1992. Extinct taxa: Accounting for 99.999...% of the earth's biota; pp. 1–16 *in* M. J. Novacek and Q. D. Wheeler (eds.), *Extinction and Phylogeny.* New York: Columbia University Press.

O'Leary, M. 1999. Parsimony analysis of total evidence from extinct and extant taxa and the cetacean-artiodactyl question (Mammalia, Ungulata). *Cladistics* 15:315–330.

Padian, K. 1995. Form versus function: the evolution of a dialectic; pp. 264–277 *in* J. J. Thomason (ed.), *Functional Morphology in Vertebrate Paleontology.* Cambridge: Cambridge University Press.

Patterson, C. 1980. Cladistics. *Biologist* 27:234–240

———. 1981. Significance of fossils in determining evolutionary relationships. *Annual Review of Ecology and Systematics* 12:195–223.

Polly, D. and C. Sidor (conveners). 1999. Beyond the cladogram [symposium]. *Journal of Vertebrate Paleontology* 19:8A.

Pough, F. H., C. M. Janis, and J. B. Heiser 1999. *Vertebrate Life,* 5th edition. Upper Saddle River, NJ: Prentice-Hall, Inc.

Rinehart, L. F. and S. G. Lucas. 2001. A statistical analysis of a growth series of the Permian nectridean *Diplocaulus magnicornis* showing two-stage ontogeny. *Journal of Vertebrate Paleontology* 21:803–806.

Rosen, D. E., P. L. Forey, B. Gardiner, and C. Patterson. 1981. Lungfishes, tetrapods, paleontology and plesiomorphy. *Bulletin of the American Museum of Natural History* 167:159–276.

Ross, C. F. 1999. How to carry out functional morphology. *Evolutionary Anthropology* 7:217–222.

Ross, H. H. 1974. *Biological Systematics.* Reading, MA: Addison-Wesley Publishing.

Russell, E. S. 1916. *Form and Function: A Contribution to the History of Animal Morphology.* London: John Murray.

Sereno, P. C. 1997. The origin and evolution of dinosaurs. *Annual Review of Ecology and Systematics* 25:435–489.

Shubin, N. H. and C. R. Marshall. 2000. Fossils, genes, and the origin of novelty. *Paleobiology* 26:324–340.

Sidor, C. A. and J. A. Hopson. 1998. Ghost lineages and "mammalness": assessing the temporal pattern of character acquisition in the Synapsida. *Paleobiology* 24:254–273.

Smith, A. B. and D. T. J. Littlewood. 1994. Paleontological data and molecular phylogenetic analysis. *Paleobiology* 20:259–273.

Stanley, S. M. 1973. An explanation for Cope's Rule. *Evolution* 27:1–26.

Taquet, P. 1994. Georges Cuvier, ses liens scientifiques européens. *Montbéliard Sans Frontières, Colloque international de Montbéliard:* 287–309.

Thomason, J. J. 1985. Estimation of locomotory forces and stresses in the limb bones of Recent and extinct equids. *Paleobiology* 11:209–220.

Van Valkenburgh, B. 1985. Locomotor diversity within past and present guilds of large predatory mammals. *Paleobiology* 11:406–428.

Vogel, S. 1998. *Cat's Paws and Catapults: Mechanical Worlds of People and Nature.* New York: W. W. Norton and Co.

———. 2003. *Comparative Biomechanics: Life's Physical World.* Princeton, NJ: Princeton University Press.

Weishampel, D. B. 1995. Fossils, function, and phylogeny; pp. 34–54 *in* J. J. Thomason (ed.), *Functional Morphology in Vertebrate Paleontology.* Cambridge: Cambridge University Press.

Wiley, E. O., D. Siegel-Causey, D. R. Brooks, and V. A. Funk. 1991. The compleat cladist. A primer of phylogenetic procedures. *University of Kansas Museum of Natural History, Special Publication* 19:1–158.

Witmer, L. M. 1995. The Extant Phylogenetic Bracket and the importance of reconstructing soft tissues in fossils; pp. 19–33 *in* J. J. Thomason (ed.), *Functional Morphology in Vertebrate Paleontology.* Cambridge: Cambridge University Press.

Zelditch, M. L., H. D. Sheets, and W. L. Fink. 2003. The ontogenetic dynamics of shape disparity. *Paleobiology* 29:139–156.

PART ONE New Fossils and Phylogenies

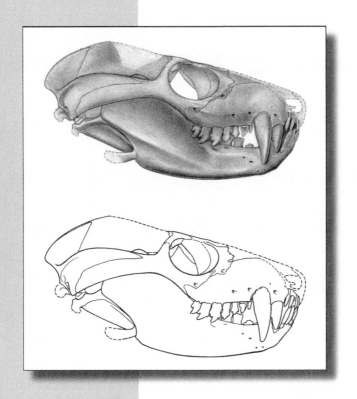

Plate 1. Two reconstructions of the skull of *Lumkuia fuzzi*, a cynodont, in right lateral view. Illustration by Claire Vanderslice, from: J. A. Hopson & J. W. Kitching (2001), A probainognathian cynodont from South Africa and the phylogeny of nonmammalian cynodonts, *Bulletin of the Museum of Comparative Zoology* 156(1): 5–35.

2

The Mandible of *Whatcheeria deltae*, an Early Tetrapod from the Late Mississippian of Iowa

R. Eric Lombard
and John R. Bolt

Introduction

It is a pleasure to dedicate this study to Dr. James Hopson, who has been our friend and colleague for many years. Although our participation in this volume is thus entirely appropriate on personal and collegial grounds, our taxic suitability might be questioned. Jim, after all, is an authority on early amniotes, whereas we have mostly concentrated on amphibian grade tetrapods. We at first thought that this problem could be solved by contributing a paper on *Whatcheeria* noting that, in our original description, we had suggested that it might be "the first outgroup to Anthracosauria," and therefore could be claimed to at least have its nasal region under the amniote tent. However, recent cladistic studies have put this notion in doubt (below). At this point all we can do is throw ourselves on the mercy of the court of collegial opinion. We point out that our organisms of interest and amniotes are all tetrapods, and we ask: what, after all, are a few branching points among friends?

Since 1985, more new early-tetrapod (pre-Pennsylvanian) localities have been discovered, and more species have been named, than in the preceding 100 years. *Whatcheeria deltae* (Lombard & Bolt, 1995) from the Mississippian of Iowa is one of the important recent additions to this roster of early tetrapods. *Whatcheeria* is important in general for its age and the number of specimens recovered and in particular for its combination of primitive morphology and interesting autapomorphies. At present, the systematic relations of *Whatcheeria* are uncertain, as it is characterized by many primitive features and, so far, is not known to possess synapomorphies that clearly associate it with groups other than the recently described sister-taxon *Pederpes finneyae* Clack (2002). The most recent and most inclusive cladistic studies indicate that *Whatcheeria* may be a stem tetrapod (Clack, 2002; Ruta et al., 2002, 2003) or of uncertain affinity as a branch of a polytomy forming the tetrapod crown group (Ruta et al., 2002). These are not novel hypotheses as both have been suggested in prior studies as well as a relationship to anthracosaurs, a group thought to

21

be related to the origin of amniotes (Coates, 1996; Ahlberg & Clack, 1998; Laurin, 1998; Paton et al., 1999—as reanalyzed by Ruta et al., 2002—and Carroll, 2001). All these studies agree, however, that *Whatcheeria* is a quite primitive tetrapod.

The jaw of *Whatcheeria* is represented by more than thirty specimens preserving both internal and external views and representing a range of sizes. Because mandibles are durable, much of the other newly discovered early-tetrapod material consists of some portion of that element. And, because mandibles are morphologically complex, they figure in both systematic and functional-morphological studies.

Our knowledge of the mandible in the earliest tetrapods was recently summarized and extended by Ahlberg and Clack (1998). The mandibles of the earliest tetrapods are clearly directly comparable to those in related fish-grade sarcopterygians. There is, however, enough diversity among the taxa that the general trends of morphological change during the transition to a more terrestrial existence are now reasonably clear. The picture continues to improve with new morphological descriptions and new taxa (Daeschler, 2000; Bolt & Lombard, 2001; Clack, 2001, 2002; Ruta et al. 2002).

In this paper, we provide a description of the jaw based on the best preserved specimens available for *Whatcheeria*. This description is used as a basis for a reconstruction of the jaw in internal and external views. The description allows us to revise some character state designations attributed to *Whatcheeria* in the recent literature and to add state observations that previously have been coded as unknown. Finally, the morphology of the adsymphysial in *Whatcheeria* as described here, together with recent observations on that element in other taxa, enables us to suggest an extension of the present scenario for jaw evolution in the earliest tetrapods.

Materials and Methods

Table 2.1 lists the specimens used in this study. All known *Whatcheeria* specimens come from a single locality near the town of Delta in southeastern Iowa. The Delta locality has produced numerous fossil vertebrates from two adjacent collapse structures. The age of the fills in these structures is apparently Early Chesterian of the Mississippian of North American nomenclature, correlative with the Viséan V3b (Asbian) of the Lower Carboniferous of Europe (Bolt et al., 1988; Witzke et al., 1990). The fills include limestone conglomerates, shales, and lime mudstones. All appear to represent deposition in a nonmarine aquatic environment that might

Table 2.1. *Whatcheeria deltae* mandible specimens studied for this paper

Specimen no. and side(s)	Approx. length tip to tip (cm)	Description
PR 1634L & R	15.5E	Associated with skull, and only partly visible; see fig. 1A in Lombard and Bolt (1995).
PR 1642R		Prepared almost completely free, but sutures difficult to see.
PR 1644L* & R*		Associated with some skull elements in limestone block; very well preserved, relatively uncrushed; left complete, visible in lateral and partially in medial view; right missing posterior one-third, visible in occlusal and lateral views.
PR 1646L & R	17.0	Associated with partial skull in shale, strongly crushed; left partly exposed in lateral and medial views, right exposed in lateral view but only partly exposed in medial view.
PR 1665L*	19.0	Crushed especially posteriorly, but generally well preserved; prepared free.
PR 1666		Anterior one-quarter of mandible, prepared free; crushed dorsoventrally, sutures obscure but shows dentition well.
PR 1692L	25.0E	Exposed in medial view in limestone slab. Missing a section near symphysis, so length estimate more uncertain than usual. Relatively uncrushed, including articular region.
PR 1700L & R	17.5E	Associated with type skull; posterior one-quarter of left is exposed in medial view, badly damaged; right is mostly covered by skull, so less than half of lateral surface is visible.
PR 1705L & R	18.0E	In shale; both badly crushed, anterior portion of left bent upward at a right angle, neither one fully exposed.
PR 1755L* & R	16.5	In limestone; exposed in lateral view, crushed and deformed, except left symphysial region is well preserved and prepared free.
PR 1792L* & R*	14.0	Associated with partial skull in shale; left is exposed in lateral view, right in medial view.
PR 1809L* & R*		Associated with large partial skull in shale; only the symphysial regions of these mandibles are useful, but both are well preserved, the left exceptionally so.
PR 1813L* & R*		Associated with reasonably complete but badly crushed skull that has been prepared free; strongly crushed but otherwise reasonably well preserved; left is prepared free, right is in situ and thus partly obscured by skull.
PR 1814L	17.0	Associated with partial skull in shale block; entire mandible, in lateral view.
PR 1816L & R	19.0	Associated with partial skull plus postcranium in shale block; both mandibles in medial view, damaged and incomplete; right includes only articulated angular plus surangular.

(continued)

Table 2.1. *(continued)*

Specimen no. and side(s)	Approx. length tip to tip (cm)	Description
PR 1819R*	18.0E	Medially exposed in limestone block; well preserved, but missing both the surangular and the articular portion of Meckelian bone.
PR 1888R		Small mandible laterally exposed in shale, associated with partial skull; twisted and crushed, missing articular region.
PR 1955R*	25.0	Medially exposed in shale, lateral side mostly unprepared; most of the area of exomeckelian fenestra is missing, but otherwise fairly well preserved.
PR 1988R*	22.0E	Medially exposed in limestone block; generally well preserved, but missing part of symphysial region, and some of coronoid area is damaged.
PR 2105L*		Small mandible, anterior portion embedded in Castolite and sawn into twenty-two sections at 1.5-mm intervals.
PR 2245L*		Articulated angular and surangular, exposed in medial view in limestone block.
PR 2246L* & R		Associated on same surface of limestone block with right suspensorial region of skull, which is catalogued under same number; both mandibles incomplete and crushed; anterior one-quarter of right side is crushed dorsoventrally and prepared free, shows dentition reasonably well, especially area around dentary fang.

Note: This is not a complete list of *Whatcheeria* specimens, as those considered to contribute no significant information are not included. All listed specimens are at least major portions of the mandible, unless otherwise noted. All specimens are in the collections of the Field Museum. In view of the distortion and/or damage all specimens have undergone, measurements should be considered approximate; therefore, they are given only to nearest 0.5 cm. Those that are less certain are identified as estimates by an "E" immediately after the number. When parts of both left and right jaws are available for a single individual, they are catalogued under the same number. Side (left or right) is indicated with a capital letter suffix. For example, PR 1809L refers to the left mandible of individual PR 1809. The "L" or "R" in such cases is only an indicator of side and not part of the Field Museum catalogue number. An asterisk indicates specimens that were particularly informative.

have ranged from fresh to brackish. The vertebrate fauna include both tetrapods and a variety of "fish." The tetrapods consist of *Whatcheeria,* by far the commonest, as well as rare embolomeres, colosteids, and very incomplete specimens that currently are not assignable.

The degree of completeness of *Whatcheeria* specimens ranges from a few nearly complete skeletons to individual bones. A few specimens are essentially undistorted, but most are more or less crushed. Deformation was sometimes by plastic flow, with little obvious breakage. In the specific case of the mandible, almost the only undistorted examples are from the symphysial region. Most other mandibular specimens are severely crushed, mostly mediolaterally but occasionally dorsoventrally. However, we were able to extract useful information from many of the large number of available mandibular specimens, enabling us to describe the dentition and most sutures with considerable confidence. A few sutures remain dubious because they appear to be closed in all specimens available, or they run through areas that are always heavily damaged. Our reconstruction of the entire mandible is necessarily rather schematic, as no specimen provides all the topographic information that would be critical to a detailed reconstruction of the posterior portion.

All specimens were prepared mechanically using a needle and pin vise, sometimes also with handheld pneumatic air hammers. All cited specimens are in the collections of the Field Museum, Chicago, Illinois.

Description

Organization of Descriptive Section

The description below is organized alphabetically by body part—i.e., by bone (because bones are the only body parts preserved). This is done for consistency with the PRESERVE character protocol. We have published an introduction to this protocol (Lombard & Bolt, 1999) and used it in that and other recent publications that contain character lists (Bolt & Chatterjee, 2000; Bolt & Lombard, 2001). Although we do not provide here a complete list of mandibular characters for *Whatcheeria,* we plan to incorporate one into a future description of the *Whatcheeria* skull. In the text we note as far as practical the specimen(s) on which each descriptive statement is based.

Protocol for Dentition Descriptions

Dentitions of early tetrapods and osteolepiform "fish" are usually described in terms of size and/or locations of various classes of teeth. Assignment of teeth to "fang" and "denticle" classes traditionally has been subjective, and fangs in particular have variously been identified based on relative size, position, or some combination of the two. Characters based on such subjectively identified but similarly named entities often do not deal with comparable things and therefore can lead to erroneous results in phylogenetic analysis. We here follow the convention advocated by Bolt and Lombard (2001, 1032) for unambiguously assigning teeth to classes, with minor modification.

1. Marginal teeth of the skull and mandible represent the size standard for "tooth" of the upper and lower dentition, respectively. In most cases this will be practically identical, but in some groups such as Colosteidae up-

per and lower teeth differ significantly in size. Thus the relative-size-based definition of fangs and denticles may differ between palatal and mandibular dentitions.

2. "Fangs" are by definition [at least] twenty-five percent greater in maximum basal diameter and/or height than the average of adjacent marginal teeth.

3. "Denticles" by definition have twenty [10 in Bolt & Lombard, 2001] percent or less of the average maximum basal diameter and/or height of adjacent marginal teeth.

General Features

In general shape and in the distribution of the individual bones, the jaw of *Whatcheeria* is directly comparable to those of the large Devonian stem tetrapods—i.e., it is remarkably primitive. In lateral and medial views the mandible has the curved, overall shape common to early tetrapods (fig. 2.1). There is no elevated surangular crest. The adductor fossa opens mostly dorsally, as in other early tetrapods and the osteolepiform

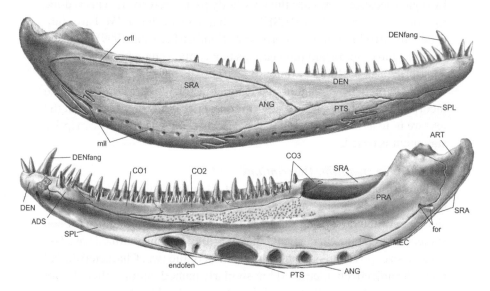

Figure 2.1. *Whatcheeria deltae,* schematic reconstruction of right jaw. (Top) Lateral view. (Bottom) Medial view. Abbreviations for this and subsequent figures: ADS, adsymphysial; ADSteeth, adsymphysial teeth; ANG, angular; ART, articular region of Meckelian bone; CO1, CO2, CO3, coronoids 1, 2, and 3; DEN, dentary; DENfang, dentary fang; DENppr, dentary postdental process; endofen, endomeckelian fenestra(e); for, foramen; MAN, mandible; MEC, Meckelian ossification; MECfen, endomeckelian fenestra; mll, indication of mandibular lateral line; orll, indication of oral lateral line; PEA, prearticular; PTS, postsplenial; SPL, splenial; SRA, surangular. Drawing by M. Donnelly.

sarcopterygians related to them. The articulating surface of the articular is oriented directly dorsally, as usual in tetrapods and contrasting with its more posterior orientation in osteolepiform sarcopterygians. Evidence of the lateral line is composed of a combination of open sulci and rows of foramina, the latter indicating that the lateral line ran partly within the bone. The partially covered course of the lateral line makes it difficult to determine its maximum extent. The mandibular lateral line begins in the surangular near the jaw joint and runs anteriorly across the surangular, angular, postsplenial and splenial. The oral sulcus branches off the mandibular lateral line in the surangular and crosses that bone in the direction of the dentary, which it appears not to reach.

The ten dermal bones primitive for tetrapods are present: dentary with a single tooth row and an anterior fang pair; four infradentaries; three coronoids, each of which bears a row of teeth; a prearticular with denticles; and a large, toothed adsymphysial that forms part of the symphysis. These bones are unsculptured except in the largest specimens (e.g., PR 1809), and even in that specimen sculpturing is weakly developed and confined to the anterior part of the mandible and the midventral surface of the angular. There is a single elongate "Meckelian" fenestra. Following Bolt and Lombard (2001), we refer to this as an exomeckelian fenestra, as it is developed in dermal bone; Meckelian fenestrae developed in endochondral bone are referred to as endomeckelian fenestrae. The elongate exomeckelian fenestra in *Whatcheeria* is actually open posteriorly. The splenial forms the anterior dorsal and ventral borders of the fenestra. The remainder is bounded by the prearticular dorsally and the postsplenial and angular ventrally. In large specimens, the Meckelian is ossified. Large mandibles thus have a continuous ventromedial exposure of Meckelian bone from the anterior border of the exomeckelian fenestra to the posterior tip of the articular, which in *Whatcheeria* is really a region in the extensive Meckelian bone rather than a local element. This type of exomeckelian fenestra, and such an extensive Meckelian ossification, is generally believed to be primitive for sarcopterygians, and the occurrence of both in *Whatcheeria* is unique for the post-Devonian tetrapods known at present.

Adsymphysial

The best-preserved adsymphysial is that in the left mandible of PR 1809 (fig. 2.2). We begin by describing this example. However, PR 1809 is a very large individual, and there appears to be significant, possibly size-related, variation in adsymphysial morphology. We therefore supplement

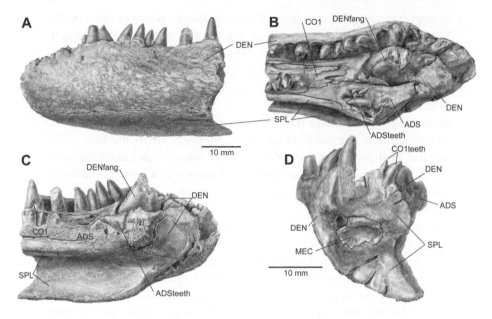

Figure 2.2. *Whatcheeria deltae* PR 1809, uncrushed anterior end of a large left jaw, prepared free of limestone. Enhanced digital photograph. Specimen shows relationships of bones around adsymphysial as well as large dentary fang in pit. (A) Lateral view. (B) Dorsal (occlusal) view. (C) Medial view. (D) Cross section at level of coronoid 1. Drawing by M. Donnelly.

the description of PR 1809 with a description of differing conditions in smaller individuals.

The adsymphysial of PR 1809 has an extensive dorsal and medial exposure. It is bounded anteriorly and anteromedially by the dentary, while a robust posterior extension is wedged between coronoid 1 laterally and the splenial medially. Its suture with coronoid 1 is modestly interdigitated, whereas the surface expression of the sutures with the dentary and splenial shows no interdigitation. The adsymphysial may participate in the symphysis. Its apparent symphysial surface is in the plane of the rugose symphysial contribution of the dentary and has a texture reminiscent of the dermal ornament found on many temnospondyl and other amphibians [see, e.g., *Greererpeton* (Bolt & Lombard, 2001), although *Greererpeton* is not a temnospondyl]. Assuming that this is a symphysial surface, the adsymphysial constitutes about nineteen percent of the total symphysial area by our measurements. This is far less than the fifty percent contribution found in *Greererpeton* (Bolt & Lombard, 2001), but well above the contribution seen in very early tetrapods such as *Acanthostega*

(Ahlberg & Clack, 1998) and in some osteolepiform sarcopterygians, in which the adsymphysial plate sometimes has no part in the symphysis at all (*Cladarosymblema;* Fox et al., 1995).

The dorsal surface of the adsymphysial bears teeth immediately above and posterior to its symphysial area. There is an outer row of three small functioning teeth, which is continued anteriorly by two empty tooth loci. Immediately medial to the outer row is a marginal-sized tooth, behind which there are two empty tooth loci. The middle one of these has been left mostly matrix filled to support the single functioning tooth.

In the smallest mandible with a reasonably well-preserved adsymphysial (left side of PR 1755, not illustrated), none of the sutures bounding the adsymphysial is interdigitated and the posterior process is much more slender than that of large specimens. The same is true of the somewhat larger right mandible of PR 1644 (fig. 2.3). The texture of the flat symphysial surface in PR 1755 does not differ from that of adjacent areas of the bone, but that of PR 1644 is somewhat ornamented. The outer tooth row in PR 1755 consists of only two teeth separated by a large empty tooth locus belonging to the inner series. There is one functioning tooth in the inner series, whose basal diameter is smaller than that of the adjacent dentary marginal teeth but larger than that of the outer adsymphysial row.

The adsymphysial never carries fangs in *Whatcheeria*. The number of adsymphysial teeth varies greatly, although there is a general pattern of small teeth in an outer row and larger teeth in a more medial (inner) position. The lowest number of adsymphysial teeth found is in PR 1705, which has only a single large although submarginal-sized tooth. On the other extreme, PR 1692 has at least fifteen teeth in total (the dentigerous area is partly matrix covered). In that specimen, teeth are arranged in two irregular rows, with at least eight teeth plus empty loci in the outer row and seven teeth plus loci in the inner row. The inner row includes at its midpoint two larger but submarginal-sized teeth.

Angular

In larger specimens, the angular is one of the most strongly ornamented bones of the mandible; in those individuals, there are numerous closely spaced pits concentrated in its midventral portion. However, the pits are not surrounded by the netlike system of ridges that constitute a true ornament. On the external surface, the angular is bounded anteriorly by the dentary, dorsally by the surangular, and anteroventrally by the postsplenial. Posteroventrally as well as on its ventromedial surface the angular is bounded by the Meckelian ossification or cartilage. Sutures with all

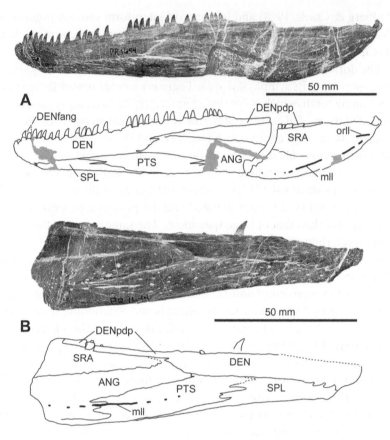

Figure 2.3. *Whatcheeria deltae* PR 1644, right and left jaws from a small individual, preserved in limestone. Gray shading represents damaged areas that have been repaired with epoxy. (A) Photo (above) and outline drawing (below) of left jaw in lateral view. (B) Photo (above) and outline drawing (below) of right jaw in lateral view. This specimen is incomplete posteriorly, and partly covered anteriorly by a clavicle, which has been edited from the photograph. The dentary has been tilted inward, so no dentary teeth are visible except one, which has been broken off at the base and thus was not displaced inward with the others.

these bones are straight to gently sinuous in exterior appearance, with no fine interdigitation. The ventral parts of the angular-postsplenial and angular-surangular sutures have a zigzag pattern. The suture with the Meckelian is always straight. There is no medial lamina of the angular, no contact with the prearticular, and therefore no angular foramen as defined by Bolt and Lombard (2001). As noted below, however, the homologue of this foramen may be present in *Whatcheeria*. In the angular, the course of the mandibular lateral line is marked by alternating foramina and open

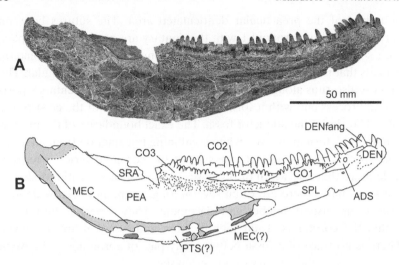

Figure 2.4. *Whatcheeria deltae* PR 1665. Medium-sized left jaw prepared completely free of matrix. (A) Photo of jaw in medial view. (B) Line drawing of (A). Area between the ventral margin of the prearticular and ventral margin of the jaw is the exomeckelian fenestra. It contains the endochondral Meckelian bone, which surrounds a number of endomeckelian fenestrae as a series of arches. Light gray indicates Meckelian bone and articular. Darker gray indicates broken bases of the Meckelian bone's arches.

sulci. The relative proportions of these vary greatly from specimen to specimen.

Coronoid Series

The coronoid series comprises coronoids 1–3, although the adsymphysial can be considered an anterior extension of the series. On most specimens, the coronoid-dentary suture was usually on an axis of rotation when the jaw was laterally crushed and so is always visible and often more or less open. Its surface expression is a straight line, and in three dimensions the suture appears to be a simple butt joint. The coronoid-prearticular and coronoid-splenial sutures are not so clear, and in some specimens are untraceable even in areas that appear well preserved. However, because of the number of specimens available the general course of these sutures can be reconstructed with confidence (fig. 2.1). The coronoid 1-splenial suture is not interdigitated; the suture of each of the coronoids with the prearticular may have one or two long interdigitations, judging from PR 1665 (fig. 2.4). In areas of individual specimens where the coronoid-prearticular sutures cannot be traced, the boundary between bones can be assumed to coincide approximately with the lateral boundary of the prearticular denticulated area. The sutures between coronoids and of coronoid 1 with the dentary are moderately complex, with some large-scale interdigitation. Coronoid 3 has a posterolateral process that is applied to the dorsomedial surface of the surangular. The process also runs along part of the medial surface of the dentary's post-dental process and with it forms the anterior portion of the crest of the lateral border of the adductor fossa. The exact boundaries of the process cannot be determined, even when combining information from all available specimens. The dorsal exposure of the coronoid series is narrow, little more than the width of the bases of the coronoid teeth. Medial to the tooth bases there are often several long, deep, and narrow depressions of inconstant number and placement. They appear to be formed within the coronoids. Their identification and function are unknown. There is no trace of a fossa between any pair of coronoids or between coronoid 1 and the dentary or adsymphysial.

The coronoid dentition consists of a single row of teeth, somewhat smaller than the marginal teeth and present on each coronoid. There are no fangs. In a minor departure from the single-rowed condition, PR 1819, PR 1955, and PR 1665 each has a single instance where coronoid 1 bears two teeth in medial-lateral juxtaposition. The coronoid teeth are attached primarily in acrodont fashion, although coronoid 2 in particular has a somewhat elevated lateral wall. However, there is no sign of the high tooth-bearing vertical lamina described by Ahlberg and Clack (1998) in *Panderichthys* and *Acanthostega*. There is often a diastema between coronoids of one to three or more teeth. Coronoid teeth show evidence of active replacement in the form of empty tooth loci that can be in any position in the row. We have not observed, however, any replacement pits in functioning teeth.

Dentary

On the lateral surface of the jaw, the dentary has the usual contacts with the infradentary bones. Anteriorly, the dentary forms most of the lateral surface of the mandible and makes contact with the splenial and postsplenial. Passing posteriorly the dentary makes a relatively abrupt transition where it makes contact with the angular, to about one-half or less of the height of the lateral exposure seen more anteriorly. This abrupt transition at the angular-dentary contact is apparently unique to *What-cheeria*. In *Panderichthys* and Devonian tetrapods where the appropriate anatomy is known in detail, the dentary begins a gentle taper at the anterior end of its contact with the angular (illustrations in Ahlberg & Clack, 1998). A slightly more abrupt transition appears to occur in the Lower

Carboniferous *Crassigyrinus* (Clack, 1998), and a very abrupt transition occurs in its contemporary *Greererpeton* but at the dentary-surangular contact (Bolt & Lombard, 2001).

Posterior to the transition point, the dentary tapers to an end as an edentulous postdental process that is embedded in a shallow groove in the surface of the surangular. The process terminates at the posterior one-third of the adductor fossa, thus somewhat more anteriorly than that in many Devonian tetrapods. At the anterior end of the adductor fossa the postdental process, together with the posterolateral process of coronoid 3, forms the dorsal border of the surangular crest. Posteriorly the post-dental process is bounded both dorsally and ventrally by the surangular.

The dentary overlaps the infradentary bones at its joints with those elements. Most of the sutures with infradentary bones are noninterdigitated, although there is modest interdigitation in the angular-dentary suture and the anterior end of the dentary-splenial suture. The dentary-splenial suture is difficult to trace in a number of specimens, although fortunately it is clear on both mandibles of the relatively small PR 1644 (fig. 2.3) and in the symphysial region of the large PR 1809L (fig. 2.2).

Ornament on the dentary is most strongly developed in the symphysial area and in larger specimens, although it does not attain the degree of relief seen in many other early tetrapods. The specimens available to us show little sign of ornamentation on the dorsal portion of the dentary, although we do not have a well-exposed lateral surface on any of the largest jaws. None of our specimens shows a lateral longitudinal ridge dividing the dentary into dorsal and ventral areas, as has been described for *Acanthostega* and other early tetrapods (Ahlberg & Clack, 1998; Ahlberg et al., 1994).

In medial view, the dentary is seen to form most of the symphysis (fig. 2.2C). The entire symphysial area is in the same plane, so there is no sign of the interdentary gap seen in *Greererpeton* (Bolt & Lombard, 2001). In no specimen can the dentary-splenial suture within the symphysis be definitely determined. The reconstruction of this suture in figure 2.2C is therefore conjectural. Most of the rugose anterior area of the symphysis is composed of dentary. Figure 2.2C shows a broad but poorly defined channel ventrally in the rugose symphysial area. Anterodorsally, this is within the dentary; posteroventrally, it is within the splenial. It is present in other jaws as well, sometimes better developed than in PR 1809L. Its function is unknown.

In dorsal view, the dentary consists mostly of a medial shelf and a lateral wall, to both of which the teeth attach. There is a single marginal tooth row, with no accessory teeth. The marginal tooth row is continuous

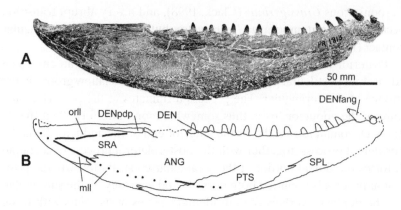

Figure 2.5. *Whatcheeria deltae* PR 1813, medium-sized right jaw preserved in occlusion with a skull, both prepared free of shale. Jaw is partly obscured posteriorly by cheek region of skull and at its anterior tip by a partial pterygoid. Skull and pterygoid fragment have been edited from the photograph and are not drawn in the sketch. (Photo of jaw in lateral view.)

from the symphysis to the anterior margin of the adductor fossa. Teeth are bluntly pointed, with somewhat mesiodistally compressed bases. The anteriormost tooth is often strongly procumbent, but not always—see, e.g., PR 1813L (fig. 2.5) and PR 1665 (fig. 2.4). There are no markedly enlarged marginal teeth, unlike those that occur in the upper dentition (Lombard & Bolt, 1995). The dentary teeth are approximately the same size as opposing upper jaw teeth. The presence of numerous gaps in the tooth row, associated with empty tooth loci with tooth bases in various stages of erosion, is evidence of active replacement. Resorption pits in the bases of functioning teeth are rare, although we have observed a few. Several jaws preserve a sufficiently complete tooth row to permit estimation of the number of functioning teeth plus empty loci. Estimated counts (table 2.1) range from thirty-five to forty-two, which suggests that, within the size range of available specimens, tooth counts are not strongly correlated with mandible length.

Medial to the marginal tooth row are loci for two large fangs, of which we have seen only a single fully functioning one in any specimen. The fangs are obviously alternately replaced; aside from the evidence of empty loci, there is a developing crown within an otherwise empty locus in PR 1809L, and PR 1705L (not illustrated) has a partly emplaced fang anterior to the functioning one. Fangs are very deeply embedded in the dentary; sectioned specimen PR 2105 shows a fang base extending to the bottom of the Meckelian canal. Because the fang is anchored by bone of attachment, this position must be original. The fangs plus their bone of attachment in fact occupy so much space that there seems to be none available for Meckelian bone, which therefore could not have reached the anterior tip of the jaw. The fangs are apparently borne entirely on the dentary; this is the usual condition and would not require comment but for the exceptional thinness of the dentary medial to the fang bases. We originally considered that the fangs might be anchored medially to the adsymphysial, which would be a unique condition. That this is not the case is suggested by two observations. In some specimens (e.g., 1644L and PR 1809L), a thin wall of dentary can be seen medial to the base of the fang. In others (e.g., PR 1644R and PR 1775R), the adsymphysial has been displaced upward presumably by medial-lateral pressure, with little damage. This seems unlikely if the adsymphysial were extensively bound to the jaw by the bone of attachment. The large symphysial fangs of *Caerorhachis bairdi*, attributed to the adsymphysial by Ruta et al. (2002), might plausibly be attributed to the dentary instead as in *Whatcheeria*. The *Caerorhachis* material is difficult to interpret in this region and, lacking some of the excellently preserved specimens of *Whatcheeria*, we would certainly have assigned its dentary symphysial fangs to the adsymphysial.

Meckelian and Articular

In mature individuals of *Whatcheeria*, the evidence from cross-sectional breaks and from physical continuity in medial view shows that a single Meckelian ossification extends at least from the anterior extremity of the exomeckelian fenestra to the articular. Thus we here consider all parts of the Meckelian, including the articular, under a single heading. In medial view, from anterior to posterior, the Meckelian bone appears first as a series of arches delineating endomeckelian fenestrae within the exomeckelian fenestra, then as a strip of bone between the ventral margin of the prearticular and medial edge of the angular, and then as the articular forming the jaw joint posterodorsally where it is bounded medially by the prearticular and laterally by the surangular (figs. 2.4, 2.6, 2.7). This morphology is generally comparable to that in *Panderichthys* and some Devonian tetrapods where the Meckelian is similarly extensively ossified (Ahlberg & Clack, 1998), but it differs from all known contemporaneous and later tetrapods.

We cannot determine from the specimens available how far an ossified Meckelian may extend anterior to the exomeckelian fenestra. No contribution to the symphysis is discernible on the surface (fig. 2.2C). In the cross-sectional view of PR 1809L (fig. 2.2D), a poorly preserved ossification within the Meckelian canal can be distinguished. The extent of the Meckelian ossification anterior to this point is unknown.

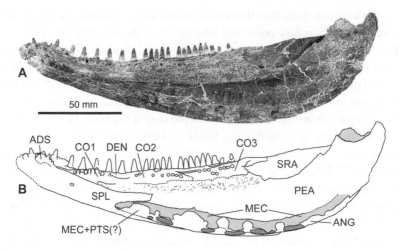

Figure 2.6. *Whatcheeria deltae* PR 1988. Large right jaw preserved in limestone. Symphysis is damaged but Meckelian bone is well-preserved. (A) Photo of jaw in medial view. (B) Line drawing of (A). The area between the ventral margin of the prearticular and ventral margin of the jaw is the exomeckelian fenestra. It contains the endochondral Meckelian bone, which surrounds a number of endomeckelian fenestrae as a series of arches. Light gray indicates Meckelian bone and articular. Darker gray indicates broken bases of the Meckelian bone's arches.

When ossified in the region of the exomeckelian fenestra, the Meckelian is pierced by several endomeckelian fenestrae, ranging from six to nine in the specimens available (figs. 2.4, 2.6). Because of crushing, the arches forming the boundaries of the fenestrae are broken off ventrally in all available specimens, always near the base that is usually visible as a broken surface that may be slightly elevated. This manner of breakage suggests that the bases may themselves be part of the Meckelian bone. It does not necessarily follow, however, that the Meckelian bone floors the Meckelian canal in the area of the exomeckelian fenestra; clear sutural evidence is lacking. The sparse comparative evidence available from the porolepiform sarcopterygian *Laccognathus* and the elpistostegalian *Panderichthys* suggests considerable morphological diversity in the extent of Meckelian ossification (Gross, 1941) and the possibility that the Meckelian sometimes formed the medial wall of the Meckelian canal ventral to the prearticular but did not floor it. This may be the case in *Whatcheeria*. In PR 1665 and PR 1988, there is a short exposure of what could be a Meckelian-postsplenial suture about midlength in the jaw (figs. 2.4, 2.6). In neither case, however, can this suture be traced far enough anteriorly or posteriorly to make it certain that the floor of the exomeckelian fenestra is floored by Meckelian bone.

As best we can determine, in the region of the adductor fossa where the Meckelian lies deep to the prearticular, it partially floors the fossa and has a plate-like dorsal extension adjoining the internal surface of the prearticular. PR 1955 is broken in an approximately frontal plane through the adductor fossa at two different anteroposterior levels. These cross sections show the Meckelian bone directly underlying the prearticular and extending for about half the dorsoventral height of that bone. This morphology is comparable to that seen in similar sections in *Acanthostega* (Ahlberg & Clack, 1998, fig. 8) and in some osteolepiform and porolepiform sarcopterygians (Gross, 1941; *Glyptolepis,* fig. 5F; *Laccognathus,* fig. 14H). We are unable to distinguish any other component of the Meckelian bone in the cross-sectional breaks available.

As noted above, the *Whatcheeria* material includes specimens in which it is likely that only the articular region of the Meckelian is ossified.

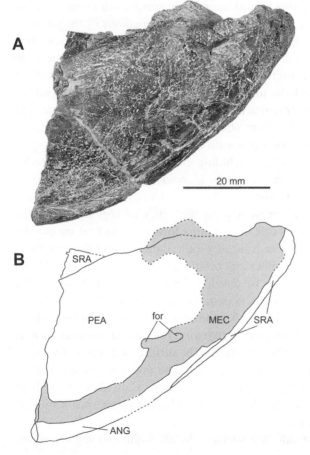

Figure 2.7. *Whatcheeria deltae* PR 1955. Posterior one-quarter of large jaw. (A) Photo of posterior extremity in medial view. (B) Outline drawing of (A). Two foramina are indicated, one (more dorsal) developed partly in the prearticular but penetrating the Meckelian bone, the other developed entirely within the Meckelian.

We have described conditions in PR 1792R; PR 1644L (fig. 2.3), a slightly larger jaw, shows the same slight degree of Meckelian ossification, although there the anterior part of the postsplenial-splenial suture is partly obscured by damage. PR 1646R (not illustrated) is slightly larger again and also shows Meckelian ossification only in the articular region. On the other hand, numerous examples show that by the time a jaw has attained 19–20 cm in overall length, the Meckelian is as fully ossified in the region of the exomeckelian fenestra as it is in the largest available specimens.

The joint surface of the articular is directed dorsally, a tetrapod character (Ahlberg & Clack, 1998). It always has the appearance of unfinished bone and presumably was cartilage-covered. The surface is saddle-shaped, with a low longitudinal ridge separating shallow glenoid depressions. In the only jaw where the articular is relatively uncrushed (PR 1692), these depressions are of markedly different sizes, with the lateral depression being several times larger than the medial. In occlusal view, this specimen shows a substantial anteromedial extension of the articular, which is apparently clasped laterally and medially by the prearticular. Indications of this condition can be seen in other specimens, such as PR 1665, but, without the evidence of PR 1692, these crushed specimens would be practically uninterpretable in this area. This configuration of the articular appears to be unique to *Whatcheeria*. On the medial side of the jaw, the Meckelian-prearticular suture is usually difficult to trace. The combination of evidence from numerous specimens, however, makes it possible to reconstruct its course with confidence. It is clear that the posterior part of the Meckelian, including the articular, has a large medial exposure. Further observations on the glenoid region are given in the section on the surangular below.

A short distance anteroventral to the articular region, the medial surface of the jaw shows one or two large foramina. As best we can determine from material that is usually quite damaged in this area, these foramina vary in both number and position. In some cases (PR 1692, not illustrated), they appear to be developed mostly or entirely within the prearticular (although doubtless penetrating the Meckelian deep to the prearticular). In PR 1955 (fig. 2.7), on the other hand, the prearticular is notched for the upper foramen; otherwise, both are developed entirely within the Meckelian and connected on the surface of that bone by a shallow groove (see reconstruction, fig. 2.1). The upper of these foramina may be the chorda tympani foramen; the more ventral one is perhaps homologous to the angular foramen described in *Greererpeton* by Bolt and Lombard (2001).

On the ventral mandibular surface, the Meckelian-surangular suture can be traced well enough in several specimens (e.g., PR 1955, fig. 2.7) to show that the ventral exposure of the Meckelian extends only to approximately the ventral midline. On the lateral surface of the jaw the Meckelian-surangular suture is usually obscure, perhaps in part because the unfinished articular surface of the jaw joint is difficult to prepare. Several lines of evidence show, however, that the articular is not exposed laterally and that in fact the surangular not only covers the articular in lateral view but itself forms part of the jaw joint. There is first the negative evidence that no specimen shows an articular-surangular suture in lateral view. More convincing are two other observations. PR 2245 is a disarticulated posterior portion of the jaw, consisting of an articulated surangular and angular preserved in medial view on a limestone block. This specimen shows clearly that the surangular formed the continuation of the lateral part of the articular surface of the jaw joint. The surangular portion of the joint surface is heavily pitted but does not have the rugose, unfinished surface of the articular portion itself. Ventral to the surangular part of the glenoid is an extensive and heavily striated region on the inner side of the surangular. We interpret this to be an attachment site for the articular. PR 1644L is a small *Whatcheeria* mandible in a limestone block (fig. 2.3). It is sufficiently immature that only the articular portion of the Meckelian is ossified. Due partly to postmortem displacement, this articular is well separated from the surangular, whose articular region can be seen both in both lateral and dorsal views. As in PR 2245, the surangular has a small articular surface with a rugose texture very similar to that of the true, Meckelian-derived articular region.

There is no retroarticular process in *Whatcheeria,* and the postglenoid region is small. Except for the small lateral portion formed by the surangular, the entire postglenoid region is formed by the articular. On the medial side of the postglenoid area, several specimens show a shallow upwardly facing pit (PR 1665, fig. 2.4) or an approximately dorsoventrally oriented groove (PR 1692, not illustrated, and PR 1955, fig. 2.7). In the latter two, which are large specimens, the medial wall of the groove shows a smooth-surfaced boss. We have referred to a similar structure in *Greererpeton* as a retroarticular pit and suggested that it marks the site of insertion of jaw depressor musculature (Bolt & Lombard, 2001). This remains a plausible interpretation, although alternative functions for this pit cannot be ruled out.

Postsplenial

The postsplenial forms part of the lateral and ventral walls of the mandible. It has no medial lamina. A series of grooves and openings indicate the course of the mandibular lateral line running along the ventral surface. There is no surface sculpture. On the lateral and ventral surface, the postsplenial-splenial suture generally has a single pronounced inter-digitation visible laterally (fig. 2.5). The suture with the angular is straight over most of its course but has some pronounced interdigitations along its midlength (fig. 2.5). The suture with the dentary presents a straight ap-pearance at the surface. In medial view, the postsplenial forms part of the ventral edge of the exomeckelian fenestra (figs. 2.3, 2.6). In larger indi-viduals, the Meckelian in this region contacts the postsplenial as a series of arch bases between endomeckelian fenestrae (figs. 2.3, 2.6). In other post-Devonian tetrapods, the postsplenial articulates with the prearticu-lar on the medial surface of the jaw. In *Whatcheeria,* the presence of the Meckelian precludes this contact.

Prearticular

The prearticular covers most of the medial side of the mandible, ex-tending from the articular area forward to about midlength of coronoid 1. It borders the adductor fossa, where its dorsal margin is concave upward. It also forms all but the anteriormost dorsal border of the exomeckelian fenestra.

In all specimens where it is extensively exposed, the prearticular has been more or less flattened by crushing, sometimes so severely that the entire bone lies in the same plane. PR 1644L and -R, which unfortunately could be only very incompletely prepared and therefore are not illus-trated in medial view, apparently preserve the prearticular in something approximating its original three-dimensional shape. From this and other specimens (e.g., PR 1988, fig. 2.6), it can be seen that the medial aspect of the prearticular has distinct dorsal and ventral faces, sharply separated from one another at a low crest. This crest originates just ventral to the dorsal border of the adductor fossa and about midlength of the fossa. From there it runs forward to the prearticular-splenial suture, anterior to which it is continued on the splenial. At its posterior and anterior ends the crest is rounded; in midlength its summit is a sharp edge. Posterior to the origin of the crest, most of the prearticular in the articular region is at or ventral to the level of the crest. Therefore, we consider the entire area to form part of the ventral face. The ventral face follows the overall curvature of the jaw, but its surface is otherwise smooth, with no large foramina or other prominent features except in some individuals for the one or two foramina described above under the Meckelian bone.

The ventral border of the ventral face is always at least slightly above the endomeckelian fenestrae and, consistent with this, is never notched. This is also reported in *Ichthyostega* but differs from the notched condition in *Acanthostega gunnari* (Ahlberg & Clack, 1998). There is no indication that the prearticular in *Whatcheeria* is a composite bone, such as Ahlberg and Clack (1998) described in the elpistostegalian *Panderichthys* and *Acanthostega* and is known as well in a number of osteolepiforms (see references in Ahlberg & Clack, 1998).

On the ventral face, the prearticular-splenial suture is visible in a number of specimens, at about midlength of coronoid 2, and its continuation onto the dorsal face is also well shown in several specimens. The coronoid-prearticular suture is often difficult to trace, as noted in the description of the coronoids. However, it can be seen clearly in a few specimens, which enables us to determine that the prearticular dorsal face gradually narrows anteriorly to end as a blunt point at the level of the anterior part of coronoid 1. Other than the prearticular-splenial suture, the prearticular has no contact with the infradentary bones because of intervention of the large exomeckelian fenestra.

The prearticular ventral face is edentulous. The dorsal face carries a denticle shagreen, which covers most of that surface up to the prearticular-coronoid suture. Denticles are absent only from the area adjacent to the adductor fossa. Denticles are borne on a surface that is slightly convex upward as preserved and thus might have been more convex before crushing. The orientation and morphology of the prearticular dentigerous area seems quite similar to that of *Acanthostega* (illustrations in Ahlberg & Clack, 1998), despite Ahlberg and Clack's description of the *Acanthostega* prearticular as bearing a "broad longitudinal band of shagreen" on a "raised dorsomesially facing ridge" (p. 20). The distinction appears to be between forms such as *Acanthostega* and *Whatcheeria*, in which the transition between dorsal (denticle-bearing) and ventral faces is abrupt, and those in which it is continuous (such as *Greererpeton;* Bolt & Lombard, 2001).

Splenial

When oriented with the dentary teeth directed dorsally, uncrushed specimen PR 1809 indicates that the splenial was primarily ventral and medial on the jaw with less lateral exposure (fig. 2.2D). The external surface lacks conventional dermal ornament even in the largest specimens, although there are a number of large and small foramina, some of which we presume to be elements of the lateral line system. The mandibular lateral line is partly open and partly enclosed. It is difficult to trace on most specimens, so its exact anterior extent is uncertain. The medial lamina forms the anterior one-third of the medial surface of the jaw. The posterior portion of that lamina forms the converging dorsal and ventral borders of the exomeckelian fenestra, defining its anterior extent. Approximately halfway between the anterior end of the exomeckelian fenestra and the anterior tip of the jaw the splenial is pierced by a foramen, which we consider a possible homologue of the medial parasymphysial foramen (fig. 2.6; the foramen is present in PR 1809L but hidden under the dorsal part of the splenial in fig. 2.2C).

In three-dimensional specimens, the anteroventral border of the splenial is drawn out into a free ventral flange. This is comparable to that in *Acanthostega,* although it does not extend as far medially (compare fig. 2.2D with Ahlberg & Clack, 1998, fig. 3B). On the medial surface, the splenial sutures dorsally with the dentary, adsymphysial, coronoid 1, and prearticular successively from anterior to posterior. The sutures with the latter three are readily distinguishable but that with the dentary at the symphysis is not, so its contribution to the symphysis can only be estimated (figs. 2.2C, 2.6). On the external surface, the splenial-postsplenial suture is readily traceable and runs from anterodorsal to posteroventral. This suture is also visible internally in the floor of the exomeckelian fenestra.

Surangular

The surangular extends anteriorly from the glenoid for half the length of the lateral surface of the mandible (figs. 2.3, 2.5). Its dorsal margin is relatively straight, and nowhere does the surangular rise above the level of the jaw joint. For about half the length of the adductor fossa, it forms most of the lateral wall and rim of the fossa. Anteriorly, however, the postdental process of the dentary plus coronoid 3 form the rim of the fossa for a short distance. The postdental process soon sinks below the rim of the adductor fossa to run in a shallow groove on the lateral face of the surangular. The angular-surangular suture is mostly straight, with a few large-scale interdigitations near its posterior terminus. Posteriorly, the surangular reaches the ventral midline of the mandible where it sutures to the Meckelian bone in larger individuals and presumably lay on the Meckelian cartilage in smaller specimens. This morphology precludes contact between the surangular and the prearticular. The details of this joint are described above in the section on the Meckelian and articular. The surangular is devoid of external surface sculpture. It participates in the lateral part of the glenoid (described above under Meckelian and articular). A surangular contribution to the jaw joint is thought to be uncommon among early tetrapods, although it was proposed by Smithson (1982) for the colosteid *Greererpeton burkemorani*. That the surangular does form part of the joint in *Whatcheeria* suggests that this condition may be more widespread among early tetrapods than is generally recognized.

A discontinuous series of open sulci and foramina indicate that the mandibular lateral line entered the bone near the glenoid and ran anteroventrally partly within the surangular to continue in the angular. Just anterior to the lateral line's entry point, the oral lateral line branched off dorsally to run anteriorly toward the dentary. Like the mandibular, the oral lateral line ran partly within the bone and partly in open sulci. Unlike the mandibular, however, it apparently did not communicate with the surface through foramina. We have not been able to follow the course of the oral lateral line beyond the surangular, and it may have been confined to that bone.

Discussion

Characters of the Jaw in **Whatcheeria**

In their 1998 review, Ahlberg and Clack published a data matrix for fifty characters of the mandible in early tetrapods and the elpistostegalian fish-grade sarcopterygian *Panderichthys*. In a broader study, Ruta et al. (2003) scored forty-six characters for the lower jaw (of a total 319) for ninty taxa: eighty-eight tetrapods and the fish grade sarcopterygians *Panderichthys* and *Eusthenopteron*. For characters of the jaw, the two studies overlapped on nineteen so that, between the two publications, seventy-eight jaw characters are scored for the taxa examined in common. In these works, character states are coded as putatively primitive or derived a priori, with primitive states scored as 0 (zero) by convention. We retain that convention for comparative purposes in our remarks below.

Ahlberg and Clack were able to assign a state to *Whatcheeria* for thirty-five of their fifty characters. Ruta et al. (2003) were able to assign a state to *Whatcheeria* for thirty-two of their forty-six characters for the lower jaw. As a consequence of the work presented in this paper, we are able to assign a state for all seventy-eight unique characters that can be applied to that genus, including both new observations as well as revisions of earlier assignments. Table 2.2 presents a summary of our character state assignments for *Whatcheeria*. We found one of Ahlberg and Clack's characters (24, width of Meckelian bone exposure in the middle part of the jaw relative to the width of coronoids) difficult to interpret. We left their scoring of this character unchanged for comparative purposes during analysis but otherwise do not recommend its use.

Table 2.2. States for jaw characters in *Whatcheeria deltae* as determined in this paper (L&B), compared with determinations by Ahlberg and Clack (1998) (A&C) and Ruta et al. (2003) (RCQ)

	Character number						State	
#	A&C	RCQ	Part	Feature	State	L&B	A&C	RCQ
1	27	178	ADS	bone	*present*	0	0	0
2		181	ADS	denticles	*absent*	1		1
3	32	179	ADS	fang	*absent*	0	0	0
4	30		ADS	irregular teeth	*present*	1	1	
5	31	180	ADS	tooth row	*present*	1	1	1
6		194	ANG	bone	*present*	0		0
7		196	ANG	joint with prearticular	*absent*	**1**		?
8	1		ANG	joint with prearticular	*absent*	**1**	?	
9	3	195	ANG	medial lamina	*absent*	**0**	?	?
10	2	197	ANG	reaches posterior end of jaw	*absent*	0	0	0
11		202	CO1	bone	*present*	0		0
12		204	CO1	denticles	*absent*	1		0
13	13	203	CO1	fangs	*absent*	1	1	1
14	11		CO1	fangs medial to tooth row	*absent*	1	1	
15	5	189	CO1	joint with splenial	*present*	**1**	?	?
16		205	CO1	tooth row	*present*	0		0
17		206	CO2	bone	*present*	0		0
18		208	CO2	denticles	*absent*	1		1
19	14	207	CO2	fangs	*absent*	1	1	1
20	12		CO2	fangs medial to tooth row	NA	1	1	
21	6	190	CO2	joint with splenial	*absent*	**0**	?	?
22		209	CO2	tooth row	*present*	0		0
23		210	CO3	bone	*present*	0		0
24		212	CO3	denticles	*absent*	1		1
25		215	CO3	exposure in lateral view	*present*	1		1
26	15	211	CO3	fangs	*absent*	**1**	?	?
27	7	214	CO3	posterodorsal process	*present*	**1**	0	0
28		213	CO3	tooth row	*present*	0		0

Table 2.2. *(continued)*

	Character number						State	
#	A&C	RCQ	Part	Feature	State	L&B	A&C	RCQ
29	20	182	DEN	accessory tooth row	*absent*	1	1	1
30	21	183	DEN	anterior fang pair	*present*	0	0	0
31		186	DEN	teeth	*present*	0		0
32	19		DEN	tooth attachment	*homodont*	0	0	
33		185	DEN	u-shaped notch anteriorly	*absent*	0		0
34	17	184	DEN	ventral edge morphology	*not chamfered*	0	0	0
35	10		MAN	accessory coronoid teeth	*absent*	**0**	1	
36		216	MAN	coronoid 3 posterodorsal process contributes to surangular crest	NA	NA		?
37	8		MAN	coronoid denticles	*present*	0	0	
38	9		MAN	coronoid tooth row	*all coronoids*	0	0	
39		221	MAN	dentary teeth longer than maxillary teeth	*absent*	0		0
40	25		MAN	dorsal margin of Meckelian foramina/ fenestrae bounded by	*meckelian*	**0**	1	
41	18		MAN	furrow in dentary-splenial suture	*absent*	0	0	
42	26		MAN	height of meckelian foramina/ fenestrae relative to adjacent prearticular	*less*	0	0	
43	16		MAN	joint with infradentaries an overlap and loose	*present*	0	0	
44		200	MAN	lateral exposure of surangular relative to that of the angular	*greater than fifty percent*	**0**		?

(continued)

Table 2.2. *(continued)*

	Character number					State		
#	A&C	RCQ	Part	Feature	State	L&B	A&C	RCQ
45	28		MAN	lateral parasymphysial foramen	*absent*	0	0	
46		187	MAN	length of dentary relative to length of skull	*greater than fifty percent*	0		0
47		188	MAN	length of medial lamina of the splenial relative to distance between anterior tip of jaw and anterior margin of adductor fossa	*greater than fifty percent*	**1**		?
48	45		MAN	mandibular lateral line	*present*	0	0	
49	46		MAN	mandibular lateral line morphology	*mostly open groove, short sections enclosed and opening through pores*	2	2	
50	29		MAN	medial parasymphysial foramen	*present*	**1**	0	
51	4		MAN	number of coronoid bones	3	0	0	
52	50	217	MAN	orientation of adductor fossa opening	*dorsal*	0	?	0
53	49		MAN	pattern of dermal ornament	NA	**NA**	1	
54		222	MAN	teeth cusp chisel-shaped	*absent*	0		0
55		219	MAN	teeth cusp number	*monocuspid*	0		0
56		223	MAN	teeth dimpled	*absent*	0		0
57		218	MAN	teeth pedicillate	*absent*	0		0
58		224	MAN	teeth short and conical	*absent*	0		0
59		220	MAN	teeth with two cuspules arranged labio-lingually	*absent*	0		0

Table 2.2. *(continued)*

#	Character number A&C	RCQ	Part	Feature	State	State L&B	A&C	RCQ
60	24		MAN	width of Meckelian in exomeckelian fenestra relative to width of coronoids	*equal or greater*	0	0	
61	22		MEC	exposed in precoronoid fossa	NA	**<u>NA</u>**	1	
62	23		MEC	ossified in exomeckelian fenestra	*present*	**<u>0</u>**	?	
63	41		PEA	denticles	*present*	**<u>0</u>**	?	
64	42		PEA	denticle distribution	*dorsal band*	**<u>1</u>**	?	
65	40	201	PEA	joint with splenial	*present*	**<u>0</u>**	?	?
66	39		PEA	joint with surangular	*absent*	**<u>0</u>**	?	
67	36		PEA	location of surface pattern center of radiation	NA	NA	?	
68	37		PEA	longitudinal ridge	*present*	1	1	
69	38		PEA	medially projecting flange on dorsal edge at adductor fossa	*absent*	0	0	
70	33	191	PTS	bone	*present*	0	0	0
71	35		PTS	joint with prearticular	*absent*	**<u>0</u>**	?	
72	34	192	PTS	medial lamina	*absent*	**<u>0</u>**	?	?
73	48	193	PTS	pit line	*absent*	1	1	1
74	43		SPL	ventral flange	*present*	**<u>0</u>**	?	
75		198	SRA	bone	*present*	**<u>0</u>**		?
76	44		SRA	crest	*absent*	**<u>0</u>**	?	
77	47		SRA	mandibular lateral line	*present*	0	0	
78		199	SRA	pit line	*absent*	**<u>1</u>**		?

Note: We retained the character and state numbers used in those works but otherwise have restated the characters to conform to PRESERVE format, which required rewording (Lombard and Bolt, 1999). NA = not applicable; refers to features that are logically impossible. For instance, there is no precoronoid fossa in *Whatcheeria*, so a character related to the construction of this fossa (Character 61) is meaningless. L&B character states in bold and underlined are new or corrected observations made in this paper.

Incorporating our additions/revisions to the data for all seventy-eight characters from these studies provides a rough indication of overall primitiveness for the jaw—i.e., the more zeroes assigned to a taxon, the more primitive it is. Seventy-four of the seventy-eight characters can be assigned a numerical score for *Whatcheeria*. For these characters, *Whatcheeria* has the putatively primitive state in fifty of seventy-four, or 67 percent. Similarly, combining the data for Devonian stem tetrapods from the two previous studies indicates that *Ventastega* has fifty-eight of seventy-seven, or 75 percent *Ichthyostega* has fifty-four of seventy-three, or 74 percent; and *Acanthostega* has fifty-two of seventy-seven, or 68 percent putatively primitive states for jaw characters. With obvious caveats, these quick calculations indicate that the mandible of *Whatcheeria* overall is quite primitive—a surprising finding given its relatively late age but consistent with its position in recent systematic studies (Clack, 2002; Ruta et al., 2002, 2003).

Ahlberg and Clack (1998) performed a phylogenetic analysis of their jaw data using PAUP 3.1 (Swofford, 1993). The result was an incompletely resolved consensus tree of more than 2,500 equally parsimonious trees. Refinements resulted in 308 equally parsimonious trees. In both results, *Whatcheeria* was positioned as a stem tetrapod between the Devonian genera *Acanthostega* and *"Tulerpeton."* Given this result, we expected that our additional data for *Whatcheeria* would not improve resolution significantly. That most of our additions and revisions assigned putatively primitive states to *Whatcheeria* also indicates an expectation of little change. This proved to be the case when we used their data matrix with the changes provided by our new observations in the same PAUP analysis they performed. We did not perform a similar analysis with the data of Ruta et al. (2003), as their work was undergoing revision as we completed this manuscript. None of the other observations presented in the present work that have not been translated into formal characters appear to provide strong evidence for relationships of *Whatcheeria.*

Evolution of the Mandible in Early Tetrapods

On the basis of their morphological descriptions and a cladogram constructed from characters of lower jaws alone, Ahlberg and Clack (1998) suggested a working hypothesis for the sequence in which mandibular characters evolved in early tetrapods. Bolt and Lombard (2001) added to this an evolutionary scenario focusing on the Meckelian bone/cartilage and the exo- and endomeckelian fenestrae. Several trends in the distribution of individual jaw elements are evident during the early evolution of tetrapods, with exceptions and possible reversals in individual cases. The dentary increases its dorsoventral contribution to the lateral surface from about a third to well over half. Concomitantly the lateral exposure of the infradentary bones decreases, but their contribution to the medial surface increases significantly. This expansion onto the medial surface begins anteriorly with the splenial and is followed by the angular, surangular, and postsplenial. As this expansion takes place, the prearticular retreats posteriorly from its primitive condition where it reaches the symphysis. As the prearticular retreats and the infradentary bones expand medially, the exomeckelian fenestra becomes reduced in relative size and may subdivide into several fenestrae. The Meckelian bone, which formed endomeckelian fenestrae within the exomeckelian fenestra, ceases to ossify except posteriorly (as the articular). After the Devonian, with the present exception of *Whatcheeria,* the Meckelian is no longer a visible element in the midregion of fossilized jaws. With the loss of this ossification, the angular and surangular become extensively sutured to the prearticular. On the external surface, the mandibular and oral lateral lines are at first entirely enclosed in bony canals, which communicate with the surface via foramina. Later, some segments lose the bone covering and come to lie in open sulci; eventually none of the lateral line is covered, so that the evidence for its presence is one or two (if both oral and mandibular lateral lines are indicated) continuous open sulci on the surface of the infradentary bones. There are equally clear changes in the dentition, primarily progressive loss of the coronoid and adsymphysial dentitions.

This study suggests some additions to this general working scenario:

1. The adsymphysial in osteolepiform sarcopterygians is superficial, almost scale-like, and has very small teeth or denticles. In many early tetrapods, it is larger than in osteolepiforms, carries larger teeth, and is more of a structural element of the mandible. In a number of the earliest tetrapods, the adsymphysial has posterior and ventral extensions that give it a significant area of contact with the dentary, splenial, and coronoid 1. It also forms a considerable part of the symphysis—calculated by us as about nineteen percent in *Whatcheeria* and fifty percent in the colosteid *Greererpeton.* A more complete integration of the adsymphysial into the jaw is plausibly interpreted as a result of its increased size and acquisition of larger teeth compared with osteolepiforms. Given the position of the adsymphysial at the anterior end

of the lower jaw, its greater participation in the symphysis is not surprising. The robust early-tetrapod type of adsymphysial persists into the Pennsylvanian. The adsymphysial in the Middle Pennsylvanian *Megalocephalus pachycephalus* described by Ahlberg and Clack (1998) is very similar to that of *Greererpeton* (Bolt & Lombard, 2001). The jaw from the Pennsylvanian of Nova Scotia described by Godfrey and Holmes (1989) has a large toothed adsymphysial, although its participation in the symphysis cannot be assessed because of poor preservation. However, the Early Permian *Archeria crassidisca* is the only post-Pennsylvanian tetrapod known to us to even possess an adsymphysial. As described by Holmes (1989), the large, tooth-bearing adsymphysial of *Archeria* also forms part of the symphysis, although the extent of its participation is unclear beyond Holmes's statement (p. 181) that it "appears to form the posterior part of the symphysis." The rise and fall of the adsymphysial in early tetrapods would be unpredictable from the only evidence available until recently: the minor element in osteolepiform sarcopterygians and its supposed absence in Permo-Pennsylvanian tetrapods.

2. The symphysis appears to occupy an increasingly large area in early tetrapods, compared with osteolepiform sarcopterygians. We cannot quantify this at present, as symphysial area is rarely given explicitly in publications, and indeed the borders of the symphysis can be somewhat difficult to establish in some cases such as *Whatcheeria*. Nonetheless, the trend appears clear from illustrations of Ahlberg and Clack (1998) and authors they cite.

Cranial Kinesis and the Whatcheeria *Lower Jaw*

We have suggested elsewhere that the mandible of *Greererpeton* had a mobile joint at the symphysis (Bolt & Lombard, 2001), which would have accommodated mediolateral movements of the kinetic skull table-cheek joint. Morphology of the skull table-cheek suture provides morphological evidence for movement of this joint, independent of assumptions about the history of kinetism in the "fish"-tetrapod transition. Morphological concomitants of a mobile intersymphysial joint in *Greererpeton* included an unusually coarsely rugose symphysial surface on the adsymphysial (brassicate structure) and an interdentary gap anterior to the symphysial portion of the adsymphysial. None of these indications of kinesis is present in *Whatcheeria*. The skull table-cheek suture appears immobile, the adsymphysial lacks a brassicate symphysial surface, and there is no apparent interdentary gap. Some degree of mobility at the mandibular symphysis cannot be ruled out, even though *Whatcheeria* lacks the specialized symphysis seen in *Greererpeton* (and the baphetid *Megalocephalus;* Ahlberg & Clack, 1998).

Acknowledgments

We thank L. Barber, L. Bergwall, R. Masek, H. Rittenhouse, W. Simpson, and A. Smith for preparing the specimens used in this study. We are grateful to the numerous members of field crews who worked at the Delta site and to the Iowa Geological Survey Bureau for their generous assistance in collecting at the site. Figures 2.1 and 2.2 were done by M. Donnelly. This work was supported by National Geographic Society grant #3303-86 to J.R.B. and by National Science Foundation grants DEB-9207475 and DEB-9306294 to both authors.

Literature Cited

Ahlberg, P. E. and J. A. Clack. 1998. Lower jaws, lower tetrapods—a review based on the Devonian genus *Acanthostega. Transactions of the Royal Society of Edinburgh, Earth Sciences* 89:11–46.

Ahlberg, P. E., E. Luksevics, and O. Lebedev. 1994. The first tetrapod finds from the Devonian (Upper Famennian) of Latvia. *Philosophical Transactions of the Royal Society of London (B)* 343:303–328.

Bolt, J. R. and S. Chatterjee. 2000. A new temnospondyl amphibian from the Late Triassic of Texas. *Journal of Vertebrate Paleontology* 74:670–683.

Bolt, J. R. and R. E. Lombard. 2001. The mandible of the primitive tetrapod *Greererpeton,* and the early evolution of the tetrapod lower jaw. *Journal of Paleontology* 75:1016–1042.

Bolt, J. R., R. M. McKay, B. J. Witzke, and M. P. McAdams. 1988. A new Lower Carboniferous tetrapod locality in Iowa. *Nature* 333:768–770.

Carroll, R. L. 2001. The origin and early radiation of terrestrial vertebrates. *Journal of Paleontology* 75:1202–1213.

Clack, J. A. 1998. The Scottish Carboniferous tetrapod *Crassigyrinus scoticus* (Lydekker)—cranial anatomy and relationships. *Transactions of the Royal Society of Edinburgh, Earth Sciences* 88:127–142.

———. 2001. *Eucritta melanolimnetes* from the Early Carboniferous of Scotland, a stem tetrapod showing a mosaic of characteristics. *Transactions of the Royal Society of Edinburgh: Earth Sciences* 92:75–95.

———. 2002. An early tetrapod from "Romer's Gap." *Nature* 418:72–76.

Coates, M. I. 1996. The Devonian tetrapod *Acanthostega gunnari* Jarvik: postcranial anatomy, basal tetrapod interrelationships and patterns of skeletal evolution. *Transactions of the Royal Society of Edinburgh: Earth Sciences* 87:363–421.

Daeschler, E. B. 2000. Early tetrapod jaws from the Late Devonian of Pennsylvania, USA. *Journal of Paleontology* 74:301–308.

Fox, R. C., K. S. W. Campbell, R. E. Barwick, and J. A. Long. 1995. A new osteolepiform fish from the lower Carboniferous Raymond Formation, Drummond Basin, Queensland. *Memoirs of the Queensland Museum* 38:97–221.

Godfrey, S. J. and R. B. Holmes. 1989. A tetrapod lower jaw from the Pennsylvanian (Westphalian A) of Nova Scotia. *Canadian Journal of Earth Sciences* 26:1036–1040.

Gross, W. 1941. Über den Unterkiefer einiger devonischer Crossopterygier. *Abhandlungen der Preussischen Akademie der Wissenschaften, Mathematisch-naturwissenschaftliche Klasse Nr.* 7:1–51.

Holmes, R. B. 1989. The skull and axial skeleton of the Lower Permian anthracosauroid amphibian *Archeria crassidisca* Cope. *Palaeontographica Abteilung A* 207:161–206.

Laurin, M. 1998. The importance of global parsimony and historical bias in understanding tetrapod evolution. Part I. Systematics, middle ear evolution, and jaw suspension. *Annales des Sciences Naturelles, Zoologie* 13e série 19:1–42.

Lombard, R. E. and J. R. Bolt. 1999. A microsaur from the Mississippian of Illinois and a format for morphological characters. *Journal of Paleontology* 73:908–923.

Lombard, R. E. and J. R. Bolt. 1995. A new primitive tetrapod, *Whatcheeria deltae,* from the Lower Carboniferous of Iowa. *Palaeontology* 38:471–494.

Paton, R. I., T. R. Smithson, and J. A. Clack. 1999. An amniote-like skeleton from the Early Carboniferous of Scotland. *Nature* 398:508–513.

Ruta, M., A. R. Milner, and M. I. Coates. 2002. The tetrapod *Caerorhachis bairdi* Holmes and Carroll from the Lower Carboniferous of Scotland. *Transactions of the Royal Society of Edinburgh: Earth Sciences* 92:229–261.

Ruta, M., M. I. Coates, and D. L. J. Quicke. 2003. Early tetrapod relationships revisited. *Biological Reviews* 78:251–345.

Smithson, T. R. 1982. The cranial morphology of *Greererpeton burkemorani* Romer (Amphibia: Temnospondyli). *Zoological Journal of the Linnean Society* 76:29–90.

Swofford, D. L. 1993. *PAUP: Phylogenetic Analysis Using Parsimony, Version 3.1.* Computer program distributed by the Illinois Natural History Survey, Champaign.

Witzke, B. J., R. M. McKay, B. J. Bunker, and F. J. Woodson. 1990. Stratigraphy and Paleoenvironments of Mississippian Strata in Keokuk and Washington Counties, Southeast Iowa. *Iowa Department of Natural Resources, Guidebook Series* No. 10:1–105.

Theropod Dinosaurs from the Early Jurassic of Huizachal Canyon, Mexico

Regina C. Munter *and*
James M. Clark

Introduction

In this chapter, we describe the first known theropod dinosaur material from the Early Jurassic deposits of Huizachal Canyon, Mexico. The La Boca Formation in Huizachal Canyon has produced a rich and diverse vertebrate assemblage of small therapsid, mammalian, and reptilian fossils (Clark et al., 1994, 1998; Fastovsky et al., 1995, in press). Zircons from a pyroclastic layer disconformably underlying the fossil-bearing mudstones have been given a preliminary Pb-Pb radiometric date of 186 ± 2 million years before present (Fastovsky et al., 1998, 2005), placing the assemblage as no older than the early part of the Toarcian marine invertebrate stage late in the Early Jurassic (Gradstein et al., 1995). (An earlier attempt at dating the same layer by the U-Pb method was reported as giving a younger age [Fastovsky et al., 1995], placing it in the early Middle Jurassic, but these results were not duplicated with the Pb-Pb method.) The first discovery from this site was the tritylodontid therapsid *Bocatherium mexicanum* (Clark & Hopson, 1985), now represented by four articulated skulls. Three species of amphilestid mammals were recovered here as well (Montellano et al., 1998). The most commonly preserved fossils in the La Boca Formation are sphenodontid rhynchocephalians, including two named monotypic genera closely related to the living *Sphenodon—Cynosphenodon* (Reynoso, 1996) and the dwarf form *Zapatadon* (Reynoso & Clark, 1998). Other taxa from this formation include the only specimens of the enigmatic burrowing diapsid *Tamaulipasaurus morenoi* (Clark & Hernandez, 1994), an exceptionally well-preserved partial skeleton of the pterosaur *Dimorphodon weintraubi* (Clark et al., 1998) and two taxa of crocodylomorphs (Clark et al., 1994).

Dinosaurs previously were known from Huizachal Canyon based only on fragmentary remains. Ornithischian dinosaurs were identified from isolated teeth collected at the Dinosaur National Monument South locality (Clark et al., 1994). Large bone fragments from the Rene's Roost locality were tentatively identified as those of sauropodomorphs, although this has not yet been confirmed.

The two specimens described here were found approximately ¾km apart, but are similar in size and the possibility they are from the same individual cannot be entirely ruled out. The first comprises an articulated pelvis and sacrum that bears several derived similarities with coelophysoids. The second comprises several pieces of a skull preserving a pneumatized paroccipital process, a feature characteristic of birds and other derived theropod dinosaurs. The two specimens are treated separately below.

Methods

The specimens were prepared (by R.C.M.) manually with carbide and steel needles under a Wild MZ7 microscope. An air-powered grinder and airscribe were also used away from the specimen. The combination of silicic and hematitic cement in the matrix around the specimen did not allow chemical preparation.

The pelvic specimen's phylogenetic relationships were examined utilizing the data matrix of Carrano et al. (2002) with an additional character and the data matrix of Rauhut (2003) (see appendix A). The cladistic analysis of the Carrano et al. (2002) matrix was conducted using the branch-and-bound search of PAUP* (Swofford, 1998), using equally weighted characters. The larger data set was analyzed with the same program using the heuristic search with 100 replicates of random stepwise addition and a branch-and-bound search of a subset of the taxa.

The following institutional abbreviations are used: IGM, Instituto de Geología, Universidad Nacional Autónoma de Mexico, Ciudad Universitaria, Mexico; QG, National Museum of Natural History, Bulawayo, Zimbabwe (formerly Queen Victoria Museum).

Systematic Paleontology

Dinosauria Owen, 1842
Saurischia Seeley, 1887
Theropoda Marsh, 1881
Coelophysoidea Welles, 1984

COMMENT: The taxonomy of the suprageneric groups within Theropoda to which *Coelophysis* belongs is complicated by uncertainties about the validity of some higher taxa (e.g., the Ceratosauria, including *Coelophysis* and *Ceratosaurus;* Holtz, 2000) and of type species based on incomplete material. A long debate over the validity of *Coelophysis bauri* (Cope, 1889) due to the fragmentary nature of the holotype was resolved by the International Commission on Zoological Nomenclature (ICZN) in 1996 with the validation of this species and the erection of a neotype. However, the family Coelophysidae (Welles, 1984) may be a junior synonym of the Podokesauridae Huene (1914) (this applies to other names in the family group, such as Coelophysoidea, following the ICZN's Principle of Coordination). The incomplete type material of *Podokesaurus holyokensis* Talbot, 1911, was destroyed in a fire, and several authors recommend that the validity of this species (and any higher taxa based on it) be restricted to the type material (e.g., Rowe & Gauthier, 1990). Adding to the confusion, the coelophysoid genus *Syntarsus* Raath (1969) was recently recognized as a preoccupied name and replaced with *Megapnosaurus* by Ivie et al. (2001). The most recent published phylogenetic analysis of basal theropods (Carrano et al., 2002) recognized a group comprising the genera *Coelophysis, Megapnosaurus,* and *Liliensternus* as the Coelophysidae, and the study of Rauhut (2003) additionally included the genera *Shuvosaurus, Gojirasaurus,* and *Segisaurus* in the larger group Coelophysoidea. Both studies placed *Dilophosaurus, Ceratosaurus,* and Abelisauridae closer to tetanurans, contra Holtz (2000).

IGM 6624 (figs. 3.2–3.4)

Material. The right half of an articulated sacrum and pelvis of an adult theropod dinosaur, previously illustrated before preparation by Fastovsky et al. (1995, fig. 14), discovered by the junior author on April 12, 1994.

Locality and Horizon. The fossil is from the upper, epiclastic unit of the La Boca Formation (Strater, 1993; Fastovsky et al., 1995) in Huizachal Canyon, Mexico (fig. 3.1). This unit is considered late Early Jurassic based on unpublished radiometric dates from below the fossiliferous horizon (Fastovsky et al., 1998). The specimen was found within a single block of eroded mudstone in a streambed at the Casa de Fidencio sublocality in the western end of Huizachal Canyon, in the Ejido de Huizachal (precise specimen locality data at the IGM available to qualified researchers).

The locations of the individual specimens indicate that IGM 6624 and the cranial specimen could not have originated from the same site but may derive from the same horizon. It is therefore unlikely, though not impossible, that they are from the same individual. The cranial specimen was found in situ on the north side of a hill about ¾km west of where IGM 6624 was found in a nodule lying free in a small wash on the east side of

Figure 3.1. Location of Huizachal
Canyon in Mexico. From
Fastovsky et al. (1995).

the hill. From the drainage pattern on this hill, it is clear that the pelvis could not have washed down from the same site as the skull. The piece of mudstone surrounding the pelvis exhibited the effects of substantial water transport, and because it was not found in situ it is possible that the two specimens were originally entrapped in the same stratigraphic layer at different sites on the hill. The disarticulated nature of most specimens in this deposit and their occurrence in debris flows deposited over an area of great topographic relief (Fastovsky et al., 1995) also allow for the possibility of parts of the same animal being widely distributed.

The specimens were encased in a matrix composed of illite mudstone cemented with hematite and silica. The bones were permineralized and replaced with apatite, quartz, calcite, and possibly chlorite (Fastovsky et al., 1995: 569); they may also be replaced with vivianite (Ortlam, 1967; A. Ortlam, personal communication), making the specimens appear blue-green against the red matrix. Like most of the specimens from these deposits, there is little evidence of distortion or crushing.

Description. The specimen consists of an articulated partial pelvis and associated vertebrae, including an incomplete right ilium, ischium, and pubis (fig. 3.2), and the right halves of two dorsal and five sacral vertebrae (fig. 3.3). The anterior process of the iliac blade is missing and the anterior-most preserved portion ends dorsal to the anterior periphery of the pubic peduncle (fig. 3.4). The ilium, pubis, and ischium form a complete and perforate acetabulum (fig. 3.2). The proximal ends of the pubis and ischium are preserved, and much more of the pubis is preserved than of the ischium. All seven vertebrae are eroded so that the neural canal and the internal structure of the vertebrae are exposed (fig. 3.3). The erosion is not consistent throughout the specimen and the left wall of the neural canal is preserved on the fourth and fifth sacral vertebrae. The right anterior half of the first preserved dorsal vertebra is missing in addition to the eroded portion. The right half of the second preserved dorsal vertebra is almost complete but is missing the anterior part of its ventral border. All five sacral vertebrae are otherwise preserved in their entirety except for the anteroventral margin of the fourth sacral vertebra, which has experienced minor cracking.

The middle of the centrum is constricted on all seven vertebrae. The constriction is most pronounced on the two preserved dorsal vertebrae where the ventral surface is strongly arched in lateral view (fig. 3.2). Bulges at the cranial and caudal ends of each centrum represent fused intervertebral joints in the sacral vertebrae and enhance the constriction of the centra. Large pleurocoels are absent on the lateral surface of the vertebrae, although the presence of small pleurocoels cannot be ruled out

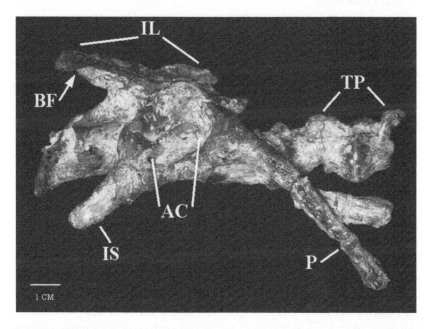

Figure 3.2. IGM 6624 in right lateral view. Abbreviations: IL, ilium; SAC, supra-acetabular crest; TP, transverse process; P, pubis; AC, acetabulum; IS, ischium; BF, brevis fossa.

because of the specimen's state of preservation. The fifth sacral centrum has a gentle concave depression in the middle of its posterior surface. The articulating surfaces of the other vertebrae are obscured by their co-ossification.

The right anterior half of the first preserved dorsal vertebra is missing. The transverse process extends anterodorsolaterally from the center of the neural arch. Posterior to the level of the transverse process, the proximal portion of a neural spine is preserved extending dorsally from the midline and anteriorly to the broken end of the vertebra. Posterior to the neural spine, a short postzygapophysis nearly vertical in orientation extends posterolaterally from near the midline.

The prezygapophysis of the second preserved dorsal vertebra is much longer than the postzygapophysis of the first. The preserved portion lies close to the midline and extends directly anteriorly, traversing the intervertebral articulation. On the lateral surface of the second preserved dorsal vertebra, a transverse process extends dorsolaterally, and an incomplete postzygapophysis lies anterior to the intervertebral articulation. At the base of the transverse process on the lateral surface of the second dorsal vertebra lies a low ridge near where the neurocentral suture presumably extended.

The dorsal and sacral vertebrae are not oriented in a straight line, and the dorsal vertebrae project anterodorsally from the sacrum at a 10° angle (fig. 3.3). The individual vertebrae of the dorsal region are more robust than those of the sacrum, although the last sacral centrum is almost as large as the dorsal centra. Distinguishing individual vertebrae becomes increasingly difficult posteriorly within the sacrum because constriction of the centra and expansion at intervertebral contacts are less prominent.

The sacrum consists of five fused sacral vertebrae, indicating the individual was probably an adult or late subadult. The individual vertebrae decrease in length from the cranial to the caudal ends of the sacrum (see table 3.1). Although sutures between vertebrae are not evident, small ventral bulges between the gently convex ventral surface of the centra indicate where fusion between individual vertebrae has occurred. Constriction of the centra is more prominent on the first, second, and fifth sacral vertebrae. The sacrum is probably complete, because the last centrum is expanded posteriorly in the same way in *Megapnosaurus rhodesiensis, Megapnosaurus kayentakatae* (R. Tykoski, personal communication), and *Coelophysis*. Furthermore, the last sacral vertebra lies posterior to the ilium (figs. 3.2, 3.3). Much of the lateral surface of the sacrum cannot be described because it is obscured by the pelvic elements.

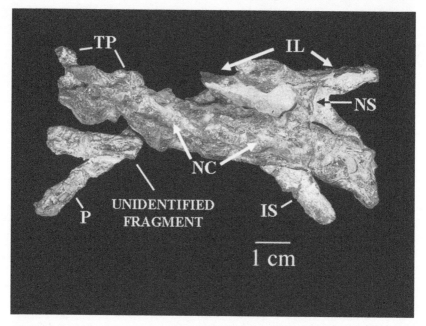

Figure 3.3. IGM 6624 in medial view. Erosion exposed the medial surface of the dorsal and sacral vertebrae and neural canal. An unidentified fragment is attached to the medial surface of the pubic shaft by matrix. Abbreviations: NS, neural spine; NC, neural canal; other abbreviations as in figure 3.2.

An anterior expansion on the first sacral vertebra where it articulates with the last dorsal vertebra strongly resembles the same contact in the two species of *Megapnosaurus* and in *Coelophysis*. This expansion is significantly larger than those between any other vertebrae in the sacrum. A very small and complete prezygapophysis extends anterodorsally from near the midline of the neural arch. On the lateral surface, the proximal end of the transverse process from the first sacral vertebra is preserved and strongly resembles the transverse processes of the dorsal vertebrae.

The second sacral rib or transverse process is twisted, and apparently what was once the dorsal surface of the distal end is now directed anteriorly. This sacral rib has experienced additional distortion and projects anterodorsally from the dorsolateral surface of the second sacral vertebra, contacting the ventral and medial surfaces of the ilium opposite the anterior end of the acetabulum. Dorsal to the posterior end of the second sacral vertebra on the ventral surface of the ilium, a triangular remnant of the second neural spine contacts the ilium with a subquadrangular articulation and extends ventrally toward the sacrum (fig. 3.3). Inside the

Table 3.1. Measurements (mm); vertebrae are measured from centrum to centrum

IGM 6624 (pelvis and sacrum)	
First preserved dorsal vertebra	13.1
Second preserved dorsal vertebra	20.0
Sacral vertebra 1	16.2
Sacral vertebra 2	12.7
Sacral vertebra 3	9.6
Sacral vertebra 4	9.6
Sacral vertebra 5	13.4
Length of entire sacrum	61.5
Length of incomplete ilium	52.1
Mediolateral width of brevis fossa (measured across ventral surface of ilium)	11.0
Acetabular diameter (at its widest point)	16.2
Width of supraacetabular crest (measured from flat portion of ilium to lateral margin)	5.3
Width of dorsal wall of acetabulum	13.1
Dorsoventral length of ventral acetabular margin	7.9
Mediolateral width of ventral acetabular margin	2.79
Width of lateral process on ventral acetabular margin	3.3
Length of pubic shaft	53.1
Length of ischium	20.9
IGM 6625 (skull)	
Length of paroccipital process (incomplete)	31.2
Length of neck of occipital condyle	7.8
Diameter of occipital condyle	9.5
Diameter of floccular fossa	13.2
Length of basioccipital and basisphenoid	20.7
Width of basioccipital and basisphenoid	9.5

neural canal, a small process extends medially into the canal from the middle of the lateral wall.

A tall neural spine extends from the middle of the third sacral vertebra and contacts the ventral surface of the ilium along its dorsomedial edge. The spine is bent slightly posteriorly at its center and contacts the ilium with an expanded, subquadrangular articulation. A sacral rib or transverse process extends from the dorsolateral region of the centrum, contacting the ilium dorsal to the acetabular opening. It runs the anteroposterior length of the vertebra. A small process, similar to that on the second sacral vertebra, projects into the neural canal on the third sacral vertebra.

The fourth sacral rib or transverse process projects anteriorly from the anterolateral surface of the vertebra appearing to join the third sacral rib/process. Matrix still embedded in this region obscures whether these two structures have fused but it appears that only one large process contacts the ilium posterior to the acetabular opening. The fourth sacral vertebra is the most difficult to discern because the anterior and posterior expansions at the points of fusion between the fourth sacral vertebra to the third and fifth vertebra are not as prominent as the intervertebral expansions between other sacral vertebrae. Furthermore, a transverse vertical crack extends through its center, damaging the ventral portion of the vertebra. A small and incomplete postzygapophysis extends dorsolaterally from its posterodorsal margin.

The posterior expansion on the fifth sacral centrum is the most prominent of any on the sacral vertebrae. The posterior surface of this centrum is gently concave. A small prezygapophysis arises dorsolaterally from the anterior end of the neural arch, and the proximal remnant of a postzygapophysis projects from the posterior margin directly dorsally. Two larger processes emerge from the lateral surface of the vertebra. The first extends anterodorsolaterally from the neural arch, contacting the medial surface of the incomplete posteroventral iliac margin. The second process emanates from the lateral surface of the centrum, ventrolateral to the articulation between the first process and neural arch and is identified as the fifth sacral rib. It articulates with the incomplete posteroventral iliac margin posterior to the articulation between the first process and the ilium.

Contact between the sacral ribs or transverse processes and the medial surface of the right ilium was preserved in the second through fifth sacral vertebrae. Articulation between the first sacral vertebra and ilium probably occurred on the missing anterior process of the iliac blade. The dorsal blade of the ilium is oriented almost horizontally above the acetabulum (fig. 3.4) and slants anteroventrally relative to the sacrum (fig. 3.2). Medially, the iliac blade curves dorsally to become vertical at the midline. The medial surface of the vertical portion is rugose in a narrow area just along the edge rather than more broadly as would be produced by articulations with ribs or transverse processes, suggesting that the ilia articulated along their midline. Posteriorly, the iliac blade diverges laterally, away from the midline (fig. 3.4).

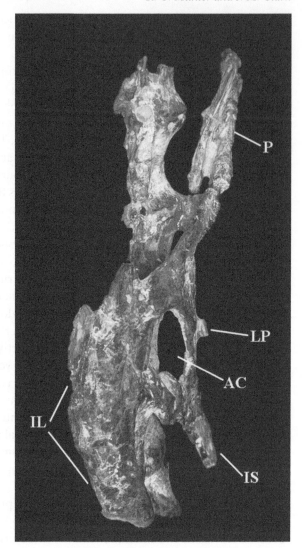

Figure 3.4. Dorsolateral view of IGM 6624. The dorsal blade of the ilium is incomplete, ending dorsal to the anterior extent of the anterior acetabular wall. It also flares laterally from the midline ending in an acute point. The medial edge of the dorsal iliac blade points dorsally and the region above the acetabulum is flat. The ischium is very thin relative to the pubis. Note the laterally directed process on the ventral acetabular border. Abbreviations: LP, lateral process; other abbreviations as in figure 3.2.

The ilium is positioned more dorsally and extends further medially than in other coelophysoids, in which the ilia do not contact, and the possibility that this position and the contact are due to postmortem distortion cannot be ruled out. Thus, the slightly dorsal orientation of the acetabulum seems unlikely to have been present in an animal with an erect hindlimb. This specimen is too poorly preserved to determine whether the shift is natural. However, the rugose medial surface of the iliac blade suggests that contact between the two ilia was indeed present, and fossils from these deposits generally show little distortion. This question cannot be decided without further material.

The ilium terminates posterolaterally in an acute process that overhangs a deep, broadly expanded brevis fossa (figs. 3.2). The preserved portion of the dorsomedial wall of the brevis fossa is formed by a thin lamina of bone and the ventromedial wall by a process that extends posteroventrally from the ilium connecting with the fifth sacral centrum. A deep notch between the dorsomedial and ventromedial walls of the ilium is probably an artifact of incomplete preservation. The brevis fossa expands in size posteriorly, as in *Coelophysis, Megapnosaurus,* and *Liliensternus liliensterni* (Rauhut, 2003). The anterior end of the ilium is missing, and its jagged anterior margin ends dorsal to the anterior periphery of the pubic peduncle (fig. 3.4). A prominent notch lies between the ventral edge of the anterior iliac blade and the pubic peduncle. It is unclear whether the anterior process of the iliac blade had a horizontal or a vertical edge, but the preserved portion is horizontal. The preserved anterior portion of the supra-acetabular crest extends dorsolaterally over the acetabulum on the lateral surface of the ilium. The posterior portion of the supra-acetabular crest is missing.

The short pubic peduncle is so tightly fused to the pubis that a suture cannot be distinguished. In *Megapnosaurus rhodesiensis* (Raath, 1969, 1977) and *Coelophysis* (Colbert, 1989) a thin subvertical sheet of bone lies ventral to the acetabulum, bound anteriorly by the pubic shaft and posteriorly by the ischiadic shaft (the puboischiadic plate of Hutchinson, 2001). This lamina of bone accommodates the obturator foramen and pubic fenestra in these taxa. In IGM 6624, the ventral edge of the pubis and ischium are thin, indicating that a thin sheet of bone may have been present on this specimen but failed to be preserved. A slight arch along the ventral border of the pubis might represent the dorsal margin of the obturator foramen and pubic fenestra that are absent in this specimen.

The pubis extends anteroventrally, curving slightly posteriorly. The pubic shaft is comma shaped in cross section, because the anterior edge of the shaft is thicker than its posteroventral margin. This is more prominent proximally, where the shaft is thicker. The pubis tapers distally and is probably incomplete. At the distal end of the pubic shaft, on the anteromedial surface, lies an anteroposterior suture that might represent where the right and left pubes joined in the pubic symphysis. A long and thin fragment lies medial to the pubis about two-thirds of the way down the pubic shaft (fig. 3.3). It may be the distal portion of the ischium, because both this fragment and the ischium are flatter than the pubis.

The ischium is also tightly fused to the ilium and no suture is apparent. At the point of articulation between the ischiadic peduncle and ischium, the proximal end of the ischium expands into the acetabulum, forming a ridge on its posterior wall. A similar expansion in *M. rhodesiensis* (fig. 3.5B) has been termed the pseudoantitrochanter (Raath, 1977), and this expansion is also present in *M. kayentakatae* (Tykoski, 1998) and *Coelophysis* (as illustrated by Colbert, 1989). The iliac and acetabular parts of the ischium converge to form a posteroventrally directed ischiadic shaft that tapers distally. The ischium is flatter and thinner than the pubis.

The ilium, pubis, and ischium form a perforate acetabulum through which the lateral surface of the sacrum is exposed (fig. 3.2). The perforation is anteroposteriorly oval rather than round, similar to the shape of the acetabulum in *M. rhodesiensis*. The head of the femur could not have filled the acetabulum, and the femur probably articulated with only the more rounded dorsal portion of the acetabulum, closer to the supraacetabular crest (Hutchinson, 2001). A posteriorly incomplete supraacetabular crest extends laterally over the acetabulum. The acetabulum is directed posterodorsally, so that the dorsal and posterior walls are visible laterally. Also in lateral view, the acetabular portion of the pubis overhangs laterally the anteroventral wall of the acetabulum. The ventral border of the acetabulum is straight and slanted posteroventrally from the pubis to the ischium. It is composed equally of the pubis and ischium and, although the two bones appear fused along their ventral contact, a suture along their dorsal contact appears to separate them. A short, laterally directed process lies posterior to the dorsal part of the suture between the pubis and ischium, but it is unclear whether this is an artifact of preservation. The ventral acetabular margin is mediolaterally thin, and the pubic portion is thicker than the ischiadic portion.

Theropoda, incertae sedis: IGM 6625 (fig. 3.5)

Material. Left posterior section of the braincase consisting of the proximal region of the paroccipital process, the prootic and floccular fossa, the occipital condyle, and a fragment of the basioccipital/basisphenoid. Another small fragment of bone may be identified as either the frontoparietal, jugal, or palatine, and a third fragment is identified as an incomplete left laterosphenoid. Many small, unidentifiable fragments of bone were found at the same site. All pieces were found in close proximity, in an area about 1 m².

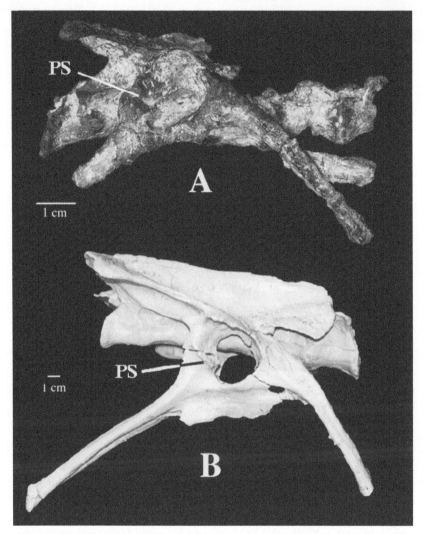

Figure 3.5. Lateral view of IGM 6624 (A) and the pelvis and sacrum of a cast of *Megapnosaurus rhodesiensis* (QG 1), reversed (B). Abbreviation: PS, pseudoantitrochanter.

Locality. From the upper, epiclastic unit of the La Boca Formation (Strater, 1993; Fastovsky et al., 1995) at the Rene's Skull sublocality of the Tierra Buena locality, western end of Huizachal Canyon.

Description. The largest of the three cranial fragments is an incomplete but articulated piece of the left posterior region of the braincase (fig. 3.6).

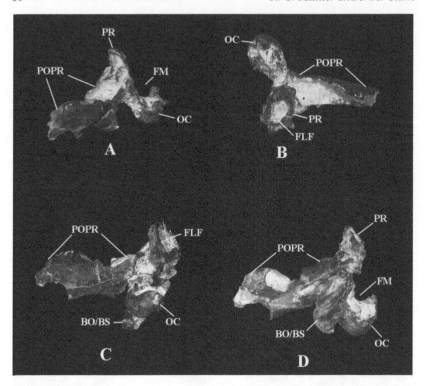

Figure 3.6. IGM 6625, partial theropod braincase. (A) posterior view; (B) dorsal view; (C) posterolateral view; (D) anterior view. Abbreviations: OC, occipital condyle; FM, foramen magnum; POPR, paroccipital process; PR, prootic; FLF, floccular fossa; BO, basioccipital; BS, basisphenoid.

It includes the occipital condyle, the proximal part of the paroccipital process, the region of the prootic that accommodates the floccular fossa, and a fragment of the basioccipital/basisphenoid. The braincase fragment was well preserved and the paroccipital process extends posterolaterally rather than laterally (fig. 3.6A). A second fragment lacks sufficient distinguishing characteristics to be identified confidently, but it may be a jugal, frontoparietal, or palatine. The third cranial fragment consists of an incomplete left laterosphenoid.

The occipital condyle is subcircular in posterior view (figs. 3.6A, C). Its dorsal surface is flat except for a longitudinal depression in the middle that extends its anteroposterior length. The anterior margin of the occipital condyle forms the ventral border of the foramen magnum. The surface of the occipital condyle is smooth except for preserved remnants of the basioccipital body along its right lateral and ventral circumference. The neck of the occipital condyle is long and shaped like a rounded triangle with three ridges protruding from a cylindrical column. The area that included the jugular foramen is present but poorly preserved.

The prootic lies anterodorsolateral to the occipital condyle (figs. 3.6A, B) and accommodates the floccular fossa, preserved as a concave depression on the prootic's medial surface. The floor of the fossa is composed of a thin sheet of bone that served as its ventral boundary. Surrounding the fossa is a thick, smooth, circular border. Unlike the smooth, solid lateral surface typical of archosaurs, the lateral surface of the prootic on this specimen is irregular with several small processes, suggesting it may have been pneumatic.

A small piece of the basioccipital or basisphenoid from the braincase anterior to the occipital condyle is preserved extending posteriorly from the anterior terminus of the condyle (fig. 3.6D). The structure is thin proximally where it contacts the braincase and flares laterally at its distal end. A suture on the ventral surface of this fragment could represent the articulation between the basioccipital and basisphenoid.

The paroccipital process is incomplete, missing its distal end and dorsal border (figs. 3.6B, C). The absence of the dorsal border exposed the hollow internal structure of the process. As preserved, the dorsal border opened dorsally at its proximal end (fig. 3.6B) and anteriorly at its distal end (fig. 3.6D). The floor and walls of the paroccipital process are thin. A hollow passage extends the length of the paroccipital process within its ventral border, as indicated by the broken distal end. The tube widens distally and is not connected to the dorsally positioned hollowed portion of the paroccipital process. The posterior surface of the paroccipital process (fig. 3.6C) lacks any indication of a posterior opening of the caudal tympanic recess, although the specimen's state of preservation does not preclude a small opening.

Among the bones of the skull, the poorly preserved second skull fragment most closely resembles a jugal, a palatine, or possibly a parietal with a portion of the frontal (fig. 3.7). The fragment is composed of three parts: a central body that extends the length of the longest straight edge, a thin oblique process that projects off of one end, and an expanded region that lies at its opposite end. Surface irregularities could be the result of poor preservation or they could represent sutures between multiple bones.

There are problems with identifying this fragment as the parietal. First, the edge that would be the medial edge of a parietal is so thin that no evidence of an articulation is present. Second, only the preserved anterior portion of the medial edge is completely straight. In both *Megapnosaurus*

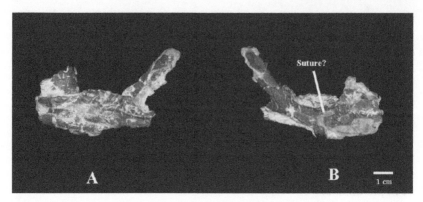

Figure 3.7. IGM 6625, unidentified element, possibly a jugal, parietal, or palatine.

(Rowe, 1989) and *Coelophysis* (Colbert, 1989), the entire medial edge of the parietal was straight. Third, a ridge is not present on the medial wall of the supratemporal fenestra in theropods, and it is unlikely that the preserved ridge is an artifact of preservation. Finally, there are no cerebral impressions on either of the fragment's surfaces.

Alternatively, the fragment could be a jugal when oriented so that the longest edge is ventral and the longest process projects posterodorsally. It exhibits the three-pronged structure characteristic of the jugal in archosaurs where the anterior process articulates with the lacrimal and maxilla, the posterodorsal process articulates with the postorbital, and the posteroventral process articulates with the quadratojugal. The fragment closely resembles a theropod jugal because the postorbital process is longer than the posterior process.

Finally, the fragment may be a palatine when oriented so that the longest process projects posterolaterally like the pterygoid ramus of the palatine. The fossa between the posterolateral process and the posterior end of the fragment might be the anterior border of the suborbital fenestra, although this region is very poorly preserved. However, the surfaces of the fragment are relatively flat, whereas the surfaces of other theropod palatines are dorsally arched.

The third cranial fragment is a left laterosphenoid. The entire dorsal portion of the bone is preserved, although the ventral edge of the fragment is incomplete, missing an articulation with the prootic and basisphenoid. The fragment consists of a central body that is shallow and elongate with three processes of the quadriradiate processes typically found on theropodan laterosphenoids. The lateral surface is convex. A prominent ridge extends the anteroposterior length of the lateral surface. The ridge slants ventrally from its anterior to posterior end. The medial surface of the fragment is smooth and concave and its ventral edge is sharply inturned ventrally toward the midline. No sutures or foramina may be identified on either of the surfaces, although the fossil's state of preservation and its small size do not preclude their existence.

At the anterior end of the fragment a process extends anteromedially similar to the anterior process in *Allosaurus* (Madsen, 1976, fig. 16) that articulates with the orbitosphenoid at the superior angle of the parasphenoid. Two processes extend posterodorsally from the dorsolateral margin of the main body. While the anterior process is shorter than the posterior one, a notch that lies between them is probably an artifact of incomplete preservation. Another process extends medially from the posterior end of the central body. It is the largest of all the processes and is flared dorsoventrally at its distal end, where its edges are jagged and incomplete. On its anterior surface a ridge extends its dorsoventral length.

The dorsal margin of the main body appears complete and is straight. A notch on the ventral edge of the lateral surface of the central body is probably one end of the opening in the laterosphenoid that accommodated the fourth cranial nerve. A second notch located on the ventral edge of the medially directed process is likely to be the opposite end of the same opening. A small fragment of bone extends ventrolaterally around the second notch, as if to form the ventral border of the opening for this cranial nerve.

Discussion

The cladistic analysis of the pelvis based on the Carrano et al. (2002) data matrix (fig. 3.8) places the specimen with the Coelophysoidea when the single character added to the analysis is included (the presence of the expansion on the proximal end of the ischium termed the pseudoantitrochanter by Raath, 1969, 1977). Without this character the placement of the pelvis is much less constrained, and the strict consensus of 560 trees is largely unresolved. Only one of the six characters unambiguously diagnosing the Coelophysoidea (124, a ventrally curved pubic shaft) is identifiable in the Huizachal pelvis. However, two other characters unambiguously place the pelvis closer than *Liliensternus* to *Coelophysis* and *Megapnosaurus* in the results from the total data set: 109 (fusion of pelvic elements in adults) and 117 (straight dorsal margin of ilium). *Coelophysis* and *Megapnosaurus* share character 87 (sacral neural spines fused), for which IGM 6624 has the plesiomorphic condition.

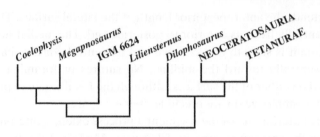

Figure 3.8. Results of a phylogenetic analysis described in appendix A, using the data matrix of Carrrano et al. (2002; appendix 3) with the addition of a 159th character. Neoceratosauria includes *Elaphrosaurus, Ceratosaurus, Masiakasaurus, Laevisuchus, Noasaurus, Genusaurus, Ilokelesia, Abelisaurus, Xenotarsosaurus, Carnotaurus,* and *Majungatholus;* Tetanurae includes *Torvosaurus, Eustreptospondylus, Afrovenator, Sinraptor, Allosaurus,* and *Ornitholestes.* Outgroups (*Herrerasaurus, Eoraptor*) not shown. Strict consensus of eighty-five trees of length 258, CI = 0.678, RI = 0.832.

The cladistic analysis of the pelvis based on the Rauhut (2003) data matrix was complicated because of the large number of taxa. With the complete matrix, only a heuristic analysis of the fifty-eight taxa was possible, resulting in a largely unresolved strict consensus. When a branch-and-bound analysis of the data matrix reduced to only twenty-two taxa was run, the strict consensus was unresolved for coelophysoids but an Adams consensus (fig. 3.9) placed the pelvis with *Megapnosaurus* and *Coelophysis.* (The basal position of *Procompsognathus* in the Adams consensus indicates that ambiguity in its relationships is responsible for the lack of resolution in the strict consensus.) The single unambiguous character supporting this node is the curved pubic shaft (character 183). These three terminals are part of a larger group with *Gojirasaurus* and *Segisaurus* supported unambiguously by one character, the position of the pubic fenestra (character 181, unknown in the Huizachal pelvis), and ambiguously by characters 116 (sacral ribs more or less continuous, reversed in the Huizachal pelvis), 202 (broad groove on distal surface of femur), and 209 (ridge on medial side of proximal end of fibula). The expanded brevis fossa (character 176) is an ambiguous synapomorphy as it is unknown in coelophysoids other than *Coelophysis, Megapnosaurus, Segisaurus,* and *Liliensternus liliensterni.*

The apparent meeting of the ilia along the midline would be a unique character for a coelophysoid, if real. A dorsomedial inclination of the ilium, and sometimes contact between them along the midline, is reported in several theropod taxa (Holtz, 1994), and Rauhut (2003) identified this condition in *Stokesosaurus* as well as several coelurosaurians (e.g., Aves).

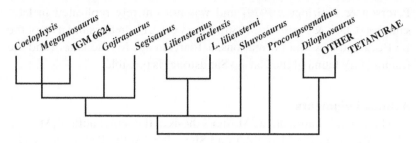

Figure 3.9. Adams consensus of 1394 trees resulting from a branch and bound analysis of the data matrix of Rauhut (2003) with twenty-two taxa (appendix A). "Other Tetanurae" includes *Magnosaurus, Monolophosaurus, Allosaurus,* Sinraptoridae, *Piatnitzkysaurus, Ceratosaurus,* and *Proceratosaurus.* Outgroups (*Herrerasaurus, Eoraptor, Staurikosaurus,* Sauropodomorpha, and Ornithischia) not shown. Tree lengths 271, CI = 0.679, RI = 0.753.

The braincase specimen is only tentatively identified as theropod, based on its general resemblance to theropod occiputs and to the presence of a caudal tympanic recess in the paroccipital process. This feature has not been reported in noncoelurosaurian theropods (Witmer, 1990), although it is reported to be present in the problematic Late Triassic theropod *Protoavis* (Chatterjee, 1991).

The presence of a coelophysoid in the Toarcian (late Early Jurassic) of Mexico is not surprising, as the fauna of the La Boca Formation shares several family level taxa with the Kayenta Formation of Arizona (Clark et al., 1994). This is the youngest well-dated coelophysoid (pending a full account of the dating of the La Boca Formation given a preliminary summary by Fastovsky et al., 1998), but it may be similar in age to *Segisaurus halli* from the Navajo Sandstone and *Megapnosaurus kayentakatae* from the Kayenta Formation. The age of the Kayenta Formation and the overlying, interfingering Navajo Sandstone are generally considered to be Sinemurian or Pliensbachian but this is poorly constrained (Clark and Fastovsky, 1986; Sues et al., 1994). The Navajo Sandstone is overlain unconformably by the Bajocian Temple Cap Sandstone and Carmel Formation (Kowallis et al., 2001) but is unlikely to be this young given its intimate relationship with the Kayenta and the fauna of the latter. However, correlations based on the Kayenta vertebrate fauna are limited by the poor age constraints on other well-known faunas of this age (e.g., the Lower Lufeng Formation of China) and the fragmentary nature of the specimens (e.g., Padian, 1989). The primary evidence for dating the Kayenta Formation as Sinemurian or Pliensbachian is a pollen sample from the underlying Moenave Formation that has never been described in detail (Cornet, in

Peterson & Pipiringos, 1979) and was not entirely replicated in later samples (Clark & Fastovsky, 1986). In any case, given the similarity of the La Boca and Kayenta Formations, a Toarcian age for the latter (and the fragmentary fauna of the Navajo Sandstone) is possible.

Acknowledgments

This paper is based on the Master's thesis of the senior author (Munter, 1999). We thank Pete Kroehler and Steve Jabo of the Paleobiology Department at the National Museum of Natural History for their help in preparing these difficult specimens. We also thank Cathy Forster (Stony Brook University) for advice and Scott Sampson (University of Utah) and the Department of Paleontology at the American Museum of Natural History for access to specimens. Reviews of the manuscript by O. Rauhut and R. Tykoski were extremely helpful, as were suggestions from M. Carrano. The fieldwork was funded by the National Geographic society and was made possible through the generosity of the people of the Ejido de Huizachal, the permission of Instituto Nacional de Antropología e Historia, and the help of the Instituto de Geologia of Universidad Nacional Autónoma de México, especially M. Montellano and R. Hernandez (discoverer of IGM 6625). The junior author thanks Jim and Sue Hopson for their many years of support (but I'm not going to go all gushy on them here).

Literature Cited

Carrano, M. T., S. D. Sampson, and C. A. Forster. 2002. The osteology of *Masiakasaurus knopfleri,* a small abelisaurid (Dinosauria: Theropoda) from the Late Cretaceous of Madagascar. *Journal of Vertebrate Paleontology* 22:510–534.

Chatterjee, S. 1991. Cranial anatomy and relationships of a new Triassic bird from Texas. *Philosophical Transactions of the Royal Society of London, Biology* 332:277–342.

Clark, J. M. and D. E. Fastovsky. 1986. Vertebrate biostratigraphy of the Glen Canyon Group in Northern Arizona; pp. 285–301 *in* K. Padian (ed.), *The Beginning of the Age of Dinosaurs.* New York: Cambridge University Press.

Clark, J. M. and R. Hernández. 1994. A new burrowing diapsid from the Jurassic La Boca Formation of Tamaulipas, Mexico. *Journal of Vertebrate Paleontology* 14:180–195.

Clark, J. M. and J. A. Hopson. 1985. Distinctive mammal-like reptile from Mexico and its bearing on the phylogeny of the Tritylodontidae. *Nature* 315:398–400.

Clark, J. M., R. Hernández, M. Montellano, J.A. Hopson, and D.E. Fastovsky. 1994. An Early or Middle Jurassic tetrapod assemblage from the La Boca Formation, northeastern Mexico; pp. 295–302 *in* N.C. Fraser and H.-D. Sues (eds.), *In the Shadow of the Dinosaurs: Early Mesozoic Tetrapods.* New York: Cambridge University Press.

Clark, J. M., J. A. Hopson, R. Hernández, D. E. Fastovsky, and M. Montellano. 1998. Foot posture in a primitive pterosaur. *Nature* 391:886–889.

Clark, J. M., M. Montellano, J. A. Hopson, R. Hernández, and V.-H. Reynoso. 1998. The Jurassic vertebrates of Huizachal Canyon, Tamaulipas; pp. 1–3 *in Avances en Investigación, Paleontología de Vertebrados.* Universidad Autonoma del Estado de Hidalgo Publicación Especial 1.

Colbert, E. H. 1989. The Triassic dinosaur *Coelophysis. Museum of Northern Arizona* 57:1–160.

Cope, E. D. 1889. On a new genus of Triassic Dinosauria. *American Naturalist* 23:626.

Fastovsky, D. E., S. A. Bowring, and O.D. Hermes. 1998. Radiometric age dates for the La Boca vertebrate assemblage (late Early Jurassic), Huizachal Canyon, Tamaulipas, México; pp. 18–25 *in Avances en Investigación, Paleontología de Vertebrados.* Universidad Autonoma del Estado de Hidalgo Publicación Especial 1.

Fastovsky, D. E., Clark, J. M., Strater, N. H., Montellano, M., R. Hernandez, and J. A. Hopson. 1995. Depositional environments of a Middle Jurassic terrestrial vertebrate assemblage, Huizachal Canyon, Mexico. *Journal of Vertebrate Paleontology* 15(3):561–575.

Fastovsky, D. E., O. D. Hermes, N. H. Strater, S. A. Bowring, J. M. Clark, M. Montellano, and R. Hernandez. 2005. Pre-late Jurassic volcanogenic redbeds of Huizachal Canyon, Tamaulipas, Mexico. *Geological Society of America Special Paper* 393:401–426.

Gradstein, F. M., F. P. Agterberg, J. G. Ogg, J. Hardenbol, P. van Veen, J. Thierry, and Z. Huang. 1995. A Triassic, Jurassic and Cretaceous time scale; pp. 95–126 *in* W. A. Berggren, D. V. Kent, M.-P. Aubry, and J. Hardenbol (eds.), *Geochronology, Time Scales, and Global Correlation. SEPM Special Publication 54.* Tulsa, OK, SEPM (Society for Sedimentary Geology).

Holtz, T. R., Jr. 1996. Phylogenetic taxonomy of the Coelurosauria (Dinosauria: Theropoda). *Journal of Paleontology* 70:536–538.

———. 2000. A phylogeny of the carnivorous dinosaurs. *Gaia* 15:5–61 [publication date of 1998 on journal incorrect.]

Hutchinson, J. R. 2001. The evolution of pelvic osteology and soft tissue on the line to extant birds (Neornithes). *Zoological Journal of Linnean Society* 131:123–168.

Ivie, M. A., S. A. Slipinski, and P. Wegrzynowicz. 2001. Generic homonyms in the Colydiinae (Coleoptera: Zopheridae). *Insecta Mundi* 15:63–64.

Kowalis, B. J., E. H. Christiansen, A. L. Deino, C. Zhang and B. H. Everett. 2001. The record of Middle Jurassic volcanism in the Carmel and Temple

Cap Formations of southwestern Utah. *Geological Society of America Bulletin* 113:373–387.

Madsen, J. H. 1976. *Allosaurus fragilis:* a revised osteology. *Utah Geological and Mineral Survey* 109:1–163.

Montellano Ballesteros, M., J. A. Hopson, J. M. Clark, D. E. Fastovsky, and R. Hernandez. 1998. Panorama de los Mammíferos Mesozoico en Mexico; pp. 12–14 *in Avances en Investigación, Paleontología de Vertebrados.* Universidad Autonoma del Estado de Hidalgo Publicación Especial 1.

Munter, R. C. 1999. Two theropod dinosaur specimens from Huizachal Canyon, Mexico. Master's thesis, Department of Biological Sciences, The George Washington University, Washington, DC.

Ortlam, D. 1967. Fossile Böden als Leithorizonte für die Gliederung des höheren Buntsandsteins im nördlichen Schwarzwald und südlichen Odenwald. *Geologisches Jahrbuch* 84:485–590.

Padian, K. 1989. Presence of the dinosaur *Scelidosaurus* indicates Jurassic age for the Kayenta Formation (Glen Canyon Group, northern Arizona). *Geology* 17:438–441.

Peterson, F. and G.N. Pipiringos. 1979. Stratigraphic relations of the Navajo Sandstone to Middle Jurassic formations, southern Utah and northern Arizona. *U.S. Geological Survey Professional Paper* 1035B:1–43.

Raath, M. A. 1969. A new coelurosaurian dinosaur from the Forest Sandstone of Rhodesia. *National Museum of Southern Rhodesia, Arnoldia* 4:1–25.

———. 1977. The anatomy of the Triassic theropod *Syntarsus rhodesiensis* (Saurischia: Podokesauridae) and a consideration of its biology. Ph.D. dissertation, Rhodes University, Salisbury, Rhodesia.

Rauhut, O. W. M. 2003. The interrelationships and evolution of basal theropod dinosaurs. *Special Papers in Palaeontology* 69:1–213.

Reynoso, V. H. 1996. *Sphenodon*-like sphenodontian (Diapsida: Lepidosauria) from Huizachal Canyon, Tamaulipas, Mexico. *Journal of Vertebrate Paleontology* 16:210–221.

Reynoso, V. H., and J. M. Clark. 1998. A dwarf sphenodontian from the Jurassic La Boca Formation of Tamaulipas, Mexico. *Journal of Vertebrate Paleontology* 18:333–339.

Rowe, T. 1989. A new species of the theropod dinosaur *Syntarsus* from the early Jurassic Kayenta Formation of Arizona. *Journal of Vertebrate Paleontology* 9:125–136.

Rowe, T. and J. Gauthier. 1990. Ceratosauria; pp. 151–168 *in* D. B. Weishampel, P. Dodson, and H. Osmólska (eds.), *The Dinosauria.* Berkeley: University of California Press.

Strater, N. H. 1993. Origins of the pre-late Jurassic strata of Huizachal Canyon (Tamaulipas) and their relationship to the tectonic evolution of northeastern Mexico. Unpublished Master's thesis, University of Rhode Island, Kingston.

Sues, H.-D., J. M. Clark, and F. Jenkins, Jr. 1994. A review of the Early Jurassic
    tetrapods from the Glen Canyon Group of the American southwest; pp.
    284–294 *in* N. C. Fraser and H.-D. Sues (eds.), *In the Shadow of the Di-
    nosaurs: Early Mesozoic Tetrapods.* New York: Cambridge University Press.
Swofford, D. L. 1998. PAUP*: Phylogenetic analysis using parsimony (*and
    other methods), Version 4.0. Sunderland, MA: Sinauer.
Tykoski, R. S. 1998. The osteology of S*yntarsus kayentakatae* and its implica-
    tions for ceratosaurid phylogeny. Unpublished Master's thesis, The Univer-
    sity of Texas at Austin.
Welles, S. P. 1984. *Dilophosaurus wetherilli* (Dinosauria:Theropoda). Osteology
    and comparisons. *Palaeontographica A* 185:85–180.
Witmer, L. M. 1990. The craniofacial air sac system of Mesozoic birds (Aves).
    Zoological Journal of the Linnean Society 100:327–378.

Appendix 3.1

The following characters from Carrano et al. (2002, appendix 2) are scored for the
pelvic specimen, IGM 6624; all other characters are unknown: 84 (1); 85 (0); 86
(1); 87 (0); 109 (1); 110 (1); 111 (0); 113 (0); 117 (1); 118 (1); 124 (1).

An additional character was added to those in Carrano et al.'s (2002) appendix
2: Character 154: Proximal end of ischium: without (0) or with (1) expansion into
acetabulum forming a ridge on its posterior wall (pseudoantitrochanter of Raath,
1969, 1977).

It was scored as follows for the twenty-three taxa in Carrano et al.'s (2002) appendix 3, in the order published there, and IGM 6624 as the 24th entry: ?011?
?00?? ????? ??0?0 0??1

The following characters from Rauhut (2003) are scored for the pelvis specimen, IGM 6624; all other characters are unknown: 113 (2), 114 (1); 115 (0);
166 (0); 167 (0); 168 (0); 171 (1); 175 (0); 176 (1); 177 (0); 179 (0); 183 (1); 188 (0);
189 (0).

To analyze the Rauhut (2003) data matrix using branch-and-bound rather than
a heuristic search, the following thirty-seven tetanurans, distant outgroups,
and poorly known basal theropod taxa were eliminated from the matrix:
*Euparkeria, Marasuchus, Poikilopleuron, Xuanhanosaurus, "Szechuanosaurus"
zigongensis, Coelurus, Compsognathus, Elaphrosaurus, Ornitholestes, Stokesosaurus, Torvosaurus, Afrovenator, Caudipteryx, Chilantaisaurus, "Chilantaisaurus" maortuensis, Ligabueino, Microvenator, Neovenator, Siamotyrannus,
Sinosauropteryx,* SMNK 2349Pal, *Avimimus, Bagaraatan, Deltadromeus, Unenla-
gia, Velocisaurus,* Abelisauridae, Aves, Baryonychidae, Carcharodontosauridae,
Dromaeosauridae, Ornithomimosauria, Oviraptorosauria, Therizinosauroidea,
Troodontidae, Tyrannosauridae.

Herpetoskylax hopsoni, a New Biarmosuchian (Therapsida: Biarmosuchia) from the Beaufort Group of South Africa

Christian A. Sidor

and Bruce S. Rubidge

Introduction

The rocks of South Africa's Beaufort Group (Karoo Supergroup) have supplied a wealth of Permo-Triassic vertebrate fossils (Kitching, 1977; Rubidge, 1995). In particular, the therapsids, or mammal-like reptiles, entombed in Karoo sediments have long been known to chronicle a critical period in premammalian history; mammalian hallmarks such as multicusped teeth and a secondary palate make their first appearance in Karoo cynodonts. In addition to the lineage that eventually gave rise to mammals, many other therapsid groups are abundantly represented in the rocks of the Beaufort Group (Sigogneau, 1970b; King, 1988; Rubidge & Sidor, 2001). Indeed, decades of collecting made it possible for most of the major therapsid clades (traditional suborders) to be recognized and reasonably well understood by the time of Broom's (1932) and Boonstra's (1972) major synthetic works.

Biarmosuchia (Hopson & Barghusen, 1986; Sigogneau-Russell, 1989) was the last major therapsid subclade to be recognized, largely because its members are known only from poorly preserved specimens. Until relatively recently, the handful of specimens known from South Africa (fig. 4.1; table 4.1) were considered to represent a family of aberrant gorgonopsians, the Ictidorhinidae (Broom, 1932; Boonstra, 1952; Sigogneau, 1970b). Sigogneau (1970b) considered ictidorhinids to be a primitive group and to have specializations also developed in other early therapsids. She also suggested that this combination of features was very reminiscent of certain Russian 'eotheriodonts', in particular *Biarmosuchus.* Mendrez-Carroll (1975) echoed these views.

Hopson and Barghusen (1986) considered the genera *Biarmosuchus, Hipposaurus, Ictidorhinus,* and *Burnetia* to be the most primitive of the post-sphenacodont synapsids and assembled them in an informally named group—Biarmosuchia. More recently, Sigogneau-Russell (1989) formally erected the Infraorder Biarmosuchia to include the families Biarmosuchidae, Hipposauridae, Ictidorhinidae, and Burnetiidae. The

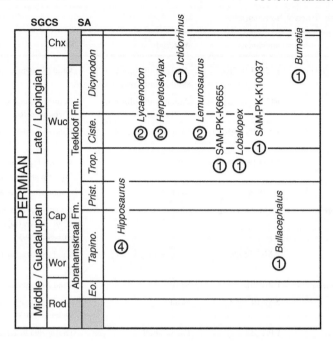

Figure 4.1. Stratigraphic distribution of South African Biarmosuchia. Circled numbers indicate the number of specimens known for each valid biarmosuchian taxon (see table 4.1). SGCS, standard global chronostratigraphic scale; SA, lithostratigraphic subdivisions of the Beaufort Group and vertebrate biozone recognized by the South African Committee for Stratigraphy (Rubidge, 1995). Assemblage Zone abbreviations: *Eo., Eodicynodon; Tapino., Tapinocephalus; Prist., Pristerognathus; Trop., Tropidostoma; Ciste., Cistecephalus.* The short interval above the *Dicynodon* Assemblage Zone represents the beginning of the primarily Triassic *Lystrosaurus* Assemblage Zone. Correlations between SGCS and SA stratigraphies are primarily based on Lucas (2002).

last taxon has been the subject of several recent papers (Rubidge & Sidor, 2002; Rubidge & Kitching, 2003; Sidor & Welman, 2003), but the Biarmosuchia as a whole remain poorly understood.

Here we describe a new genus that is referable to the Biarmosuchia. Its holotype is better preserved than many of this group's previous fossils and sheds considerable light on our understanding of basal therapsid anatomy in general. We are very pleased to offer this contribution on a group that Jim Hopson helped to initially recognize.

The following institutional abbreviations are used throughout this chapter; AMNH, American Museum of Natural History, New York; BMNH, The Natural History Museum, London; BP, Bernard Price Institute for Palaeontological Research, University of the Witwatersrand,

Table 4.1. South African Biarmosuchia

Taxon	Specimen	Material	Status
Lycaenodon longiceps	BMNH R5700	Anterior two-thirds of skull lacking lower jaws	Valid (Sigogneau-Russell, 1989)
	RC 20	Anterior two-thirds of skull lacking lower jaws	Referred specimen (Sigogneau-Russell, 1989)
Hipposaurus boonstrai	SAM-PK-8950	Skull and postcrania	Valid (Sigogneau-Russell, 1989)
	SAM-PK-9081	Skull	Referred specimen (Sigogneau-Russell, 1989)
	CGP/1/66	Skull	Undescribed specimen
Hipposaurus? brinki	SAM-PK-12252	Skull	Tentatively referred to *Hipposaurus* (Sigogneau-Russell, 1989)
Ictidorhinus martinsi	AMNH 5526	Skull lacking lower jaws	Valid (Sigogneau-Russell, 1989)
Lemurosaurus pricei	BP/1/816	Skull	Valid (Sigogneau-Russell, 1989)
	NMQR 1702	Skull	Referred specimen (Sidor and Welman, 2003)
cf. *Lemurosaurus*	BP/1/353	Partial skull	Therapsida *incertae sedis* (Sidor and Welman, 2003)
Rubidgina angusticeps	RC 55	Partial skull	*Nomen dubium* (this paper)
Herpetoskylax hopsoni	CGP/1/67	Skull with lower jaws	Valid (this paper)
	BP/1/3294	Skull with lower jaws	Referred specimen (this paper)
Burnetia mirabilis	BMNH R5397	Skull lacking lower jaws	Valid (Rubidge and Sidor, 2002)
Bullacephalus jacksoni	BP/1/5387	Skull lacking snout	Valid (Rubidge and Kitching, 2003)
Lobalopex mordax	CGP/1/61	Skull and cervical vertebrae	Valid (Sidor et al. 2004)
Burnetiamorpha indet.	SAM-PK-K6655	Anterior part of skull	Undescribed specimen
Burnetiamorpha indet.	SAM-PK-K10037	Skull with lower jaws	Undescribed specimen
Burnetiamorpha indet.	SAM-PK-K5292	Isolated pineal region	Undescribed specimen

Note: The table lists all specimens of South African biarmosuchians currently known to the authors. For each taxon, the holotype is listed first, followed by referred specimens (if any). The status column provides the most recent assessment of the taxonomic status of the taxon in the literature. The taxon *Styracocephalus platyrhinus* was previously considered to be a burnetiid (Sigogneau-Russell, 1989), but it has more recently been transferred to the Dinocephalia by Rubidge and van den Heever (1997).

Johannesburg; CGP, Council for Geosciences, Pretoria; NMQR, National Museum, Bloemfontein; PIN, Paleontological Institute, Moscow; RC, Rubidge Collection, Graaff-Reinet; SAM, South African Museum, Cape Town.

Materials

The following specimens were consulted during the course of this study and form the basis for the character codings presented in appendix 4.1: AMNH 5526 (holotype of *Ictidorhinus martinsi*), AMNH 5633, and 5634 (specimens referred to *Jonkeria*); BMNH R5700 (holotype of *Lycaenodon longiceps*); CGP/1/61 (holotype of *Lobalopex mordax*); CGP/1/66 (an undescribed specimen of *Hipposaurus*); NMQR 1702 (referred specimen of *Lemurosaurus pricei*); NMQR 3000 (holotype of *Patranomodon nyaphulii*); PIN 157/1 (holotype of *Titanophoneus potens*); PIN 157/2 (holotype of *Syodon efremovi*); PIN 1758/2, 7, 8, 19, and 255 (holotype and referred specimens of *Biarmosuchus tener*); PIN 1758/4, 327 (holotype and referred specimens of *Estemmenosuchus uralensis*); PIN 1758/6, (holotype of *Estemmenosuchus mirabilis*); RC 20 (referred specimen of *Lycaenodon longiceps*); SAM-PK-8950 (holotype of *Hipposaurus boonstrai*); SAM-PK-12213b (specimen referred to Titanosuchidae indet.); SAM-PK-K287 (specimen referred to *Titanosuchus*); SAM-PK-11884, 12030 (specimens referred to *Jonkeria*). Our codings for *Suminia getmanovi* are derived from the description of Rybczynski (2000).

Systematic Paleontology

Synapsida Osborn, 1903
Therapsida Broom, 1905
Biarmosuchia Sigogneau-Russell, 1989

Revised Definition. The most recent common ancestor of *Biarmosuchus tener* and *Hipposaurus boonstrai* and all its descendants (from Sidor, 2000).

Revised Diagnosis. Therapsids retaining the following primitive characteristics: temporal fenestra small and situated posteroventral to orbit, jaw adductor musculature does not broadly escape onto skull roof, and ventral edge of occiput moderately rotated anteriorly. Autapomorphies include orbits proportionally large, dentary with marked difference in depth between incisor and postcanine regions, coronoid region of mandible with outturned lateral shelf, incisors and postcanine teeth of approximately equal size, distal carpals 4 and 5 fused to form a single large element, distal tarsals 4 and 5 fused to form a single large element; cervical vertebrae elongate.

Herpetoskylax hopsoni gen. et sp. nov.

Holotype. CGP 1/67, well-preserved skull with lower jaws that is missing the zygomatic arch, postorbital bar, and postdentary bones on its left side.

Referred Material. BP/1/3924, transversely compressed skull with lower jaw that is approximately eighty percent of the size of the type.

Horizon and Type Locality. CGP/1/67 was collected by Mr. Alfred Chuma on the farm Matjiesfontein (235) in the Beaufort West District, Western Cape Province, South Africa, at the following coordinates: 32° 14' 05" South, 21° 39' 50" East. Based on the cohort of fossils collected with CGP/1/67, this specimen is considered to come from the *Cistecephalus* Assemblage Zone (Smith & Keyser, 1995) and thus be Late Permian in age.

Diagnosis. Distinguished from other biarmosuchians by a unique combination of primitive and derived features. Shares with *Lycaenodon* paired anterior prongs on the parietals that bound the preparietal medially. Shares with *Lycaenodon* and *Hipposaurus* a posterior extension on the postfrontal (unknown in Burnetiamorpha). Differs from *Lycaenodon* in that the postfrontal is flat, rather than convex, dorsally. *Herpetoskylax* is autapomorphic in its lack of a distinct fossa on the posterior surface of the squamosal, the degree to which the zygomatic arch appears elevated in lateral view, the facial process of the septomaxilla overlying the nasal, and the large size of the quadratojugal foramen and posttemporal fenestra.

Etymology. Herpeton, creeping animal (Greek); *skylax,* young dog or whelp (Greek). The genus name combines reptilian and mammalian roots to highlight the transitional nature of therapsid fossils. The specific name is given name in honor of Dr. James A. Hopson, for his pioneering work on biarmosuchians.

COMMENTS: We refer BP/1/3294 to *Herpetoskylax hopsoni* and not to the genus *Rubidgina* as Sigogneau (1970a) previously proposed, because we consider the latter to be a nomen dubium. Our rationale for this decision is more fully discussed in the systematic relationships section.

Description

The following description refers primarily to the holotype of *Herpetoskylax hopsoni* (CGP/1/67) with only occasional reference to its referred specimen (BP/1/3294) when the two differ. CGP/1/67 consists of a skull with articulated lower jaws and is complete apart from the absence of the left posterolateral portion (figs. 4.2–4.5). The good bone surface preservation of this specimen offers a remarkable improvement over that of most other South African biarmosuchians. Excellent sutural detail can be made out and the skull shows only minor transverse compression.

The holotype of *Herpetoskylax* retains many primitive features: the skull roof is convex in lateral view, the snout is deep, the temporal opening is small, and the lower jaw lacks a freestanding coronoid process. The orbit appears relatively large, a feature considered by Sigogneau-Russell (1989) to be a biarmosuchian synapomorphy.

Skull Roof

The anterior portion of the snout has been severely weathered, so is it impossible to determine the full extent of the premaxilla (fig. 4.2). We can infer, however, that the dorsal process of the premaxilla was relatively short, as in *Lycaenodon* (personal observation of RC 20) and *Proburnetia* (Rubidge & Sidor, 2002), because no indication of its overlap with the nasal is visible on the preserved portion of the latter element. A relatively short dorsal process of the premaxilla is in contrast to the condition present in the basal biarmosuchians *Hipposaurus* and *Biarmosuchus,* where the dorsal process of the premaxilla reaches nearly halfway between the tip of the snout and orbit. A long premaxillary dorsal process appears to be primitive for Therapsida in that this condition typifies early dinocephalians and anomodonts as well (Hopson & Barghusen, 1986). The remnants of four premaxillary teeth are preserved on either side. We hypothesize the presence of an additional, more medial, tooth in life because of a gap between the most medial preserved tooth and the midline. Serrations could be detected only on the posterior margin of the second to last tooth on the left. The absence of serrations on the other incisors is probably due to wear.

A septomaxilla is preserved on both sides. The more complete, right septomaxilla presents a well-preserved subnarial process that extends between the maxilla and external naris. Posterodorsally, the facial process of the septomaxilla extends onto the surface of the nasal bone. This configuration is unlike that observed in other biarmosuchians (and early therapsids in general), where the septomaxilla slots between the nasal

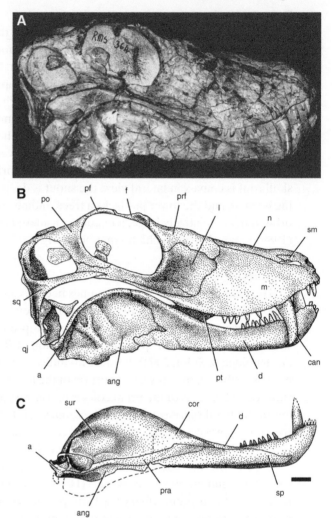

Figure 4.2. *Herpetoskylax hopsoni,* gen. et sp. nov. (CGP/1/67), in lateral view (A, B) and medial view (C) of right lower jaw (reversed). Incisor region of (C) has been reconstructed. Abbreviations: a, articular; ang, angular; can, canine tooth; cor, coronoid; d, dentary; f, frontal; j, jugal; l, lacrimal; m, maxilla; n, nasal; pf, postfrontal; po, postorbital; pra, prearticular; prf, prefrontal; qj, quadratojugal; sm, septomaxilla; sp, splenial; sq, squamosal; sur, surangular. Scale bar applies to (B) and (C) and equals 1 cm.

and maxilla (Sidor, 2003). The ventral margin of this process is paralleled by an elongate septomaxillary foramen.

The maxilla is a large, smooth-surfaced bone that forms most of the lateral antorbital region of the skull (fig. 4.2). The nature of the contact between the maxilla and premaxilla is unclear because of poor preservation. Anterodorsally, the maxilla forms a short contact with the septomaxilla just below the external aperture of naris, whereas dorsally it has a long, straight contact with the nasal. The posterior margin of the body of the maxilla forms a near vertical contact with the anterior edge of the prefrontal, lacrimal, and jugal bones. More posteriorly, the maxilla sends a long tongue of bone along the ventral margin of the skull below the jugal to a point under the orbit. *Herpetoskylax* lacks precanine maxillary teeth. A single, well-developed canine is present, posterior to which is a long diastema. The canine bears coarse serrations along its posterior margin. Several conical postcanine teeth project orthogonal to the convex ventral margin of the maxilla. There are five postcanine teeth preserved on the right and six on the left. On both sides the last two postcanines are the smallest, with the rest being roughly the same size. All the postcanine teeth bear serrations along their posterior carinae. The presence of serrations along the anterior carinae is unclear.

The paired nasals have a large exposure on the skull roof (fig. 4.3). Postmortem transverse compression of the skull has resulted in the nasals being more sharply convex in frontal section than was probably true in life. In dorsal view, the nasomaxillary sutures are medially concave, with the nasals being narrowest about halfway between the anterior margin of the prefrontal and the tip of the snout. No indication of the suture between the nasals and premaxillae can be observed. Posteriorly, the nasals are constricted between the prefrontals and contact the frontals.

The frontal is an elongate, flat bone that stretches between the nasals and parietals (fig. 4.3). Each of these paired elements contributes to half of a deep, inverted V-shaped suture that incises deeply into the posterior margin of the nasals. Midway along its total length, the frontal contributes to a large portion of the orbit's dorsal margin, although a distinct orbital lappet is not formed on the frontal, as in sphenacodontids and edaphosaurids (Reisz et al., 1992; Modesto, 1995). Posteriorly, the frontal forms a slender wedge that is bounded by the postfrontal laterally and the parietal and preparietal medially. The posterior-most tip of the frontal wedge stops at a point in line with the posterior margin of the parietal foramen.

The prefrontal forms the anterodorsal margin of the orbit in lateral view (fig. 4.2). Above the orbit, the bone is a horizontally oriented flat element, whereas anterior to the orbit it takes on a vertical orientation. Between these two surfaces, the prefrontal creates a rounded right angle in frontal section. Anteriorly, the prefrontal contacts the maxilla along a near vertical suture before contacting the nasal on its dorsal surface. Posteriorly, the prefrontal is in contact with the frontal along the supraorbital margin. The prefrontal and lacrimal contact one another along a subhorizontal suture that dips ventrally as it approaches the orbital margin.

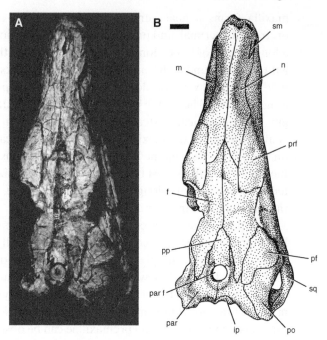

Figure 4.3. Photograph (A) and illustration (B) of *Herpetoskylax hopsoni*, gen. et sp. nov. (CGP/1/67), in dorsal view. Abbreviations: f, frontal; ip, interparietal; m, maxilla; n, nasal; par, parietal; par f, parietal foramen; pf, postfrontal; po, postorbital; pp, preparietal; prf, prefrontal; sm, septomaxilla; sq, squamosal. Scale bar applies to (B) and equals 1 cm.

The postfrontal is a simple bone that forms the posterior third of the orbital margin in dorsal view (fig. 4.3). In contrast to the condition in burnetiamorph biarmosuchians, the postfrontal in *Herpetoskylax* is unpachyostosed and has a relatively flat dorsal surface. Its most interesting feature is a narrow lappet of bone that extends parasagittally between the frontal and postorbital. A similar condition occurs in *Lycaenodon* (RC 20), which was accurately portrayed by Broom (1940, fig. 6), and in *Hipposaurus* (CGP/1/66).

In lateral view, the lacrimal has a large exposure anterior to the orbit and forms the dorsal half of a prominent antorbital depression. The degree to which this depression is an artifact of the transverse compression suffered by the specimen is unknown. On the left, a dorsoventrally elongated lacrimal foramen is visible. An almost horizontal suture extends forward from the anteroventral margin of the orbit and marks the ventral contact of the lacrimal with the jugal.

The morphology of the jugal is seen most clearly in *Herpetoskylax* among biarmosuchians. In lateral view (fig. 4.2), it is an elongate bone that extends from the anteroventral corner of the temporal fenestra to the posterior margin of the body of the maxilla and possesses little, if any, of a postorbital process. The jugal deepens slightly anteriorly, where it contacts the lacrimal and maxilla. The jugal forms most of the ventral margin of the orbit. The portion of the jugal forming the ventral margin of the skull is greatly constricted between the posterior process of the maxilla and the zygomatic process of the squamosal. The jugal barely enters into the temporal fenestra and narrowly separates the squamosal and postorbital in lateral view.

The postorbital is a complex bone forming the bulk of the postorbital bar in addition to a large posterior, or supratemporal, ramus. Ventrally, it wraps under the orbit and has a relatively long sutural contact with the jugal. The postorbital bar is relatively thin transversely and shows no indication of jaw adductor musculature escaping onto its dorsal surface, as in *Biarmosuchus* (Ivachnenko, 1999) and eutherapsids. The postorbital contacts the postfrontal along the posterodorsal margin of the orbit. In dorsal view, the postorbital forms the lateral portion of the broad intertemporal skull roof. It contacts the parietal medially and tabular and squamosal posteriorly.

The squamosal is a triradiate element that forms much of the posterior margin of the skull in lateral view (fig. 4.2). Its anterior, or zygomatic, ramus is a narrow rod of bone that extends under the temporal fenestra and reaches a point under the posterior third of the orbit. *Herpetoskylax* appears unique among biarmosuchians in the degree to which the zygomatic arch is elevated and infratemporal fossa is exposed in lateral view. However, it is possible that transverse compression of the skull has accentuated these features. The dorsal ramus of the squamosal forms the posterior margin of the skull and sutures to the postorbital near the posterodorsal corner of the temporal fenestra. This ramus appears relatively thin in lateral view but can be seen to widen ventrally when viewed from behind. Inside the temporal fenestra, the anterior surface of the squamosal is smooth and undoubtedly served as the origination area for jaw adductor musculature. In posterior view, the ventral, or quadrate, ramus of the squamosal is broadly convex from side to side. In contrast to the condition in most other therapsids, a concavity, the homologue of the external auditory meatus (Allin & Hopson, 1992), is not clearly present on the posterior surface of the squamosal above the temporomandibular joint.

In dorsal view (fig. 4.3), the base of a median, arrowhead-shaped preparietal can be seen to form the anterior margin of the raised parietal foramen. This element is in contact with the frontals and parietals along its lateral margins. The parietal foramen is positioned near the back of the skull roof, roughly in line with the postorbital bar. The preparietal has a complex distribution among therapsids and it seems unlikely that this ossification evolved only once (Sidor, 2001). This neomorphic ossification is present in some biarmosuchians, most anomodonts (except venyukovioids and some Late Triassic forms), and most gorgonopsians (except in some large-skulled rubidgines). When it occurs in the first two groups, the preparietal forms the anterior margin of the parietal foramen and extends between the parietals. In the latter group, the preparietal is interposed between the frontals and parietals and so is excluded from the parietal foramen.

The parietal is a paired bone that forms much of the posterior skull table. Just lateral to the parietal foramen, the parietal extends anteriorly as a thin prong of bone to slot between the preparietal and frontal. A similar condition is visible in *Lycaenodon* (Broom, 1940). Immediately behind the parietal foramen, the parietal meets it mate along a short longitudinal suture. The lateral edge of the parietal trends posterolaterally and sutures to the frontal, postfrontal, and postorbital. In occipital view, the parietal has a relatively large exposure and is positioned between the interparietal, tabular, and postorbital.

Palate

The lateral portions of the palate are obscured because the jaws are tightly occluded. In addition, the palatal ramus of the premaxilla and the anterior portion of the vomer and internal nares are difficult to observe because of this area's inaccessibility to preparation.

In ventral view, the vomer is an elongate, longitudinally troughed bone (fig. 4.4). A midline suture could not be detected and so this normally bilateral element must be regarded as fused. Fusion is complete in *Lycaenodon* (Sidor, 2003) and *Ictidorhinus* (Sigogneau, 1970b) but is only partial in *Burnetia* (Rubidge & Sidor, 2002). As far as can be ascertained, the lateral edges of the vomer are downturned only in its middle portion. Anteriorly, these ridges appear to blend with the body of the vomer and form a flat-to-convex ventral surface. The downturned edges of the vomer stop more abruptly in the posterior direction.

The palatines are large bones that form most of the visible palate (fig. 4.4). Each forms the lateral and posterolateral margins of the choanae. Anteriorly, the contact between the palatine and maxilla is obscured by the mandible. The posterior portion of the palatine expands laterally and ventrally to form a prominent, anteriorly rounded palatal boss. This anterior margin of this boss is studded with the roots of small conical teeth that are arranged in an anteriorly pointing V. This boss continues

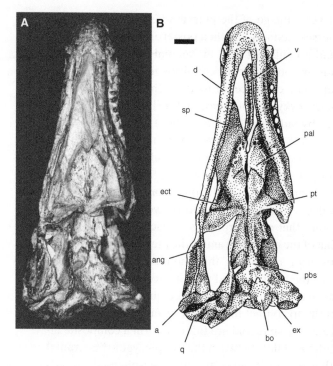

Figure 4.4. Photograph (A) and illustration (B) of *Herpetoskylax hopsoni*, gen. et sp. nov. (CGP/1/67), in ventral view. Abbreviations: a, articular; ang, angular; bo, basiocciptal; d, dentary; ect, ectopterygoid; ex, exoccipital; pal; palatine; pbs, parabasisphenoid; pt, pterygoid; q, quadrate; sp, splenial; v, vomer. Scale bar applies to (B) and equals 1 cm.

posteriorly onto the pterygoid. The left and right palatal bosses have been compressed together to such a degree that only a narrow slit is present in between them. As a consequence, it is impossible to determine whether the vomer contacted the palatine or pterygoids in this region.

The ectopterygoid is relatively large and well preserved on the left side; its medial margin is semicircular, whereas its lateral edge is straight. This element forms much of the anterior face of the transverse flange of the pterygoid and is in contact with the pterygoid posteriorly and the palatine anteromedially.

The contact between the palatine and pterygoid is clearly visible as an anteromedially directed suture positioned midway along the palatal boss. Although transverse compression has drastically decreased the distance between the left and right palatal bosses, it is clear that these structures were relatively large and that the pterygoid contribution is slightly larger in area than that of the palatine. The pterygoid contribution to the palatal boss is ventrally convex and bears two obliquely oriented ridges that converge posteriorly. The roots of small teeth are present on the pterygoid portions of the palatal bosses; there are seven teeth on the left boss and ten on the right. The transverse flange of the pterygoid is thick laterally but thins markedly as it courses medially and then posteromedially. Its ventral surface is ridge-like and is confluent with the medial ridge of the ptery-goid's basicranial ramus. In contrast to the condition in the referred spec-imen (BP/1/3294), there is no evidence for teeth on the transverse flange of the pterygoid in *Herpetoskylax*'s holotype (CGP/1/67). This observa-tion parallels that given by Sidor and Welman (2003) for *Lemurosaurus*: pterygoid flange teeth are retained in the smaller, presumably more juve-nile, specimen but are absent in the larger, more adult specimen. As de-scribed for burnetiamorphs (Sidor, 2000; Rubidge & Sidor, 2002; Sidor & Welman, 2003), the basicranial ramus of the pterygoid in *Herpetoskylax* bears a parasagittal ridge. A trough is formed between the basicranial rami, although it appears that an interpterygoid vacuity is absent. Be-tween this ridge and that of the quadrate ramus, the pterygoid forms a hor-izontal shelf. The posterior margin of this shelf is contacted by the basipterygoid process of the basisphenoid and then opens posteriorly as the cranioquadrate passage. The vertically oriented quadrate ramus ridge continues posterolaterally to contact the medial surface of the quadrate.

The parasphenoid and basisphenoid are fused to form a composite el-ement, which is triangular and slots between the two pterygoid basicranial rami (fig. 4.4). On either side of its apex, the parabasisphenoid extends dorsally and laterally to form basipterygoid processes that suture to the medial side of the pterygoid's quadrate ramus. As in other therapsids, the basicranial joint shows no evidence of being mobile in life. The paraba-sisphenoid broadens posteriorly to become triangular in ventral view. The lateral margins of the triangle are raised and form the anterior margin of the fenestra ovalis. The central portion of the parabasisphenoid bears a shallow fossa that is bisected by a slight midline ridge. The basisphenoid sutures to the opisthotics laterally and the basioccipital posteriorly.

Occiput

The occiput of *Herpetoskylax* (fig. 4.5) is almost square in outline and is rotated slightly anteroventrally so that its ventral edge lies just anterior to its dorsal edge. The left side of the occiput is missing the lateral portion of the paroccipital process, the tabular, the squamosal, the quadrate, and the quadratojugal. In addition, both stapes are missing. The endochon-dral portion of the occiput is solidly united and has disarticulated slightly from the dorsal, dermal elements. As a consequence, a substantial gap separates the left portion of the supraoccipital from the tabular and postparietal.

The unpaired interparietal (or postparietal) is a relatively large, sub-rectangular element that is missing much of its left dorsolateral margin. It is positioned with the parietal above and supraoccipital below. Ventrolaterally it contacts the tabular, whereas dorsolaterally it contacts the parietal. The posterior surface of the interparietal bears the remnants of a broad median ridge.

Only the right tabular is present. As preserved, it is a triangular element that narrows ventrally and extends well below the level of the posttemporal fenestra. The dorsal margin of the tabular is eroded and was probably taller in life. Medially, the tabular contacts the interparietal,

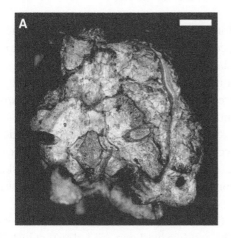

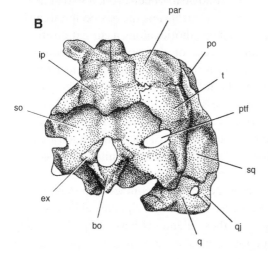

Figure 4.5. Photograph (A) and illustration (B) of *Herpetoskylax hopsoni,* gen. et sp. nov. (CGP/ 1/67), in posterior view. Abbreviations: bo, basiocciptal; ex, exoccipital; ip, interparietal; par, parietal; po, postorbital; ptf, posttemporal fenestra; q, quadrate; qj, quadratojugal; so, supraoccipital; sq, squamosal; t, tabular. Scale bar applies to (A) and equals 1 cm.

supraoccipital, and paroccipital process of the opisthotic and forms the lateral margin of the posttemporal fenestra. The tabular is bordered laterally by the squamosal. Separation between the braincase and temporal region makes it difficult to determine whether the tabular contacted the postorbital as well.

The supraoccipital is a transversely broad bone that appears completely fused to the opisthotic in posterior view (fig. 4.5). It forms the dorsal margin of the foramen magnum and posttemporal fenestra. Immediately above the former, the supraoccipital is swollen to form a broad, median ridge. The latter are large, elliptical openings that angle slightly dorsolaterally. Because of the braincase's slight disarticulation from the remainder of the skull, sutural facets for the tabular are visible along the free edge of the supraoccipital.

Small, paired exoccipitals form the lateral margins of the foramen magnum. Below the lateral process of the exoccipital, two foramina are present. The larger and more laterally placed jugular foramen is positioned between the exoccipital and opisthotic, whereas the smaller hypoglossal foramen resided entirely within the exoccipital. As in *Proburnetia,* the exoccipitals fail to contact one another on the dorsal surface of the occipital condyle (Rubidge & Sidor, 2002). This arrangement is a derived feature unlike that of pelycosaur-grade synapsids (Romer & Price, 1940), dinocephalians (Orlov, 1958), and anomodonts (Rybczynski, 2000), where the exoccipitals contact and form the upper portion of the occipital condyle.

Below the foramen magnum, the basioccipital exclusively forms the occipital condyle. Unfortunately, the condyle has been eroded so that its full shape cannot be described. In posteroventral view, the basioccipital can be seen to be overlapped by the paired opisthotics along a curved suture.

Only the right paroccipital process of the opisthotic is complete (fig. 4.5). It has a gently concave posterior surface and deepens laterally, where it is overlapped by the tabular for approximately half its height. An edge separates the paroccipital process into distinct posterior and ventral faces. In ventral view, the ventral face expands laterally, where it contacts the quadrate and forms the posterior border of the cranioquadrate passage. Sidor (2000) and Sidor et al. (2004) noted a similar expansion in *Lobalopex.*

Only the right quadrate is preserved. It is in contact with the quadratojugal laterally and both emerge from below the squamosal in posterior view (fig. 4.5). It is possible that these elements have slipped ventrally from their position in life and therefore show too great an exposure. The quadratojugal foramen is very large. Medially, the quadrate is in contact with the anterior side of the paroccipital process of the opisthotic and makes a short contact with the quadrate ramus of the pterygoid. In ventral view, the articular surface of the quadrate is oriented anteromedially and bears two condyles. These condyles are anteroposteriorly narrow, convex surfaces that are separated by an obliquely oriented trough. Most of the dorsal process of the quadrate is obscured by the squamosal and is inaccessible to preparation. An unidentified element lies anterior to the quadrate in the right infratemporal fossa (fig. 4.2).

Lower Jaw

The left ramus of the lower jaw is missing its postdentary portion, but the right ramus is essentially complete and well preserved. In lateral view (fig. 4.2), the lower jaw is deep in the region of the incisors and canine, shallows posteriorly to its thinnest point below the last maxillary postcanine tooth, and then deepens again in the coronoid region. Although the jaws are tightly occluded, a great deal of the mandible's medial surface can be observed (fig. 4.2C).

The dentary forms most of the lateral surface of the jaw and, on the right side, can be seen to bear at least eight postcanine teeth. All these teeth occlude lingual to the upper postcanines, but it is not possible to determine whether additional teeth are present anteriorly because of the upper canine. A single lower canine is positioned anteromedial to the upper canine and supports serrations along its posterior carina. Although its tip cannot be observed, this tooth appears to be smaller than the upper canine. Three incisor teeth are preserved in the left ramus and there is space for an additional tooth more posteriorly. The lateral surface of the dentary is relatively flat, but, in the region below the orbit, the dorsal margin swells to form an overhanging ledge that continues onto the surangular. A similar structure is present in other biarmosuchians in which this region is preserved (Sigogneau, 1970a; Rubidge & Sidor, 2002; Sidor & Welman, 2003). The dentary reaches its deepest point just anterior to the midpoint of the orbit. From there, the dentary contacts the surangular and angular along a Z-shaped suture, although the precise contact between the latter two elements is difficult to make out. The effect of this suture is to deeply incise into the posterior margin of the dentary. Although not figured by Sigogneau (1970a), BP/1/3924 shows the same configuration.

The angular is a large bone that forms the ventral half of the mandible's lateral surface posterior to the dentary. The reflected lamina is exceptionally large and its posterior margin closely approaches the posterior margin of the lower jaw. On the lamina's lateral surface, a characteristic set of ridges and fossae are present, the posterodorsal branch of which bifurcates into a distinctive cleft. Between the posterodorsal and posteroventral ridges, the lateral surface of the reflected lamina forms a shallow fossa. The exact location of the angular's lateral contacts with the surangular and articular is difficult to discern. In ventral view, the angular can be seen to slot between the dentary and splenial.

The surangular forms the dorsal border of the lower jaw in the coronoid region. Here, the surangular is thickened into a moderate shelf that overhangs its smooth, mildly concave lateral surface. The position of the contact between the surangular and angular is difficult to determine. In medial view, the surangular contacts the articular and prearticular along its ventral margin.

The splenial is a thin, flat bone confined to the medial aspect of the mandible (fig. 4.2C). It extends from the symphysis to just anterior to the level of the transverse flange of the pterygoid. The splenial makes contact with the anterior ramus of the prearticular and the very anterior tip of the coronoid along its posterodorsal border.

The (posterior) coronoid is a relatively large bone in *Herpetoskylax*. It is visible in medial view as a triangular element that tapers anteriorly from its point of emergence from behind the transverse flange of the pterygoid. In contrast to the condition in pelycosaur-grade synapsids, the coronoid is not visible in lateral view. No trace of an anterior coronoid is detectable.

The prearticular is a thin, strap-shaped bone that, in medial view, courses posteroventrally from its contact with the splenial. Near its posterior contact with the articular, the prearticular twists about its long axis to achieve a more horizontal orientation. In addition, when viewed from below, the prearticular widens transversely so that it matches the width of the articular cotyles. Anterior to the transverse flange of the pterygoid, the prearticular sutures to the coronoid above, splenial below, and dentary laterally. Posterior to the pterygoid's transverse flange, the prearticular contacts the surangular above, angular below, and articular posteriorly. A Meckelian foramen on the medial aspect of the mandible is absent.

The articular of *Herpetoskylax* is the best preserved in any biarmosuchian specimen of which we are aware. In lateral view, the articular sutures to the surangular immediately lateral to the articular cotyles. This suture is difficult to trace but appears to dip ventrally so that the articular has only a small exposure on the lateral surface of the mandible, just posterior to the body of the angular. In ventral view, the articular continues its contact with the angular for a short distance before contacting the prearticular along an anteromedially oriented suture. In dorsomedial view, the articular sutures to the prearticular anterior to the articular cotyles. The articular region consists of two concave facets that are transversely oriented. Behind the lateral cotyle, a small dorsal process is present. Although originally described as a gorgonopsian feature (Parrington, 1955; Hopson & Barghusen, 1986), this structure has since been noted in several biarmosuchians and so might be more widely distributed (Sidor, 2000; Sidor et al., 2004). Ventromedial to the dorsal process, the base of an incomplete retroarticular process is present.

Phylogenetic Relationships of *Herpetoskylax*

Biarmosuchian Terminal Taxa

Before assessing the position of *Herpetoskylax* among biarmosuchians, it is necessary to discuss the rationale behind our recognition of some biarmosuchian genera. These taxonomic decisions are part of an ongoing revision of basal therapsids from South Africa (Sidor, 2000; Rubidge & Sidor, 2002; Rubidge & Kitching, 2003; Sidor, 2003; Sidor & Welman, 2003; Sidor et al. 2004). We focus here on nonburnetiid biarmosuchians, as several previous analyses have demonstrated the monophyly of Burnetiamorpha (Rubidge & Kitching, 2003; Sidor & Welman, 2003).

Biarmosuchus tener Tchudinov, 1960

Holotype. PIN 1758/2, deformed skull with lower jaws and partial postcranial skeleton.

Referred Material. PIN 1758/7, PIN 1758/8, PIN 1758/19, 1758/255.

Recorded Temporal Range. Early Kazanian (Middle Permian); Ezhovo locality, Ocher District, Russia.

Diagnosis. Distinct fossa on postorbital at anterodorsal margin of lateral temporal fenestra; postorbital bar transversely broad; postcanine teeth with basal constriction.

COMMENTS: In his review of the genus, Ivachnenko (1999) synonymized *Biarmosuchus antecessor* (PIN 1758/7), *Eotitanosuchus olsoni* (PIN 1758/1, PIN 1758/85, PIN 1758/319), *Ivantosaurus ensifer* (PIN 1758/292), and *Chthomaloporus lenocinator* (PIN 1758/17) under *Biarmosuchus tener. Biarmosuchus tagax,* which is known from several specimens collected from the Arkhangelsk Region (Mezen District), and *Biarmo-*

suchus tchudinovi, from Udmurtia (Zav'yalovo District), were considered specifically distinct (Ivachnenko, 1990). Although we have not had the opportunity to study all the supposed *Biarmosuchus* material synonymized by Ivachnenko (1999), our examination of the *Eotitanosuchus olsoni* holotype (PIN 1758/1) leads us to believe that it should be considered taxonomically distinct. Thus, for this analysis, only the following specimens were used: PIN 1758/2, 1758/8, 1758/255 (all originally assigned to *B. tener*) and PIN 1758/7 (holotype for *B. antecessor*).

Hipposaurus boonstrai Haughton, 1929

Holotype. SAM-PK-8950, skull, lower jaw, and the majority of the postcranial skeleton.

Referred Material. SAM-PK-9081, skull with lower jaws; CGP/1/66, skull with lower jaws.

Recorded Temporal Range. Middle Permian; *Tapinocephalus* Assemblage Zone, Abrahamskraal Formation, Beaufort Group, South Africa.

Diagnosis. Jugal with tall postorbital process; palatal teeth numerous; quadrate ramus of squamosal long.

COMMENTS: Sigogneau-Russell (1989) provided the most recent review of this genus and concluded that *H. boonstrai* represents its only valid species. In contrast, Sigogneau-Russell only tentatively considered "*Hipposaurus brinki*" (SAM-PK-12252) to belong to the same genus. Our examination of all the material suggests that a revision of the genus is in order. For the purpose of this analysis, we consider only those specimens conforming to the diagnosis of the type species (SAM-PK-8950, SAM-PK-9081; CGP/1/66).

Herpetoskylax hopsoni Sidor & Rubidge, 2003

Holotype. CGP/1/67, nearly complete skull and lower jaws.

Referred Material. BP/1/3294, transversely compressed skull and lower jaws.

Recorded Temporal Range. Late Permian; *Cistecephalus* Assemblage Zone, Teekloof Formation, Beaufort Group, South Africa.

Diagnosis. See section on Systematic Paleontology.

COMMENTS: Two specimens are assigned to *H. hopsoni:* the holotype (CGP/1/67) and a specimen (BP/1/3294) previously referred to as *Rubidgina* sp. by Sigogneau (1970a) and Sigogneau-Russell (1989). Sigogneau-Russell (1989) presumably refrained from assigning the latter specimen to the type species of *Rubidgina, Rubidgina angusticeps,* because of the holotype's poor preservation and several differences between the two specimens. Indeed, our examination of the holotype of *R. angusticeps* (RC 55), indicates no autapomorphies by which to definitively assign new specimens. In addition, the physical incompleteness of RC 55 probably will preclude additional future preparation from revealing diagnostic features. Therefore, we declare *R. angusticeps* to be a nomen dubium. For the purpose of this analysis, our character coding for *Herpetoskylax* stems primarily from its holotype, because it represents the larger and presumably more adult individual.

Lycaenodon longiceps Broom, 1925

Holotype. BMNH R5700, anterior half of skull without lower jaws.

Referred Material. RC 20, anterior two-thirds of skull without lower jaws.

Recorded Temporal Range. Late Permian; Teekloof and Balfour formations, *Cistecephalus* Assemblage Zone, Beaufort Group, South Africa.

Diagnosis. "Differs from other biarmosuchian therapsids in (1) premaxilla possessing scroll-like choanal process; (2) unpaired vomer widest at level of upper canine and tapering both anteriorly and posteriorly; (3) choanal and post-choanal portions of vomer connected by posterolateral lamina; and (4) parietals with short contact between preparietal and pineal foramen. Shares with *Ictidorhinus* (5) convex postfrontal with shallow grooves on external surface" (from Sidor, 2003).

COMMENTS: The taxonomic history of *Lycaenodon* most recently was reviewed by Sidor (2003). RC 20 was designated the holotype of "*Hipposaurus rubidgei*" by Broom (1940). Boonstra (1952) suggested that this specimen did not show the characteristic features of the genus and therefore transferred it to a new genus, *Hipposauroides.* Later, Sigogneau (1970b) transferred RC 20 to *Lycaenodon longiceps,* a taxon previously described by Broom (1925). We agree with Sigogneau (1970b) and find the common possession of a distinct postfrontal morphology in both specimens to be a convincing synapomorphy. The incompleteness of both specimens makes detailed comparison difficult, although they resemble one another and should be considered to belong to the same taxon until the RC 20 is further prepared and redescribed (Sidor, 2003). Codings for *Lycaenodon* in appendix 4.1 are based on both BMNH R5700 and RC 20.

Ictidorhinus martinsi Broom, 1913

Hypodigm. AMNH 5526, small skull lacking lower jaws.

Recorded Temporal Range. Late Permian; Balfour Formation, *Dicynodon* Assemblage Zone, Beaufort Group, South Africa.

Diagnosis. Supraorbital border very convex and thick (from Sigogneau-Russell, 1989), snout deep with dorsal process of premaxilla vertical or slightly anterodorsal in orientation.

COMMENTS: Since Broom's (1913) original description, the holotypic specimen appears to have been subjected to acid preparation, with the result that the outer surface of the skull is smooth and preserves little detail. For example, the presence of a preparietal was discussed by Broom (1913) and figured by Boonstra (1935), although we cannot confidently identify this element today. As noted above, we consider *Ictidorhinus* to be a valid taxon pending a thorough redescription of its anatomy or the discovery of additional material.

The possibility exists that *Ictidorhinus* might represent a juvenile of *Lycaenodon* (the extremely large orbit and short snout being growth-related features). Evidence in support of this potential synonymy is that AMNH 5526 shows the same convex postfrontal morphology, albeit in a more extreme form (possibly caused by diagenetic deformation). However, considering that these taxa are found in different biozones, we believe these genera should be treated as discrete until more specimens of *Ictidorhinus* and *Lycaenodon* are discovered.

Lemurosaurus pricei Broom, 1949

Holotype. BP/1/816, small skull with lower jaws.

Referred Material. NMQR 1702, skull with lower jaws.

Recorded Temporal Range. Late Permian; Middleton and Balfour formations, *Cistecephalus* Assemblage Zone, Beaufort Group, South Africa.

Diagnosis. "Therapsid differing from all other burnetiamorphs in the retention of the following plesiomorphic features: skull roof relatively un-pachyostosed; median nasal boss absent; posteriorly-directed squamosal 'horns' absent. Autapomorphic in its possession of a nubbin-like boss at the dorsal apex of the lateral temporal fenestra, the low, ridge-like form of the median frontal crest, and the very coarse serrations present on the posterior margin of the postcanine teeth" (from Sidor and Welman, 2003).

COMMENTS: Sidor and Welman (2003) provide a recent review of this genus. They recognize one species of *Lemurosaurus, L. pricei,* despite the presence of minor differences between the holotype and the newly re-ferred specimen. They also suggest that Sigogneau's (1970b) assignment of BP/1/353 as cf. *Lemurosaurus* is unfounded because of the pro-nounced morphological difference in the supraorbital region between that specimen and the holotype. Because NMQR 1702 is much more complete and well preserved, in addition to being the larger and presum-ably more adult specimen of *Lemurosaurus,* we use it exclusively for cod-ing this genus in appendix 4.1.

Lobalopex mordax Sidor, Hopson, Keyser, 2004

Holotype. CGP/1/61, skull and lower jaws, with the first four cervical vertebrae in articulation.

Recorded Temporal Range. Early Late Permian; *Tropidostoma* As-semblage Zone, Teekloof Formation, Beaufort Group, South Africa.

Diagnosis. "Therapsid with cranial bosses more moderately developed than in other burnetramorphs; midline nasal boss low and ridge-like; supraorbital ridges weakly developed; squamosals forming weakly devel-oped, posteriorly projecting supratemporal 'horns.' Autapomorphic in possession of parietals with shallow fossac on their dorsal surface just lat-eral to parietal foramen; vomer deeply troughed ventrally and narrow; palatine teeth forming two longitudinal rows separated from a single row of pterygoid teeth by a short gap; laterally projecting knob present at junction of pterygoid and palatine bosses; transverse flange of pterygoid edentulous, with sharply, ridged ventral surface; ectopterygoid with thin lamina connecting lateralmost portion of transverse flange and primary palate" (from Sidor et al., 2004).

COMMENTS: Sidor (2000) provided a preliminary description of CGP/1/61. A full description was published by Sidor et al. (2004). We

have used this specimen and *Lemurosaurus* as representative burnetiamorphs because they are among the morphologically least derived
members of this clade (Sidor, 2000; Sidor and Welman, 2003).

Methods

Our assessment of the phylogenetic position of *Herpetoskylax* builds
on the cladistic analyses of Sidor (2000), Sidor and Welman (2003), and
Rubidge and Kitching (2003). We used fifty-six cranial characters for this
analysis, thirty-nine of which pertained to the skull, eight to the lower jaw,
and nine to the dentition. No postcranial characters were used because
most of the taxa in this analysis are based solely on cranial material. Our
data matrix is presented in appendix 4.1, and our characters and character states are listed in appendix 4.2. The primary branch-and-bound
PAUP (Swofford, 1993) analysis was run with all characters weighted
equally and unordered (see appendix 4.2 for details).

The ingroup included thirteen terminal taxa, seven of which are biarmosuchians. The relatively speciose biarmosuchian subclade Burnetiamorpha was represented by *Lemurosaurus* and *Lobalopex,* as these are
relatively basal taxa and burnetiids proper form a robust clade (Sidor
et al., 2004). Although testing the position of biarmosuchians among
therapsids was not of primary interest to us, we included six representative nontheriodont therapsids (two anteosaurian dinocephalians, two
tapinocephalian dinocephalians, and two anomodonts) in the ingroup
because Ivachnenko (1999) has suggested an intimate relationship between *Biarmosuchus* and anteosaurians. *Dimetrodon limbatus* and *Haptodus garnettensis* were used as consecutive outgroups and coded exclusively from the literature using Romer and Price (1940) and Laurin
(1993), respectively.

Results

Our analysis yielded nine minimum-length trees, consensus trees of
which are presented in figure 4.6. Dinocephalia and Anomodontia each
formed a clade in all trees, as did post-*Biarmosuchus* biarmosuchians.
The nine trees primarily differed in the position of *Biarmosuchus* as well
as in the relative position of *Herpetoskylax* and *Lycaenodon.*

In most trees (fig. 4.6B), Anomodontia was recovered as the sister-
taxon to Biarmosuchia (including *Biarmosuchus*), with Dinocephalia positioned as the most primitive therapsid clade. In three trees, Anomodontia was interposed on the stem between *Biarmosuchus* and the
remaining biarmosuchians included in this analysis. In three other trees,
Anomodontia was positioned as the most basal group of therapsids, with

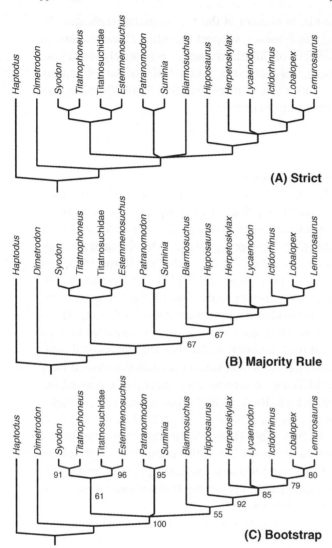

Figure 4.6. Results of cladistic analysis of biarmosuchian relationships. Strict (A) and majority rule (B) consensus cladograms based on nine minimum length trees. Tree statistics: tree length = 104 steps, CI = 0.66, RI = 0.77. Percentage occurrences are given for clades present in the majority of trees in (B). Bootstrap percentages, out of 500 replicates, are given above branches receiving support of fifty percent or greater in (C). Data matrix and characters are presented in appendices 4.1 and 4.2.

Dinocephalia and Biarmosuchia as sister taxa. Most previous analyses have suggested that Anomodontia and Dinocephalia are more closely related to one another than either is to Biarmosuchia (Hopson & Barghusen, 1986; Gauthier et al., 1988).

Figure 4.6C presents the results of bootstrap analysis. Relatively well-supported clades include Therapsida, Anteosauria, Tapinocephalia, Anomodontia, and post-*Biarmosuchus* biarmosuchians, each of which appears in more than ninety percent of the 500 bootstrap replicates. As with the primary analysis, however, support including *Biarmosuchus* as the most primitive member of the Biarmosuchia is relatively low. These results are not surprising, however, given the small size of the data matrix and the incomplete nature of most biarmosuchian fossils (including an almost complete absence of postcrania).

Discussion

Previous Analyses of Biarmosuchia

Hopson and Barghusen (1986, 88) were tentative in their recognition of the taxon Biarmosuchia, stating that, "This group is characterized only by the possession of primitive therapsid characters; therefore, we do not know whether it is monophyletic or paraphyletic." Despite their cautionary statement, this group was included in the analyses of Rowe (1986) and Gauthier et al. (1988) as a (presumably monophyletic) terminal taxon.

Sigogneau-Russell (1989) formally named Biarmosuchia as an infraorder of Therapsida and considered it to represent a natural group. However, numerous features in her diagnosis are clearly primitive for therapsids (e.g., intertemporal roof definitely wider than the interorbital roof, absence of supratemporal, three sacral vertebrae) and therefore do not provide compelling evidence for biarmosuchian monophyly. Among the characters listed by Sigogneau-Russell (1989) that appear to be synapomorphic for Biarmosuchia is the possession of proportionally large orbits. Sigogneau-Russell (1989, table 1) provided a graph in support of this potential synapomorphy and our analysis of this feature corroborates her initial findings (fig. 4.7). Except for one burnetiamorph, all biarmosuchians display an orbit-to-skull length relationship that is greater than would be expected for therapsids generally.

Hopson (1991, 649) reviewed the evidence for biarmosuchian monophyly and proposed three synapomorphies for the clade: (1) postcanine teeth with basal swelling and coarsely serrated margins, (2) distal carpals 4 and 5 fused to form a single large element, and (3) distal tarsals 4 and 5 fused to form a single large element. The latter two features are observable only in *Biarmosuchus* and *Hipposaurus* and thus can be considered synapomorphies of Biarmosuchia only when these taxa are placed into a broader phylogenetic framework. Likewise, we find the first character proposed by Hopson (1991) not to be compelling evidence for biarmosuchian monophyly. Although *Lemurosaurus* has long been known to possess a specialized postcanine dental morphology (Broom, 1949), we consider the

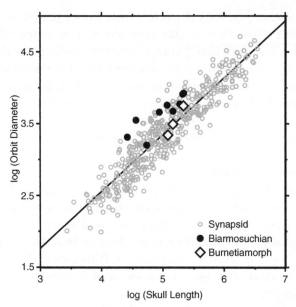

Figure 4.7. Plot of orbit diameter against maximum skull length for nonmammalian synapsids. Biarmosuchian data points are highlighted, with filled circles denoting nonburnetiamorph taxa (i.e., *Biarmosuchus* + traditional "ictidorhinids") and open diamonds denoting burnetiamorphs. As predicted by Sigogneau-Russell (1989), nonburnetiamorph biarmosuchians are not randomly distributed about the synapsid regression but instead lie uniformly above it. From left to right, filled circles denote the following specimens: *Rubidgina* (RC 55); *Ictidorhinus* (AMNH 5526), *Herpetoskylax* (BP/1/3294), *Herpetoskylax* (CGP/1/67), *Biarmosuchus* (PIN 1758/255), *Biarmosuchus* (PIN 1758/2), *Hipposaurus* (SAM-PK-8950), *Biarmosuchus* (PIN 1758/8), and *Hipposaurus* (CGP/1/66). From left to right, open diamonds denote the following specimens: *Lobalopex* (CGP/1/61), *Burnetia* (BMNH R5697), and *Proburnetia* (PIN 2416/1). Regression equation for nonbiarmosuchian synapsids: log (orbit diameter) = $-0.602 + 0.791 \times$ log(skull length). $R^2 = 0.857$. $N = 436$. Data from Sidor (2000).

condition present in the remaining biarmosuchians to be not sufficiently similar to warrant its recognition as a distinctive character state.

In his unpublished dissertation, Sidor (2000) provided the first cladistic appraisal of biarmosuchian monophyly and this clade's position within Therapsida. His primary analysis found nine characters to support biarmosuchian monophyly, but this clade was unexpectedly positioned in a relatively advanced position, as the sister-taxon to Theriodontia. Battail (1992) found a similar result. Sensitivity analyses, in the form of decay indices, bootstrapping, and various alternative weighting schemes, however, showed that, whereas biarmosuchian monophyly was reasonably robust, its placement within Therapsida was much less so (Sidor,

2000, figs. 4.4, 4.5). Several characters from Sidor's (2000) analysis were used in the present analysis (see appendix 4.2), although others were excluded because they were found to be problematic. Sidor (2000) proposed an additional postcranial synapomorphy for Biarmosuchia: the presence of elongated cervical vertebrae. As with the postcranial characters proposed by Hopson (1991), this feature can be observed only in selected taxa (e.g., *Biarmosuchus, Lobalopex*) and thus does not provided unambiguous support for biarmosuchian monophyly.

Biarmosuchian Monophyly

The present analysis did not find unambiguous support for biarmosuchian monophyly. However, two characters can be viewed as diagnostic of Biarmosuchia when plotted on the Adams and majority-rule consensus trees (characters 40 and 50 in appendix 4.2). In addition, Sigogneau-Russell's (1989) observation that a large orbit appears to characterize the group seems valid, although it is not included as a cladistic character in our analysis (fig. 4.7).

Compared with other early therapsids, the biarmosuchian mandible shows a pronounced difference in the depth of the dentary along its length (character 40). The anterior portion of the dentary, which houses the lower incisors and canine, is very deep and rises superiorly to parallel the margin of the maxilla and premaxilla. In contrast, the portion of the dentary housing the postcanine dentition is quite shallow. Anteosaurian dinocephalians (e.g., *Syodon, Titanophoneus, Anteosaurus*) also enlarge the anterior depth of the dentary, undoubtedly to accommodate the roots of their massive incisors but more or less maintain this thickness of bone in the postcanine region.

The second feature found to diagnose biarmosuchians is the possession of incisors and postcanines of approximately equal size (character 50). In *Haptodus, Dimetrodon*, toothed anomodonts, and dinocephalians, the incisor teeth are enlarged relative to the more posterior, postcanine teeth. Our interpretation that large incisors are primitive for Therapsida (and probably Sphenacodontia) and that relatively small incisors are a derived feature is novel. Although toothed anomodonts lack a definitive canine, tooth size clearly decreases posteriorly in these taxa from a maximum in the incisor region.

Even though our analysis of cranial features produced only moderate support for biarmosuchian monophyly (fig. 4.6), we believe the addition of postcranial features to a future analysis would strengthen this support; that is, given the general cladogram topology found here and the distribution of postcranial features discussed by Sidor (2000), the three postcranial characters previously proposed by Hopson (1991) and Sidor (2000) undoubtedly would be understood as additional biarmosuchian synapomorphies.

Biarmosuchian Interrelationships

Post-*Biarmosuchus* biarmosuchians: The clade including *Hipposaurus, Lycaenodon, Ictidorhinus, Herpetoskylax,* and the burnetiamorphs *Lemurosaurus* and *Lobalopex* essentially conforms to the taxon Ictidorhinidae sensu Mendrez-Carroll (1975). This group is primarily known from southern Africa, although two burnetiamorphs (*Proburnetia* and *Niuksenitia*) not included in this analysis have been recovered from Russia (Ivachnenko et al., 1997).

Hipposaurus shares with other more advanced biarmosuchians several characters of the temporal region, including the presence of a preparietal bone (character 24), a long posterior extension on the postfrontal (character 16), and the lack of fossa on the dorsolateral surface of the postorbital for the attachment of jaw adductor musculature (character 13). Furthermore, because of missing data, the following characters are synapomorphic for this clade under DELTRAN optimization: the basipterygoid articulation located at the level of the basicranial ramus of the pterygoid (character 35), the stapedial foramen is absent (character 39), the articular bears a dorsal process along its posterior margin (character 45), and the anterior incisors intermesh (character 49).

Jaw musculature had its origin on the lateral surface of the postorbital bone in *Biarmosuchus,* dinocephalians, and anomodonts. This character distribution suggests that therapsids primitively possessed this configuration and that its absence must be regarded as a reversal among post-*Biarmosuchus* biarmosuchians.

The presence of a dorsal process on the articular in some biarmosuchian therapsids is unexpected, as this has long been thought to be an exclusively gorgonopsian feature (Parrington, 1955; Hopson & Barghusen, 1986). Nonetheless, in those specimens in which the jaw articulation is well preserved (CGP/1/61; CGP/1/66; CGP/1/67), a distinct process can be seen to emanate from the posterodorsal aspect of the articular. The condition is unknown in *Biarmosuchus* and so this feature may be a biarmosuchian synapomorphy instead. As noted by Parrington (1955), this process would have served to prevent the jaws from being opened excessively.

Post-*Hipposaurus* biarmosuchians: In contrast to *Biarmosuchus, Hip-*

posaurus, and most other early therapsids, the remaining biarmosuchians included in this analysis possess, or are inferred to possess, a relatively short dorsal process of the premaxilla (character 2). In addition, the posterior margin of each frontal is V-shaped and extends backward to at least the level of the parietal foramen (character 11). On the palate, the palatine dentition is restricted to a small patch (character 29). Finally, the posterior margin of the dentary is deeply incised by angular (character 41) in lateral view, as in shown in figure 4.2 for *Herpetoskylax.*

Position of Biarmosuchia within Therapsida

The morphology of the temporal region (temporal roof, temporal fenestra, and its associated jaw adductor musculature) has figured prominently in discussions of therapsid evolution (Broom, 1932; Hopson, 1987; Laurin & Reisz, 1996; Rubidge & Sidor, 2001). Indeed, this complex was chief among the characteristics used by Hopson and Barghusen (1986) to argue for the relatively basal position of Biarmosuchia within Therapsida. Our research supports the idea that Biarmosuchia is a monophyletic group, but placing this group within the broader scheme of therapsid evolution remains difficult to do with confidence.

Although the temporal region of biarmosuchians most closely resembles the condition of pelycosaur-grade synapsids, the excellent preservation of *Herpetoskylax* has permitted the recognition of several unexpectedly derived features as well. In addition to the presence of a dorsal process on the articular discussed above, the exoccipitals do not contact on the floor the foramen magnum in *Herpetoskylax.* This character has previously been recognized only in theriodonts (i.e., gorgonopsians, therocephalians, and cynodonts) among therapsids. In addition, the vomer is unpaired (which is also characteristic of gorgonopsians) and the lower canine fits into the choana (which is true of both gorgonopsians and therocephalians). This suite of derived features, some or all of which might have been gained in parallel with theriodonts, suggests that questions remain to be answered regarding the phylogenetic position of biarmosuchians. This is especially true considering the low support values for the cladogram produced here and the fact that Sidor (2000) and Battail (1992) have, at least tentatively, placed biarmosuchians as the sister-taxon to advanced, theriodont therapsids.

Summary and Future Directions

The past fifteen years have produced fundamental insights into the evolution of early therapsids on the southern continents (Rubidge, 1990;

Rubidge & Hopson, 1990; Modesto et al., 1999; Langer, 2000; Rubidge & Sidor, 2001). *Herpetoskylax* is the best understood non-burnetiid biarmosuchian recovered thus far and casts considerable new light on the cranial morphology of this group. It, along with other well-preserved biarmosuchian fossils from South Africa, has provided strong character evidence in support of the monophyly of a group including "ictidorhinids" and burnetiamorphs (Rubidge & Kitching, 2003; Sidor, 2003; Sidor & Welman, 2003). Future research should strive to incorporate South African fossils into a more global picture of early therapsid evolution. A fuller understanding of this important period in synapsid evolution will be possible only with increased international collaboration and the search for new Permian localities outside the traditional collecting areas.

Acknowledgments

Jim Hopson has been a wonderful mentor, colleague, and friend to both of us, and we are very pleased to be able to offer this contribution in his honor. His remarkable breadth of therapsid experience and knowledge has been generously shared with us over the years, and for this we are very grateful. We thank Drs. André Keyser and Johann Neveling (CGP) for the loan of the holotype of *H. hopsoni* and the following museum personnel for access to comparative material: Mark Norell, Gene Gaffney (AMNH), Angela Milner, Sandra Chapman (BMNH), Johann Welman (NMQR), Michael Ivachnenko (PIN), and Roger Smith and Sheena Kaal (SAM). We are indebted to the late Gert Modise (BP) for his excellent preparation of CGP/1/67 and to Ernst du Plessis for the illustrations. Research leading to this contribution has been supported by the National Research Foundation of South Africa, a fellowship from the Smithsonian Institution, and the New York College of Osteopathic Medicine. Our manuscript benefited from the careful reviews provided by Drs. Robert Reisz and Sean Modesto.

Literature Cited

Allin, E. F. and J. A. Hopson. 1992. Evolution of the auditory system in Synapsida ("mammal-like reptiles" and primitive mammals) as seen in the fossil record; pp. 587–614 *in* D. B. Webster, R. R. Fay, and A. N. Popper (ed.), *The Evolutionary Biology of Hearing.* New York: Springer-Verlag.

Battail, B. 1992. Sur la phylogénie des thérapsides (Reptilia, Therapsida). *Annales de Paléontologie* 78:83–124.

Boonstra, L. D. 1935. On the South African gorgonopsian reptiles preserved in the American Museum of Natural History. *American Museum Novitates* 772:1–14.

————. 1952. Die gorgonopsiër-geslag, *Hipposaurus,* en die familie, Ictidorhinidae. *Tydskrif vir Wetenskap en Kuns* 12:142–149.

————. 1972. Discard the names Theriodontia and Anomodontia: a new
classification of the Therapsida. *Annals of the South African Museum*
59:315–338.

Broom, R. 1905. On the use of the term Anomodontia. *Records of the Albany
Museum* 1:266–269.

————. 1913. On some new carnivorous therapsids. *Bulletin of the American
Museum of Natural History* 32:557–561.

————. 1925. On some carnivorous therapsids. *Records of the Albany Museum*
25:309–326.

————. 1932. *The Mammal-like Reptiles of South Africa and the Origin of Mammals.* London: Witherby.

————. 1940. Some new Karroo reptiles from the Graaff-Reinet District. *Annals of the Transvaal Museum* 20:71–87.

————. 1949. New fossil reptile genera from the Bernard Price collection. *Annals of the Transvaal Museum* 21:187–194.

Gauthier, J., A. Kluge, and T. Rowe. 1988. Amniote phylogeny and the importance of fossils. *Cladistics* 4:105–209.

Haughton, S. H. 1929. On some new therapsid genera. *Annals of the South African Museum* 28:55–78.

Hopson, J. A. 1987. The mammal-like reptiles: a study of transitional fossils.
The American Biology Teacher 49:16–26.

————. 1991. Systematics of the nonmammalian Synapsida and implications
for patterns of evolution in synapsids; pp. 635–693 *in* H.-P. Schultze and
L. Trueb (eds.), *Origins of the Higher Groups of Tetrapods: Controversy
and Consensus.* Ithaca: Comstock Publishing Associates.

Hopson, J. A. and H. Barghusen. 1986. An analysis of therapsid relationships;
pp. 83–106 *in* N. Hotton, P. D. MacLean, J. J. Roth, and E. C. Roth (eds.),
The Ecology and Biology of the Mammal-like Reptiles. Washington, DC:
Smithsonian Institution Press.

Ivachnenko, M. F. 1990. Late Paleozoic tetrapod faunal assemblage from the
Mezen' River basin. *Paleontological Journal* 81–90.

————. 1999. Biarmosuches from the Ocher Faunal Assemblage of Eastern
Europe. *Paleontological Journal* 33:289–296.

Ivachnenko, M. F., V. K. Golubev, Y. M. Gubin, N. N. Kalandadze, I. V. Novikov,
A. G. Sennikov, and A. S. Rautian. 1997. *Permskie i Triasovye tetrapody Vos-
tochnoi Evropy (Permian and Triassic Tetrapods of Eastern Europe).* Moscow: GEOS.

King, G. M. 1988. Anomodontia; pp. 1–174 *in* P. Wellnhofer (ed.), *Encyclopedia
of Paleoherpetology, Part 17C.* Stuttgart: Gustav Fischer.

Kitching, J. W. 1977. The distribution of the Karroo vertebrate fauna. *Bernard
Price Institute for Palaeontological Research Memoir* 1:1–131.

Langer, M. C. 2000. The first record of dinocephalians in South America: Late Permian (Rio do Rasto Formation) of the Paraná Basin, Brazil. *Neues Jahrbuch für Geologie und Paläontologie, Abhandlungen* 215:69–95.

Laurin, M. 1993. Anatomy and relationships of *Haptodus garnettensis,* a Pennsylvanian synapsid from Kansas. *Journal of Vertebrate Paleontology* 13:200–229.

Laurin, M. and R. R. Reisz. 1996. The osteology and relationships of *Tetraceratops insignis,* the oldest known therapsid. *Journal of Vertebrate Paleontology* 16:95–102.

Lucas, S. G. 2002. Tetrapods and the subdivision of Permian time; pp. 479–491 *in* L. V. Hills, C. M. Henderson, and E. W. Bamber (eds.), *Carboniferous and Permian of the World.* Canadian Society of Petroleum Geologists.

Mendrez-Carroll, C. 1975. Comparaison du palais chez les Thérocéphales primitifs, les Gorgonopsiens et les Ictidorhinidae. *Comptes Rendus de l'Académie des Sciences à Paris D* 280:17–20.

Modesto, S. P. 1995. The skull of the herbivorous synapsid *Edaphosaurus boanerges* from the Lower Permian of Texas. *Palaeontology* 38:213–239.

Modesto, S. P., B. S. Rubidge, and J. Welman. 1999. The most basal anomodont therapsid and the primacy of Gondwana in the evolution of the anomodonts. *Proceedings of the Royal Society of London B* 266:331–337.

Orlov, Y. A. 1958. Khishchnye deinotsefaly fauny Isheeva (Titanozukhi). *Akademiya Nauk SSSR Trudy Paleontologicheskogo Instituta* 72: 1–114.

Osborn, H. F. 1903. The reptilian subclasses Diapsida and Synapsida and the early history of the Diaptosauria. *Memoirs of the American Museum of Natural History* 1:265–270.

Parrington, F. R. 1955. On the cranial anatomy of some gorgonopsids and the synapsid middle ear. *Proceedings of the Zoological Society of London* 125:1–40.

Reisz, R. R., D. S. Berman, and D. Scott. 1992. The cranial anatomy and relationships of *Secodontosaurus,* an unusual mammal-like reptile (Synapsida: Sphenacodontidae) from the Early Permian of Texas. *Zoological Journal of the Linnean Society* 104:127–184.

Romer, A. S. and L. I. Price. 1940. Review of the Pelycosauria. *Geological Society of America Special Paper* 28:1–538.

Rowe, T. 1986. Osteological diagnosis of Mammalia, L. 1758, and its relationship to extinct Synapsida. Unpublished Ph.D. thesis, University of California, Berkeley.

Rubidge, B. S. 1990. A new vertebrate biozone at the base of the Beaufort Group, Karoo Sequence (South Africa). *Palaeontologia Africana* 27:17–20.

———. 1995. *Biostratigraphy of the Beaufort Group (Karoo Supergroup).* Pretoria: Government Printer.

Rubidge, B. S. and J. A. Hopson. 1990. A new anomodont therapsid from South Africa and its bearing on the ancestry of Dicynodontia. *South African Journal of Science* 86:43–45.

Rubidge, B. S. and J. W. Kitching. 2003. A new burnetiamorph (Therapsida: Biarmosuchia) from the lower Beaufort Group of South Africa. *Palaeontology* 46:199–210.

Rubidge, B. S. and C. A. Sidor. 2001. Evolutionary patterns among Permo-Triassic therapsids. *Annual Review of Ecology and Systematics* 32:449–480.

Rubidge, B. S. and C. A. Sidor. 2002. On the cranial morphology of the basal therapsids *Burnetia* and *Proburnetia* (Therapsida: Burnetiidae). *Journal of Vertebrate Paleontology* 22:257–267.

Rubidge, B. S. and J. A. van den Heever. 1997. Morphology and systematic position of the dinocephalian *Styracocephalus platyrhinus. Lethaia* 30:157–168.

Rybczynski, N. 2000. Cranial anatomy and phylogenetic position of *Suminia getmanovi,* a basal anomodont (Amniota: Therapsida) from the Late Permian of Eastern Europe. *Zoological Journal of the Linnean Society* 130:329–373.

Sidor, C. A. 2000. Evolutionary trends and relationships within the Synapsida. Unpublished Ph.D. dissertation, University of Chicago, Chicago.

———. 2001. Simplification as a trend in synapsid cranial evolution. *Evolution* 55:1419–1442.

———. 2003. The naris and palate of *Lycaenodon longiceps* (Therapsida: Biarmosuchia), with comments on their early evolution in the Therapsida. *Journal of Paleontology* 77:153–160.

Sidor, C. A. and J. Welman. 2003. A second specimen of *Lemurosaurus pricei* (Therapsida: Burnetiamorpha). *Journal of Vertebrate Paleontology* 23:631–642

Sidor, C. A., J. A. Hopson, and A. W. Keyser. 2004. A new burnetiamorph therapsid from the Teekloof Formation, Permian, of South Africa. *Journal of Vertebrate Paleontology* 24:938–950.

Sigogneau, D. 1970a. Contribution à la connaissances des Ictidorhinides (Gorgonopsia). *Palaeontologia Africana* 13:25–38.

———. 1970b. *Révision Systématique des Gorgonopsiens Sud-Africains.* Paris: Cahiers de Paléontologie.

Sigogneau-Russell, D. 1989. Theriodontia I; pp. 1–127 *in* P. Wellnhofer (ed.), *Encyclopedia of Paleoherpetology, Part 17B.* Stuttgart: Gustav Fischer.

Smith, R. M. H. and A. W. Keyser. 1995. Biostratigraphy of the *Cistecephalus* Assemblage Zone; pp. 23–28 *in* B. S. Rubidge (ed.), *Biostratigraphy of the Beaufort Group (Karoo Supergroup).* Pretoria: Council for Geosciences.

Swofford, D. L. 1993. *PAUP: Phylogenetic Analysis Using Parsimony.* Program distributed by the Illinois Natural History Survey, Champaign.

Tchudinov, P. K. 1960. Upper Permian therapsids from the Ezhovo locality (Verkhnepermskie terapsidy ezhovskogo mestonakhozhdenip). *Paleontologicheskii Zhurnal* 1960(4):81–94.

Appendix 4.1

Data matrix used in the cladistic analysis of burnetiamorph interrelationships. A complete list of specimens used in this analysis is provided in the Materials section (table A4.1). *Estemmenosuchus* includes character information from both *Estemmenosuchus uralensis* and *Estemmenosuchus mirabilis.* Titanosuchidae includes character information from both *Titanosuchus* and *Jonkeria.* Characters and character state definitions are provided in appendix 4.2. "?" denotes missing data. Multistate taxa denote polymorphism.

Appendix 4.2

The following is a list of the characters and character states used to construct the cladogram in figure 4.6. The number preceding the character definition corresponds to that of the columns in table A4.1. Following the last character state for each character is a citation for previous uses of the character in the literature. When an asterisk follows the citation, it denotes that the character definition has been modified or that an additional character state(s) has been added (or deleted). Citations are in the following form: (author: character number), except those of Hopson and Barghusen (1986), which are (author: table.clade.character number), King (1988), which are (author: suite.character number), and Sidor (2000), which are (author: appendix number.character number). Abbreviations for authors are as follows: GKR, Gauthier et al. (1988); HB, Hopson and Barghusen (1986); MRW, Modesto et al. (1999); R, Rybczynski (2000), S, Sidor (2000); SH, Sidor and Hopson (1998); SW, Sidor and Welman (2003).

1) Length of vomerine process of premaxilla: short (0); long, approaching level of upper canine posteriorly (1); or absent in ventral view (2) so that vomer abuts body of premaxilla. (S:3.2.1; S:4.2.6; SW:1*)

2) Length of posterodorsal process of premaxillae: short (0), or long, reaching to a level posterior to that of the upper canine (1). (HB:1.2.7*; K:E.9*; R:4*; SH:1*; S:4.2.1; SW:2)

3) Alveolar margin of premaxilla subhorizontal (0) or greatly upturned (1). (SH:2*)

4) Antorbital region long (0) or short (1). (HB:1.6.5; MRW:6*)

5) Septomaxilla contained within external naris (0) or escapes to have facial exposure (1). (GKR:3*; HB:1.2.2; MRW:8*; SH:6; SW:3)

6) Septomaxillary facial process short (0) or long (1). (MRW:8*; R:7*)

7) Lateral surface of lacrimal bears one or more deep fossae: absent (0), present (1). (S:3.2.5*; S:4.2.13; SW:4)

8) Maxilla contacts prefrontal: absent (0), present (1). (HB:1.2.9*; GKR:27; SH:8; S4.2.9; SW:5)

9) Shape of dorsal surface of nasals: flat (0), with median boss (1). (S:3.2.3; S:4.2.14; SW:6)

10) Shape of dorsal surface of frontals: flat (0), or with low, broad ridge (1). (S:4.2.16; SW:14*)

11) Frontal without (0) or with (1) long posterolateral process that extends to the level of pineal foramen midpoint.

12) Supraorbital margin thin (0) or moderately to greatly thickened (1). (S:3.2.7; S:4.2.22*; SW:7*)

13) Adductor musculature originates on ventral surface of postorbital only (0) or on ventral and lateral surfaces (1). (SH:15*,16*; S:4.2.21*)

14) Thickness of postorbital bar: thin, A-P length much less than orbital or temporal fenestra diameter (0), thickened, so that A-P length is approximately half the diameter of the temporal fenestra (1). (S:4.2.18)

15) Postorbital bar narrow (0) or broadened in transverse plane (1).

16) Postfrontal without (0) or with (1) posterior extension along its medial contact with the frontal.

17) Jaw adductor musculature extends onto dorsal surface of postfrontal: absent (0), present (1). (S:4.2.20)

18) Shape of dorsal surface of parietal: flat (0), low and diffuse swelling (1), or forms well-defined chimney (2). (HB:1.2.8*; GKR:18*; SH:21*; MRW:18*; R:17*; S:4.2.24*; SW:16)

19) Temporal fenestra small (0) or expanded posterodorsally (1) so that adductor musculature origination on squamosal is visible in dorsal view. (HB:1.3.1; SH:14*)

20) Intertemporal region wider (0) or narrower (1) than interorbital region. (HB:1.4.1*; SH:18*)

21) Ventrolateral surface of zygomatic arch and suborbital bar: smooth (0) or with ventrolaterally projecting bosses (1). (S:3.2.9*; SW:9*)

22) Zygomatic arch elevated above margin of upper tooth row so as to fully expose quadrate and quadratojugal in lateral view: absent (0), present (1). (HB:1.6.1; GKR:59*; K:A.3*)

23) Length of squamosal's anterior ramus: stops under temporal fenestra (0) or under orbit (1). (MRW:11*)

24) Preparietal: absent (0), present (1). (MRW:18*; R:18*; S:4.2.23; SH:48; SW:17)

25) Supratemporal: present (0), absent (1). (GKR:23; HB:1.2.3; S1.1.2.16; SH:22; SW:18)

26) Tabular contacts paroccipital process of opisthotic (0) or restricted dorsally (1). (MRW:21*)

27) Vomer: paired (0) or unpaired (1). (HB:1.8.12; SH:26)

28) Choanal and postchoanal portions of vomer meet at similar level on palate (0) or choanal portion is offset ventrally from postchoanal portion (1).

29) Palatine dentition broadly distributed (0), restricted to small area (1), or absent (2). (GKR:121; MRW:25*; S:4.2.45; SH:36*; SW:23*)

30) Teeth on transverse flange of pterygoid: present (0), absent (1). (S:3.2.17; SW:21*)

31) Position of transverse flange of pterygoid: under posterior half of orbit (0), under anterior half of orbit (1), or preorbital (2). (S:4.2.47; SH:73*; SW:24)

32) Pterygoid without (0) or with (1) shelf posterior to its transverse flange. (HB:2.15.5; S:4.2.50)

33) Dentition on palatal ramus of pterygoid: present (0), absent (1). (SH:37; MRW:26*)

34) Basicranial rami of pterygoids broadly separated (0), narrowly separated by sagittal trough (1), or broadly contacting anterior to basicranium (2). (HB:1.2.10*; MRW:28*; SW:25*)

35) Basipterygoid articulation located high above primary palate (0), just dorsal to basicranial ramus of pterygoid (1), or at level basicranial ramus (i.e., suture visible in ventral view) (2).

36) Ectopterygoid teeth: present (0), absent (1). (SH:39; SW:22)

37) Shape of postparietal: wider than tall (0), approximately rectangular (1), or taller than wide (2). (SW:29)

38) Forward rotation of occiput: none (0), moderate (1), pronounced (2). (S:4.2.55; SH:42; SW:26)

39) Stapedial foramen: present (0), absent (1). (S:4.2.39; SH:76; SW:27)

40) Dentary height in canine versus anterior postcanine regions: nearly equivalent (0), or shows pronounced difference (1). (S:4.2.58; SW:30)

41) Dentary-angular suture runs diagonally across lateral surface of mandible (0) or posterior margin of dentary deeply incised (1). (S:4.2.62*; SW:31)

42) Angular with pattern of ridges and fossae on its lateral surface: absent (0), present (1). (S:4.2.68; SH:98*; SW:32)

43) Sharp lateral ridge on dorsal margin of lower jaw in postcoronoid eminence region: absent (0), or present (1). (SW:34*)

44) Foramen between prearticular and angular (sometimes bordered by splenial as well) on medial surface of lower jaw: absent (0), present (1). (S:4.2.69)

45) Articular dorsal process: absent (0), present (1). (HB:1.8.6; S:4.2.73; SW:33)

46) Lateral mandibular fenestra: absent (0), present (1). (HB:1.6.4*; K:A.6*; GKR:87*; SH:94*; B:57*; MRW:36)

47) Coronoid (posterior): present (0), absent or greatly reduced (1). (K:A.4*; MRW:38*)

48) Differentiation of upper tooth row: clear separation into incisors, canine, and postcanines (0), or teeth decrease in size posteriorly, upper canine barely differentiated (1). (HB:2.17.2*; HB:2.18.1*; GKR:111*,114*; MRW:1*)

49) Degree of intermeshing between upper and lower incisors: none (0), anterior incisors only (1), complete intermeshing (2). (HB:1.1.2*; S:4.2.81*; SH:105*)

50) Upper incisors much larger (0) or roughly equivalent in size to postcanines (1). (GKR:119*; S:4.2.80)

51) Lower canine fits into choana (0), or into fossa roofed by premaxilla and maxilla (1), or passes anterior and external to upper canine (2). (HB:1.8.4*; HB:1.10.5*; S2.2:81*; S:3.2.2; SW:37)

52) Upper and lower canines with heels absent (0) or small heels present (1). (HB:2.18.2*)

53) Number of upper postcanines: twelve or greater (0), fewer than 12 (1). (GKR:116; SH:112; SW:38)

54) Lower postcanine teeth inset from lateral margin of dentary: absent (0), present (1). (S:4.2.89)

55) Upper postcanine teeth confluent with upper incisor row medial to canine: absent (0), present (1). (HB:2.15.4)

56) Postcanine teeth with triangular crown bearing coarse, megaserrations along both anterior and posterior carinae: absent (0), present (1). (GKR:155*; HB:2.15.3; R:3*; SH:113*)

Table A4.1. Data matrix for cladistic analysis of biarmosuchian relationships

	12345	1 67890	11111 12345	11112 67890	22222 12345	22223 67890	33333 12345
Haptodus	00000	?0000	00000	00000	00000	00000	00000
Dimetrodon	00000	?0000	00000	00000	00000	00000	00000
Patranomodon	00?1?	00100	00100	00010	01111	10121	00111
Suminia	01011	00100	00100	?0210	01101	11121	00111
Syodon	21101	00100	00101	01211	00001	0001?	10021
Titanophoneus	21101	10100	01101	01211	00001	00010	10121
Titanosuchidae	21001	10100	01110	00111	00101	00001	21121
Estemmenosuchus	21001	00?1?	01?10	00210	00?01	?1000	2102?
Biarmosuchus	?1001	10100	00101	00200	00101	0?100	1002?
Hipposaurus	?1001	10100	00000	10200	?0?11	0?100	10012
Lycaenodon	10001	10100	10000	102?0	???1?	?1111	1001?
Herpetoskylax	?0001	10100	10000	10200	0(01)111	0111(01)	10012
Ictidorhinus	1?001	101?1	?1000	?02?0	00???	?11??	?0???
Lemurosaurus	??001	11101	?1000	?0200	101??	0?110	1001?
Lobalopex	1?00?	?1111	11000	?0200	10??1	0?111	10012

Table A4.2. Data matrix

	33334 67890	44444 12345	44445 67890	55555 12345	5 6	% Missing
Haptodus	0?000	00000	00000	?0000	0	5.4
Dimetrodon	00000	00000	00000	10000	0	1.8
Patranomodon	10010	01100	111??	??100	0	10.7
Suminia	11010	11010	111?0	??100	1	7.1
Syodon	12100	10010	00020	11100	0	1.8
Titanophoneus	12100	0001?	00020	11100	0	1.8
Titanosuchidae	1?2?0	10010	00020	21011	1	3.6
Estemmenosuchus	??200	1001?	0?020	20011	1	17.9
Biarmosuchus	11101	011??	0?0?1	?0100	0	14.3
Hipposaurus	11111	(01)1?11	00011	?0100	0	10.7
Lycaenodon	1????	?????	??0?1	001??	?	39.3
Herpetoskylax	111?1	11101	00011	?0100	0	5.4
Ictidorhinus	11???	?????	??0?1	0????	?	55.4
Lemurosaurus	11111	1110?	0?011	?0100	0	19.6
Lobalopex	11111	11101	000?1	00100	0	14.3

The Postcranial Skeleton of *Kayentatherium wellesi* from the Lower Jurassic Kayenta Formation of Arizona and the Phylogenetic Significance of Postcranial Features in Tritylodontid Cynodonts

Hans-Dieter Sues *and*
Farish A. Jenkins Jr.

Introduction

The Tritylodontidae are a clade of derived, presumably herbivorous eucynodonts with a superficially somewhat rodent-like dentition, comprising enlarged, procumbent incisors and, separated from the latter by a prominent diastema, multi-cusped, multiple-rooted molariform teeth. They are currently known from the Lower and Middle Jurassic of Europe, the Lower Jurassic of western North America and southern Africa, the Lower or Middle Jurassic of Mexico, the Lower to Upper Jurassic of China and the Lower Cretaceous of Japan and Russia (Hopson & Kitching, 1972; Kemp, 1982; Kermack, 1982; Clark & Hopson, 1985; Sun & Li, 1985; Lewis, 1986; Sues, 1986a, b; Tatarinov & Matchenko, 1999; Matsuoka & Setoguchi, 2000). Initially considered basal mammals related to multituberculates (e.g., Simpson, 1928), the Tritylodontidae subsequently have been interpreted as derived gomphodont cynodonts by most authors (Watson, 1942; Young, 1947; Kühne, 1956; Crompton & Ellenberger, 1957; Crompton, 1972; Hopson & Kitching, 1972, 2001; Sues, 1985; Hopson & Barghusen, 1986; Hopson, 1991; Sidor & Hopson, 1998).

Kemp (1982, 1983) enumerated numerous derived characters in the postcranial skeleton shared by Tritylodontidae and Mammalia and proposed a sister-group relationship between these two groups. Sues (1985) argued that some of these features are also present in the traversodont eucynodont *Exaeretodon* from the Upper Triassic Ischigualasto Formation of northwestern Argentina. He followed Crompton (1972) in regarding Tritylodontidae and "Traversodontidae" as sister-taxa and suggested that tritylodontids are yet another example of the extensive parallelism among derived synapsids. Rowe (1988) attempted to reconcile these conflicting points of view by placing *Exaeretodon* as the sister-taxon of a clade containing Tritylodontidae and Mammaliaformes (= Mammalia *sensu* Luo et al., 2002), but he left the affinities of the other traversodonts unresolved. Hopson (1991, 697) cited *Exaeretodon* as providing "the best evidence for a relationship between tritylodontids and more primitive gomphodonts . . . [demonstrating] the likelihood that the mammal-like features of tritylodontids evolved independently of those of mammals within the gomphodont clade." Wible (1991, 20), acknowledging that analyses of the craniodental evidence have provided no clear resolution to the controversy, observed that "overwhelming support for a Tritylodontidae/ Mammaliaformes clade comes thus far from the postcranial skeleton."

Mammalian features of the tritylodontid postcranial skeleton thus have been used as synapomorphies to support the hypothesis that Tritylodontidae are the sister-group of Mammalia, and alternatively have been interpreted as evidence for significant homoplasy in cynodont evolution.

Recent studies have not favored the view that tritylodontids and mammals are closely related. With the exception of similarities in the structure of the orbital wall, Luo (1994, 104) concluded that "support from other areas of the skull and dentition for the tritylodontid-mammal hypothesis is, at best, very weak and inconclusive." A comprehensive cladistic analysis of Cynodontia by Hopson and Kitching (2001) hypothesized a dichotomy of eucynodonts into Probainognathia and Cynognathia, with *Morganucodon* nested in the former and Tritylodontidae in the latter. Interestingly, this analysis incorporated numerous postcranial features that, in studies by other authors, have been alleged to be mammal-like in tritylodontids. Hopson and Kitching (2001, 27) cited two Late Triassic cynodonts that are "very mammallike postcranially" as "critical evidence supporting the probainognathian-cynognathian dichotomy" (cf. Bonaparte & Barberena, 2001) "and the occurrence of a truly extraordinary amount of homoplasy in eucynodont evolution."

Although the tritylodontid postcranial skeleton has been important to analyses of cynodont phylogeny, the available fossil evidence in fact has been rather limited. Aside from Kühne's (1956) monographic study of *Oligokyphus,* little has been published to date on the postcranial skeleton of the Tritylodontidae. Kühne's work was based on hundreds of dissociated skeletal remains from Early Jurassic fissure-fillings in southwestern England. A bimodal size distribution led Kühne to distinguish two species, *Oligokyphus major* and *Oligokyphus minor,* but the lack of skeletal association limits the utility of this material to comparisons of individual bones. Young (1947) described several limb bones referable to *Bienotherium* from the Lower Jurassic Lower Lufeng Formation of Yunnan (China). Fourie (1962) provided a brief account on the postcranial skeleton of *Tritylodontoideus maximus* (probably a subjective junior synonym of *Tritylodon longaevus;* Hopson & Kitching, 1972) from the Lower Jurassic Clarens Formation of South Africa; unfortunately, this specimen is preserved only as a natural mold on two slabs of sandstone. Sun and Li (1985) reported on some postcranial remains of *Bienotheroides wanhsiensis* from the Middle or Upper Jurassic Shaximiao Formation of Sichuan (China). Winkler et al. (1991) commented on a partial postcranial skeleton of an indeterminate tritylodontid from the Lower Jurassic Navajo Sandstone of Arizona.

We present here a description of tritylodontid postcranial remains from the Lower Jurassic Kayenta Formation of northeastern Arizona, most notably a well-preserved partial skeleton of *Kayentatherium wellesi* Kermack, 1982 (including *Nearctylodon broomi* Lewis, 1986 as a subjective junior synonym; Sues, 1986a,b). Our intent is not only to place the structure of the postcranial axial and appendicular skeleton on record but also to explore the functional and phylogenetic inferences that may be drawn from the available material. We specifically review the issue of mammal-like features in the tritylodontid postcranial skeleton, an assessment that different authors have used to reach very different phylogenetic conclusions.

Sues et al. (1994) discussed the diversity, stratigraphic context, and distribution of the fossil vertebrates from the Kayenta Formation of northeastern Arizona, central and southern Utah. The first tritylodontid remains to be recovered from the Kayenta Formation were collected in the early 1950s from outcrops 2.5 to 3 m below the top of the formation ("typical facies"), just below the presumed contact with the overlying Navajo Sandstone, on Comb Ridge, 11.4 km airline, bearing 069° true, from Kayenta, Arizona (Lewis, 1986). The tritylodontid material described by Sues (1986a) and in this study was collected from exposures of the "silty facies" of the Kayenta Formation along the Adeii Eechii Cliffs on Ward Terrace, from the region of Dinnebito Wash to Tuba City, on lands of the Navajo Nation in northeastern Arizona.

The following institutional abbreviations are used: BMNH, The Natural History Museum [formerly British Museum (Natural History)], London; MCZ, Museum of Comparative Zoology, Harvard University, Cambridge, MA; MNA, Museum of Northern Arizona, Flagstaff; USNM, National Museum of Natural History (formerly United States National Museum), Washington, DC.

Material

The tritylodontid specimens that form the basis for this study were collected by joint field-parties from the Museum of Comparative Zoology and the Museum of Northern Arizona, led by F.A.J., Jr. between 1977 and

1982. Sues (1986a) provided a detailed anatomical account on the skull and dentition of two tritylodontid taxa, *Kayentatherium wellesi* and *Dinnebitodon amarali*, and Sues (1986b) briefly reviewed the phylogenetic relationships and biostratigraphic significance of the tritylodontids from the Kayenta Formation.

The principal specimen used for the osteological description is MCZ 8812, a well-preserved partial skeleton of a large individual of *K. wellesi*, which includes the pectoral girdle and sternum, both forelimbs, and vertebral column (with most of the ribs in association) from the atlas-axis to the posterior dorsal region. The specimen was preserved lying on its left side, and the articular context of the bones was largely undisturbed. The posterior portion of the skeleton had already weathered out at the time of discovery, but screening of the talus below the site yielded pieces of the left ilium, several caudal centra, and unidentifiable fragments of bone. Additional postcranial material used in this study and referable to *K. wellesi* on the basis of associated dentitions comprises MCZ 8811 and MNA V3141, V3224, and V3235. MNA V3141 includes the glenoid regions of both scapulae, the right humerus, left ulna and radius, partial right manus, incomplete left femur, and the proximal ends of both tibiae. This material, which was collected as surface float, includes the distal end of a second right humerus and thus represents the intermingled remains of two individuals of similar body size. MNA V3224 and V3235 include fragmentary axial and appendicular elements. We tentatively refer several specimens to *K. wellesi* mainly on the basis of large size: MCZ 8832, 8835, and 8838. The ischium is based on a right element associated with a partial maxilla and dentary of *D. amarali* (MNA V3232). Most of the bones show signs of crushing and plastic distortion, limiting their utility for comparative purposes.

H.-D.S. has viewed the holotype of *Nearctylodon broomi* (USNM 317201; Lewis, 1986), a nearly complete articulated skeleton of a small, probably juvenile individual of *K. wellesi*, which is now on public display at the National Museum of Natural History (Washington, DC). The preliminary report by Lewis (1986) provided some details about the dentition of this and other specimens collected from the same quarry on Comb Ridge. This collection has yet to be fully described. These apparently immature tritylodontid skeletons, although useful for showing the overall body proportions of juveniles, are poorly preserved, and none appears to be fully ossified (Lewis, 1986, 299).

On several occasions both authors have examined the postcranial elements of *Oligokyphus* that were described by Kühne (1956) and are housed in The Natural History Museum, London. Most of these bones are fragmentary and variously abraded, which accounts for Kühne's use of multiple specimens to produce composite reconstructions for individual skeletal elements. Comparison of *Oligokyphus* and *Kayentatherium* elucidates various important postcranial features in these tritylodontids and necessitates modification of some of Kühne's descriptions and reconstructions.

Description

Postcranial Axial Skeleton

The vertebral column is most completely documented in MCZ 8812, which preserves an articulated series of vertebrae from the atlas (including a possible fragment of a proatlas) to the posterior dorsal region, along with most of the dorsal ribs and a number of isolated caudal vertebrae. Most vertebrae are crushed and often considerably distorted. Well-preserved, dissociated vertebral centra from other specimens referable to *Kayentatherium* were also available for examination.

Atlas. The atlas is composed of paired, unfused atlantal neural arches and a centrum (fig. 5.1). The atlantal arch is composed of a thin dorsal plate, representing the lamina, and a robust ventral base that bears articular facets for the centrum and occipital condyles and represents the pedicle. The thin medial margins of the laminae appear not to have been in contact along the midline (fig. 5.1B), although they were probably ligamentously joined in life. The dorsolateral, or external, surface of each lamina is almost flat and bears no indication of any muscular process. The base of the lamina near its junction with the pedicle is thickened (fig. 5.1B). The two facets at the base of the neural arch (pedicle) are of unequal size, with the facet for the occipital condyle being more deeply concave and considerably larger. Each arch base supports a ventrolaterally projecting transverse process. A groove appears just behind the transverse process and continues as a deep sulcus passing dorsally between the process and the lateral aspect of the occipital facet. This groove, which extends between the anterior margin of the lamina and the dorsal rim of the occipital facet, presumably marked the course of the vertebral artery. The atlantal arches and intercentrum were presumably ligamentously joined.

The atlantal and axial centra are not suturally attached to each other, unlike the condition in *Oligokyphus* (BMNH R 7315; Kühne, 1956, fig. 43) and *Morganucodon* (Jenkins & Parrington, 1976, fig. 1G, H). A short

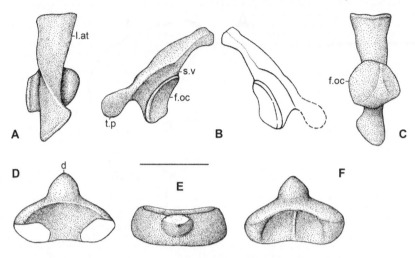

Figure 5.1. Atlantal elements of Tritylodontidae indet., MCZ 8839. (A–C) Right atlantal arch in (A) lateral, (B) anterior (with addition of left element), and (C) medial views. (D–F) Atlantal centrum with dens in (D) dorsal, (E) anterior, and (F) ventral views. Scale bar equals 1 cm. Abbreviations: d, dens; f.oc, facet for contact with occipital condyle; l.at, lamina of atlantal arch; s.v, vertebrarterial sulcus; t.p, transverse process.

dens, which projects from the dorsal half of the anterior face of the centrum, is stout at its base but attenuates on all sides to a distinct apical point (figs. 5.1D–F, 5.2A). An obliquely inclined, anteroventrally facing facet for articulation with the first intercentrum is situated ventral to the dens. The contact areas for the atlantal arches are set off from the remainder of the dorsal surface of the centrum, which forms a broad concavity divided by a low median ridge.

Axis. The axial centrum is constricted at midlength and relatively short (fig. 5.2A). As in *Oligokyphus* (Kühne, 1956, fig. 43), the anteroposterior lengths of the atlantal and axial centra are about equal. The robust transverse process of the axis projects posterolaterally and has an expanded distal end. The prezygapophyses were apparently small. The large postzygapophyses bear ventrolaterally facing facets. As on all other vertebrae, anapophyses are absent. The tall, hatchet-shaped neural spine is mediolaterally compressed and thin except for anterior and posterior terminal tuberosities. The dorsal margin of the process is distinctly convex in lateral view (fig. 5.2A). The lateral face of the process, broad and gently concave, provided ample surface for suboccipital muscles, possibly M. obliquus capitis.

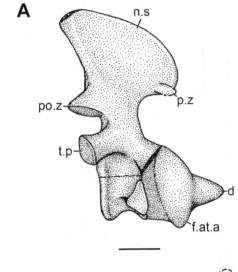

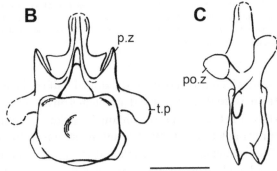

Figure 5.2. *Kayentatherium wellesi,* MCZ 8812. (A) Atlas-axis complex in right lateral view. Scale bar equals 1 cm. (B, C) Postaxial cervical vertebra in (B) anterior and (C) right lateral views. Scale bar equals 2 cm. Abbreviations: d, dens; f.at.a, facet for contact with atlantal arch; n.s, neural spine; p.z, prezygapophysis; po.z, postzygapophysis; t.p, transverse process.

Postaxial Cervical Vertebrae. MCZ 8812 has five postaxial cervical vertebrae (C3–7), for a total of seven cervicals. The disk-like, amphicoelous centra are about three times as wide as long (fig. 5.2B, C), unlike the cervicals of *Oligokyphus* in which the width of the centrum only slightly exceeds its length (Kühne, 1956). The centra bear midventral keels as do those of the atlas and axis. Their anterior and posterior articular surfaces have sharp ventral rims. The short, dorsally projecting neural spines taper toward slightly thickened apices. The prezygapophyses project more dorsally than those in *Oligokyphus* (cf. fig. 5.2B and Kühne, 1956, fig. 41C) and thus do not present the great transverse width seen in dorsal view of the cervicals in *Megazostrodon* (Jenkins & Parrington, 1976, fig. 6A). The pre- and postzygapophyseal facets are shallowly concave and convex, respectively, along a transverse plane and, as a whole, are inclined to within

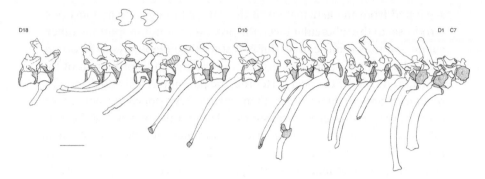

Figure 5.3. *Kayentatherium wellesi,* MCZ 8812. Presacral vertebral column from the seventh cervical (C7) to the eighteenth dorsal (D18) vertebra, in right lateral view. Outlines of apices of neural spines above dorsal vertebrae thirteen and fourteen. Hatching indicates broken surfaces, stippling unfinished bony surfaces. Scale bar equals 4 cm.

about 30° of the vertical plane. The pedicles of the slender neural arches are narrow anteroposteriorly and enclose a very wide neural canal. The transverse processes are short and stout. The parapophyseal and diapophyseal facets are well separated. The parapophyses are positioned more ventrally on anterior cervical centra and are successively displaced dorsally on posterior cervicals; on C7, the parapophysis is located halfway up on the anterior rim of the centrum. As in *Massetognathus* (Jenkins, 1970) and *Oligokyphus* (Kühne, 1956), the cervical ribs were not intervertebral in position. Although Kühne (1956, 141) interpreted the cervical ribs in *Oligokyphus* as fused, his illustrations show that the diapophyses and parapophyses are broken (Kühne, 1956, fig. 44A–C). In *Kayentatherium,* the cervical ribs are not fused to the vertebrae.

Dorsal Vertebrae. The first dorsal vertebra (fig. 5.3) differs from the last cervical in having a slightly longer centrum and an anteroposteriorly broader neural spine, which is more or less triangular in transverse section.

The neural spines of the first three dorsal vertebrae are tall, mediolaterally narrow, and distinctly inclined posterodorsally (fig. 5.3). The neural spine of the third thoracic has a slightly thickened apex. The neural arch of the first dorsal bears a small process, which, on the second dorsal, becomes a prominent tubercle and, starting at the third dorsal, a distinct, dorsally and slightly laterally projecting process. On the posterior dorsal vertebrae, the process changes into a more or less horizontal, increasingly laminar feature posteriorly. Kühne (1956, 102) noted that anterior dorsal vertebrae of *Oligokyphus* bear "a rather strong diapophysis, which is well

separated from the articulation of the rib." Although situated just posterodorsal to the tubercular facet, the process does not support the tuberculum and thus is musculotendinous rather than diapophyseal in origin.
The robust prezygapophyses project beyond the anterior margin of the
centrum. As on the cervical vertebrae, the parapophyseal and diapophyseal facets are confluent rather than separate. All dorsal vertebrae have
distinct, irregular neurocentral sutures that join the pedicles of the neural arches to the centra. Intervertebral foramina for the exit of spinal
nerves are largely intravertebral in position; deep notches are formed
along the posterior margins of the pedicles, but corresponding notches
on the anterior margins of successive pedicles are only slightly developed.
The centra are constricted at mid-length and bear ventral keels.

A gradient in neural spine structure begins in the region of the fourth
to sixth dorsal vertebrae (fig. 5.4A): the neural spines decrease in height
posteriorly, and lateral exostoses appear on their apices. Starting at the
sixth dorsal vertebra, the spines become more sharply inclined posteriorly,
the lateral exostoses are prominent, and the posterior margins of the
apices are extended as a pronounced lappet, producing a trident appearance in dorsal view (figs. 5.3, 5.4B). The posterior lappets presumably
represent attachment sites for robust supraspinous and/or interspinous
ligaments. Starting at the eleventh dorsal vertebra, the apices of the neural spines become much expanded and flattened to form distinct "spine
tables" (fig. 5.3), and the neural spines become broad anteroposteriorly.
Lewis (1986, 297) noted apical expansion of the neural spines on the dorsal vertebrae of the Comb Ridge specimens of *Kayentatherium,* where
"transverse expansion [is] greatest in the posterior dorsals, especially in

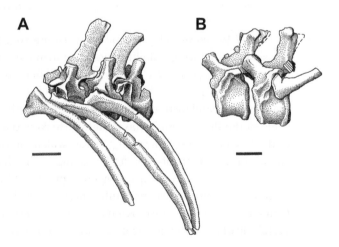

Figure. 5.4. *Kayentatherium
wellesi,* MCZ 8812. (A) Dorsal
vertebrae four to six with
associated ribs in left lateral
view. (B) Dorsal vertebrae fifteen
and sixteen with associated ribs
in left lateral view. Scale bars
equal 2 cm.

lumbars." A similar differentiation of the dorsal neural spines is present in *Exaeretodon,* where the first four vertebrae have tall, slender spines, and the following twelve have spines with much expanded dorsal apices (Bonaparte, 1963).

Both pairs of zygapophyseal facets of the first dorsal vertebra and the prezygapophyseal facets of the second dorsal vertebra are steeply inclined at angles comparable to those of the cervical series. The postzygapophyseal facets of the second dorsal vertebra, however, appear to be more or less horizontal, a condition that persists through the ninth dorsal vertebra. A gradational return to an inclined orientation appears to occur from the tenth through twelfth dorsal vertebrae. The zygapophyseal facets of the posterior dorsal vertebrae are steeply inclined at about 40° from the vertical plane.

No abrupt structural transition exists between the thoracic and lumbar regions. Although a gradational change in zygapophyseal orientation is apparent, neither an anticlinal nor a diaphragmatic vertebra is present. All dorsal neural spines are posterodorsally inclined. The centra of the dorsal vertebrae become increasingly longer and more massive toward the posterior end of the series (fig. 5.3). Starting at about the tenth dorsal vertebra, the parapophyseal facets shift increasingly posterodorsally.

Lewis (1986, 297) reported a count of nineteen dorsal vertebrae for the Comb Ridge specimens of *Kayentatherium,* of which "the last two could be called lumbar," but did not specify any criteria for this distinction. The last "lumbar," or nineteenth dorsal vertebra, thus appears to be missing in MCZ 8812. Jenkins (1971) distinguished thirteen thoracic and seven lumbar vertebrae in *Thrinaxodon,* with the thoracolumbar transition gradational and to some extent arbitrarily determined on the basis of costal structure.

Caudal Vertebrae. Comparison of the numerous isolated caudal centra with those of *Oligokyphus* demonstrates that the caudal vertebral column displays a similarly pronounced proximodistal gradation. On the basis of material available, caudal vertebrae can be described only by distinguishing proximal and distal centra. The tail in *Kayentatherium,* which comprised at least nineteen vertebrae (Lewis 1986), was apparently long and well developed.

Proximal caudal vertebrae (fig. 5.5A, B) have short centra and apparently robust neural arches, the pedicles of which are positioned anteriorly on the centrum (as in *Oligokyphus;* Kühne, 1956, fig. 47). On proximal centra, the ventral half of the posterior articular surface is recessed and

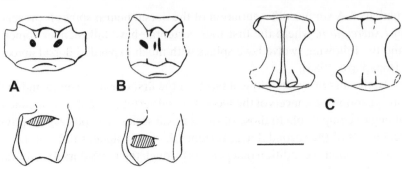

Figure 5.5. *Kayentatherium wellesi,* MCZ 8812. (A, B) Proximal caudal vertebrae in ventral (top) and right lateral (bottom) views. (C) More distal caudal vertebra in dorsal (left) and ventral (right) views. Scale bar equals 2 cm.

faces posteroventrally. The horizontal transverse processes arise variably from the sides of the centra, on some centra from near the neurocentral junction (fig. 5.5A) and on others from mid-centrum (fig. 5.5B). A longitudinal ventral groove presumably carried coccygeal vessels (Kühne, 1956). Distal caudal centra are more elongate. The neural spines and various other processes evidently became progressively smaller and eventually disappeared toward the distal end of the tail (fig. 5.5C). In these vertebrae, the neural canal is reduced to a narrow sulcus. Many caudal centra bear vascular foramina of various shapes and positions, especially on their ventral surfaces.

Ribs. The dorsal ribs lack the imbricating plates or processes present in many basal eucynodonts (fig. 5.3). In the more anterior dorsal ribs, the articular region of the proximal end is long, narrow, and relatively flat, with the capitular and tubercular facets separated by a mere constriction. The tuberculum is usually situated above and behind the capitulum. The shaft of a typical rib is anteroposteriorly flattened. Proximally, the shaft arches laterally and somewhat posteriorly; the distal half is straighter and directed ventrally. A distinct projection or rugosity is located a short distance distal to the tuberculum on the dorsal margin of the shaft and may mark the insertion of parts of longissimus dorsi. Distally, the shaft gradually attenuates and then expands again to support a ventrally facing, oval facet for a costal cartilage. The gentle curvature of the distal portion of the more anterior dorsal ribs indicates a rather deep thorax. The first dorsal rib differs from others in the great width of its shaft. The shafts of the seventh cervical ribs, which are distinctly more slender, are preserved bilaterally along the anterior margins of the first thoracic ribs in MCZ 8812.

Ribs nine and twelve on the left side as well as rib seven on the right side in MCZ 8812 exhibit fractures with nearthrotic formations (fig. 5.3). In each instance, both ends of the fractured shaft became expanded through callus formation and developed irregular articular facets, forming a secondary joint. One posterior dorsal rib bears a massive nodular growth just distal to its proximal head.

Pectoral Girdle

Scapula. The scapula is structurally complex. Viewed from medial aspect, the scapular blade without the acromion is more or less triangular in outline (figs. 5.6B, 5.8B), flaring broadly toward the convex vertebral margin. The smooth medial surface of the scapular blade has three confluent surfaces of different orientation; these surfaces are delimited by lines of pronounced flexure, which have been accentuated in MCZ 8812 by crushing during fossilization. An anteriorly facing surface is separated from a large medially facing area by a line of curvature extending from the anterior angle to the anteromedial region of the supraglenoid buttress. The posterior border of the scapula is thickened and supports a slightly concave, posteriorly and slightly laterally facing postscapular fossa (fig. 5.6C) below the posterior angle; in *Cynognathus,* a comparable fea-

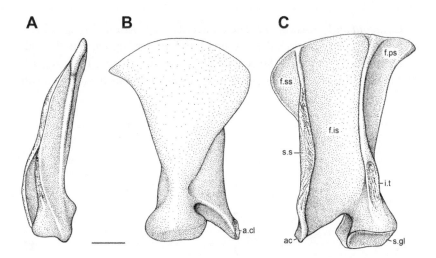

Figure 5.6. *Kayentatherium wellesi,* MCZ 8812. Left scapula in (A) posterior, (B) medial, and (C) lateral views. The anteroposterior width of the scapular blade has been somewhat exaggerated by crushing during fossilization. Scale bar equals 2 cm. Abbreviations: a.cl, area of attachment of clavicle; ac, acromion; f.is, infraspinous fossa; f.ps, postscapular fossa; f.ss, supraspinous fossa; i.t, insertion of M. triceps brachii (caput scapularis); s.gl, scapular glenoid facet; s.s, scapular spine.

ture was interpreted as representing the origin of M. teres major by Gregory and Camp (1918, 471).

Viewed from lateral aspect, the anterior and posterior borders of the scapula are reflected laterally, enclosing a broad infraspinous fossa between them (fig. 5.6C). A groove above the supraglenoid buttress probably marks the origin of the caput scapularis of M. triceps brachii (fig. 5.6C; cf. Jenkins, 1971, fig. 18D). The ventral end of the posterior border is markedly thickened, where it forms a broad posterolateral area above the glenoid facet. The anterior border, which is homologous to the mammalian scapular spine, bears an expanded, rugose margin for muscular attachment, most likely M. deltoideus. The robust, anterolaterally projecting acromion bears a distinct clavicular facet (fig. 5.6B). It is undercut anteroventrally along its junction with the scapular base. An incipient, biplanar supraspinous fossa is formed between adjacent surfaces of the scapular blade and spine at the anterior angle of the scapula (fig. 5.6C).

The scapular base supports a large, rounded, concave glenoid facet and bears narrow articular surfaces for the coracoid and procoracoid. The base forms a short process for contact with the procoracoid. The scapular facet of the glenoid faces ventrally rather than posterolaterally, as in other nonmammalian cynodonts. In contrast to more basal eucynodonts, the dorsal margin of the glenoid facet is more defined and overhangs the glenoid socket more extensively.

Coracoid and Procoracoid. The coracoid and procoracoid (fig. 5.7) are comparable to those in other nonmammalian cynodonts, but they are much smaller relative to the scapula (fig. 5.7; cf. Jenkins, 1971, fig. 17). The coracoid facet of the glenoid has a rather bulbous posterior termi-

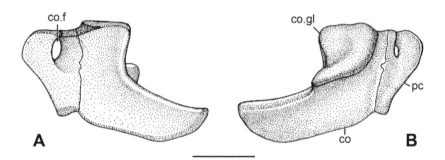

Figure 5.7. *Kayentatherium wellesi,* MCZ 8812. Right coracoid and procoracoid in (A) medial and (B) lateral views. The relative orientation of bones differs in the two illustrations. Scale bar equals 2 cm. Abbreviations: co, coracoid; co.f, coracoid foramen; co.gl, glenoid facet of coracoid; pc, procoracoid.

nus (fig. 5.7B) and apparently faced posterolaterally and only somewhat dorsally. The procoracoid does not contribute to the formation of the coracoid facet of the glenoid, unlike the condition in most other known nonmammalian cynodonts with the exception of *Exaeretodon* (Bonaparte, 1963).

Clavicle. The clavicle is distinctly angular (fig. 5.9A, B), a feature that is also present in the derived probainognathian *Prozostrodon* (Bonaparte & Barberena, 2001). Its medial two-thirds comprise a slender shaft that expands proximally into a spatulate plate for articulation with the interclavicle. The plate bears prominent striations parallel to the long axis of the bone, principally on its dorsal surface; it is thickest along its long axis and is delimited by sharp margins, especially anteriorly. The lateral third of the clavicle, which is offset from the medial two-thirds at an almost right angle, is a fairly robust rod with a posterodorsal and slightly lateral orientation. A recess on the distal end, which is partially delimited by a distinct ridge, represents the articular contact with the acromion.

Interclavicle. The interclavicle, which is thin and dorsoventrally flattened, bears broad, anterolaterally projecting clavicular processes (fig. 5.9C). These processes are flat and were overlapped ventrally by the expanded proximal ends of the clavicles. The short posterior ramus is slightly expanded posteriorly. A ventromedial projection may mark the origin of a portion of M. pectoralis.

Sternum. As in *Bienotheroides* (Sun & Li, 1985, fig. 4), the manubrium of *Kayentatherium* comprises a pair of ossifications that presumably were

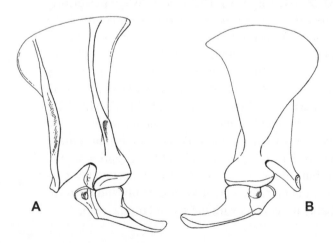

A   **B**

Figure 5.8. *Kayentatherium wellesi*, MCZ 8812. Left scapulocoracoid in (A) lateral and (B) medial views. Crushing of the scapular blade during fossilization has not been corrected.

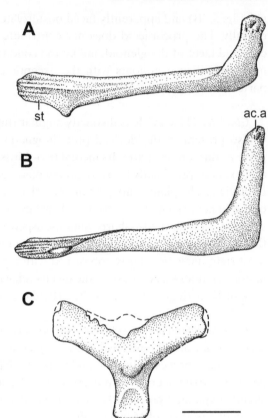

Figure 5.9. *Kayentatherium wellesi,* MCZ 8812. Left clavicle in (A) dorsal and (B) anterior views. (C) Interclavicle, ventral view. Scale bar equals 2 cm. Abbreviations: ac.a, area for contact with acromion; st, striated articular surface.

joined along the midline by cartilage in life. Only the left manubrial element is preserved in MCZ 8812 (fig. 5.10). The bone is thicker posteriorly than anteromedially. Its lateral margin is anteroposteriorly concave, the ventral surface is flat, and the dorsal surface is distinctly concave. The anterior margin is thin and rounded. A large, raised convex facet at the anterolateral corner faces anterolaterally. Kühne (1956, 113) interpreted the large facet on the manubrium in *Oligokyphus* as contacting the coracoid and a smaller one as contacting the first thoracic rib. In MCZ 8812, there is no evidence of facet development along the ventromedial margin of the coracoid that would indicate an articulation with the manubrium; a more plausible interpretation is that the large manubrial facet represents the articulation with the robust first thoracic rib. A second facet on the posterolateral corner of the manubrial element faces dorsally and posterolaterally and was probably for contact with the second thoracic rib.

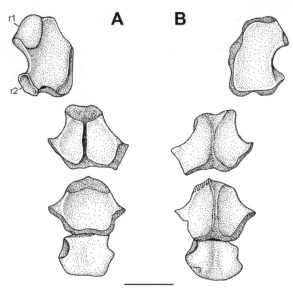

Figure 5.10. *Kayentatherium wellesi,* MCZ 8812. Associated left manubrial element and three sternebrae in (A) dorsal and (B) ventral views. Scale bar equals 2 cm. Abbreviations: r1, r2, contact facets for first and second thoracic ribs.

In addition to the manubrium, the sternum of MCZ 8812 comprises three unfused sternebrae (fig. 5.10). The anterior two sternebrae are wider than thick and flare posteriorly; costal facets are situated on the posterior half of the lateral margins and face posterolaterally. The posterior sternebra is a massive, rounded element with deep anterolateral concavities for rib contact. The sternum of *Oligokyphus* is formed by four fused sternebrae (Kühne, 1956, fig. 51B).

Forelimb

Humerus. The humerus (fig. 5.11) is robust, with flaring proximal and distal ends and a proportionately short diaphysis, which is triangular in transverse section. The greatest width across the epicondyles is about sixty percent of humeral length in MCZ 8812 [compared with forty percent in *Massetognathus,* about fifty percent in *Thrinaxodon,* fifty-four percent in *Exaeretodon,* and fifty-eight percent in *Chiniquodon* (Jenkins, 1971); however, the ratio in *Megazostrodon* is only about thirty-two percent (Jenkins & Parrington, 1976, fig. 7d)]. The planes of the proximal and distal ends are offset relative to one another (cf. fig. 5.11A, C), with the result that the humerus has a twisted appearance. The angle between the planes is approximately 40°, comparable to that in other nonmammalian cynodonts (Jenkins, 1971).

The proximal end of the humerus is a continuous surface of unfinished, presumably cartilage-covered bone that extends from the areas of medial

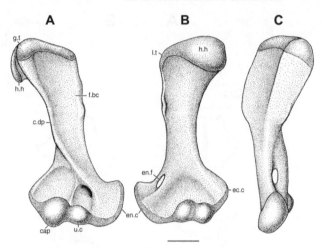

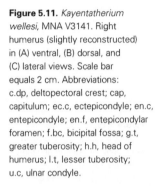

Figure 5.11. *Kayentatherium wellesi*, MNA V3141. Right humerus (slightly reconstructed) in (A) ventral, (B) dorsal, and (C) lateral views. Scale bar equals 2 cm. Abbreviations: c.dp, deltopectoral crest; cap, capitulum; ec.c, ectepicondyle; en.c, entepicondyle; en.f, entepicondylar foramen; f.bc, bicipital fossa; g.t, greater tuberosity; h.h, head of humerus; l.t, lesser tuberosity; u.c, ulnar condyle.

and lateral tuberosities. The bulbous articular surface of the central area, representing the head of the humerus, is directed anteromedially as well as dorsally. The head is raised above the adjacent dorsal surface of the shaft; ventrally, the head is delineated from the ventral surface by a slight ridge. The articular surface of the humeral head is confluent with adjoining areas where medial and lateral tuberosities would be expected; the texture of these areas, however, suggests that they were covered by cartilage and incorporated into the joint. A distinct lesser tuberosity is not developed. At the proximomedial end of the humerus, where the insertion of M. subscapularis might be expected (lesser tuberosity), the bulbous and somewhat expanded articular surface intersects with a broad surface that extends distally to about the middle of the shaft (fig. 5.11C). A distinct projection near its distal end may represent an attachment of M. triceps (as in monotremes; Howell, 1937, fig. 2G).

There is no distinct demarcation for a greater tuberosity. The proximolateral end of the humerus, where this tuberosity might be expected, is an angular and slightly dorsally reflected intersection between the proximal articular surface and the deltopectoral crest. The length of the robust deltopectoral crest is about fifty-three percent of the total humerus length, compared with about forty-one percent in *Thrinaxodon* (Jenkins, 1971, fig. 28E) and thirty-five percent in *Megazostrodon* (Jenkins & Parrington, 1976, fig. 7d). Set at a distinct angle to the adjacent proximal portion of the shaft, the crest borders a wide bicipital depression and is thickest distally. In MCZ 8812, the deltopectoral crest bears distinct striations. The medial edge of the deltopectoral crest continues posterodistally as a low, rounded ridge that forms a bar across the large entepicondylar foramen. There is no ectepicondylar foramen.

The entepicondyle is thicker and broader than the ectepicondyle. The former is farther from the radioulnar articulation than the latter, an asymmetry that increases the relative moment arm of the antebrachial flexors over the extensors (Jenkins, 1973). The broad ulnar condyle wraps around the distal end and extends onto both the dorsal and ventral aspects of the humerus, although the condyle is less extensive ventrally than dorsally. The radial condyle (*sensu* Jenkins, 1973) also wraps around the distal end. The intercondylar groove is relatively broad. The surficial texture of the bone on both condyles extends medially and laterally across the entire distal end of the humerus. Thus, the free margins of the epicondyles appear to have been cartilaginous.

Ulna. The olecranon process is long relative to the total length of the ulna (fig. 5.12). Measured from the center of the well-defined semilunar

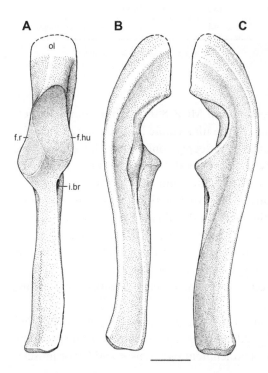

Figure 5.12. *Kayentatherium wellesi*, MNA V3141. Left ulna in (A) anterior, (B) medial, and (C) lateral views. Scale bar equals 2 cm. Abbreviations: f.hu, facet for contact with humerus; f.r, facet for contact with radius; i.br, insertion of M. brachialis; ol, olecranon.

Table 5.1. Selected measurements (in cm) of bones of the pectoral girdle and forelimb in two specimens of *Kayentatherium wellesi* (MCZ 8812 and MNA V3141)

Measurement	MCZ 8812		MNA V314
	Left	Right	
Scapula			
Height (dorsoventral)	13.7	13.4	
Glenoid facet length	3.1	3.1	2.6
Glenoid facet width	1.9	1.9	2.3
Humerus			
Length	14.2	14.2	12.7
Proximal width	6.3	6.3e	5.1
Distal width	8.8*	8.7*	6.3
Radius			
Length	12.1	11.9	10.2
Maximum diameter of head	2.6	3.5*	2.5
Distal width	3.2	3.9*	3.3
Ulna			
Length	16.2	16.2*	14e
Olecranon length	4.1	4.3*	3.8e

Note: Asterisk indicates measurement affected by plastic deformation; e denotes estimate. Olecranon length measured from midnotch to proximal end.

notch, the length of the olecranon in MCZ 8812 is about thirty-six percent that of the distal portion of the ulna (table 5.1). The large facet for the ulnar condyle of the humerus is deeply concave proximodistally and gently concave transversely. The proximal end of this facet is situated more laterally than the more distal region and also appears less concave. The facet for the radial condyle is not very distinct and is situated lateral to the proximal portion of the facet for the ulnar condyle. A facet for the proximal head of the radius, corresponding to the mammalian incisura radialis, is developed distal to the facet for the radial condyle of the humerus. Just distal to the large facet for the ulnar condyle is a deep, elongate pit on the medial edge of the ulnar shaft, possibly representing the insertion of M. brachialis.

The gently sigmoidal shaft of the ulna is transversely narrow. Either side of the shaft bears a longitudinal groove, especially near the proximal end, representing the areas of attachment of extensor and flexor muscles

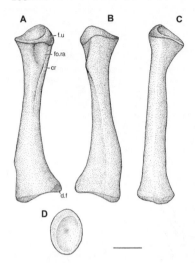

Figure 5.13. *Kayentatherium wellesi,* MCZ 8812. Left radius in (A) posteromedial, (B) anterolateral, and (C) medial views. (D) Proximal articular surface of left radius. Scale bar equals 2 cm. Abbreviations: cr, crest; d.f, facet for distal contact with ulna; f.u, facet for proximal contact with ulna; fo.ra, radial fossa.

(Jenkins, 1971). There is a well-developed anteromedial crest on the ulna for anchoring the radioulnar interosseous ligament. The distal portion of the ulnar shaft is flat anteriorly. The distal end of the ulna has its largest diameter in an anteroposterior direction. The distal articular facet is convex, with no trace of a styloid process.

Radius. The radius (fig. 5.13) has a slightly sigmoidal curvature, comparable to that of the ulna. The expanded proximal and distal ends are linked by a cylindrical shaft. The radial head bears an oval, concave articular facet (fig. 5.13D). The posterior margin of the head extends more proximally than does the anterior margin, with the result that the facet is inclined slightly anteriorly (fig. 5.13A). On the posteromedial aspect of the radial head is a broad, anteriorly tapering facet for contact with the ulna (fig. 5.13A). The extent of the facet indicates only modest capability for antebrachial pronation and supination (*contra* Kühne, 1956). A distinct ridge extends from the ulnar facet distally and posteromedially along the shaft; this feature is associated with a medial depression (fig. 5.13A). The ridge may represent the attachment of the radioulnar interosseous ligament; alternatively, a distinct crest on this ridge (fig. 5.13A) may represent a bicipital tuberosity, the point of attachment for M. biceps brachii.

The distal end of the radius is widest mediolaterally. The concave distal articular facet is delimited by sharp bony edges. The radius of MNA V3141 preserves a small but distinctly protuberant facet on the posterolateral margin of the distal end that articulates with the ulna. The size of

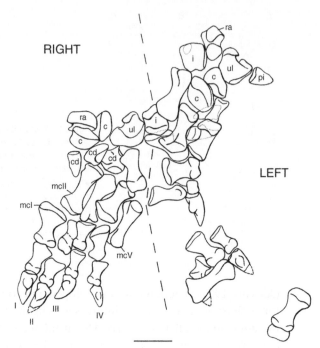

Figure 5.14. *Kayentatherium wellesi*, MCZ 8812. Dissociated left and right manus as preserved, mostly in ventral view. Lines divide the bones of the left manus from those of the right. Scale bar equals 3 cm. Abbreviations: c, centrale; cd, distal carpal; i, intermedium; mcI–V, metacarpal I–V; pi, pisiform; ra, radiale; ul, ulnare. Roman numerals denote manual digits.

this facet and the opposing area on the ulna, which is correspondingly narrow, indicate that movement at the distal radioulnar joint was modest and involved pivoting of the radius on the ulna. A styloid process is absent.

Manus. Both forefeet are preserved in MCZ 8812 (fig. 5.14). The bones were slightly disarticulated, especially those of the left manus, and some elements are missing. MNA V3141 also includes a partial manus.

The carpus comprises four proximal elements, two centralia, and at least three distal carpals (preserved on the right side of MCZ 8812).

The ulnare (fig. 5.15A), the largest proximal carpal, has a convex lateral margin and a medial margin that bears two raised facets separated by a notch for a perforating vessel. The more proximal facet, which is large and triangular, contacted the intermedium; the smaller distal facet presumably articulated with the lateral centrale. The ulnare is anteroposteriorly longer along its lateral margin than along its medial edge, and its dorsoventral thickness is greater medially than laterally. Its proximal end bears a large facet for articulation with the distal end of the ulna. The ventral surface of the ulnare is concave, whereas the dorsal face is saddle shaped.

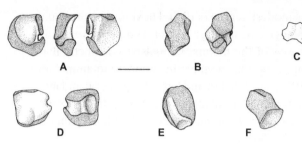

Figure 5.15. *Kayentatherium wellesi,* MCZ 8812. Carpal bones. (A) Left ulnare in (from left to right) ventral, medial, and dorsal views. (B) Left radiale in (left) proximal and (right) distal views. (C) Left pisiform in ventral view. (D) Left intermedium in (left) dorsal and (right) ventral views. (E) Left medial centrale. (F) Left lateral (proximal) centrale. Scale bar equals 2 cm.

The thick intermedium (fig. 5.15D) has a saddle-shaped dorsal surface and a notched lateral margin that corresponds to the notched medial margin of the ulnare. The nearly straight medial margin of the inter-medium forms an extensive facet for contact with the radiale. The proximal and distal facets for the ulna and lateral centrale, respectively, are well developed. On the ventral surface of the intermedium, a proxi-modistal ridge separates two concavities.

The rather flat distal face of the nodular radiale (fig. 5.15B) is divided by a groove into two distinct facets. The rounded proximal surface contacted the distal end of the radius.

The pisiform is represented only by a fragmentary element on the left side (fig. 5.15C). The more or less triangular bone is flat and thin except for the area that bears an articular facet for the ulnare.

The two centralia (fig. 5.15E, F) may have been aligned transversely to the long axis of the manus as reconstructed for *Thrinaxodon* by Jenkins (1971, 128). One of the bones is slightly flattened dorsoventrally and possibly represents the medial centrale. The other, with an irregularly nodular shape, may be the lateral centrale. Three bones in the right carpus represent distal carpals (fig. 5.14), but their relative positions are uncertain due to postmortem disturbance of their articular context.

The metacarpals are broadly expanded proximally and distally (fig. 5.14). Metacarpal I is the shortest and relatively widest in overall dimensions; the bone also differs from other metacarpals in having an asymmetric proximal end with the greatest expansion transversely. The proximal ends of all metacarpals are dorsoventrally deeper than the distal ends. A depression in the lateral side of the proximal end of each of metacarpals II–IV accommodates the adjoining metacarpal.

The phalanges are robust. At the interphalangeal joints, the distal condyles are dorsoventrally asymmetric; the articular facets extend further proximally onto the flexor aspect than onto the extensor surface, an asymmetry that indicates a modest degree of digital flexion. The proximal ends of the ungual phalanges bear large flexor tubercles ventrally (fig. 5.14). The distal portions of the manual unguals are distinctly flattened, unlike the tapering, cone-like unguals in other nonmammalian cynodonts (e.g., Jenkins, 1971, fig. 35A-D), and have sharp lateral and medial edges. The phalangeal formula of the manus is 2-3-3-3-3 in the Comb Ridge material referable to *Kayentatherium* (Lewis, 1986), and this is probably also the case in MCZ 8812.

Pelvic Girdle

The pelvis of *Kayentatherium* is best known from the still-undescribed Comb Ridge specimens (Lewis, 1986, fig. 2). MCZ 8812 preserves only the anterior portion of the left iliac blade. Incomplete pelvic elements, including an ilium of a small indeterminate tritylodontid (MCZ 8835) and an ischium of *Dinnebitodon,* are represented in the MCZ-MNA collections and provide the basis for the following observations.

Ilium. The elongate iliac blade (fig. 5.16A, B) was oriented anterodorsally in life (Lewis, 1986, fig. 2) and gradually expands to it greatest depth anteriorly (fig. 5.16C). A longitudinal ridge on the lateral surface (crista lateralis *sensu* Kühne, 1956) defines dorsal and ventral surfaces, which

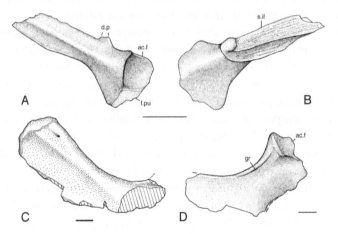

Figure 5.16. (A, B) Tritylodontidae indet. (MCZ 8835), left ilium in (A) lateral and (B) medial views. Note that the dorsal and ventral margins of the preacetabular portion are damaged. (C) *Kayentatherium wellesi* (MCZ 8812), partial left ilium in lateral view. Hatched areas denote broken surfaces. (D) *Dinnebitodon amarali* (MNA V3232), right ischium in lateral view. Scale bars each equal 1 cm. Abbreviations: ac.f, acetabular facet; d.p, posterodorsal process of ilium; f.pu, facet for contact with pubis; gr, groove; s.il, sacroiliac facet.

face dorsolaterally and ventrolaterally, respectively. In MCZ 8835, the ventral margin of the blade is a thin, slightly everted flange, and the surface between the ventral margin and the longitudinal ridge is slightly concave. The medial surface of the iliac blade is extensively marked by parallel striae and rugosities representing a well-developed sacroiliac syndesmosis; as in *Oligokyphus* (Kühne, 1956, 124), this surface is divided by a longitudinal ridge. Behind the sacroiliac articulation, the neck of the ilium is approximately triangular in transverse section (MCZ 8835). Any evidence of a pit below the lateral longitudinal ridge, which Kühne (1956, 124) interpreted as the origin of M. rectus femoris in *Oligokyphus,* has been obscured by crushing in this region in MCZ 8835. A dorsally and slightly posteriorly directed, spine-like process arises from the dorsal margin of the ilium at the level of the posterior end of the sacroiliac facet (fig. 5.16A); most of the process has been broken off in MCZ 8835, but the process is completely preserved in USNM 317201 (Lewis, 1986, fig. 2). Kühne's (1956, fig. 57) reconstruction depicts no dorsal process on the ilium of *Oligokyphus* but does show a step-like increase in the depth of the iliac blade at a location similar to that of the process. Examination of Kühne's collection of *Oligokyphus,* however, reveals two ilia in which a prominent, flange-like posterior process is more or less completely preserved (BMNH R 7445, 7448); on several other ilia (BMNH R 7442-7444, 7446, and 7447), the process has been broken off but its base is evident. The ilium of *Exaeretodon* has a posteriorly directed process (Bonaparte, 1963, fig. 9).

The large, concave acetabular facet of the ilium is oriented perpendicular to the long axis of the bone and faces posteroventrally (fig. 5.16A) rather than ventrally as in basal eucynodonts (e.g., Jenkins, 1971, figs. 45A, 46A). A triangular articular surface is situated on the ventral process for contact with the pubis. As in *Oligokyphus* (BMNH R 7441) and *Exaeretodon* (Bonaparte, 1963, fig. 9), a dorsal acetabular rim is absent, with the resulting marginal gap between the ilial and ischial facets of the acetabulum presumably filled by fibrocartilage in life. The acetabular facets of the ilium and ischium form slightly protruding lateral margins. Kühne's (1956, fig. 57) reconstruction of the pelvis of *Oligokyphus* depicts the bony acetabulum as an elongated socket with the greatest diameter in an anteroposterior direction.

Ischium. The acetabular portion of the ischium (fig. 5.16C) is separated from the plate-like, ventromedially directed ischial plate by a slight constriction. The acetabular facet faces dorsally and anterolaterally; as noted above, the dorsal margin of the acetabulum was not completed in bone.

A distinct medial facet contacted the ilium. A ridge extends from the center of the lateral rim of the acetabular facet posterodorsally, gradually attenuating toward the ischial tuberosity; a shallow sulcus is present just dorsal to the ridge. Comparable features are present in basal eucynodonts (Jenkins, 1971, figs. 45A, 47A). The thin anteroventral edge of the ischial plate contributes to the margin of an apparently large obturator foramen. The acetabular notch between the acetabular facet and the pubic process is shallow in MNA V3232 but deep in MNA V3235.

Hind Limb

The hind limb is represented in the MCZ-MNA collections only by the nearly complete left femur of an indeterminate tritylodontid (MCZ 8838) as well as parts of the left femur and the proximal ends of both tibiae in MNA V3141.

Femur. The bulbous femoral head is reflected slightly dorsomedially relative to the long axis of the shaft (fig. 5.17A, C). The articular surface as a whole is a slightly asymmetric ovoid and extends laterally onto the ridge between the head and greater trochanter (fig. 5.17D). The shape of the femoral head, which is also present in *Oligokyphus* (Kühne, 1956, fig. 58), differs from the symmetrically hemispheroidal head in *Morganucodon* (Jenkins & Parrington, 1976, figs. 12, 13e). The area of the femoral neck is only faintly defined by a slight constriction. The deep intertrochanteric fossa occupies the entire broad surface between the trochanters and the head and extends farther distally than in either basal eucynodonts (Jenkins, 1971, figs. 48B, 49B) or *Morganucodon* (Jenkins & Parrington, 1976, fig. 12c).

The robust, triangular trochanters arise from the femoral shaft in relatively ventral positions. They diverge at angles of about 40° relative to the long axis of the shaft in MCZ 8838 and are more or less aligned in the same transverse plane. The lesser trochanter is reflected dorsally and is thicker than the ridge connecting it to the femoral head.

The femur of MCZ 8838 closely resembles that of *Bienotherium* (Young, 1947, fig. 20A). Kühne's (1956, fig. 58) reconstruction of the femur of *Oligokyphus* resembles the femur of MCZ 8838 in the shape of the femoral head and the relative positions of the trochanters on contralateral sides of the shaft. However, other details shown in Kühne's reconstruction, notably a ridge on the dorsal side of the greater trochanter, the cylindrical shape of the greater trochanter, and an acute notch between the greater trochanter and femoral head, appear to represent major differences.

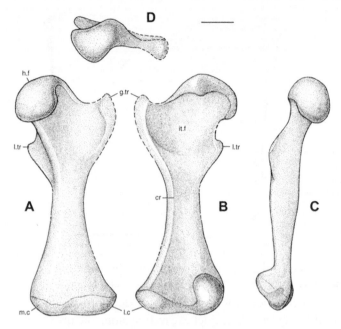

Figure 5.17. Tritylodontidae indet., MCZ 8838. Left femur (slightly reconstructed) in (A) dorsal, (B) ventral, and (C) medial views. (D) Proximal view of the proximal end. Outline of greater trochanter restored from MCZ 8832. Scale bar equals 2 cm. Abbreviations: cr, crest; g.tr, greater trochanter; h.f, head of femur; it.f, intertrochanteric fossa; l,c, lateral condyle; l.tr, lesser trochanter; m.c, medial condyle.

The shaft of the femur is oval in transverse section, with the smaller diameter between the dorsal and ventral surfaces. A crest along the posteroventral margin of the shaft (fig. 5.17B) is comparable to a ridge in basal eucynodonts (Jenkins, 1971, 172). The lateral condyle of the femur is considerably wider transversely than the medial one (fig. 5.17B) and forms a gently convex articular surface; the medial condyle has a more sharply curved articular surface and projects further ventrally than the lateral one. These features are already present in basal eucynodonts (Jenkins, 1971, 172) and are retained among some extant mammals (Jenkins & Parrington, 1976, 424). There is no distinct patellar groove.

Tibia. The narrowest dimension of the oval shaft is mediolateral rather than anteroposterior as in basal eucynodonts (Jenkins, 1971, figs. 56, 57). The proximal end, which is expanded anteriorly and laterally, bears two condylar facets that are separated by a low ridge (fig. 5.18). The slightly concave facet for the medial condyle is inclined anteromedially; the long axis of the facet extends almost transversely. The posterolateral facet for the lateral femoral condyle is elongate, and its long axis extends posteromedially. A projection on the anteromedial margin of the articular surface appears to represent a tibial tuberosity for the insertion of M. quadriceps

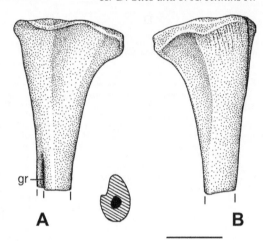

Figure 5.18. *Kayentatherium wellesi,* MNA V3141. Proximal end of right tibia in (A) posterior (with transverse section at broken distal end) and (B) anterior view. Scale bar equals 2 cm. Abbreviation: gr, groove.

femoris; similarly positioned tubercles are present in other nonmammalian cynodonts (Jenkins, 1971, 186). The cnemial crest, which extends distally from the tibial tuberosity, is not as prominently developed as in *Oligokyphus* (Kühne, 1956, fig. 60). A distinct excavation along the proximomedial aspect of the tibial shaft may mark the origin of M. tibialis anterior, as suggested for *Oligokyphus* (Kühne, 1956, 128). A prominent lateral groove on the flexor aspect possibly relates to the origin of a plantar flexor muscle. The slightly convex distal end of the tibia is irregular in outline, and its articular surface has a pitted texture suggestive of cartilaginous covering in life. An elliptical scar of uncertain identity is present on the medial aspect of the tibial shaft in MCZ 8838.

Discussion

The partial skeleton of *Kayentatherium* (MCZ 8812) and other tritylodontid postcranial remains from the Kayenta Formation offer a basis for reassessing the allegedly mammalian postcranial features that have been ascribed to this group of nonmammalian eucynodonts. The composite reconstruction of the skeleton of *Kayentatherium* (fig. 5.19), based primarily on MCZ 8812, depicts a stockily built animal with a large head, stout vertebral column, and robust girdles and limbs. Do postcranial features provide persuasive evidence in support of a sister-group relationship between Tritylodontidae and Mammalia?

Postcranial Axial Skeleton

The cervical vertebrae exhibit features that were already present, incipiently developed among basal eucynodonts, or appear to be not closely

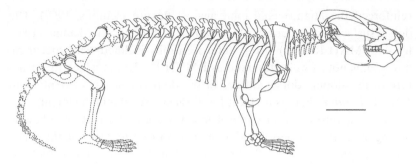

Figure 5.19. Composite reconstruction of skeleton of *Kayentatherium wellesi* in right lateral view. The skull, presacral postcranial skeleton, and individual caudal vertebrae are based on MCZ 8812. Other postcranial bones are reconstructed on the basis of additional tritylodontid specimens from the Kayenta Formation: ilium (anterior portion from MCZ 8812 and remainder from indeterminate tritylodontid, MCZ 8835), ischium (*Dinnebitodon amarali*, MNA V3232), femur (indeterminate tritylodontid, MCZ 8832 and 8838), and tibia (*Kayentatherium wellesi*, MNA V3141). Length of the tail is conjectural. Scale bar equals 10 cm.

similar to the respective condition in basal mammals. The unfused paired neural arches and the intercentrum of the atlas in *Kayentatherium* are structurally similar to those in basal eucynodonts and represent a pattern that is retained in *Morganucodon*. Although small, the axial prezygapophyses are likewise a plesiomorphic feature. The axial centrum is longer than the atlantal centrum in basal eucynodonts; these proportions are retained among mammals, but the length of the two centra is more or less equal in *Oligokyphus* (Kühne, 1956, fig. 43) and *Kayentatherium*. The presence of a distinct dens is a mammal-like feature, but, in fact, this structure originated among basal eucynodonts (Jenkins, 1971). The dens in *Kayentatherium* (MCZ 8812) is short, stout, and anteriorly attenuated. Its dorsal surface, however, is slightly inclined anteroventrally (figs. 5.1E, 5.2A) and unlike the horizontal dorsal surface of the dens in *Morganucodon*, which bears a facet for the transverse ligament of the atlas (Jenkins & Parrington, 1976, fig. 1f, g). There is no evidence of such a facet in *Kayentatherium*, and the inclination of the dorsal surface of the dens is mechanically unsuited to sustain a transverse ligament of the atlas, which is a major component of the osseoligamentous check mechanism constraining atlanto-axial flexion in mammals (Jenkins, 1969). Instead, the dens in *Kayentatherium* may have served for attachment of the apical and alar ligaments (as in mammals). Our interpretation that *Kayentatherium* lacked a transverse atlantal ligament is corroborated by the presence of axial prezygapophyses that supported the atlantal arches and effectively reinforced the atlanto-axial joint against flexure (Jenkins, 1969). The dorsal surface of the dens of *Oligokyphus* as illustrated by Kühne (1956, fig. 43; BMNH R 7315) also is anteroventrally inclined. This specimen, however, is not as well preserved as BMNH R 7316, in which the structure of the sloping dorsal surface of the dens is patently evident; as in *Kayentatherium,* there is no facet for a transverse atlantal ligament.

The structure of other cervical intervertebral articulations likewise distinguishes tritylodontids and mammals. In *Kayentatherium,* the cervical zygapophyses are dorsally directed and the facets are steeply inclined (to within about 30° of the vertical plane), which is comparable to the condition in the basal eucynodont *Thrinaxodon* (Jenkins, 1971, fig. 12C) but differs from the more horizontally inclined zygapophyses and facets in *Megazostrodon* (Jenkins & Parrington, 1976, fig. 6a, b). The only features shared by tritylodontid and mammalian cervical vertebrae are the reduced height of the neural spines and the anteroposterior narrowness of the pedicles; both may be related to an elevated (i.e., slightly extended) posture of the neck.

The dorsal vertebrae of *Kayentatherium* have almost no features that might be characterized as mammal-like. Typically in mammals, the orientation of anterior thoracic zygapophyseal facets exhibits a marked shift to a horizontal orientation at the cervicothoracic junction. In *Thrinaxodon,* the cervicodorsal boundary is delineated by a distinct zygapophyseal facet reorientation at C7; however, the shift is toward a more vertical orientation (from 55° to 15° relative to a sagittal plane through the pre- and postzygapophyses, respectively) (Jenkins, 1971, 53). In contrast, a reorientation of facet angle occurs at the second dorsal vertebra in *Kayentatherium;* the facets shift to a more or less horizontal orientation, as in mammals. The elongate and posteriorly inclined neural spines on the anterior dorsal vertebrae in *Kayentatherium* indicate a well-developed nuchal ligament and M. semispinalis for suspension of the large head (as in some mammals), and yet these features appear to be but a slight elaboration of the same fundamental pattern present in the cervicodorsal column of *Thrinaxodon* (Jenkins, 1971, fig. 12C).

An important distinguishing feature of mammals is thoracolumbar differentiation that occurs as the result of a relatively abrupt shift in the orientation of the zygapophyses and neural spines (at the diaphragmatic and anticlinal vertebrae, respectively). There is definitive evidence from a single specimen referable to *Morganucodon* that distinct thoracolumbar differentiation was already present in basal mammals (Jenkins & Parrington, 1976, fig. 6D, E). *Kayentatherium,* however, clearly lacks any such differentiation. Although the orientation of the posterior dorsal zygapophyseal facets does not preclude some axial flexion and extension in this region, strong supra- and interspinous ligaments (as indicated by the shape of the apices of the neural spines) probably would have restricted flexibility. Whereas a number of distinctive features are evident among the dorsal vertebrae and ribs, the structural transitions along the truncal series are entirely gradational.

The exostoses on the neural spines of mid- and posterior dorsal vertebrae of *Kayentatherium* are comparable to those in *Exaeretodon* (Bonaparte, 1963), but exostoses appear to be absent in *Oligokyphus*. The median exostosis, which projects posteriorly, most likely represents the site of attachment for well-developed supraspinous and/or interspinous ligaments. The posterolateral projections, commencing at the tenth dorsal and elaborated as "spine tables" more posteriorly along the dorsal column, appear to be indicative of hypertrophied epaxial musculature, particularly M. transversospinalis, which bridges adjacent vertebrae or short, contiguous series of vertebrae. Together these features most plausibly are interpreted as contributing additional rigidity to the back and are yet another example of the diverse axial features among nonmammalian cynodonts for reinforcement of the lumbar region [for discussions of this problem, see Jenkins (1971, 86–91) and Kemp (1980, 232–235).

Appendicular Skeleton

Major features of the pectoral girdle, notably the supraspinous and postscapular fossae, orientation of the glenoid, position of the scapular spine, and shape of the clavicle, invite consideration as mammal-like features, but comparisons are somewhat limited by the fact that the shoulder girdle in basal mammals is still incompletely known.

The presence of a biplanar supraspinous fossa, between an anterior extension of the scapular blade and the scapular spine, might be considered a patently mammalian feature were it not for two counterindications. The first is that a supraspinous fossa was already present among basal eucynodonts. In *Kayentatherium*, the fossa is restricted to the dorsal half of the scapula and is thus only an incomplete expression of this feature in the area where, in basal eucynodonts, there is an incipient anterior shelf (Jenkins, 1971, fig. 17A; Kemp, 1980, fig. 10). The second problem is that there is no evidence for the presence of a biplanar supraspinous fossa in basal mammals. The vertebral end of the scapula in *Megazostrodon*, albeit damaged in the only known specimen, apparently did not possess this feature, and there is no indication of a biplanar fossa on the known incomplete scapulae referable to *Morganucodon* (Jenkins & Parrington, 1976, figs. 4, 7a). Among mammals, monotremes and multituberculates (Krause & Jenkins, 1983; Sereno & McKenna, 1995) apparently never developed a biplanar supraspinous fossa, although the putative equivalent of M. supraspinatus may be identified or reconstructed as originating from the anterior surface of the laterally reflected scapular spine. Complete supraspinous fossae are known only in later Mesozoic mammals, such as Early Cretaceous triconodontids (Jenkins & Crompton, 1979; Ji et al., 1999).

A postscapular fossa is not a universal mammalian feature. It occurs principally as a fossorial specialization (e.g., in some edentates) but also in forms with a powerful forelimb stroke (such as bears and pinnipeds) in which M. teres major, a humeral retractor, is hypertrophied. A similar functional inference may be made for the postscapular fossa of *Kayentatherium*. The feature is incipiently developed in more basal eucynodonts (e.g., Jenkins, 1971, fig. 17).

The overall configuration of the glenoid and coracoid region in *Kayentatherium,* including the reduction in the relative size of the procoracoid and coracoid and the exclusion of the procoracoid from the glenoid (fig. 5.8), is comparable to that in basal mammals with one notable exception. The primarily ventral, and only slightly posterolateral, orientation of the scapular facet of the glenoid in *Kayentatherium* is closely similar to the ventral orientation generally present in more derived mammals (in which the procoracoid has been lost and the coracoid has been reduced to a small process for muscle attachment).

The scapula of *Kayentatherium* diverges from the pattern seen in basal mammals. The two most salient differences are the size and orientation of the scapular spine and the position of the infraspinous and supraspinous fossae relative to the glenoid (fig. 5.20A). In basal eucynodonts, the anterior margin of the scapula—homologous to the mammalian scapular spine—is reflected anterolaterally and situated anterior to the shoulder joint (fig. 5.20B). In basal mammals such as *Morganucodon,* the lateral reflection is relatively accentuated, so that the margin of the scapular spine is approximated to the anterior margin of the glenoid (fig. 5.20C). In more derived mammals, the scapular spine is positioned above the shoulder joint, bisecting the supra- and infraspinous fossae in a manner that provides a symmetrical approach of supra- and infraspinatus muscle fibers to the greater tuberosity of the humerus. *Kayentatherium,* in contrast, has a large, anterolaterally directed scapular spine that is situated well away from the glenoid (fig. 5.20A). Kemp (1982, 246) likewise

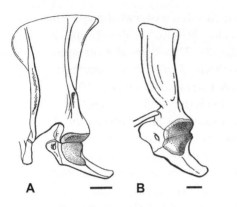

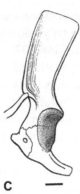

Figure 5.20. Comparison of shoulder girdles of (A) *Kayentatherium,* (B) indeterminate eucynodont (cf. *Cynognathus*), and (C) *Morganucodon* in left lateral view. Drawings are scaled to approximately the same height of the scapulocoracoid; the glenoid articular surfaces are indicated by stippling. (B) Modified from Jenkins (1971); (C) modified from Jenkins and Parrington (1976). Scale bars equal 2 cm (A, B) and 1 mm (C).

A — **B** — **C** —

recognized this unusual feature in *Oligokyphus,* which he characterized as "the extreme eversion of the acromion process of the scapula." A second unusual feature in *Kayentatherium* is the relative anterolateral displacement of the infraspinous fossa; the long axis of the central sulcus of the fossa extends anterolateral to the glenoid (fig. 5.20A) rather than dorsally through the glenoid as in basal mammals (fig. 5.20C). As a result, the supraspinatus muscle also is displaced anteriorly, away from the glenoid, rather than posteriorly and closer to the joint as in basal mammals. The development of a massive M. supraspinatus in direct line between the acromion and interclavicle may account for the distinctively angular form of the clavicle in *Kayentatherium.*

There are major differences between the humerus of *Kayentatherium* and those of basal mammals in the configuration of the tuberosities and humeral head. In *Kayentatherium,* the areas representing the greater and lesser tuberosities are confluent with the humeral head and not developed as distinct features (although they may have been more prominent as part of a cartilaginous 'epiphyseal' cap); the configuration of the bony tuberosities is comparable to that in basal eucynodonts. The articular surface of the head is ovoid with the long axis extending mediolaterally (i.e., between the tuberosities); the dorsoventral curvature is only slightly more extensive than that seen in basal eucynodonts (Jenkins, 1971, fig. 26A). In contrast, the humeri of basal mammals have distinct bony tuberosities and a hemispheroidal humeral head, a pattern retained in more derived mammals.

Kühne's (1956, fig. 55) reconstruction of the ulna in *Oligokyphus,* based on five incomplete specimens, depicts a more gracile, elongate bone than the ulna of (the much larger) *Kayentatherium.* As Kühne (1956, 120) noted, "if the humerus and the ulna [of *Oligokyphus*] are compared, the preponderance of mammalian features in the ulna is overwhelming." The basic configuration of the tritylodontid ulna, however, is already present in basal eucynodonts (Jenkins, 1971, figs. 31, 32)—sigmoidal curvature, anteroposteriorly narrow shaft, and longitudinal sulci; these features are retained in morganucodontids (Jenkins & Parrington, 1976, fig. 10).

The most widely cited feature of the ulna is the ossified olecranon process, which Hopson and Kitching (2001) scored as either present or absent in their cladistic analysis. Other authors have recognized intermediate conditions (Kemp, 1983; Rowe, 1988) or used degrees of ossification (Sidor & Hopson, 1998). Whereas tritylodontids and basal mammals both have ossified olecranon processes, more basal eucynodonts almost certainly had cartilaginous processes (Jenkins, 1971, 123). The presence of a large, ossified olecranon in *Kayentatherium* contrasts with the extensively cartilaginous 'epiphyseal' ends of other long bones, notably the humerus and femur, which represent the plesiomorphic condition for cynodonts. A large, ossified olecranon on a short ulna in a robust animal is consistent with other features (postscapular fossa, extensive deltopectoral crest on the humerus, large flexor tubercles on the ungual phalanges) that are indicative of a powerful forelimb employed in activities other than locomotion, such as digging. The relative length of the olecranon in *Kayentatherium* (MCZ 8812) is about thirty-six percent that of the distal segment of the ulna, falling well within the range of proportions (twenty to seventy-five percent) determined by Hildebrand (1995) for twenty-seven genera of present-day fossorial mammals. In *Megazostrodon,* the only basal mammal for which a complete ulna is currently known, the length of the olecranon is only about fifteen percent that of the distal segment of the ulna. We find no unambiguous phylogenetic signal in the ossification of the olecranon among eucynodonts.

The extent of the proximal radioulnar facets provides evidence for only a modest degree of antebrachial pronation and supination. With respect to the distal radioulnar articulation, Kühne (1956, 123) made the perceptive observation that in *Oligokyphus* "the distal end of the ulna bears a concavity for accommodation of the radius. In the mammals it is always the radius which has a concavity for the convex articular surface of the ulna." Although the shape of the distal ulnar articular facet in *Kayentatherium* is not known, the surface of the distal end of the ulna that contacted the radius is narrow and flat. The radial facet is also small and protuberant. Pronation-supination thus would appear to have been limited to a simple pivoting of the radius on the ulna. In typical mammals, as Kühne (1956, 123) pointed out, the radius translates about a curvilinear facet of the ulna. In *Morganucodon* (MCZ 19944), the distal end of the ulna for radial contact is distinctly rounded, and the corresponding facet on the radius is a small planar facet. Translation at the distal radioulnar joint thus appears to have been possible in basal mammals, but a comparable movement was absent in tritylodontids.

The fragmentary nature of the available tritylodontid ilia from the Kayenta Formation constrains our comparisons to relatively few observations. A juvenile skeleton of *Kayentatherium* (USNM 317201) has a dorsally and slightly posteriorly directed, spine-like process on the dorsal margin of the blade (Lewis, 1986, fig. 2); the base of a similar process is preserved on the ilia of *Kayentatherium* (MCZ 8812, fig. 5.16C) and of an indeterminate tritylodontid (MCZ 8835, fig. 5.16A, B). We interpret this process as a vestige of the more extensive posterior projection of the iliac blade in basal eucynodonts. In mammals, this process is absent or structurally modified. In *Megazostrodon,* as well as in ilia attributed to *Morganucodon,* there is no trace of a projection; the posterior half of the dorsal margin of the blade exhibits an elongate, slightly laterally reflected flange (Jenkins & Parrington, 1976, figs. 11a, 13a, d). In *Tachyglossus,* however, there are two low tuberosities along the posterior margin of the blade (Jenkins & Parrington, 1976, fig. 20d). The medial (sacral) surface of the ilium also appears to differ between tritylodontids and mammals. The incomplete anterior end of the blade of the partial ilium of MCZ 8812 reveals that the sacroiliac joint extended nearly to the anterior terminus. The incomplete ilium of MCZ 8835 (fig. 5.16B) shows that the joint surface extended posteriorly to the level of the dorsal process, in close proximity to the acetabulum. The sacroiliac joint thus occupied most of the medial surface of the iliac blade, in contrast to morganucodontids in which the articular area is situated on the anterior half of the ilium, well separated from the acetabulum by a "neck" that is circular in transverse section (Jenkins & Parrington, 1976, fig. 11b).

The anterodorsally oriented, relatively narrow iliac blade of tritylodontids has been cited as a synapomorphy shared with mammals. Kühne's (1956, fig. 57) reconstruction of the pelvis of *Oligokyphus* has been particularly influential, especially his rendition of a very narrow, rod-like, and tapering iliac blade. Examination of the original material, however, reveals that all ilia referred to *Oligokyphus* are fragmentary; in none of the specimens is the anterior end of the ilium preserved, nor are the dorsal and ventral margins complete. We found no compelling evidence in support of Kühne's (1956, fig. 57) reconstruction. Although a posterior process of the ilium is now recognized as present in *Oligokyphus* and

Kayentatherium, the anterior end of the tritylodontid ilium is still imperfectly known. On the basis of the partial ilium of MCZ 1812 (fig. 5.16C), the iliac blade appears to be somewhat spatulate, with its greatest dorsoventral dimension near the anterior terminus. Such an interpretation is consistent with the substantial evidence concerning the diversity of iliac shapes among nonmammalian eucynodonts that has been documented since the publication of Kühne's (1956) study. The ilia of more basal eucynodonts, such as *Thrinaxodon* and *Cynognathus,* have an expanded blade with both anterior and posterior projections, but even in basal forms the anterior portion of the blade has a larger surface area, indicating an anterior shift in muscular preponderance (Jenkins, 1971). In more derived eucynodonts, including the gomphodont *Exaeretodon* (Bonaparte, 1963, fig. 9) and the probainognathian *Prozostrodon* (Bonaparte & Barberena, 2001, fig. 15A), the iliac blade is composed almost entirely of an expanded anterior projection, with the posterior part reduced to a small process. Although incompletely known, the ilium of the probainognathian *Therioherpeton* has an entirely preacetabular, parallel-sided blade with no trace of a posterior process (Bonaparte & Barberena, 2001, fig. 5B). In view of the aforementioned evidence, the mammal-like pattern in tritylodontids appears far less extraordinary than it did several decades ago. In the context of the phylogenetic analysis by Hopson and Kitching (2001), which hypothesized a dichotomy between Probainognathia and Gomphodontia, a predominantly preacetabular iliac blade must have evolved independently at least twice (cf. Bonaparte & Barberena, 2001).

Tritylodontid femora, although approximating a mammalian pattern, differ in several important respects from those of basal mammals. As in more basal eucynodonts, the outline of the femoral head in MCZ 8838 is ovoid; the long axis, which passes from anterodorsal to posteroventral, is about twenty percent longer than the perpendicular short axis. In *Morganucodon,* the femoral head is very nearly if not perfectly hemispheroidal, with an axial differential estimated to be about six percent. The intertrochanteric fossa in MCZ 8838 is enormous, extending distally about one-third the length of the femur; in *Morganucodon,* the fossa is relatively restricted in extent, extending distally no more than twenty percent of the length of the femur.

The configuration of the femoral trochanters in *Kayentatherium* differs in several respects from that in basal eucynodonts and *Morganucodon.* In basal eucynodonts, the lesser trochanter is situated ventral and distal to the greater trochanter. In tritylodontids, the lesser trochanter has shifted medially but remains distal to the greater trochanter (cf. Young, 1947, fig. 20A); in terms of proximodistal placement, its apex is distal to that of the greater trochanter. In morganucodontids, the lesser trochanter has migrated proximally so that the apices of the trochanters are nearly aligned on the same transverse axis (Jenkins & Parrington, 1976, figs. 12a, 13b). The femur of MCZ 8838 thus exhibits an intermediate condition in that the trochanter is medially, but not proximally, displaced. Kühne (1956, 126) and, more recently, Hopson and Kitching (2001, 31) cited the presence of a notch separating the femoral head from the greater trochanter as an apomorphic feature; in basal eucynodonts, however, a slight indentation is present (Jenkins, 1971, fig. 48A, B), and the extent to which separation is apparent seems to reflect the degree of ossification of the proximal end. Finally, the greater trochanter in tritylodontids retains a plesiomorphic, flange-like structure, although it has a discernable apex. The greater trochanter of basal mammals, in contrast, is a proximally directed process with a pronounced apex; the proportional length of the base of the trochanter adjoining the shaft is decreased, and thus the flange-like appearance of the trochanter is reduced.

In conclusion, we find few features in the postcranial skeleton of tritylodontid cynodonts that could be interpreted as potential synapomorphies in support of a sister-group relationship between tritylodontids and mammals. In the preceding section, we demonstrated that most of the alleged postcranial similarities are only superficial in nature, and scoring them as representing the same character-state in phylogenetic analyses obscures the real structural differences in these features between tritylodontids and basal mammals. Finally, certain mammal-like features of the postcranial skeleton of the Tritylodontidae (e.g., large, ossified olecranon process) appear to represent autapomorphies for this group and thus are not useful for determining its phylogenetic relationships.

Acknowledgments

It is with the greatest pleasure that we contribute our study to a volume honoring our dear friend and colleague, James A. Hopson. Not only has Jim always been supportive of our own work in the diverse directions we have taken, but symbolically laying a large tritylodontid on his doorstep has special significance, for Jim is the veritable Dean of Cynodontia.

Charles R. Schaff discovered MCZ 8812, and William W. Amaral contributed to the preparation of this specimen. C. R. Schaff, W. W. Amaral, and the late William R. Downs collected other tritylodontid material from the Kayenta Formation. We thank Angela C. Milner and the late Alan J. Charig (The Natural History Museum, London) for access to Kühne's material of *Oligokyphus* and for many courtesies extended to us during our visits. Diane Scott converted the original drawings by H.-D.S. into electronic format. L. Laszlo Meszoly drafted figs. 5.4 and 5.16C. Fernando Abdala (University of the Witwatersrand) and Zhe-Xi Luo (Carnegie Museum of Natural History) provided constructive reviews of the manuscript. We gratefully acknowledge the Navajo Nation for granting scientific access to the region of the Adeii Eechii Cliffs and the National Science Foundation for funding the Kayenta project.

Literature Cited

Bonaparte, J. F. 1963. Descripción del esqueleto postcraneano de *Exaeretodon* sp. *Acta Geologica Lilloana* 4:5–52.

Bonaparte, J. F. and M. C. Barberena. 2001. On two advanced carnivorous cynodonts from the Late Triassic of southern Brazil. *Bulletin of the Museum of Comparative Zoology, Harvard University* 156:59–80.

Clark, J. M. and J. A. Hopson. 1985. Distinctive mammal-like reptile from Mexico and its bearing on the phylogeny of the Tritylodontidae. *Nature* 315: 398–400.

Crompton, A. W. 1972. Postcanine occlusion in cynodonts and tritylodontids. *Bulletin of the British Museum (Natural History), Geology* 21:27–71.

Crompton, A. W. and F. Ellenberger. 1957. On a new cynodont from the Molteno Beds and the origin of tritylodontids. *Annals of the South African Museum* 44:1–14.

Fourie, S. 1962. Further notes on a new tritylodontid from the Cave Sandstone of South Africa. *Navorsinge van die Nasionale Museum, Bloemfontein, South Africa* 2(1):7–19.

Gregory, W. K. and C. L. Camp. 1918. Studies in comparative myology and osteology. No. III. *Bulletin of the American Museum of Natural History* 38: 447–563.

Hildebrand, M. 1995. *Analysis of Vertebrate Structure,* 4th ed. New York: Wiley.

Hopson, J. A. 1991. Systematics of the nonmammalian Synapsida and implications for patterns of evolution in synapsids; pp. 635–693 *in* H.-P. Schultze and L. Trueb (eds.), *Origins of the Higher Groups of Tetrapods: Controversy and Consensus.* Ithaca: Comstock Publishing Associates (Cornell University Press).

Hopson, J. A. and H. R. Barghusen. 1986. An analysis of therapsid relationships; pp. 83–106 *in* N. Hotton, III, P. D. MacLean, J. J. Roth, and E. C. Roth (eds.), *The Ecology and Biology of Mammal-like Reptiles.* Washington, DC: Smithsonian Institution Press.

Hopson, J. A. and J. W. Kitching. 1972. A revised classification of cynodonts (Reptilia, Therapsida). *Palaeontologia Africana* 14:71–85.

Hopson, J. A. and J. W. Kitching. 2001. A probainognathian cynodont from South Africa and the phylogeny of nonmammalian cynodonts. *Bulletin of the Museum of Comparative Zoology, Harvard University* 156:5–35.

Howell, A. B. 1937. Morphogenesis of the shoulder architecture. Pt. V.
    Monotremata. *Quarterly Review of Biology* 12:191–205.
Jenkins, F. A., Jr. 1969. The evolution and development of the dens of the mam-
    malian axis. *Anatomical Record* 164:173–184.
———. 1970. The Chañares (Argentina) Triassic reptile fauna. VII. The post-
    cranial skeleton of the therapsid *Massetognathus pascuali* (Therapsida, Cyn-
    odontia). *Breviora* 352:1–28.
———. 1971. The postcranial skeleton of African cynodonts. *Bulletin of the
    Peabody Museum of Natural History, Yale University* 36:1–216.
———. 1973. The functional anatomy and evolution of the mammalian humero-
    ulnar articulation. *American Journal of Anatomy* 137:281–298.
Jenkins, F. A., Jr. and A. W. Crompton. 1979. Triconodonta; pp. 74–90 *in*
    J. A. Lillegraven, Z. Kielan-Jaworowska, and W. A. Clemens (eds.), *Meso-
    zoic Mammals: The First Two-Thirds of Mammalian History.* Berkeley:
    University of California Press.
Jenkins, F. A., Jr. and F. R. Parrington. 1976. The postcranial skeleton of the
    Triassic mammals *Eozostrodon, Megazostrodon* and *Erythrotherium.*
    Philosophical Transactions of the Royal Society of London, B 273:
    387–431.
Ji Q., Z.-X. Luo and S.-A. Ji. 1999. A Chinese triconodont mammal and mosaic
    evolution of the mammalian skeleton. *Nature* 398:326–330.
Kemp, T. S. 1980. Aspects of the structure and functional anatomy of the Middle
    Triassic cynodont *Luangwa. Journal of Zoology, London* 191:193–239.
———. 1982. *Mammal-like Reptiles and the Origin of Mammals.* New York:
    Academic Press.
———. 1983. The relationships of mammals. *Zoological Journal of the Linnean
    Society* 77:353–384.
Kermack, D. M. 1982. A new tritylodontid from the Kayenta Formation of Ari-
    zona. *Zoological Journal of the Linnean Society* 76:1–17.
Krause, D. W. and F. A. Jenkins, Jr. 1983. The postcranial skeleton of North
    American multituberculates. *Bulletin of the Museum of Comparative Zool-
    ogy, Harvard University* 150:199–246.
Kühne, W. G. 1956. *The Liassic therapsid* Oligokyphus. London: Trustees of the
    British Museum.
Lewis, G. E. 1986. *Nearctylodon broomi,* the first Nearctic tritylodontid;
    pp. 295–303 *in* N. Hotton, III, P. D. MacLean, J. J. Roth, and E. C. Roth
    (eds.), *The Ecology and Biology of Mammal-like Reptiles.* Washington, DC:
    Smithsonian Institution Press.
Luo, Z.-X. 1994. Sister-group relationships of mammals and transformations of
    diagnostic mammalian characters; pp. 98–128 *in* N. C. Fraser and H.-D. Sues
    (eds.), *In the Shadow of the Dinosaurs: Early Mesozoic Tetrapods.* New
    York: Cambridge University Press.
Luo, Z.-X., Z. Kielan-Jaworowska and R. L. Cifelli. 2002. In quest for a phy-
    logeny of Mesozoic mammals. *Acta Palaeontologica Polonica* 47:1–78.

Matsuoka, H. and T. Setoguchi. 2000. Significance of Chinese tritylodonts (Synapsida, Cynodontia) for the systematic study of Japanese materials from the Lower Cretaceous Kuwajima Formation, Tetori Group of Shiramine, Ishikawa, Japan. *Asian Paleoprimatology* 1:161–176.

Rowe, T. 1988. Definition, diagnosis, and origin of Mammalia. *Journal of Vertebrate Paleontology* 8:241–264.

Sereno, P. C. and M. C. McKenna. 1995. Cretaceous multituberculate skeleton and the early evolution of the mammalian shoulder girdle. *Nature* 377:144–147.

Sidor, C. A. and J. A. Hopson. 1998. Ghost lineages and "mammalness": assessing the temporal pattern of character acquisition in the Synapsida. *Paleobiology* 24:254–273.

Simpson, G. G. 1928. *A Catalogue of the Mesozoic Mammalia in the Geological Department of the British Museum.* London: Trustees of the British Museum.

Sues, H.-D. 1985. The relationships of the Tritylodontidae (Synapsida). *Zoological Journal of the Linnean Society* 85:205–217.

———. 1986a. The skull and dentition of two tritylodontid synapsids from the Lower Jurassic of western North America. *Bulletin of the Museum of Comparative Zoology, Harvard University* 151:217–268.

———. 1986b. Relationships and biostratigraphic significance of the Tritylodontidae (Synapsida) from the Kayenta Formation of northeastern Arizona; pp. 279–284 *in* K. Padian (ed.), *The Beginning of the Age of Dinosaurs: Faunal Change across the Triassic-Jurassic Boundary.* New York: Cambridge University Press.

Sues, H.-D., J. M. Clark and F. A. Jenkins, Jr. 1994. A review of Early Jurassic tetrapods from the Glen Canyon Group of the American Southwest; pp. 284–294 *in* N. C. Fraser and H.-D. Sues (eds.), *In the Shadow of the Dinosaurs: Early Mesozoic Tetrapods.* New York: Cambridge University Press.

Sun A. and Y. Li. 1985. [The postcranial skeleton of the late tritylodont *Bienotheroides.*] *Vertebrata PalAsiatica* 23:133–151. (In Chinese with English summary.)

Tatarinov, L. P. and E. N. Matchenko. 1999. A find of an aberrant tritylodont (Reptilia, Cynodontia) in the Lower Cretaceous of the Kemerovo Region. *Paleontological Journal* 33:422–428.

Watson, D. M. S. 1942. On Permian and Triassic tetrapods. *Geological Magazine* 49: 81–116.

Wible, J. R. 1991. Origin of Mammalia: the craniodental evidence reexamined. *Journal of Vertebrate Paleontology* 11: 1–28.

Winkler, D. A., L. L. Jacobs, J. D. Congleton, and W. R. Downs. 1991. Life in a sand sea: Biota from Jurassic interdunes. *Geology* 19:889–892.

Young, C. C. 1947. Mammal-like reptiles from Lufeng, Yunnan, China. *Proceedings of the Zoological Society of London* 117:537–597.

The Phylogeny of Living and Extinct Armadillos (Mammalia, Xenarthra, Cingulata): A Craniodental Analysis

Timothy J. Gaudin
and John R. Wible

Introduction

The armadillos are the only group of living mammals that bear an armored, dermal carapace. This armor typically takes the form of a flexible mosaic of dermal scutes covered by a layer of keratinous epidermal scales (Grassé, 1955). Digging specialists, the armadillos are a modestly common and diverse assemblage of small to medium-sized Neotropical insectivores and omnivores (Nowak, 1999). Moreover, the armadillos represent the most diverse assemblage within the mammalian order Xenarthra, accounting for twenty-one of the thirty living species (Wetzel, 1985a; Vizcaíno, 1995; Gaudin, 2003).

Because of the hundreds of individual bony scutes present in the carapace of each individual armadillo, their fossilization potential is quite high, and the armadillos have an extensive fossil record. Indeed, the oldest fossils of the Xenarthra are isolated armadillo scutes and tarsals recovered from the middle Paleocene deposits of Itaboraí in Brazil (Scillato-Yané, 1976; Cifelli, 1983). Nearly complete skeletons are known from Casamayoran (late Eocene; see Kay et al., 1999) age sediments, although the skeletal record for armadillos is sparse until the early Miocene (Hoffstetter, 1958, 1982; Pascual et al., 1985). Fossil armadillos (thirty-eight or thirty-nine extinct genera [Scillato-Yané, 1980; McKenna & Bell, 1997]) form part of a larger radiation of extinct armored xenarthrans, the Cingulata.

The cingulates also include large-bodied herbivorous forms, the pampatheres and glyptodonts. Pampatheres (six genera; Scillato-Yané, 1980; McKenna & Bell, 1997) were superficially armadillo-like, with a carapace composed of pectoral and pelvic bucklers joined by mobile transverse bands, but reached body lengths up to 3 m (Edmund, 1985) and were characterized by flat-crowned, bilobate teeth (Hoffstetter, 1958; Edmund, 1985). The glyptodonts (65 genera; McKenna & Bell, 1997), with their thick, immobile carapaces, short deep skulls, trilobate dentition, fused vertebral columns, elephantine limbs, and often spiked, club-like tails (Hoffstetter, 1958; Gillette & Ray, 1981), surely represent one of the most bizarre groups of mammals ever to have evolved. Moreover, these strange giants were among the most diverse and abundant groups of mammals in the late Cenozoic fossil record of South America (Hoffstetter, 1958; Pascual et al., 1985; Scillato-Yané et al., 1995).

The relationships among the living and extinct armadillos, and among the armadillos and their cingulate relatives, have been incompletely understood. Traditionally, the cingulates have been split into two groups of superfamily or family-level status. The Dasypodoidea/Dasypodidae included the living and extinct armadillos and the pampatheres, the latter typically designated as a distinct subfamily (Simpson, 1945; Grassé, 1955; Hoffstetter, 1958; Paula Couto, 1979; Scillato-Yané, 1980). The Dasypodoidea/Dasypodidae has often been considered a paraphyletic stem lineage from which other cingulates (Patterson & Pascual, 1968, 1972) or indeed other xenarthrans (Hoffstetter, 1958; Romer, 1966) are derived. The Glyptodontoidea/Glyptodontidae typically included only the glyptodonts themselves, although Patterson and Pascual (1968, 1972) suggested a shared common ancestry between glyptodonts and pampatheres.

Among the living and extinct armadillos, several subgroupings appear repeatedly in most historical classifications. Euphractine armadillos, including the living genera *Euphractus, Chaetophractus,* and *Zaedyus* and a number of purportedly related fossil genera (eight to thirteen extinct genera [Scillato-Yané, 1980; McKenna & Bell, 1997], several of which have been examined in the present study: *Macroeuphractus, Paleuphractus, Proeuphractus, Prozaedyus*), were usually recognized as a separate subfamily (Frechkop & Yepes, 1949; Grassé, 1955; Patterson & Pascual, 1968, 1972; Paula Couto, 1979; Scillato-Yané, 1980; Carlini & Scillato-Yané, 1996) or as a separate tribe within the subfamily Dasypodinae (Simpson, 1945; Hoffstetter, 1958; Wetzel, 1985b). However, this euphractine assemblage has been considered by several previous authors as a primitive stem group for other armadillos (Winge, 1941; Simpson, 1945) and hence united largely by plesiomorphic characters. The extinct eutatine armadillos and the extant genera *Priodontes* and *Cabassous* have also been recognized as distinct taxonomic groups, either as separate subfamilies (Eutatinae and Priodontinae, respectively; Patterson & Pascual, 1968, 1972) or as tribes within the subfamily Dasypodinae (Eutatini and Priodontini [Simpson, 1945; Hoffstetter, 1958; Paula Couto, 1979; Wetzel, 1985b]; Scillato-Yané [1980] recognized a Tribe Eutatini within the subfamily Euphractinae).

The placement of the remaining extant armadillo genera, *Chlamyphorus, Dasypus,* and *Tolypeutes,* as well as several well-known and distinctive extinct groups such as the peltephiline and stegotheriine armadillos, has been more controversial (Simpson, 1945; Frechkop & Yepes, 1949; Grassé, 1955; Hoffstetter, 1958; Patterson & Pascual, 1968, 1972; Paula Couto, 1979; Scillato-Yané, 1980; Wetzel, 1985b). Moreover, these earlier studies provided little insight into how the widely recognized subgroupings of living and extinct cingulates might be related to one another.

Pioneering work by Engelmann (1978, 1985) represented the first attempt to employ cladistic methodology to more clearly resolve cingulate phylogeny. He examined a wide range of cingulate taxa, among them living and extinct armadillos, pampatheres, and glyptodonts, but used a limited number of characters in his analysis. His results suggested several novel systematic arrangements. Much like previous authors, he divided the Cingulata into two groups, the Dasypoda and Glyptodonta. However, in the latter he included not only the glyptodonts and pampatheres (following Patterson & Pascual, 1968, 1972) but also the extinct eutatine armadillos. These three were united by a single derived characteristic, the presence of a core of wear-resistant osteodentine in the center of the posterior "molariform" teeth. Within Glyptodonta, Engelmann (1978, 1985) hypothesized that the eutatine armadillos (represented in his study by the Pleistocene genus *Eutatus* and the Miocene genus *Proeutatus*) were the sister-group to glyptodonts on the basis of a series of cranial and dental characters.

Engelmann (1978, 1985) divided Dasypoda into two clades, the Dasypodidae and Euphracta. Relationships within the Dasypodidae were well resolved and fairly well supported. The Dasypodidae included a monophyletic Priodontini for the extant genera *Priodontes* and *Cabassous* and a monophyletic Dasypodini for the extant genus *Dasypus* and the Miocene genus *Stegotherium*. Engelmann (1978, 1985) was less successful in elucidating the relationships within the Euphracta, with an unresolved tetrachotomy at the base of the clade. However, he did propose that the extinct genera *Peltephilus* and *Macroeuphractus* were sister taxa, and he erected a monophyletic clade to include the extant euphractans *Zaedyus* and *Chlamyphorus* and the Miocene genus *Prozaedyus*.

Engelmann (1985) referred to his results as a "framework" upon which further investigations of xenarthran relationships could be built. He was unable to fully resolve cingulate phylogeny, most notably the relationships among euphractan armadillos, and, although he generated novel hypotheses of relationship, several were weakly supported. However, few phylogenetic studies, involving either living or extinct cingulate taxa, have been published since Engelmann's (1978, 1985) work. A number of mammalian taxonomies have appeared in this period. These have either ignored Engelmann's results entirely (Carroll, 1988; Gardner, 1993) or have accepted them only in part (McKenna & Bell, 1997). Carlini and Scillato-Yané (1996) recently conducted a cladistic study of a portion of armadillo diversity. Using only morphological characters derived from the carapace and head shield, they analyzed relationships among euphractine armadillos, including the extant genera *Euphractus, Chaetophractus,* and *Zaedyus* and the extinct genera *Macroeuphractus* and *Prozaedyus,* but excluding two of Engelmann's (1978, 1985) euphractan taxa, the extant *Chlamyphorus* and the extinct *Peltephilus.* Like Engelmann (1978, 1985), Carlini and Scillato-Yané (1996) hypothesized a close relationship between *Prozaedyus* and *Zaedyus.* Unlike Engelmann (1978, 1985), Carlini and Scillato-Yané (1996) proposed that the extant genus *Euphractus* is more closely related to the extinct *Macroeuphractus* (and a cluster of other extinct euphractines not included in Engelmann's [1978, 1985] studies) than to other extant euphractines.

A study of the cingulate ear region published by Patterson et al. (1989) was more comprehensive in its taxonomic coverage of the Cingulata. They concurred with several of Engelmann's conclusions—namely, the close alliance of glyptodonts and pampatheres and of *Dasypus* and *Stegotherium*—although their systematic arrangement of other armadillos, including the extinct eutatines, differed significantly from that of Engelmann (1978, 1985). It should be noted that the two senior authors of Patterson et al. (1989), Bryan Patterson and Walter Segall, reached their conclusions independent of Engelmann (1978, 1985). Patterson and Segall conducted their study primarily in the 1940s and 1950s but left it unfinished, and it was published posthumously (see comments in preface, Patterson et al., 1989) after the appearance of Engelmann's (1978, 1985) work. In an analysis of masticatory function in eutatine armadillos by Vizcaíno and Bargo (1998), the authors noted that two of the characters employed by Engelmann (1978, 1985) to ally eutatine armadillos to glyptodonts were present in the Miocene genus *Proeutatus,* but not in the Pleistocene form *Eutatus,* and that one of the purported eutatine/ glyptodont synapomorphies was also present in pampatheres. Although Vizcaíno and Bargo (1998) assert that a close relationship between eutatines and glyptodonts is "feasible," they emphasize the need for additional study of this proposed relationship.

A recent series of molecular studies (Delsuc et al., 2001, 2002, 2003) investigating xenarthran interrelationships using three nuclear and two mitochondrial gene sequences have contributed valuable insights into the relationships among living cingulates. The first of these studies included only three cingulate taxa, but the two more recent papers incorporated all extant armadillo genera except *Chlamyphorus.* The results contradict those of Engelmann (1978, 1985) in a number of respects. They support neither the monophyly of the Dasypodidae or the Euphracta. Rather, *Dasypus* is the sister taxon to all other living armadillos, and two sub-clades are recognized among the remaining forms: Tolypeutinae, with *Tolypeutes* as the sister taxon to *Cabassous + Priodontes,* and Euphrac-tinae, with *Zaedyus* as the sister taxon to *Chaetophractus + Euphractus.* Statistical support for relationships within both Tolypeutinae and Euphractinae are weak, however.

The goal of the present study is to reanalyze the phylogenetic relationships of extant and extinct armadillos to one another and to their close relatives among the Cingulata. The analysis is based on a cladistic morphological study of the skull, lower jaw, and dentition in living and fossil armadillos, pampatheres, and glyptodonts. The present study incorporates all the cingulate taxa studied by Engelmann (1978, 1985) as well as a few additional fossil forms treated by Patterson et al. (1989) and Carlini and Scillato-Yané (1996). However, it is based on a much greater number of characters than any of these previous studies. We believe the additional character information will result in a better-resolved and more robust understanding of the relationships among living and extinct cingulates.

Materials and Methods

A total of 163 discrete craniodental characters were obtained and scored via direct observations of the specimens listed in appendix 6.1. The characters are described in appendix 6.2. Some characters could not be scored in certain fossil taxa because portions of these fossils were missing or poorly preserved. In these cases, data obtained from direct observation of specimens were supplemented by descriptions in the literature—e.g., *Eutatus seguini* (Guth, 1961; Vizcaíno & Bargo, 1998), *Propalaeohoplophorus australis* (Scott, 1903–1904), *Peltephilus pumilis* (Scott, 1903–1904; Vizcaíno & Fariña, 1997), and *Proeutatus oenophorus* (Patterson et al., 1989). In addition, three characters treating dental histology were added based on the work of Ferigolo (1985; see characters 5, 6, 7, appendix 6.2). Of the twenty-one species listed in appendix 6.1, nineteen are ingroup taxa, including a single representative species from each of the eight extant armadillo genera, nine fossil armadillos, a single well-preserved pampathere species, and one of the oldest glyptodonts known from nearly complete skeletal material, the Miocene species *Propalaeo-hoplophorus australis*. Among the nine fossil armadillos incorporated in the analysis are three species of eutatine armadillos.

   The data matrix of 163 characters and twenty-one taxa (appendix 6.3) was analyzed with the computer program PAUP (Version 4.0b10 [Swofford, 2002]). Analyses of the entire matrix were conducted by using a heuristic search with random-addition sequence and 1,000 repetitions to find the most-parsimonious trees. Characters were optimized by using PAUP's DELTRAN option in all analyses (see Gaudin [1995] for justification). In those instances in which intraspecific variation was noted for a given character in a given taxon, the taxon was coded for all relevant states and treated as polymorphic in the PAUP analyses. Of the 163 characters, fifty-seven are multistate, and thirty-three of these are ordered along numerical, positional, or structural morphoclines. Several characters proved to be parsimony uninformative in the final analyses, but all values reported for consistency index exclude uninformative characters. Following Gaudin (1995), two different weighting schemes were applied to multistate characters to assess their effect on the analysis: (1) all character state changes weighted equally, and (2) character state changes scaled so that all characters are weighted equally regardless of the number of character states. A bootstrap analysis using heuristic methods (random-addition sequence, ten repetitions for each 1,000 bootstrap replicates) was also used to evaluate the relative support for various groupings (Hillis & Bull, 1993).

   Characters were polarized via comparison to a single monophyletic outgroup, the Pilosa, the closest sister-taxon to the Cingulata (following Flower, 1882; Engelmann, 1978, 1985; Gaudin, 1993, 1995, 2004; McKenna & Bell, 1997). The Pilosa was represented by a single extant species drawn from each of its primary lineages: from the Vermilingua, the anteater *Tamandua mexicana* (appendix 6.1); and from the Tardigrada, the tree sloth *Bradypus variegatus* (appendix 6.1).

   Because of the large amount of missing data for the fossil taxa, an additional set of analyses were performed with only extant taxa, including eight species of armadillos and the two aforementioned outgroup taxa. Because of the much smaller number of taxa involved, these analyses were performed with PAUP's branch-and-bound algorithm to ensure that a globally parsimonious solution would be obtained. In all other respects, these analyses were identical to those described above for the full data matrix. Results of this second set of analyses are compared with those using the entire data matrix below.

As a by-product of this study, we were struck by the dearth of detailed descriptions and illustrations of the armadillo skull. We, therefore, embarked upon such a study of the skull of the yellow armadillo, *Euphractus sexcinctus,* which was published (Wible & Gaudin, 2004) prior to this report. Wherever possible, anatomical terminology used in this report will follow that of Wible and Gaudin (2004).

The following institutional abbreviations are used: AMNH, American Museum of Natural History, New York; CM, Carnegie Museum of Natural History, Pittsburgh; FMNH, Field Museum of Natural History, Chicago; USNM, National Museum of Natural History, Smithsonian Institution, Washington, DC; UTCM, University of Tennessee at Chattanooga Natural History Museum, Chattanooga; YPM-PU, Princeton University collection housed at Peabody Museum, Yale University.

Other abbreviations are as follows: CI, consistency index; GSL, greatest skull length; MPT, most parsimonious tree(s); RI, retention index; TL, tree length.

Results

Two different PAUP analyses have been performed with the entire data matrix: one in which the all character-state changes are all weighted equally, and a second in which character-state changes are scaled so that all characters are weighted equally. The two manipulations produce nearly identical results. The former yields four MPTs, TL = 802, CI = 0.536, and RI = 0.509. A strict consensus of these nine trees is illustrated in figure 6.1. The latter yields a single MPT, TL = 530.68, CI = 0.517, and RI = 0.520. The tree in which all character state changes are weighted equally includes two nodes that are unresolved polytomies (fig. 6.1), whereas the tree in which all characters are given equal weight is fully resolved. In addition, the two analyses differ in their phylogenetic allocation of the extant giant armadillo *Priodontes,* the latter placing this taxon as the sister-group to a clade including *Dasypus* and *Stegotherium,* the former as the sister-group to node 4, a crown group including the extant naked-tailed armadillo *Cabassous* and all more derived cingulates (fig. 6.1). Because of the large number of fossil taxa included in these analyses and the concomitantly large amount of missing data, we believe the tree illustrated in figure 6.1 is the more conservative, and subsequent discussions will focus on this tree. Characters will be referred to in these discussions according to the numeration provided in appendix 6.2. A list of the apomorphies appearing at each of the nodes on the tree illustrated in figure 6.1 is provided in appendix 6.4.

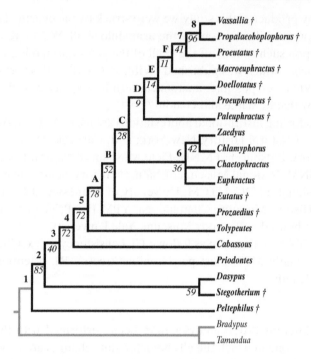

Figure 6.1. Preferred phylogeny of the Cingulata based on PAUP analysis of 163 craniodental characters in nineteen taxa, including all eight extant armadillo genera and eleven fossil armadillos, pampatheres, and glyptodonts. Characters are polarized via comparison with a single outgroup, the Pilosa (represented by the extant anteater *Tamandua* and the extant tree sloth *Bradypus*). All characters state changes are weighted equally in this analysis. The tree resulting from this analysis is a strict consensus of four MPTs (TL = 802, CI = 0.536, RI = 0.509). Numbers at nodes are bootstrap values as detailed in Materials and Methods. Boldface numbers at nodes are nodes considered in the Results and appendix 6.4. Letters at nodes are for nodes considered in appendix 6.4 only.

Node 1: Cingulata

There are thirteen unequivocal character state changes at the base of the tree and six character state changes that are equivocally optimized to occur at this node (appendix 6.4). Because the two representatives of Pilosa were constrained a priori to form a single monophyletic outgroup, it is not possible to determine the polarity of the character state changes at this node without reference to more remote outgroups. However, comparison with several generalized therian taxa (the opossum *Didelphis virginiana,* UTCM 257; the hedgehog *Atelerix algirus,* UTCM 727; and the extinct *Leptictis* [Novacek, 1986]) suggests that a number of these features are derived for the Cingulata. These include the following unambiguously assigned characters (fig. 6.2): occipital height greater than or equal to its width (character 157[2]); occipital artery traveling in a groove on the

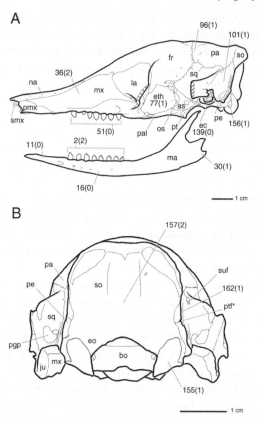

Figure 6.2. *Dasypus novemcinctus* CM 76829, skull (with zygoma cut) and mandible in lateral view (A) and skull in occipital view (B). Abbreviations: as, alisphenoid; bo, basioccipital; ec, ectotympanic; eo, exoccipital; eth, ethmoid; fr, frontal; ju, jugal; la, lacrimal; ma, mandible; mx, maxilla; na, nasal; os, orbitosphenoid; pa, parietal; pal, palatine; pe, petrosal; pgp, postglenoid process; pmx, premaxilla; pt, pterygoid; ptf*, posttemporal foramen (hidden); smx, septomaxilla; so, supraoccipital; sq, squamosal; suf, suprameatal foramen. Characters and states: 2(2), 8 lower teeth; 11(0), mandibular spout elongated; 16(0), mandible shallow; 30(1), ventral edge of angular process dorsal to lower edge of horizontal ramus; 36(2), rostrum elongated; 51(0), upper tooth row less than 30 percent skull length; 77(1), ethmoid orbital exposure large; 96(1), frontal/parietal suture at anterior edge of glenoid fossa; 101(1), postglenoid length of skull between 15 percent and 20 percent of skull length; 139(0), ectotympanic ligamentously attached or sutured to squamosal; 155(1), occipital condyle rectangular; 156(1), lateral indentation on occipital condyle; 157(2), maximum height of occiput equal or greater than maximum width; 162(1), groove for occipital artery dorsal to posttemporal foramen.

occiput extending dorsal to the posttemporal foramen (162[1]); width of the external nares much greater than their height (40[2]); external narial aperture inclined anteroventrally in lateral view (41[1]); anterior portion of the nasoturbinal lying medial to the nasal/maxillary or premaxillary suture (45[1]), and rectangular occipital condyles (155[1]). Three of these characters (40, 45, and 155) are unique to the Cingulata. We define unique binary characters as those having a CI = 1.0. In the case of multistate characters, we use a restrictive definition. Unique character states are those that are synapomorphies of a particular clade, occur in all members of that clade, and are not found in taxa outside that clade.

Node 2

The results of this analysis clearly demonstrate the unusual nature of the skull of *Peltephilus,* the famed "horned" armadillo of the Miocene Santa Cruz formation in Patagonia. *Peltephilus* is the sister-taxon to all remaining cingulates. There are at least nine unambiguous synapomorphies that support the common ancestry of all cingulates except *Peltephilus* (appendix 6.4; fig. 6.2A): eight lower teeth (2[2]); elongated mandibular spout (11[0]); angular process of the mandible lies dorsal to the ventral edge of the horizontal ramus (30[1]); rostrum 50–59 percent of GSL (36[2]); frontal/parietal suture at the level of the anterior edge of the glenoid fossa (96[1]); postglenoid length of the skull reduced, <20 percent of GSL (101[1]); ectotympanic loosely attached (139[0]); bony external auditory meatus absent (141[0]); and lateral indentation on the occipital condyles (156[1], fig. 6.2A). Only one of these characters (30) is unique to this node. A number of these features (e.g., 11, 139, 141) are reversed at higher levels of the tree.

There are eighteen additional character state changes that are equivocally assigned to node 2 (appendix 6.4). Most of these (fifteen) represent characters that were coded as unknown ("?," appendix 6.3) for *Peltephilus.* Although some of these features certainly might represent additional characters distinguishing other cingulates from *Peltephilus,* it is also possible that some are synapomorphies of the Cingulata as a whole. Indeed, several of these features are unique, present (where known) in all other cingulates (e.g., 34[2], 42[1], 52[1], 53[1], 81[1], 98[1], 99[1]).

Node 3

The extant giant armadillo *Priodontes* is allied to more derived cingulates by at least five unambiguous synapomorphies (1[4], 6[1], 97[1], 103[1], 146[1]; see appendices 6.2, 6.4). The sister-group to node 3 is a clade including the living long-nosed armadillos of the genus *Dasypus,* represented in our analysis by *Dasypus novemcinctus,* and the Miocene genus *Stegotherium.* These two genera are united by eight unequivocal synapomorphies (appendix 6.4; fig. 6.2A), including several derived resemblances noted by Engelmann (1978, 1985): a slender mandibular horizontal ramus (16[0]) and an elongated rostrum (36[3]). *Dasypus* and

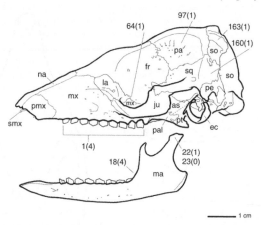

Figure 6.3. *Cabassous centralis* USNM 314577, skull and mandible in lateral view. Abbreviations: as, alisphenoid; ec, ectotympanic; fr, frontal; ju, jugal; la, lacrimal; ma, mandible; mx, maxilla; na, nasal; pa, parietal; pal, palatine; pe, petrosal; pmx, premaxilla; pt, pterygoid; smx; septomaxilla; so, supraoccipital; sq, squamosal. Characters and states: 1(4), 9 upper teeth; 18(4), angle between toothrow and anterior edge of ascending ramus 70° to 79°; 22(1), coronoid, condylar, and angular processes equidistant; 23(0), condyle greatly elevated above toothrow; 64(1), maxilla and lacrimal contact in orbit; 97(1), more than 5 foramina for rami temporales in temporal fossa of parietal; 160(1), occipital exposure of squamosal; 163(1), nuchal crest with thickened bosses.

Stegotherium are also characterized by a smooth lacrimal (49[0]), a shortened upper toothrow (51[0]), and well-developed basioccipital tubera (149[1]). There is one unique feature that serves as an unequivocal synapomorphy of *Dasypus* and *Stegotherium:* the presence of a large ethmoid exposure in the anteroventral wall of the orbit (77[1]).

Node 4

The naked tailed armadillos, represented in our analysis by the species *Cabassous centralis,* are positioned one node higher than *Priodontes,* as the sister-taxon to all more derived cingulates. There are at least nine unambiguous synapomorphies supporting this arrangement (appendix 6.4). Four of these features are unique to this node: dentition with avascular orthodentine (6[0]); a thick layer of cementum around the outside of the teeth (7[0]); orbital contact between the maxilla and lacrimal (64[1], fig. 6.3); and promontorium of the petrosal elongated anteromedially in ventral view (123[1], fig. 6.4B).

Node 5

This clade includes the extant three-banded armadillo *Tolypeutes* and all cingulates that are more derived. It is robustly supported by at least eleven unambiguous synapomorphies (appendix 6.4; fig. 6.4) including

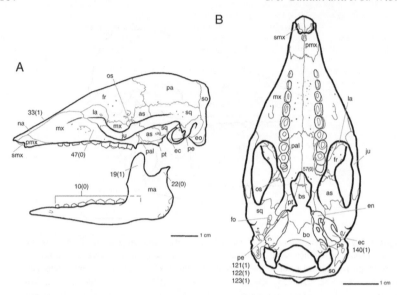

Figure 6.4. *Tolypeutes matacus* AMNH 248394, skull and mandible in lateral view, with septomaxilla based on Wegner (1922) and ectotympanic based on AMNH 246460 (A) and skull in ventral view, with ectotympanic and entotympanic missing on left side and added to right side based on AMNH 246460 (B). Abbreviations: as, alisphenoid; bo, basioccipital; bs, basisphenoid; ec, ectotympanic; en, entotympanic; eo, exoccipital; fo, foramen ovale; fr, frontal; ju, jugal; la, lacrimal; ma, mandible; mx, maxilla; na, nasal; os, orbitosphenoid; pa, parietal; pal, palatine; pe, petrosal; pmx, premaxilla; pt, pterygoid; smx; septomaxilla; so, supraoccipital; sq, squamosal. Characters and states: 10(0), lower tooth row elongated; 19(1), coronoid process elongated with anterior and posterior edges parallel; 22(0), condyle closer to coronoid than to angular process; 33(1), snout downturned; 47(0), antorbital depression; 57(0), posterior extent of palate even with last tooth; 121(1), petrosal sutured medially to basicranium; 122(1), promontorium of petrosal even with or extending slightly ventral to ventral surface of basicranium; 123(1), promontorium of petrosal elongated anteromedially; 140(1), ectotympanic medial to glenoid fossa, Glaserian fissure immediately posterior to foramen ovale.

features of the mandible (10[0], 19[1], 22[0], 27[0]), snout (33[1], 47[0]), and auditory region (121[1], 154[0]).

Node 6: *Euphracta*

The four remaining extant armadillo genera—*Euphractus, Chaeto-phractus, Zaedyus,* and *Chlamyphorus*—are grouped as a monophyletic clade in the present analysis (node 6, or what might be the equivalent of Engelmann's [1978, 1985] "Euphracta" in a restricted sense). The extant euphractans are separated from *Tolypeutes* by a pair of fossil cingulate taxa, the Pleistocene eutatine armadillo *Eutatus,* and the Miocene genus *Prozaedyus* from the Santa Cruz Formation (fig. 6.1; appendix 6.2, 6.4, nodes A, B). The extant euphractans in turn represent a sister-taxon to a

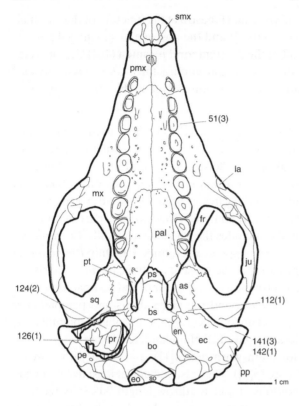

Figure 6.5. *Euphractus sexcinctus* FMNH 28350, skull in ventral view, with right auditory bulla cut. Abbreviations: as, alisphenoid; bo, basioccipital; bs, basisphenoid; ec, ectotympanic; en, entotympanic; eo, exoccipital; fr, frontal; ju, jugal; la, lacrimal; mx, maxilla; pal, palatine; pe, petrosal; pmx, premaxilla; pp, paroccipital process; pr, promontorium; ps, presphenoid; pt, pterygoid; smx, septomaxilla; so, supraoccipital; sq, squamosal. Characters and states: 51(3), upper toothrow greater than 40 percent skull length; 112(1), alisphenoid contacts ectotympanic; 124(2), processus crista facialis large, concave, and quadrangular; 126(1), epitympanic sinus; 141(3), osseous external auditory meatus elongated and tubular; 142(1), porus acousticus ovate, bordered anteriorly by a short muscular crest on ectotympanic.

crown group of extinct cingulates (fig. 6.1; appendix 6.2, 6.4, nodes C, D). This crown clade includes a number of extinct taxa (e.g., *Paleuphractus, Proeuphractus,* and *Macroeuphractus*) that traditionally have been allied with extant euphractans.

The monophyly of the extant euphractan genera themselves is supported by seven unambiguous synapomorphies (appendix 6.4, node 6). The greatest number of these are features of the auditory bulla and the ear region in general (fig. 6.5), a region recognized by Engelmann (1978, 1985) and Patterson et al. (1989) as distinctive in euphractan armadillos. The synapomorphous ear region characters in this analysis include alisphenoid contact with the ectotympanic (112[1]); an epitympanic sinus (126[1]); a large, concave, quadrangular processus crista facialis (124[2]); and an ossified, complete, elongated tubular external auditory meatus (141[3]). Character 124 is unique to extant euphractans. *Zaedyus* and *Chlamyphorus* are allied as sister taxa within Euphracta by seven unambiguous synapomorphies (appendix 6.4), all but two of which represent a reversal to a more primitive character state. The two non-reversed characters are the absence of a lacrimal fenestra at the juncture of the lacrimal, maxilla, and frontal bones (50[1]) and the absence of squamosal participation in the lateral wall of the posttemporal foramen (161[1]). The relationship among the other two extant euphractans, *Chaetophractus* and *Euphractus,* and the latter clade were unresolved in this analysis (fig. 6.1).

Node 7

Engelmann (1978, 1985) hypothesized that the eutatine armadillos were the sister-group to glyptodonts. Three eutatine armadillos were included in the present analysis: the Pleistocene genus *Eutatus,* the Pliocene *Doellotatus,* and the Miocene *Proeutatus.* As noted previously, *Eutatus* is the sister-taxon to a clade including living euphractans and crown group of extinct cingulates in the present analysis. *Doellotatus* is part of an unresolved trichotomy that also includes the Pliocene armadillo *Proeuphractus* and a crown clade (fig. 6.1; appendix 6.4, node E). Only *Proeutatus* is closely allied with glyptodonts in the present analysis, forming the sister-taxon to a monophyletic grouping of glyptodonts and pampatheres. The common ancestry of *Proeutatus* and the glyptodonts and pampatheres (node 7) is supported by eight unequivocal synapomorphies (appendix 6.4; fig. 6.6). These include a couple of dental characters: wear on anterior teeth beveled, posterior teeth worn flat (4[1]); and teeth with a wear-resistant osteodentine core (5[1]). They also include at least five reversals to more primitive conditions: upper toothrow length 30–35 percent of GSL (51[2]); promontorium of the petrosal dorsal to the basicranium (122[0]); stylomastoid foramen anteromedial or medial to the tip of the paroccipital process (127[1]); ectotympanic loosely attached to the squamosal (139[0]); and paroccipital process freestanding, its greatest width in the anteroposterior plane (134[0]).

Node 8: Glyptodonta

The close relationship of the large, herbivorous pampatheres and glyptodonts is supported by more unambiguous synapomorphies than any other node on the cladogram. There are sixteen unambiguous synapomorphies at this node (appendix 6.4), many but not all of which have to do with the distinctive masticatory apparatus of these two groups (fig. 6.7) — e.g., a deep horizontal ramus on the mandible (16[2]), a laterally directed zygomatic root (90[1]), and a transversely widened glenoid fossa (117[2]). As noted by Patterson et al. (1989), there are also many resemblances in bony anatomy of the auditory region of these two groups (fig. 6.7), including the lack of a medial attachment between the petrosal and the basicranium (121[0]), the position of the fossa incudis between the squamosal

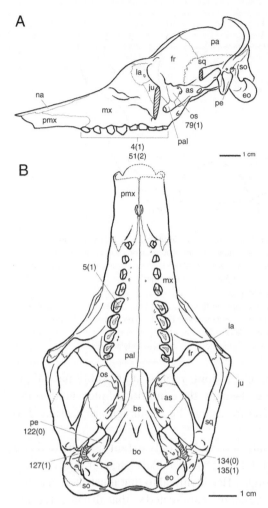

Figure 6.6. *Proeutatus oenophorus,* FMNH P13197, skull in lateral view (with zygoma cut) (A) and in ventral view (B). Abbreviations: as, alisphenoid; bo, basioccipital; bs, basisphenoid; eo, exoccipital; fr, frontal; ju, jugal; la, lacrimal; mx, maxilla; na, nasal; os, orbitosphenoid; pa, parietal; pal, palatine; pe, petrosal; pmx, premaxilla; so, supraoccipital; sq, squamosal. Characters and states: 4(1), anterior teeth with beveled wear, posterior flat; 5(1), tooth center with elevated core of osteodentine; 51(2), upper toothrow between 35 percent and 40 percent skull length; 79(1), optic caval recessed; 122(0), promontorium of petrosal dorsal to ventral surface of basicranium; 127(1), stylomastoid foramen anteromedial or medial to paroccipital mastoid process; 134(0), paroccipital process with greatest width in anteroposterior plane; 135(1), paroccipital and paracondylar processes closely approximated.

and petrosal (125[1]), the separation of the ectotympanic from the glenoid by a distinctive horizontal postglenoid fossa with a centrally located post-glenoid foramen (143[1]), and the presence of a large paracondylar process (154[2]). Finally, four of the sixteen unambiguous synapomorphies are unique to this node, including character 143 as well as the presence of a distinct fossa on the anterodorsal surface of the mandibular condylar process (29[1]), the presence of pterygoids in line with or lateral to the toothrow, framing narrow U-shaped choanae (59[2], fig. 6.7), and the presence of a common fossa for the sphenopalatine foramen and the confluent sphenorbital fissure/foramen rotundum (72[1]). The rugose pterygoids

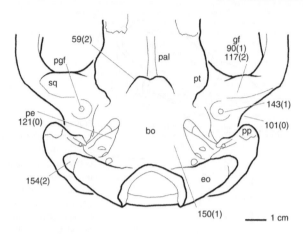

Figure 6.7. *Vassallia maxima* (= *Plaina* cf. *subintermedia*), FMNH P14424, basicranium in ventral view (redrawn from Patterson et al., 1989, fig.15B). Abbreviations: bo, basioccipital; eo, exoccipital; gf, glenoid fossa; pal, palatine; pp, paroecipital process; pt, pterygoid; pe, petrosal; pgf, postglenoid foramen. Characters and states: 59(2), posterior margin of palate forms a narrow U with pterygoids in line with or slightly lateral to toothrow; 90(1), posterior root of zygoma laterally directed; 101(0), postglenoid length of skull greater than 20 percent of skull length; 117(2), glenoid fossa wide and short; 121(0), petrosal unattached medially; 143(1), porus acousticus far posterior to glenoid fossa, separated by horizontal postglenoid fossa in which postglenoid foramen is centrally located; 150(1), rectus capitis fossae weakly indicated; 154(2), paracondylar process well developed.

(104[1]) of pampatheres and glyptodonts represent an additional unique feature of this node. However, because the condition of pterygoids is unknown in *Proeutatus*, this feature cannot be unambiguously assigned to node 8.

A branch-and-bound analysis including only extant taxa yielded two MPTs (TL = 522, CI = 0.721, RI = 0.584; fig. 6.8) if character state changes were weighted equally. A single MPT identical to that illustrated in figure 6.8A resulted if character state changes are scaled so that all characters are weighted equally. These two MPTs differ in a number of respects from the arrangement of extant taxa obtained using the entire matrix.

In the tree illustrated in figure 6.8A, the genera *Priodontes* and *Cabassous* form a monophyletic group united by six unambiguous synapomorphies: the presence of foramina on the frontal in the vicinity of the dorsal midline (94[1]); the presence of five or more vascular foramina for rami temporales in the temporal surface of the parietal (97[1], fig. 6.3); the stylomastoid foramen positioned posteromedial to the tip of the paroccipital process (127[2]); the presence of a distinct swelling at the juncture of the mallear neck and manubrium (146[1]); a substantial occipital exposure of the squamosal (160[1], fig. 6.3); and the presence of distinct, posteriorly thickened dorsal bosses on the nuchal crest (163[1], fig. 6.3). This "priodontine" clade forms the basal-most branch of the tree, whereas, in the analysis of the complete data set, *Dasypus* is a more basal taxon. The monophyly of living euphractans is obtained in both the complete data set and the tree shown in figure 6.8A, but the inter-relationships differ substantially. In the latter *Chaetophractus* and *Euphractus* are sister taxa, united by five unambiguous synapomorphies: ten lower teeth (2[4]); lacrimal fenestra present (50[0]); upper toothrow elongated, its length > 40 percent of GSL (51[3], fig. 6.5); an ovate porus acousticus bordered anteriorly by a short muscular crest on the ectotympanic (142[1], fig. 6.5), and the participation of the squamosal in the lateral wall of the posttemporal foramen (161[0]). The other extant euphractans, *Zaedyus* and *Chlamyphorus,* form successive sister taxa to this latter clade.

In the tree illustrated in figure 6.8B, the relationships among living euphractans are identical to those depicted in figure 6.8A. However, a monophyletic group including *Tolypeutes, Dasypus, Cabassous,* and *Priodontes* is obtained, reminiscent of Engelmann's (1985) clade "Dasypodidae." This relationship is supported by ten unambiguous synapomorphies: persistent piriform fenestra in the adult (109[1]); presence of a

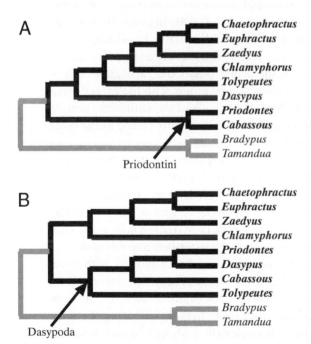

Figure 6.8. Phylogeny of the Cingulata based on PAUP analysis of 163 craniodental characters in 8 extant armadillo genera. Characters are polarized via comparison to a single outgroup, the Pilosa (represented by the extant anteater *Tamandua* and the extant tree sloth *Bradypus*). When character state changes are weighted equally, two MPTs (A) and (B) result (TL = 522, CI = 0.721, RI = 0.584). When character state changes are scaled such that all characters receive equal weight, a single MPT identical to (A) is produced. See Results for a discussion of these trees.

single large suprameatal foramen (120[1]); absence of an epitympanic sinus (126[0]); ectotympanic crura either ligamentously attached or sutured to the squamosal (139[0], fig. 2A); absence of an ossified auditory meatus (141[0]); anteroposteriorly compressed mallear head (144[2]); the presence of a distinct swelling at the junction of the mallear neck and manubrium (146[1]); malleus with a large lamina and small lateral process (147[2]); height of occiput roughly equivalent to or greater than width (157[2], fig. 6.2B); and the presence of a groove for the occipital artery dorsal to the posttemporal foramen (162[1], fig. 6.2B). As in Engelmann's (1985) clade, *Tolypeutes* is the basal member of this "dasypodid" clade. Contrary to both Engelmann's results and the tree illustrated in figure 6.8A, the monophyly of the Priodontini is not supported. Rather, *Dasypus* and *Priodontes* are allied as sister taxa, united by thirteen unequivocal synapomorphies (6[1], 7[1], 18[1], 22[2], 23[1], 31[1], 64[0], 96[2], 101[1], 105[0], 122[0], 123[0], 136[0]), a number of which seem to be related either to the weakening of the masticatory apparatus (6, 7, 18, 23) or the absence of an ossified auditory bulla (105, 136) in these taxa.

Discussion

The results of the various analyses performed in the present study vary somewhat in their details. Nevertheless, a consistent pattern of phylogenetic relationships appears in all (figs. 6.1, 6.8). In every analysis, the four living genera of "euphractan" armadillos (*sensu* Engelmann, 1978, 1985) form a monophyletic clade, even in those analyses that include fossil taxa. In all but one analysis, the living "dasypodid" armadillos (*sensu* Engelmann, 1978, 1985) are a paraphyletic group, with the extant genus *Tolypeutes* being the most derived. In analyses of the entire data matrix, the Santacrucian armadillo *Peltephilus* is consistently placed as the sister genus to all other cingulates. Moreover, despite large amounts of missing data the position of the fossil armadillos is remarkably stable irrespective of character weighting schemes employed. Among the most notable results, the eutatine armadillos form a polyphyletic assemblage, and the armadillos taken as a whole are paraphyletic, with a clade including glyptodonts and pampatheres serving as the most derived node in the tree. The Santacrucian eutatine *Proeutatus* is the sister-taxon to this node.

We compared our hypothesized scheme of relationships with those proposed by Paula Couto (1979), Scillato-Yané (1980), Engelmann (1978, 1985), and McKenna and Bell (1997), by constraining the PAUP analysis to produce the MPTs consistent with these hypotheses (using the same outgroup and other settings as those described above). In each instance, the scheme of relationships hypothesized by these authors was substantially longer than our MPT (from fig. 6.1). A tree consistent with the classification of Paula Couto (1979) was seventy-seven steps longer than our MPT; a tree consistent with the classification of Scillato-Yané (1980) was thirty-nine steps longer than our MPT; a tree consistent with the cladograms of Engelmann (1978, 1985) was seventy-four steps longer; and a tree consistent with the classification of McKenna and Bell (1997) was forty-nine steps longer. In addition, we constrained PAUP to produce the MPT consistent with the phylogenetic hypotheses of Delsuc et al. (2003), using only living taxa and assuming that *Chlamyphorus,* which is not included in Delsuc et al.'s (2003) study, would group with the other "euphractan" taxa. The MPT tree produced by this analysis was only nine steps longer than our MPT (from fig. 6.8A).

The position of *Peltephilus* as the sister-taxon to all other cingulates is one of the most robust nodes on our cladogram. It is supported by at least nine unequivocal synapomorphies. That the relationship between *Peltephilus* and other cingulates is a remote one is not a novel suggestion. Several previous authors have placed *Peltephilus* in a family separate from (but related to) all other armadillos (Simpson, 1945; Romer, 1966; Paula Couto, 1979), and at least two previous classifications have implied that the family Peltiphilidae is at least as distinct from other armadillos as are the glyptodonts (Hoffstetter, 1958; Barlow, 1984). Peltephilid cranial morphology has long been recognized as unusual in a number of respects (Scott, 1903–1904; Vizcaíno & Fariña, 1997). *Peltephilus* is best known as the only "horned" armadillo, with a pair of elongated, slightly recurved dermal scutes from the cranial portion of the cephalic shield forming "horns" on top of the snout. Additionally, it has an unusually short, broad skull and mandible for a cingulate, with a fused symphysis and a ventrally directed angular process. The teeth differ markedly from other cingulates. They are pointed and laterally compressed, and their function in mastication has been the subject of debate (Vizcaíno & Fariña, 1997). *Peltephilus* is also the only well-known xenarthran with an anteriorly closed dentition.

Patterson et al. (1989) and Engelmann (1985) used the possession of a completely ossified auditory bulla to ally this taxon to other "euphractan" armadillos. The possession of such an "advanced" character is troubling if *Peltephilus* occupies a basal position within cingulates, as our results suggest. However, we note that among sloths an ossified auditory bulla is apparently gained and lost several times (Gaudin, 1995, 2004). *Peltephilus* is the only Santacrucian armadillo with a completely ossified bulla, and is the oldest cingulate known to exhibit such a characteristic (Patterson et al., 1989). Moreover, the bulla of *Peltephilus* shows a number of morphological features not present in other bullate armadillos. The external auditory meatus is ossified, but directed largely anterolaterally, in contrast to the largely dorsal orientation of other armadillos (Patterson et al., 1989). This external auditory meatus obscures the glenoid fossa in lateral view, the latter being placed farther ventrally on the skull than in any other cingulate (Patterson et al., 1989). There are also a number of unusual features of the posterior crus of the ectotympanic, including its wide separation from the paroccipital process and its participation in the floor of the external auditory meatus (Patterson et al., 1989). The unusual morphology of the osseous bulla of *Peltephilus* may be indicative of an origin independent from that of the euphractan armadillos.

The four extant genera of "dasypodid" armadillos (*sensu* Engelmann, 1978, 1985) are paraphyletic in all but one of our analyses (figs. 6.1, 6.8A). In both the full data set and two of three analyses comprising only extant taxa, *Tolypeutes* is more closely allied with euphractan armadillos than with any of the other dasypodid armadillos. This is a fairly robust result. It is supported by eleven unambiguous synapomorphies drawn from several different regions of the skull and lower jaw, and it receives fairly high bootstrap support (fig. 6.1). To our knowledge the present study is the first to suggest close ties between *Tolypeutes* and euphractans, although previous authors have recognized the distinctiveness of *Tolypeutes* by placing the genus in a separate, monotypic tribe or subfamily (Simpson, 1945; Hoffstetter, 1958; Paula Couto, 1979; Scillato-Yané, 1980; Wetzel, 1985b).

Although the preponderance of evidence suggests an alliance between *Tolypeutes* and euphractans, the allocation of *Tolypeutes* as a basal "dasypodid" armadillo cannot be ruled out. Our own analysis suggests that a monophyletic "Dasypodidae" (*sensu* Engelmann, 1978, 1985) is equally as parsimonious as the alliance of *Tolypeutes* and euphractans if only living taxa are considered. This dasypodid clade, with *Tolypeutes* as its basal most member (fig. 6.8B), is supported by at least ten unequivocal synapomorphies.

The phylogenetic position of the giant armadillo *Priodontes* is perhaps least resolved of all the taxa considered here. In the analyses of the full data set, if all characters are weighted equally, *Priodontes* is the sister-taxon to a clade including *Dasypus* and *Stegotherium*. *Priodontes* is also the closest relative of *Dasypus* in one of the analyses involving only extant taxa (see Results; fig. 6.8B). In the MPT shown in figure 6.1, derived from the full data set and weighting all character state changes equally,

Priodontes is the sister-taxon to a crown group incorporating *Cabassous* and all more derived cingulates. In the MPT shown in figure 6.8A based solely on extant taxa, *Priodontes* and *Cabassous* form a monophyletic group at the base of Cingulata. Given the strong ecological, behavioral, and postcranial similarities of *Cabassous* and *Priodontes* (Eisenberg, 1989; Redford & Eisenberg, 1992; Eisenberg & Redford, 1999; Wetzel, 1985a, 1985b; Nowak, 1999), previously published armadillo classifications have invariably placed the two taxa in the same subfamily and often in the same tribe (Simpson, 1945; Hoffstetter, 1958; Paula Couto, 1979; Wetzel, 1985b). Engelmann's (1978, 1985) cladistic analyses also ally the two as sister taxa, as does the molecular study of Delsuc et al. (2003; *contra* Delsuc et al., 2002) Our inability to recover this clade with the full data matrix is an unexpected result. It is possible that including more species of *Cabassous* would result in our recovering a monophyletic "Priodontini" (*sensu* Simpson, 1945 and others). *Cabassous* is among the most variable of the extant armadillo genera in its cranial anatomy (e.g., see Wetzel, 1980; Patterson et al., 1989). It also seems likely that the inclusion of postcranial characters in subsequent analyses might recover such a grouping. Nevertheless, it is clear that *Cabassous* shares a number of derived cranial features with *Tolypeutes* and other crown group cingulates and that the node (node 4) joining it to these taxa is fairly robust, with high bootstrap support (fig. 6.1). This relationship is supported by nine unambiguous synapomorphies, including four unique features.

The alliance of *Dasypus* and *Stegotherium* is consistent with the phylogeny of Engelmann (1978, 1985), the study of the auditory region by Patterson et al. (1989), and the classifications of Scillato-Yané (1980) and McKenna and Bell (1987). It is supported by eight unambiguous synapomorphies, including features associated with a reduced masticatory apparatus noted by Engelmann (1985), as well as the unique presence of a large ethmoid exposure forming much of the anteroventral orbital wall.

The monophyly of the four extant genera of euphractan armadillos is supported in analyses of the entire data matrix as well as those analyses including only living taxa. In the former instance, the clade including extant euphractans excludes all fossil taxa. This result is surprising given that almost every published phylogeny or classification that includes both fossil and extant armadillos since Simpson (1945) placed the fossil genera *Prozaedyus, Proeuphractus, Paleuphractus,* and *Macroeuphractus* in the same subfamily or tribe as the extant genera. These would include the phylogenies of Engelmann (1978, 1985) and Carlini and Scillato-Yané (1996) and the classification of McKenna and Bell (1997). As noted in the introduction, this euphractan assemblage is viewed by several earlier authors (e.g., Winge, 1941; Simpson, 1945) as an armadillo "stem group" and hence may be based largely on plesiomorphic characters.

In our preferred tree (fig. 6.1), *Prozaedyus* is basal to living euphractans. The basal position of *Prozaedyus* is consistent with what Engelmann (1985, 56) termed its "disturbingly primitive" anatomy. However, according to both Engelmann (1978, 1985) and Carlini and Scillato-Yané (1996), *Prozaedyus* is considered not a basal euphractan but a close relative of the extant *Zaedyus*. More surprisingly, in our results the closest relatives of three extinct taxa—*Paleuphractus, Proeuphractus,* and *Macroeuphractus*—are not the living euphractan armadillos or *Prozaedyus* but rather a crown group assemblage including two eutatine armadillos and the glyptodonts and pampatheres (fig. 6.1). We fail to recover any of the euphractine subclades supported by Carlini and Scillato-Yané's (1996) cladistic analysis of dermal carapace morphology. For example, in our results (fig. 6.1) *Macroeuphractus* is closely related to *Proeutatus,* glyptodonts, and pampatheres rather than the euphractine *Proeuphractus* (Carlini & Scillato-Yané, 1996), and *Paleuphractus* is a basal member of the crown group (node D, fig. 6.1) rather than a close relative of the extant genus *Euphractus* (Carlini & Scillato-Yané, 1996). However, none of the basal nodes (nodes D–F) in our crown group assemblage is particularly well supported based either on numbers of synapomorphies or on bootstrap values (fig. 6.1; appendix 6.4). We believe our results speak to the broad similarities among armadillos traditionally placed in the Euphractinae and Eutatinae and the difficulty of sorting them out phylogenetically.

Although the monophyly of the living euphractans is supported in our analyses, the relationships among them are not consistently resolved. When only extant taxa are considered, *Chaetophractus* and *Euphractus* form a monophyletic clade with *Zaedyus* and *Chlamyphorus* as successive sister groups (fig. 6.8). These results are fully consistent with the results of Delsuc et al.'s (2003) molecular analysis. However, when fossil taxa are added to the analysis, *Zaedyus* and *Chlamyphorus* become a monophyletic group, with either *Chaetophractus* as its sister-group (when all characters are weighted equally) or with an unresolved basal trichotomy for extant euphractans (when all characters state changes are weighted equally; fig. 6.1). The latter result is similar to the scheme of phylogenetic relationships hypothesized for these taxa by Engelmann (1978, 1985), with the exception that Engelmann (1978, 1985) includes the Santacrucian genus *Prozaedyus* as the closest sister-taxon to *Zaedyus* and *Chlamyphorus* based on dental formula and several postcranial characters (although we found that the dental formula in these taxa was variable and did not match the pattern described by Engelmann [1978, 1985]; see appendices 6.2, 6.3). In the end these four living genera are extremely similar to one another both morphologically and ecologically (Eisenberg, 1989; Redford & Eisenberg 1992; Eisenberg & Redford, 1999; Nowak, 1999), and it is difficult to find characters that join any particular subgroup to the exclusion of the others. It is worth nothing in this context that the results of Delsuc et al. (2002) and Delsuc et al. (2003) are at odds with one another over the relationships among the three euphractan taxa considered in their analyses, with different gene sequences supporting different relationships. None of these purported relationships received strong statistical support, and Delsuc et al. (2003) suggest that the living genera may have been produced by a rapid evolutionary diversification. *Chlamyphorus* is distinctive from other euphractans in a number of respects, but most of these distinctive features appear to be autapomorphous characteristics related to its fossorial habitus. They do not help to ally it with the other euphractan genera.

The eutatine armadillos traditionally have been allied with one another on the basis of similarities in the carapace, skull, and dentition — in the case of the latter, the possession of obliquely oriented teeth with multiple dentine layers of differential hardness (Hoffstetter, 1958). In our results, however, the eutatines form a polyphyletic assemblage. The large Pleistocene taxon *Eutatus* is basal to the living euphractans, whereas the long-snouted Pliocene genus *Doellotatus* lies in an unresolved trichotomy at node E, and the Miocene *Proeutatus* is the sister-taxon to the node containing glyptodonts and pampatheres. It should be noted here that several of the dental features that purportedly diagnose eutatines are not present in all members. For example, in *Doellotatus* we found neither obliquely oriented teeth nor the presence of multiple layers of dentine. Vizcaíno and Bargo (1998, 380) assert that the multiple layers of dentine "are not very evident" in the Oligo-Miocene eutatine *Stenotatus*. Additionally, based on their analysis of masticatory function, these authors identify at least three different feeding types in the group: an herbivorous type including *Eutatus,* an omnivorous type with specializations for herbivory including *Proeutatus,* and an omnivorous type with specializations for insectivory including *Stenotatus* (Vizcaíno & Bargo, 1998). This would make eutatines the most trophically diverse of all the cingulate subgroups but is also consistent with our findings that these taxa may not be closely related to one another.

Engelmann (1978, 1985) proposed that eutatines were allied with glyptodonts and pampatheres on the basis of their shared possession of teeth with a wear-resistant osteodentine core. He further suggested that eutatines were the sister-group to glyptodonts, united by at least five synapomorphies, including teeth with an outer layer of wear-resistant compact dentine, the elevation of the basicranium and mandibular condyle above their respective toothrows, a triangular promontorium, and a mandibular ascending ramus that is anteriorly inclined. Vizcaíno and Bargo (1998) noted that the last feature is characteristic of *Proeutatus* but not *Eutatus* and that the elevated basicranium and mandibular condyle are also present in pampatheres. Our findings echo the comments of Vizcaíno and Bargo (1998) in the sense that not all eutatines show similarity to glyptodonts, and those that do are not necessarily more closely related to glyptodonts than are the pampatheres. In our analyses, only *Proeutatus* is closely related to glyptodonts. It is the sister-taxon to a clade including glyptodonts and pampatheres (fig. 6.1). In our preferred tree, the node (node 7) uniting glyptodonts, pampatheres, and *Proeutatus* is supported by eight unambiguous synapomorphies. Only three of these eight features are also found in *Eutatus* [teeth with osteodentine core, 5(1); optic foramen recessed within the fossa for the sphenorbital fissure/foramen rotundum, not visible in lateral view, 79(1); and paroccipital and paracondylar processes closely approximated, 135(1)], and none is present in *Doellotatus*. The relationship between *Proeutatus*, glyptodonts, and pampatheres is supported regardless of the character weighting scheme used.

Between the publication of Simpson's (1945) classification and the phylogenetic studies of Engelmann (1978, 1985), most cingulate classifications separated glyptodonts from armadillos at the family or superfamily level. Where pampatheres were considered, they were included as a separate family or subfamily of armadillos. Patterson and Pascual (1968, 1972) were the first to advocate a close relationship between pampatheres and glyptodonts, based largely on the earlier work of Patterson and Segall on the cingulate auditory region [published posthumously by Turnbull; see preface to Patterson et al. (1989)]. Engelmann (1978, 1985) also advocated a close relationship between these groups based on a single synapomorphy: teeth with an osteodentine core. Subsequent taxonomists have often ignored this conclusion (Barlow, 1984; Carroll, 1988), although it is incorporated in McKenna and Bell's (1997) classification. Our results link pampatheres and glyptodonts into a monophyletic group supported by sixteen unequivocal synapomorphies, four of which are unique to this node. Most (but not all) of the similarities at this node are derived either from the auditory region or from the masticatory apparatus. The latter almost certainly result from similar adaptations in the two groups for processing a herbivorous diet (Fariña, 1985; Vizcaíno et al., 1998).

Whereas the hypothesis of a close common ancestry of glyptodonts and pampatheres is not novel, our placement of that clade as the crown group among cingulates to our knowledge is without precedent. The effect of this placement is to make the armadillos paraphyletic, with glyptodonts and pampatheres derived from within the clade including euphractan and eutatine armadillos. It has long been acknowledged that glyptodonts and armadillos share a common ancestry, a relationship that can be diagnosed by a substantial number of cranial and postcranial synapomorphies (Engelmann, 1978, 1985). Moreover, most authors have hypothesized that glyptodonts evolved from early armadillos, noting that the carapaces of the earliest well-known glyptodonts bear rudiments of the transverse, mobile bands characteristic of armadillo carapaces (Scott, 1903–1904; Hoffstetter, 1958, 1982; Patterson & Pascual, 1968, 1972). However, most authors regard the split between glyptodonts and armadillos as one of the most basal splits within Cingulata (e.g., see Hoffstetter, 1958; Patterson & Pascual, 1968, 1972). Engelmann (1978, 1985) divides the Cingulata into two sister groups, a Dasypoda that includes most armadillos and a Glyptodonta that includes eutatines, pampatheres, and glyptodonts. However, the only derived feature that he cites in support of the dasypodan clade is the presence of teeth that are oval in cross-section. That even this feature is derived rests on the rather tenuous assumption that lobate teeth of some kind are primitive for Cingulata; other authors have considered the ovate teeth of extant armadillos to represent the primitive condition (Hoffstetter, 1958, 1982). In the end there are few if any unique derived features that are characteristic of armadillos to the exclusion of glyptodonts and pampatheres.

A consideration of our phylogenetic scheme in the light of stratigraphic data demonstrates how poor the early Cenozoic (and perhaps, the late Mesozoic) fossil history of cingulates is and consequently how little is understood about the early evolution of the Cingulata. The oldest genera incorporated in the present study are Santacrucian in age (late early to early middle Miocene; Flynn & Swisher [1995]). These would include a taxon from the crown group, the glyptodont *Propalaeohoplophorus*. This in turn necessitates that the split among armadillos, pampatheres, and glyptodonts, as well as the more basal splits among various armadillo lineages, must predate the Santacrucian LMA (Land Mammal Age). However, undoubted glyptodont remains in the form of isolated scutes are known from the middle Eocene Mustersan LMA (Pascual et al., 1985). Isolated tarsals from deposits at Itaboraí in Brazil have also been identified as glyptodont-like, suggesting that glyptodonts may be recognizably distinct as early as the late Paleocene (Cifelli, 1983). The oldest pampathere remains include isolated scutes and teeth from the late Eocene Casamayoran LMA (Patterson & Pascual, 1968, 1972; Pascual et al., 1985). If glyptodonts and pampatheres were in fact distinct by the Paleocene or Eocene, our results would necessitate that virtually the entire cingulate radiation take place before this time. Yet to date not a single scrap of identifiable xenarthran fossil remains predates the late Paleocene Itaboraian LMA (Pascual et al., 1985), despite the preserve of several well characterized and well preserved mammalian faunas from the early Paleocene of South America (e.g., Muizon, 1991; Bonaparte et al., 1993). Clearly there is much to be gained in terms of understanding early cingulate (and early xenarthran) evolution by improving our knowledge of the early Cenozoic and late Mesozoic fossil record of South America.

Conclusions

To summarize, our cladistic investigation of the phylogenetic relationships among Cingulata based on craniodental morphology provides strong support for a number of phylogenetic conclusions. These include the remote position of the horned armadillo *Peltephilus* as the sister-taxon to all other cingulates, the paraphyly of the armadillos as a whole, and the crown group position of a monophyletic clade including pampatheres and glyptodonts. In addition, our results provide evidence that the four extant "dasypodid" armadillo genera *Cabassous, Dasypus, Priodontes,* and *Tolypeutes* may form a paraphyletic assemblage at the base of Cingulata. *Tolypeutes* is likely the most derived "dasypodid," being more closely related to euphractan and eutatine armadillos than to other extant dasypodids. The position of the extant giant armadillo *Priodontes* is not well resolved. There is only equivocal support for the traditional alliance of *Priodontes* and *Cabassous,* as the latter has a number of derived cranial features not present in *Priodontes* or *Dasypus.* Our analyses provide tentative support for the monophyly of a clade including the four extant genera of euphractan armadillos *Chaetophractus, Chlamyphorus, Euphractus,* and *Zaedyus* to the exclusion of any fossil taxa included in the analysis. However, we were unable to unambiguously resolve the relationships among these four genera. Lastly, the results of our analysis suggest that the eutatine armadillos are a paraphyletic assemblage, with the Miocene genus *Proeutatus* representing the sister-taxon to pampatheres and glyptodonts.

The results of this analysis are startling in that they neither support the monophyly of armadillos as a whole, not offer unequivocal support for any of the traditional subframilial groupings within armadillos. However, it must be remembered that the present study is based entirely on craniodental data and that the inclusion of postcranial characters or soft tissue data might substantially alter our phylogenetic conclusions. Moreover, there remains much missing data on many of the fossil taxa included in our analyses. In many instances, these fossils are well preserved but in dire need of additional preparation and study. Finally, our results highlight the poor early Cenozoic (and late Mesozoic) record of cingulates, suggesting that our understanding of basal evolution within this group could be much improved by the discovery of more fossil material from this critical time period in South American history.

Acknowledgments

We are indebted to the following institutions and individuals for access to the specimens that formed the basis of this study: Meng Jin, Department of Vertebrate Paleontology and Ross MacPhee, Department of Mammalogy, American Museum of Natural History, New York; Larry Heaney, Bruce Patterson, and Bill Stanley, Division of Mammals, and John Flynn and Bill Simpson, Department of Geology, Field Museum of Natural History, Chicago; Richard Thorington and Linda Gordon, National Museum of Natural History, Smithsonian Institution, Washington, DC; and Jacques Gauthier, Peabody Museum, Yale University, New Haven, Connecticut. This manuscript was greatly improved by the comments of Gerry DeIuliis, Tim Koneval, and Sergio Vizcaíno. We thank Julia Morgan Scott (fig. 6.6) and Gina Scanlon (figs. 6.2–6.5, 6.7) for their skillful work in preparing the illustrations for this study. Tim Gaudin's work on this project was supported in part by NSF RUI Grant DEB 0107922.

It is with real pride and personal pleasure that we dedicate this paper to Jim Hopson. Jim's seemingly boundless knowledge of and enthusiasm for vertebrate paleontology has inspired former students and postdocs alike to explore multidinous and varied branches on the vertebrate phylogenetic tree, even that weird and wonderful corner occupied by the edentate mammals. We have both benefited greatly from Jim's scientific insight and friendship over the years and are glad of this opportunity to honor him and his contributions to our discipline.

Literature Cited

Barlow, J. C. 1984. Xenarthrans and pholidotes; pp. 219–239 *in* S. Anderson and J. Knox Jones, Jr. (eds.), *Orders and Families of Recent Mammals of the World.* New York: Wiley.

Bonaparte, J. F., L. M. Van Valen, and A. Kramartz. 1993. La fauna de Punta Peligro, Paleoceno Inferior, de la provincia del Chubut, Patagonia, Argentina. Evolutionary Monographs 14: 1–60.

Carlini, A. A. and G. J. Scillato-Yané. 1996. *Chorobates recens* (Xenarthra, Dasypodidae) y un análisis de la filogenia de los Euphractini. *Revista del Museo de La Plata (N.S.), Paleontología IX* 59:225–238.

Carroll, R. L. 1988. *Vertebrate Paleontology and Evolution.* New York: Freeman and Co.

Cifelli, R. L. 1983. Eutherian tarsals from the late Paleocene of Brazil. *American Museum Novitates* 2761:1–31.

Delsuc F., F. M. Catzeflis, M. J. Stanhope, and E. J. P. Douzery. 2001. The evolution of armadillos, anteaters and sloths depicted by nuclear and mitochondrial phylogenies: implications for the status of the enigmatic fossil *Eurotamandua. Proceedings of the Royal Society of London B* 268:1605–1615.

Delsuc F., M. Scally, O. Madsen, M. J. Stanhope, W. W. de Jong, F. M. Catzeflis, M. S. Springer, and E. J. P. Douzery. 2002. Molecular phylogeny of living xenarthrans and the impact of character and taxon sampling on the placental tree rooting. *Molecular Biology & Evolution* 19(10): 1656–1671.

Delsuc F., M. J. Stanhope, and E. J. P. Douzery. 2003. Molecular systematics of armadillos (Xenarthra, Dasypodidae): contribution of maximum likelihood and Bayesian analyses of mitochondrial and nuclear genes. *Molecular Phylogenetics and Evolution* 28: 261–275.

Edmund, G. 1985. The fossil giant armadillos of North America (Pampatheriinae, Xenarthra = Edentata); pp. 83–93 *in* G. G. Montgomery (ed.), *The Ecology and Evolution of Armadillos, Sloths, and Vermilinguas.* Washington, DC: Smithsonian Institution Press.

Eisenberg, J. F. 1989. *Mammals of the Neotropics. Volume 1. The Northern Neotropics: Panama, Colombia, Venezuela, Guyana, Suriname, French Guiana.* Chicago: University of Chicago Press.

Eisenberg, J. F. and K. H. Redford. 1999. *Mammals of the Neotropics. Volume 3. The Central Neotropics: Ecuador, Peru, Bolivia, Brazil.* Chicago: University of Chicago Press.

Engelmann, G. 1978. The logic of phylogenetic analysis and the phylogeny of the Xenarthra. Ph.D. dissertation, Columbia University, New York.

———. 1985. The phylogeny of the Xenarthra; pp. 51–64 *in* G. G. Montgomery (ed.), *The Ecology and Evolution of Armadillos, Sloths, and Vermilinguas.* Washington, DC: Smithsonian Institution Press.

Fariña, R. A. 1985. Some functional aspects of mastication in Glyptodontidae (Mammalia). *Fortschritte der Zoologie* 30:277–280.

Ferigolo, J. 1985. Evolutionary trends of the histological pattern in the teeth of Edentata (Xenarthra). *Archives of Oral Biology* 30(1):71–82.

Flower, W. H. 1882. On the mutual affinities of the animals composing the order Edentata. *Proceedings of the Zoological Society of London* 358–367.

Flynn, J. J. and C. C. Swisher, III. 1995. Cenozoic South American Land Mammal Ages: Correlation to global geochronologies. *Geochronology Time Scales and Global Stratigraphic Correlation, SEPM Special Publication* 54:317–333.

Frechkop, S. and J. Yepes. 1949. Étude systématique et zoogéographique des dasypodidés conservés a l'institut. *Bulletin de l'Institut royal des Sciences naturelles de Belgique* 25(5):1–56.

Gardner, A. L. 1993. Order Xenarthra; pp. 63–68 *in* D. E. Wilson and D. M. Reeder (eds.), *Mammal Species of the World,* 2nd ed. Washington, DC: Smithsonian Institution Press.

Gaudin, T. J. 1993. The phylogeny of the Tardigrada (Xenarthra, Mammalia) and the evolution of locomotor function in the Xenarthra. Ph.D. dissertation, University of Chicago.

———. 1995. The ear region of edentates and the phylogeny of the Tardigrada (Mammalia, Xenarthra). *Journal of Vertebrate Paleontology* 15(3):672–705.

———. 2003. Phylogeny of the Xenarthra (Mammalia) *in* R.A. Fariña, S. F. Vizcaíno, and G. Storch (eds.): Morphological studies in fossil and extant Xenarthra (Mammalia). Senckenbergiana Biologica 83: 27–40.

———. 2004. Phylogenetic relationships among sloths (Mammalia, Xenarthra, Tardigrada): the craniodental evidence. Zoological Journal of the Linnean Society 140: 255–305.

Gillette, D. D. and C. E. Ray. 1981. Glyptodonts of North America. *Smithsonian Contributions to Paleobiology* 40:1–255.

Grassé, P.-P. 1955. Ordre des Édentés; pp. 1182–1266 *in* P.-P. Grassé (ed.), *Traité de Zoologie, Vol. 17, Mammifères.* Paris: Masson et Cie.

Guth, C. 1961. La région temporale des Édentés. Ph. D. dissertation, L'Université de Paris.

Hillis, D. M. and J. J. Bull. 1993. An empirical test of bootstrapping as a method for assessing confidence in phylogenetic analysis. *Systematic Biology* 42(2): 182–192.

Hoffstetter, R. 1958. Xenarthra; pp. 535–636 *in* J. Piveteau (ed.), *Traité de Paléontologie, Vol. 2, no. 6, Mammifères Évolution.* Paris: Masson et Cie.

———. 1982. Les édentés xénarthres, un groupe singulier de la faune Neotropical; pp. 385–443 *in* E. M. Gallitelli (ed.), *Paleontology, Essential of Historical Geology.* Modena, Italy: STEM Mocchi Modena Press.

Kay, R. F., R. H. Madden, M. G. Vucetich, A. A. Carlini, M. M. Mazzoni, G. H. Re, M. Heizler, and H. Sandeman. 1999. Revised geochronology of the Casamayoran South American Land Mammal Age: climatic and biotic implications. *Proceedings of the National Academy of Sciences USA* 96: 13235–13240.

McKenna, M. C. and S. K. Bell. 1997. *Classification of Mammals Above the Species Level.* New York: Columbia University Press.

Muizon, C. de. 1991. La fauna de mamíferos de Tiupampa (Paleoceno Inferior, Formacion Santa Lucia), Bolivia; *in* Suarez-Soruco, R. (ed.), Fósiles y Facies de Bolivia – Vol. 1, Vertebrados. *Revista Técnica de YPFB* 12(3–4): 575–624.

Novacek, M. J. 1986. The skull of leptictid insectivorans and the higher-level classification of eutherian mammals. *Bulletin of the American Museum of Natural History* 183:1–112.

Nowak, R. L. 1999. *Walker's Mammals of the World,* 6th ed. Baltimore: Johns Hopkins University Press.

Pascual R., M. G. Vucetich, G. J. Scillato-Yané, and M. Bond. 1985. Main pathways of mammalian diversification in South America; pp. 219–247 *in* F. G. Stehli and S. D. Webb (eds.), *The Great American Biotic Interchange.* New York: Plenum Press.

Patterson, B. and R. Pascual. 1968. Evolution of mammals on southern continents. *Quarterly Review of Biology* 43:409–451.

———. 1972. The fossil mammal fauna of South America; pp. 247–309 *in* A. Keast, F. C. Erk, and B. Glass (eds.), *Evolution, Mammals, and Southern Continents.* Albany: State University of New York Press.

Patterson, B., W. Segall, and W. D. Turnbull. 1989. The ear region in xenarthrans (= Edentata, Mammalia). Part I. Cingulates. *Fieldiana, Geology, New Series* 18:1–46.

Paula Couto, C. de. 1979. *Tratado de Paleomastozoologia.* Rio de Janeiro: Academia Brasileira de Ciências.

Redford, K. H. and J. F. Eisenberg. 1992. *Mammals of the Neotropics. Volume 2. The Southern Cone: Chile, Argentina, Uruguay, Paraguay.* Chicago: University of Chicago Press.

Romer, A. S. 1966. *Vertebrate Paleontology.* Chicago: University of Chicago Press.

Scillato-Yané, G. J. 1976. Sobre un Dasypodidae (Mammalia, Xenarthra) de Edad Riochiquense (Paleoceno superior) de Itaboraí, Brasil. *Anais, Academia Brasileira de Ciências* 48:527–530.

———. 1980. Catálogo de los Dasypodidae fósiles (Mammalia, Edentata) de la República Argentina. *Actas del Segundo Congreso Argentino de Paleontología y Biostratigrafía y Primer Congreso Latinoamericano de Paleontología* III:7–36.

Scillato-Yané, G. J., A. A. Carlini, S. F. Vizcaíno, and E. Ortiz Jaureguizar. 1995. Los Xenartros; pp. 183–209 *in* M. T. Alberdi, G. Leone, and E. P. Tonni (eds.), *Evolución Biológica y Climática de la Región Pampeana Durante los Últimos Cinco Millones de Años. Un Ensayo de Correlación con el Mediterráneo Occidental.* Madrid: Monografías, Museo Nacional de Ciencias Naturales.

Scott, W. B. 1903–1904. Mammalia of the Santa Cruz Beds. Part 1: Edentata. *Reports of the Princeton Expeditions to Patagonia* 5:1–364.

Simpson, G. G. 1945. The principles of classification and a classification of mammals. *Bulletin of the American Museum of Natural History* 85:1–350.

———. 1948. The beginning of the age of mammals in South America. Part 1. introduction. Systematics: Marsupialia, Edentata, Condylarthra, Litopterna, and Notioprogonia. *Bulletin of the American Museum of Natural History* 91:1–227.

Swofford, D. L. 2002. *PAUP: Phylogenetic Analysis Using Parsimony, Version 4.0b10.* Sunderland, MA: Sinauer Associates.

Vizcaíno, S. F. 1995. Identificación específica de las "mulitas," genero *Dasypus* L. (Mammalia, Dasypodidae), del noroeste Argentino. Descripción de una nueva especie. *Mastozoología Neotropical* 2(1):5–13.

Vizcaíno, S. F. and M. S. Bargo. 1998. The masticatory apparatus of the armadillo *Eutatus* (Mammalia, Cingulata) and some allied genera: paleobiology and evolution. *Paleobiology* 24(3):371–383.

Vizcaíno, S. F. and R. A. Fariña. 1997. Diet and locomotion of the armadillo *Peltephilus:* a new view. *Lethaia* 30:79–86.

Vizcaíno, S. F., G. De Iuliis, and M. S. Bargo. 1998. Skull shape, masticatory apparatus and diet of *Vassallia* and *Holmesina* (Mammalia: Xenarthra: Pampatheriidae): when anatomy constrains destiny. *Journal of Mammalian Evolution* 5(4):291–322.

Wegner, R. N. 1922. Der Stützknochen, Os nariale, in der Nasenhöhle bei den Gürteltieren, Dasypodidae, und seine homologen Gebilde bei Amphibien, Reptilien und Monotremen. *Gegenbaurs Morphologisches Jahrbuch* 51: 413–492.

Wetzel, R. 1980. Revision of the naked-tailed armadillos, genus *Cabassous* McMurtrie. *Annals of Carnegie Museum* 49(20):323–357.

———. 1985a. The identification and distribution of recent Xenarthra (= Edentata); pp. 5–21 *in* G. G. Montgomery (ed.), *The Ecology and Evolution of Armadillos, Sloths, and Vermilinguas.* Washington, DC: Smithsonian Institution Press.

———. 1985b. Taxonomy and distribution of armadillos, Dasypodidae; pp. 23–46 *in* G. G. Montgomery (ed.), *The Ecology and Evolution of Armadillos, Sloths, and Vermilinguas.* Washington, DC: Smithsonian Institution Press.

Wible, J. R., and T. J. Gaudin. 2004. On the cranial osteology of the yellow armadillo *Euphractus sexcinctus* (Dasypodidae, Xenarthra, Placentalia). *Annals of Carnegie Museum* 73(3): 117–196.

Winge, H. 1941. *The Interrelationships of the Mammalian Genera. Vol. 1: Mono-tremata, Marsupialia, Insectivora, Chiroptera, Edentata.* Copenhagen: C. A. Reitzels Forlag.

Appendix 6.1

Table A6.1. Specimens examined in connection with the present study

Cabassous centralis	FMNH 121185, 121224; USNM 1494, 314577; CM 1044
Cabassous unicinctus	FMNH 47960 (ear ossicles)
Chaetophractus villosus	FMNH 24333, 60467, 63865
Chlamyphorus retusus	CM 2139; USNM 283134
Dasypus novemcinctus	FMNH 18758 (ear ossicles), 34353, 55664; CM 76829; UTCM 15
Doellotatus prominens[†]	FMNH P14351, P14358, P14526
Euphractus sexcinctus	FMNH 21403, 28350, 28354, 34338, 54325 (ear ossicles); CM 6399
Eutatus seguini[†]	AMNH 11231
Macroeuphractus retusus[†]	FMNH P14493, PM 8404 (cast), 8405 (cast)
Peltephilus pumilis[†]	YPM-PU 15391
Paleuphractus argentinus[†]	FMNH P14412, P14442
Proeuphractus scalabrinii[†]	FMNH P14360, P14784, P15435
Proeutatus oenophorus[†]	FMNH P13197, P13199, P13203
Propalaeohoplophorus australis[†]	YPM-PU 15007, 15291
Prozaedyus exilis[†]	YPM-PU 15567, 15579
Priodontes maximus	FMNH 25271, 72913; CM 20942
Stegotherium tesselatum[†]	YPM-PU 15565, 15566
Tolypeutes matacus	FMNH 21407, 28339, 28340, 28341, 121540; AMNH 246460, 248394
Vassallia maxima[†1]	FMNH P14424
Zaedyus pichiy	FMNH 23809, 129838; CM 2364, 21053, 86390
Outgroups	
Bradypus variegatus	FMNH 20132 (ear ossicles), 30738, 60164; CM 1987, 4457
Tamandua mexicana	FMNH 15163, 22397, 25261, 74915, 93095; CM 76827

Note: † denotes extinct taxa.

[1]This specimen is described by Patterson et al. (1989) under the name *Plaina subintermedia*. G. DeIuliis and A. G. Edmund have since restudied the specimen in detail. They placed this specimen in the genus *Vassallia* (G. DeIuliis, personal communication; see also Vizcaíno et al., 1998).

Appendix 6.2

Listing of characters (dental, mandibular, and cranial, from anterior to posterior) and character states. * denotes multistate characters; ** denotes multistate and ordered characters.

**1. Number of upper teeth: 5 (0), 6 (1), 7 (2), 8 (3), 9 (4), or more than 9 (5).

**2. Number of lower teeth: 4 (0), 7 (1), 8 (2), 9 (3), 10 (4), or more than 10 (5).

3. Premaxillary teeth: present (0), or absent (1).

*4. Tooth wear: occlusal surface beveled anteroposteriorly on all teeth or on all but the first and/or last tooth (0), anterior portion of the toothrow exhibiting beveled wear, posterior teeth worn flat (1), occlusal surface worn flat on all teeth (2), occlusal surface slanted lingually on the upper teeth, labially on the lowers (3), or occlusal surface slanted posteriorly on the upper teeth, anteriorly on the lowers (4).

5. Histology of central region of tooth: tooth center composed of modified orthodentine (0), or tooth center composed of an elevated core of osteodentine (1) (Ferigolo, 1985).

**6. Modified orthodentine core of teeth: avascular (0), poorly vascularized (1), well-vascularized (2), or highly vascularized, almost resembling a vasodentine (3) (Ferigolo, 1985).

7. Layer of cementum forming outer layer of teeth: thick (0), or thin (1) (Ferigolo, 1985).

*8. Orientation of long axis of teeth relative to long axis of toothrow: all teeth oriented parallel to the long axis of the toothrow (0), posterior teeth oriented obliquely to the long axis of the toothrow (1), anterior teeth oriented obliquely to the long axis of the toothrow (2), or all teeth oriented obliquely to the long axis of the toothrow (3).

*9. Shape of teeth in lower toothrow: all teeth with ovate cross-section (0); anterior teeth strongly compressed labiolingually, posterior teeth ovate (1); anterior teeth ovate, posterior teeth slightly bilobate (2); anterior teeth ovate, posterior teeth strongly bilobate (3); posterior teeth ovate, anterior teeth slightly bilobate (4); anterior and posterior teeth may be slightly bilobate, intervening teeth ovate (5); or anterior teeth ovate or bilobate, posterior teeth trilobate (6).

10. Length of lower toothrow: elongated, occupies more than half the total length of the mandible (0); or short, occupies less than half the total length of the mandible (1).

**11. Length of mandibular spout: elongated, greater than or equal to the length of the mandibular symphysis (0); reduced, half the length of the symphysis or less (1); or spout absent (2).

12. Mandibular symphysis delineated from mandibular horizontal ramus by distinct angle in lateral view: clear angle present (0); or clear angle absent, ventral edge of the symphysis grades imperceptibly into the ventral edge of the ramus (1).

13. Position of posterior end of mandibular symphysis relative to anteriormost mental foramen: posterior to (0), or anterior to (1).

14. Anterior termination of lower toothrow: posterior to the mandibular symphysis (0), or at least two lower teeth dorsal to the mandibular symphysis (1).

**15. Number of mental foramina: 1 (0), 2 (1), or 3 or more (2).

**16. Depth of mandible: shallow, maximum depth of the horizontal ramus < ten percent of the maximum mandibular length (0); of moderate depth, maximum depth of the horizontal ramus ≥ ten percent, <twenty percent of the maximum mandibular length (1); or deep, maximum depth of the horizontal ramus ≥ twenty percent of the maximum mandibular length (2).

**17. Position of last lower tooth relative to ascending ramus of mandible: fully hidden in lateral view (0), partially hidden in lateral view (1), or located well anterior to (2).

**18. Angle between toothrow and anterior edge of ascending ramus of mandible: <40° (0), 40°–49° (1), 50°–59° (2), 60°–69° (3), 70°–79° (4), 80°–89° (5), or ≥90° (6).

*19. Coronoid process shape: elongated, flares distally (0); elongated, with the anterior and posterior edges roughly parallel to one another (1); elongated, with the anterior and posterior edges strongly divergent ventrally (2); or small, with the anterior and posterior edges strongly divergent ventrally (3).

20. Distal extent of coronoid process relative to condylar process: dorsal to (0), or ventral to (1).

21. Distinct lateral ridge crosses coronoid process anteroventrally somewhat proximal to its tip: absent (0), or present (1).

**22. Relative distance between coronoid, condylar, and angular processes of mandible: condyle closer to the coronoid than to the angle (0), three processes essentially equidistant (1), or condyle closer to the angle than to the coronoid (2).

23. Mandibular condyle: greatly elevated above the toothrow, vertical distance from the condyle to the toothrow more than twice the height of the mandible measured at the last tooth (0), or not greatly elevated above the toothrow (1).

**24. Shape of mandibular condyle in dorsal view: narrow, greatest anteroposterior length more than one and a half times greater than the greatest width (0); greatest width and the anteroposterior length roughly equivalent (1); or wide, greatest width more than one and a half times the greatest anteroposterior length (2).

25. Lateral expansion of mandibular condyle in dorsal view: present (0), or absent (1).

26. Medial expansion of mandibular condyle in dorsal view: absent (0), or present (1).

*27. Dorsal surface of mandibular condyle in posterior view: concave transversely (0), flat (1), convex transversely (2), or concavo-convex (3).

**28. Orientation of mandibular condyle in lateral view: faces posterodorsally (0), dorsally (1), or anterodorsally (2).

29. Fossa on anterodorsal surface of condylar process of mandible, immediately anterior to articular condyle: absent (0), or present (1).
30. Ventral edge of angular process: ventral to the lower edge of the horizontal ramus (0), or dorsal to the lower edge of the horizontal ramus (1).
31. Position of mandibular foramen: less than two tooth lengths (defined here as maximum anteroposterior diameter of the largest tooth) behind the last molar (0), or more than five tooth lengths behind the last molar (1).
32. Strong mylohyoid groove on anterior medial surface of mandibular horizontal ramus: absent (0), or present (1).
33. Orientation of snout in lateral view: horizontal (0), or downturned anteriorly (1).
*34. Septomaxilla shape: absent (0); present, lacking distinct lamina palatina or processus intrafenestralis (1); or present, possessing distinct lamina palatina and processus intrafenestralis (2).
*35. Shape of processus ascendens and processus intrafenestralis of septomaxilla: processes slender, circumference fairly uniform throughout (0); processus intrafenestralis flared distally (1); or both processes elaborate, the processus ascendens bearing a posterior prong and the processus intrafenestralis a broad lateral flange (2).
**36. Length of rostrum to anterior orbital rim: short, <forty percent of the GSL (0); ≥forty percent, <fifty percent of the GSL (1); ≥fifty percent, <sixty percent of the GSL (2); or elongated, ≥sixty percent of the GSL (3).
**37. Length of nasal bone: short, <twenty-five percent of the GSL (0); ≥twenty-five percent, <thirty percent of the GSL (1); ≥thirty percent, <thirty-five percent of the GSL (2); or elongated, ≥thirty-five percent of the GSL (3).
38. Length of premaxillary/nasal suture: elongated, >twelve percent of the condylobasal skull length (0); or short, <ten percent of the condylobasal skull length (1).
39. Well-defined notch on anteroventral edge of premaxilla: absent (0), or present (1).
**40. External naris shape: height > width (0), width = height (1), or width > height (2).
41. Inclination of external narial aperture in lateral view: directed anteriorly or anterodorsally (0), or anteroventrally (1).
42. Vomer overlaps premaxilla dorsally within external narial aperture: absent (0), or present (1).
43. Anterior edge of nasal in dorsal view: with distinct medial and lateral processes separated by an indentation or notch (0); or without distinct processes, typically forming an evenly convex curve (1).
44. Anterior extent of nasal bone relative to nasoturbinal: posterior to (0), or anterior to anterior tip of Nasoturbinal (1).

45. Position of anterior tip of nasoturbinal: at the lateral edge of the nasal (0); or medial to the nasal/maxillary and nasal/premaxillary sutures (1).
46. Position of infraorbital foramen relative to zygomatic process of maxilla: on lateral face of, visible in ventral view (0); or anterior to, not visible in ventral view (1).
47. Antorbital depression on lateral surface of maxilla: present (0), or absent (1).
48. Shape of facial exposure of lacrimal: triangular (0), or quadrangular (1).
49. Surface of lacrimal dorsal to lacrimal foramen (in position of lacrimal tubercle): smooth (0), or rough (1).
50. Lacrimal fenestra at the juncture of the lacrimal, maxilla, and frontal bones: present (0), or absent (1).
**51. Length of upper toothrow: short, <thirty percent of the GSL (0); ≥thirty percent, <thirty-five percent of the GSL (1); ≥thirty-five percent, <forty percent of the GSL (2); or elongated, ≥forty percent of the GSL (3).
52. Incisive foramen size: large, maximum diameter equal to or greater than half the anteroposterior length of the premaxilla measured at the midline (0); or small, maximum diameter less than half the anteroposterior length of the premaxilla (1).
53. Maxilla participation in posterolateral margin of incisive foramina: present (0); or absent, incisive foramen completely or almost completely encircled by the premaxilla (1).
*54. Shape of maxilla/premaxilla palatal suture in ventral view: U- or V-shaped, midpoint directed posteriorly (0); approximately flat (1); V-shaped, midpoint directed anteriorly (2); or W-shaped, midline apex directed anteriorly (3).
*55. Maxillary/palatine suture shape: rectangular, with angular anterolateral corners (0); U-shaped with rounded anterolateral corners (1); or M-shaped, with the midline apex directed posteriorly and the anterolateral corners rounded (2).
**56. Anterior extent of horizontal process of palatine relative to anterior root of zygoma: posterior to or even with the posterior edge of zygomatic root (0), anterior to the posterior edge of the zygomatic root (1), or anterior to the anterior edge of the zygomatic root (2).
57. Posterior extent of palate relative to tooth row: even with the last tooth or not extending posteriorly beyond the toothrow by more than one anteroposterior diameter of the longest tooth (0), or extended posteriorly beyond the toothrow by more than one anteroposterior diameter of the longest tooth (1) (from Wetzel, 1985b).
**58. Median palatine suture: flat (0), raised to form a midline crest along the posterior half of the suture (1), or raised to form a midline crest along the entire length (2).

*59. Posterior margin of palate: forms a narrow v or u, choanae narrow, pterygoids inset medially versus the toothrow (0); forms a broadly open v or u, choanae broad, pterygoids in line with the toothrow (1); or forms a narrow u, choanae narrow, but pterygoids in line with or slightly lateral to the toothrow (2).

60. Pterygoid exposure on posterior edge of hard palate: absent (0), or present (1).

61. Maximum width of skull at zygomatic arches: wide, >fifty percent of the GSL (0); or narrow, <forty-five percent of the GSL (1).

**62. Width of interorbital area: narrow, <twenty percent of the GSL (0); ≥twenty percent, <twenty-five percent of the GSL (1); ≥twenty-five percent, <thirty percent of the GSL (2); or wide, ≥thirty percent of the GSL (3).

63. Maxillary foramen: not visible in ventral view (0), or visible in ventral view (1).

64. Contact between orbital process of maxilla and lacrimal in medial wall of orbit: absent (0), or present (1).

65. Shape of palatine orbital exposure: tall and narrow, maximum height greater than the anteroposterior length (0); or long and low, maximum anteroposterior length much greater than the height (1).

66. Palatine/orbitosphenoid contact: absent (0), or present (1).

67. Palatine/frontal contact: absent (0), present (1).

68. Position of orbital wing of palatine's dorsal edge relative to sphenorbital fissure and/or foramen rotundum: ventral to (0), or at the level of or extending dorsal to (1).

69. Posterior process of palatine orbital exposure extending between alisphenoid and pterygoid: absent (0), or present (1).

70. Position of orbital exposure of pterygoid's dorsal edge relative to foramen ovale: ventral to (0), or at the level of or extending dorsal to (1).

71. Sphenopalatine and caudal palatine foramina: confluent (0), or separate (1).

72. Sphenopalatine foramen position versus sphenorbital fissure/foramen rotundum: separate (0), or in common fossa (1).

73. Groove connecting sphenorbital fissure/foramen rotundum to sphenopalatine foramen: absent (0), or present (1).

74. Orbitosphenoid participation in sphenopalatine/caudal palatine foramen: present (0), or absent (1).

**75. Number of ethmoid foramina: 1 (0), 2 (1), or 3 (2).

76. Orbitosphenoid participation in lowermost ethmoid foramen: absent (0), or present (1).

77. Ethmoid orbital exposure: rudimentary or absent (0), or large (1).

78. (Anterior opening of posttemporal canal) cranioorbital foramen: absent (0), or present (1).

79. Optic foramen position: visible in lateral view (0); not visible in lateral view, recessed within the fossa for the sphenorbital fissure/foramen rotundum (1); or, not visible in lateral view but anterior to sphenorbital fissure, hidden by large muscular process (ossified ala hypochiasmata of Wible and Gaudin [2004])

*80. Position of optic foramen within orbitosphenoid: dorsocentral (0), central (1), or posteroventral (2).

81. Foramen rotundum and sphenorbital fissure: separate (0), or confluent (1).

82. Orbital opening for vidian nerve: within recess for the sphenorbital fissure/foramen rotundum (0), or ventral to the recess for the sphenorbital fissure/foramen rotundum (1).

*83. Orbital muscular crest: absent (0); present but weak, connected to the medial edge of the glenoid fossa, situated only on the squamosal (or squamosal plus alisphenoid) in a position dorsal to the optic carval (1); present, connected to the medial edge of the glenoid fossa, extends dorsal to the optic carval across the squamosal, alisphenoid and frontal (2); or present, unconnected to the medial edge of the glenoid fossa, occurs approximately at the level of the optic carval, crossing the alisphenoid, parietal and frontal (3).

*84. Zygomatic process of maxilla: weak or absent (0); present, excluded from the anterior margin of the infratemporal fossa by the jugal/lacrimal contact (1); or present, forms the anterior margin of the infratemporal fossa (2).

*85. Depth of jugal (posterior to descending process): uniform depth throughout its length (0), maximum depth in the anterior half, narrower posteriorly (1), or maximum depth in posterior half, narrower anteriorly (2).

*86. Ventral process on zygomatic arch: absent (0), present as ventral boss on the zygomatic arch at the maxillary/jugal suture (1); present as elongated, anteroposteriorly compressed process on the anterior half of the jugal (2); or present as elongated, mediolaterally compressed process on the posterior half of the jugal (3).

87. Dorsal edge of zygomatic process of squamosal: terminates anterior to the external auditory meatus/porus acousticus (0), or elongated posteriorly, extending dorsal or posterodorsal to the external auditory meatus/porus acousticus (1).

88. Postorbital process of jugal/squamosal: absent (0), or present (1).

*89. Shape of jugal/squamosal suture: suture absent, zygoma incomplete (0); suture oriented obliquely relative to the long axis of the skull (1); or suture horizontal relative to the long axis of the skull, may be L-shaped with long horizontal component (2).

90. Orientation of posterior root of zygoma: directed anteriorly to anterolaterally (0), or laterally (1).

91. Shape of braincase: elevated, parietal with extensive lateral exposure (braincase typically cylindrical with domed parietals) (0); or dorsoventrally compressed, parietals appear relatively flat (1).

**92. Position of temporal lines: lines meet in the dorsal midline to form a sagittal crest (0), lines approximate midline but do not meet to form a sagittal crest (1), or lines are laterally situated, do not approximate midline (2).

93. Shape of frontal in dorsal view: relatively narrow transversely, unexpanded (0); or expanded transversely into a distinct forehead "shield" (1).

94. Multiple foramina on dorsal surface of frontal, around midline: absent (0), or present (1).

95. Vascular foramina situated in posterolateral region of frontal (within temporal fossa): present (0), or absent (1).

**96. Position of frontal/parietal suture at its lateral-most point relative to anterior edge of glenoid fossa: well anterior to (0), at the level of (1), or posterior to (2).

97. Number of foramina for rami temporales in temporal fossa of parietal (near parietal/squamosal suture): 5 or less (0), or more than 5 (1).

98. Alisphenoid/parietal contact: present (0), or absent, excluded by frontal/squamosal contact (1).

99. Supraoccipital exposure on skull roof: present (0), or absent (1).

100. Shape of nuchal crest in dorsal view: lacking distinct posterior extensions lateral to the midline, roughly C-shaped (0); or W-shaped, with distinct posterior extensions situated lateral to the midline (1).

**101. Length of postglenoid region of skull: ≥twenty percent of the GSL (0); reduced, <twenty percent of the GSL (1); or very reduced, <fifteen percent of the GSL (2).

102. Form of vomerine alae (wings): fused posteriorly (0); or separate, exposing the overlying presphenoid or ethmoid (1).

103. Endocranial surface of presphenoid forms raised median crest anterior to optic foramina: absent (0), or present (1).

104. Entopterygoid processes: gracile (0), or rugose (1).

*105. Pterygoid contact with bulla (pterygoid tympanic process): absent (0); present, with the ectotympanic only (1); or present, with the entotympanic only (2).

**106. Vidian nerve (nerve of pterygoid canal) pathway: internal, contained in a canal within the pterygoid or between the pterygoid and sphenoid (0); partially exposed along a groove at the base of the medial surface of the pterygoid, partially enclosed by a canal within the pterygoid (1); or fully exposed along a groove at the base of the medial surface of the pterygoid (2).

*107. Position of posterior foramen for vidian nerve: between vomer and pala-
tine (0), within palatine (1), within pterygoid (2), between pterygoid
and alisphenoid (3), between pterygoid and palatine (4), or between
entotympanic, alisphenoid and pterygoid (5).

108. Distinct groove for eustachian (auditory) tube on basisphenoid: absent
(0), or present (1).

109. Piriform fenestra: absent in adult (0), or persists in adult (1).

110. Foramen ovale/piriform fenestra relationship: separate (0), or
confluent (1).

111. Large vascular foramen (transverse canal foramen of Wible and Geudin
[2004]) located immediately anterior or ventral to foramen ovale: absent (0), or present (1).

*112. Alisphenoid contact with bulla (alisphenoid tympanic process): absent
(0), with ectotympanic only (1), with entotympanic only (2), or with
both ectotympanic and entotympanic (3).

113. Alisphenoid contribution to margin of Glaserian fissure: absent, fissure
lateral to the alisphenoid, between the ectotympanic and squamosal
(0); or present (1).

114. Squamosal/pterygoid contact: absent (0), or present (1).

*115. Entoglenoid process: absent (0), present as a distinct ridge near the
squamosal/alisphenoid suture (typically contacting the ectotympanic)
(1), or present as in state (1) but distinctly inflated (2).

*116. Shape of glenoid fossa: concave (0), flat (1), or convex (2).

**117. Maximum width of glenoid fossa relative to anteroposterior length: gle-
noid long and narrow, ratio of width to length < 1.0 (0); width of glenoid roughly equivalent to length, ratio ≥ 1.0, < 1.5 (1); or glenoid
wide and short, ratio of width to length ≥ 1.5 (2).

118. Position of glenoid fossa relative to optic foramen: at the level of or ventral to (0), or dorsal to (1).

119. Distinct depression accommodating postglenoid foramen situated immediately posterior to glenoid fossa: absent (0), or present (1).

*120. Suprameatal foramen: absent (0), single large foramen present (1), or
multiple small foramina present (2).

121. Petrosal contact with basicranial axis: petrosal unattached (0), or sutured
medially to basicranial axis (1).

122. Position of promontorium of petrosal relative to ventral surface of basicranium: dorsal to (0), or even with or extending slightly ventral to (1).

123. Shape of promontorium of petrosal: globose (0), or elongated, extended
anteromedially (1).

*124. Shape of processus crista facialis (=?tegmen tympani): flat, rather small,
rectangular (0); concave, rather small, ovate (1); large, concave, quadrangular (2); convex lump, small, ovate (3); concave, greatly enlarged,
tripartite (4); or small, concave, elongated anteromedially (5).

125. Fossa incudis: formed entirely by the petrosal (0), or formed between the squamosal and petrosal (1).
126. Epitympanic sinus: absent (0), or present (1).
**127. Position of stylomastoid foramen: directly anterior to the paroccipital process (0), anteromedial or medial to the paroccipital process (1), or posteromedial to the paroccipital process (2).
*128. Medial extension and distal expansion of tympanohyal: absent (0); medial extension present, but unexpanded distally (1); or medial extension present, flares distally (2).
129. Vagina processus hyoidei: absent (0), or present (1).
130. Position of vagina processus hyoidei: visible in posterior view (0), or displaced anteriorly, not visible in posterior view (1).
*131. Distinct stapedius fossa below elongated ridge originating on promontorium between fenestra vestibuli and fenestra cochleae: fossa and ridge absent (0), fossa and ridge present (1), or fossa well-defined but ridge absent (2).
132. Groove posterior to fenestra vestibuli for stapedius muscle tendon: absent (0), or present (1).
133. Digastric fossa: absent or only weakly indicated (0), or well-developed depression present (1).
134. Paroccipital process (see Wible and Gaudin [2004]; equals "mastoid process" of Patteraon et al. [1989]. Gaudin [1995] et al.) shape and position: greatest width in the anteroposterior plane, freestanding (0); or greatest width in the mediolateral plane, appressed to the posterior surface of the ectotympanic/external auditory meatus (1).
135. Distance from distal tip of paroccipital process to distal extremity of paracondylar process (or exoccipital crest, if latter is poorly developed): paroccipital and paracondylar processes well-separated, distance between them ≥ five percent of the GSL (0); or processes closely approximated, distance between them < five percent of the GSL (1).
136. Auditory bulla: with substantial membranous or cartilaginous contribution, with horseshoe-shaped ectotympanic that may exhibit either a small anteroventral contact with a single entotympanic or both an antero- and posteroventral contact with rostral and caudal entotympanics (0); or bony, formed by broad fusion of the entotympanic and ectotympanic (1).
137. Entotympanic participation in tympanic cavity floor: small or nonexistent (0), or forming ½ or more (1).
138. Anteroposterior length of entotympanic relative to ectotympanic: longer (0), or shorter (1).
139. Ectotympanic/squamosal attachment: ectotympanic crura either ligamentously attached or sutured to the squamosal (0), or ectotympanic crura fused to the squamosal (1).

*140. Position of ectotympanic and fissura Glaseri: ectotympanic posterior to the glenoid fossa, Glaserian fissure well posterior of the foramen ovale (0); ectotympanic medial to the glenoid, Glaserian fissure immediately posterior to the foramen ovale (1); or ectotympanic medial to glenoid fossa, Glaserian fissure lateral to foramen ovale (2).

**141. Lateral surface of ectotympanic: no ossified external auditory meatus present (0), forms incomplete external auditory meatus (1), forms complete but short external auditory meatus (2), or forms complete, tubular, elongated external auditory meatus (3).

**142. Shape of porus acousticus and form of ectotympanic: porus ovate, ectotympanic lacking a vertical muscular crest externally (0); porus ovate, ectotympanic with a short vertical muscular crest anterior to the porus, crest not extending ventrally onto the external surface of the bulla (1); or porus drawn into a narrow V ventrally, sharp muscular crest present on the outer surface of the ectotympanic extending ventrally across the external surface of the bulla (2).

143. Relationship of porus acousticus to glenoid fossa: porus situated close to the glenoid, postglenoid fossa absent or if present, either situated medial to the glenoid or posterior to the glenoid but overlapped ventrally by the ectotympanic (0); or porus far posterior to the glenoid, separated by horizontal postglenoid fossa in which the postglenoid foramen is centrally located (1).

**144. Shape of mallear head: elongated anteroposteriorly (0), bulbous, almost spherical (1), or anteroposteriorly compressed (2).

145. Angle between incudal facets on mallear head: >90° (0), or <90° (1).

146. Distinct swelling at junction of mallear neck and manubrium: absent (0), or present (1).

*147. Size of lamina and lateral process of manubrium of malleus: both small (0), lamina small but the lateral process enlarged (1), or lamina large but the lateral process small (2).

148. Position of cochlear cavaliculus (aqueduct): opens within the jugular foramen (0), or opens within the cranial cavity, dorsal to the jugular foramen (1).

149. Basioccipital tubera: rudimentary or absent (0), or well-developed (1).

**150. Rectus capitis/longus capitis fossae on basioccipital: absent (0), faintly indicated (1), or well-developed (2).

151. Articular facet on odontoid notch of basioccipital: absent (0), or present (1).

152. Hypoglossal foramen position: at the level of or dorsal to the jugular foramen (0), or ventral to the jugular foramen (1).

**153. Number of hypoglossal foramina: 1 (0), 2 (1), or 3 (2).

**154. Paracondylar process of exoccipital: absent (0), weakly indicated (1), or present (2).

155. Occipital condyle shape in ventral view: roughly triangular, narrowing laterally (0), or roughly rectangular (1).
156. Lateral indentation in occipital condyle: absent (0), or present (1).
**157. Shape of occiput: low and wide, maximum width is ≥1.5 times the maximum height, (0); moderately wide, maximum width ≥ 1.25, <1.5 times the maximum height (1); or maximum height is roughly equivalent to (<1.25) or greater than the maximum width (2).
**158. Inclination of occipital midline in lateral view: anterodorsal (0), approximately vertical (1), or posterodorsal (2).
159. Width of foramen magnum: narrow, maximum width less than or equal to one half the maximum width of the occiput (0); or wide, maximum width greater than one half the maximum width of the occiput (1).
160. Occipital exposure of squamosal: absent (0), or present (1).
161. Squamosal participation in lateral wall of posttemporal foramen: present (0), or absent (1).
162. Groove for occipital artery continues dorsal to posttemporal foramen: absent (0), or present (1).
163. Nuchal crest: of uniform thickness (0), or bearing distinct, posteriorly thickened dorsal bosses (1).

Appendix 6.3

Table A6.3. Data matrix

Cabassous										
42100	00001	01A0A	1C431	0101A	11101	00022	22102	11111	A1A10	11110
11IA0	03111	11011	00AAI	A01A0	10022	10120	02110	11111	01102	0201A
A2001	10101	01131	0221A	10A00	1AA01	00021	1200I	A10C1	12001	A?01A
Chaetophractus										
44000	00B00	11012	1AH10	10020	10201	0A120	2E002	11111	00110	3112A
00200	0E010	1A1A1	A00AC	A0100	11222	11120	11110	11111	01102	CA0?0
1CA01	2111C	1?120	10010	00010	11011	31010	01102	A1001	10101	010
Chlamyphorus										
32100	??100	00000	10631	00020	10201	00120	22002	11111	00101	2112?
?0?00	03010	100?0	00010	00000	11221	11121	10111	?0?11	01102	05000
01101	21110	1????	?0?10	??010	11012	30010	01102	11001	00101	100
Dasypus										
E2100	21001	01A0C	02010	02110	11101	10021	3E112	11111	00001	0112A
11A11	1D101	00111	A0A1A	A110A	11220	01110	0A10A	20111	C10A0	BJA10
10001	10001	00011	0110?	10001	0??01	00021	0201B	01A01	12C10	11A

(continued)

Table A6.3. *(continued)*

Doellotatus

G4000	??000	00001	114??	?????	????1	01???	?????2	1?111	0011?	?112?
?0200	??01?	?????1	0?0??	??10?	1?22?	?1120	?01??	01111	??1??	2?110
12101	20111	011??	?0?11	??01?	10011	120??	??102	01?11	1??01	010

Euphractus

44000	00000	11A12	1AG10	A0020	10201	00120	2E002	11111	00A10	31121
10200	0D010	1A111	0001C	A0100	11221	11120	11110	21111	11002	C2000
1CA01	21112	11120	10010	0A010	11011	31010	01100	A1A01	A0201	01A

Eutatus

33101	??001	0A?01	10310	10020	102?1	??1??	30002	1?1??	001??	1??20
00200	0C111	00010	00110	00?1?	10221	10120	111?0	21A11	2??02	0?100
13101	2?111	0????	?0?10	??011	11011	12010	0??10	01001	10100	010

Macroeuphractus

34?00	??240	1111C	10200	?0120	10201	010??	1??02	1?1??	00???	3????
?0020?	000??	?????1	0????	??10?	1?2?0	11120	001??	?1??1	1????	??1?0
1???1	22111	?????	?0???	??010	11010	120??	???02	11?11	1110?	?10

Paleuphractus

44000	??000	11011	114?0	??0?0	??2?1	0?1??	1?1?2	1?1??	00110	3????
??200	0201?	?????1	?????	??1??	??020	11120	121?0	21111	1????	??0?0
?3001	20111	111?0	00???	0?010	11011	120??	?????	01001	?0201	010

Peltephilus

21040	??211	210??	22130	?2120	1?0?0	??0??	12002	1?111	01?10	01001
1??0	0100?	?20?1	?10?1	?????	??0?	?202?	?1???	00111	00??1	0210?
0??0?	??1??	??001	02002	?????	?1???	??010	11010	310??	??000	011

Priodontes

55130	11011	0100C	12A3A	02101	11101	10021	2E112	11111	110A0	C1120
11A10	12101	11001	A011A	10011	10112	00120	02111	21111	11100	A?01A
10002	0010C	00041	022A1	11100	00A00	00021	1?000	11I21	11101	A11

Proeuphractus

4?000	??20?	?????	?0???	?????	?????	?10??	12002	1????	00010	31120
10200	?101?	?????1	?????	?????	1?22?	?1?2?	101??	01?11	1????	??1?0
13101	2?111	1????	???11	??010	11011	120??	???01	01?01	1010?	??0

Proeutatus

44011	??120	0101C	12610	10020	10201	010??	230?2	111??	1011?	21121
10200	01010	101?1	0001?	?0?10	11221	11020	11110	11111	11???	24100
A??02	2111C	101?0	0100?	00001	0??01	000??	??012	01001	11001	010

Table A6.3. *(continued)*

Propalaeohoplophorus

32121	??26?	11012	206?1	?0020	13211	0?0??	0?102	01101	01111	1113?
0022?	000??	00??1	?10??	???1?	10121	20121	?01?1	11111	0??1?	0E110
1??01	2210A	00101	0120?	??001	???00	??1??	???01	11021	11001	110

Prozaedyus

E4100	??010	101?C	02410	10020	10101	001??	22002	1?1??	00010	11?20
1020?	010?1	??011	?000?	??10?	1?121	11120	11100	1A111	1??02	??0?0
1?001	10101	101??	?0???	??010	01001	100??	?2?01	11001	12100	??0

Stegotherium

AC130	??001	000?1	02A31	02101	12101	100??	33112	1?1??	01001	11111
01121	10201	101?1	10?11	?0100	0110?	1?012	30020	01100	20111	12001
1?00B	??010	0B002	00112	000?1	0100?	??000	0??00	00021	??011	?11

Tolypeutes

43100	00000	01A0I	10K10	00010	10101	00120	20112	11111	000AA	31110
00A00	13A1A	00111	A01AC	00101	1A020	00020	00101	0AA10	A110B	03010
AB001	2010C	11101	0110?	10000	AAA01	00021	1201C	01001	A2110	010

Vassallia

43?11	??230	10010	20410	11120	10211	011??	13?02	11111	00010	21??2
00220	01010	10100	01110	00010	11122	10111	00110	11111	01110	20110
1?001	22111	00151	01???	20001	???? 0	??1??	??110	01021	01101	100

Zaedyus

F3A00	00000	11A12	12410	00020	10201	00120	2C002	11111	00A11	21121
10200	0E010	1AA11	000A1	A0100	11120	11120	1A110	21111	11102	CE000
1JA01	2111C	11120	10?10	00010	11011	30010	0?102	A1A01	A110A	100

Bradypus

00100	31351	1001I	2CE20	00011	12000	0000?	00?00	00010	01011	10000
00B10	03A00	110A0	00010	A0012	01302	31A00	02101	0000A	00002	2A000
00011	0000I	10001	11010	10100	11010	20000	01011	00020	00000	10A

Tamandua

?????	?????	201?0	0????	??101	02200	?011?	12101	00000	01100	?0000
21A11	10101	11001	A0110	00021	A0300	00000	02001	00000	01001	0?000
01010	00000	10001	1100?	10000	10110	20001	0000A	00010	01000	000

Note: The symbol "?" represents missing data; the symbol "n" is used in cases in which a character is not applicable to a given taxon. The following symbols are used to represent character states in polymorphic taxa: A = (0, 1); B = (0, 2); C = (1, 2); D = (1, 3); E = (2, 3); F = (3, 4); G = (4, 5); H = (5, 6); I = (0, 1, 2); J = (1, 2, 3); K = (3, 4, 5). Characters and character states are described in appendix 6.2.

Appendix 6.4

Distribution of apomorphies on the consensus tree illustrated in figure 6.1. Characters are numbered according to the scheme provided in appendix 6.2, with numbered character states shown in parentheses. Unambiguous synapomorphies are boldface.

Node 1. Cingulata: 1(2), 2(1), **12(1),** 19(3), 22(2), **40(2), 41(1), 43(1), 45(1), 83(0), 84(2),** 100(1), **114(0),** 120 (2), **152(1), 155(1), 157(2), 158(1), 162(1),** 163(1) (see figs. 6.2, 6.3).

Node 2. **2(2), 11(0),** 15(1), 28(1), **30(1),** 34(2), **36(2),** 42(1), 52(1), 53(1), 54(2), 56(1), 58(1), 81(1), 89(2), **96(1),** 98(1), 99(1), **101(1),** 107(1), 109(1), 121(0), 126(0), **139(0), 141(0),** 144(2), **156(1)** (see fig. 6.2).

Node 3. **1(4), 6(1),** 27(1), 88(1), **97(1), 103(1),** 111(1), 118(1), 128(2), **146(1)** (see fig. 6.3).

Node 4. **6(0), 7(0), 18(4), 22(1), 23(0),** 54(1), **64(1),** 69(1), 78(1), **122(1), 123(1), 140(1),** 147(2) (see figs. 6.3, 6.4).

Node 5. 2(3), **10(0), 19(1), 22(0),** 25(0), **27(0), 33(1), 47(0), 57(0),** 59(0), **67(0),** 92(1), **121(1), 154(0), 163(0)** (see fig. 6.4).

Node 6. **15(2),** 35(0), 55(1), **75(1),** 83(2), 109(0), **112(1), 117(1), 124(2), 126(1),** 128(0), **141(3),** 147(1) (see fig. 6.5).

Node 7. **4(1), 5(1),** 37(3), **51(2), 79(1),** 122(0), **127(1),** 129(0), **134(0), 135(1), 139(0)** (see fig. 6.6).

Node 8. **2(3), 16(2),** 17(0), **29(1), 56(0), 59(2), 72(1), 83(1), 87(0), 90(1), 101(0),** 104(1), 117(2), **121(0), 125(1),** 140(0), **143(1), 150(1), 154(2), 161(1)** (see fig. 6.7).

Node A. **21(1), 24(2), 38(0),** 54(2), **58(2),** 61(0), 85(1), 86(1), **91(1),** 95(0), 105(2), 116(1), 120(1), **127(0), 134(1), 141(1).**

Node B. 17(1), **28(2), 48(1), 96(2),** 112(3), **113(1),** 116(2), **119(1),** 129(1), **139(1), 142(1),** 144(1), 145(0), 146(0), **157(0).**

Node C. 2(4), **3(0),** 11(1), **14(1), 51(3),** 63(0), **65(0), 68(1),** 73(0), 80(0), **82(1),** 87(1), 94(1), **106(1),** 107(2), 125(0), 131(0), 148(1), **150(2), 160(1).**

Node D. **36(1),** 142(2).

Node E. 32(1), **33(0), 62(1),** 83(2), **92(0), 96(1),** 106(2), **108(1),** 130(1).

Node F. 8(2), **117(1), 157(1).**

Dasypus + Stegotherium: **16(0),** 31(1), **36(3),** 39(1), **49(0), 50(1), 51(0),** 60(1), **67(0),** 69(1), **77(1),** 78(1), 92(1), 96(2), 136(0), **149(1).**

Zaedyus + Chlamyphorus: **2(3), 21(0), 50(1), 51(2), 142(0), 161(1), 162(0).**

PART TWO   Large-Scale Evolutionary Patterns

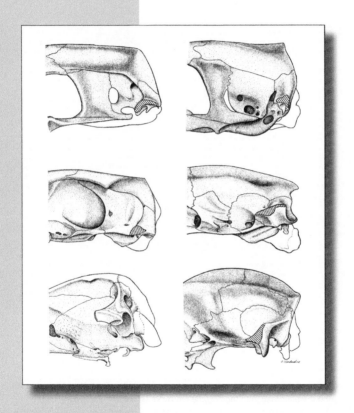

Plate 2. Transformation of the basicranium during mammalian evolution. Each illustration is in left lateral view, showing the braincase with the zygoma removed. Upper left: *Pachygenelus monus*. Upper right: *Morganucodon oehleri*. Center left: *Chulsanbataar vulgaris*. Center right: *Vincelestes neuquenianus*. Lower left: *Ornithorhynchus anatinus*. Lower right: *Didelphis marsupialis*. Illustration by Claire Vanderslice, from: J. R. Wible & J. A. Hopson (1993), Basicranial evidence for early mammal phylogeny; pp. 45–62 *in* F. S. Szalay, M. J. Novacek and M. C. McKenna (eds.), *Mammal Phylogeny. Mesozoic Differentiation, Multituberculates, Monotremes, Early Therians and Marsupials* (New York: Springer-Verlag).

7

The Origins of High Browsing and the Effects of Phylogeny and Scaling on Neck Length in Sauropodomorphs

J. Michael Parrish

Introduction

The giant sauropods of the Late Jurassic are among the most familiar dinosaurs, by virtue of their immense sizes and their improbably elongated necks. Although many other vertebrates, including plesiosaurs, tanyostropheids, and extant giraffes, also possess very long necks, the large size, abundance, and diversity of sauropods has made their striking body form a favorite subject for functional speculations since the earliest reasonably complete sauropod fossils were discovered in the latter part of the nineteenth century (e.g., Marsh, 1878). The juxtaposition of herbivory and the elongate neck in sauropodomorphs has naturally led to adaptationist arguments that suggest that, over time, continued extension of the neck among sauropodomorphs occurred to enable them to reach vegetation at greater and greater distances above the ground.

The purpose of this study is to evaluate the adaptive roles that body form and limb posture played in the evolution of herbivorous adaptations within the Sauropodomorpha. This question is considered from several different perspectives. First, maximum browsing heights of herbivorous tetrapods are estimated from the origin of terrestrial vertebrate herbivory through the appearance of the first sauropodomorphs in the Late Triassic to assess whether the elongate neck of sauropodomorphs might have conferred a selective advantage on the earliest members of the clade. Next, after an overview of sauropodomorph phylogeny, trends in relative neck length and cervical vertebral count are evaluated among sauropodomorphs to test the hypothesis that relative elongation of the neck among members of the clade is, as a general trend, size related and indicative of a positively allometric relationship between neck length and body size. Next, recent studies of feeding adaptations in sauropodomorphs are surveyed, and the possibility is considered that trends toward neck elongation throughout the sauropods may not be primarily adaptive for increased browsing height but instead may reflect a phylogenetic constraint that is expressed more strongly with phyletic size increase. Finally, the phylogenetic distribution of limb proportions among the basal members of the various dinosaur clades is considered to test the hypothesis that the basal locomotor pattern for sauropodomorphs was bipedal rather than quadrupedal.

Materials and Methods

To test the hypothesis that the synapomorphic elongation of the neck in sauropodomorphs could have conferred a selective advantage on them relative to other herbivores, maximum browsing heights were estimated for herbivorous tetrapods from their first appearance up to the appearance of large sauropodomorph in the Norian (fig. 7.1). In determining browsing heights of Paleozoic and Mesozoic herbivorous vertebrates, the relative size, inferred locomotor pattern, and potential mode of feeding were taken into account. The value for a given interval reflects the inferred maximum feeding height for any of the potentially herbivorous vertebrates known from that time period.

Next, to consider the relationships between neck length and body size, trends in neck length, vertebral count, and body size were examined for sauropodomorphs as well as for several dinosaur and archosaur outgroups. For this study, skeletal proportional data were collected from genera for which all or most of the presacral vertebral column and/or limb lengths are known. The quantities measured were the lengths of the cervical and dorsal sections of the vertebral column, diameter of the femur at midshaft, and combined lengths of humerus + ulna and femur + tibia.

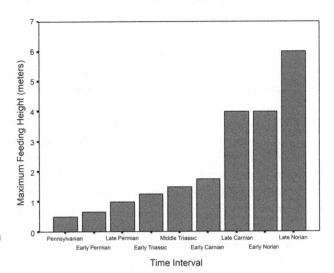

Figure 7.1. Distribution of herbivorous tetrapod browsing heights versus stratigraphic intervals.

The ratio of the summed forelimb and hind limb lengths is intended as an index of potential bipedal capability. In addition to a variety of sauropodomorphs, data were compiled for a basal archosaur (*Euparkeria*), basal ornithodirans (*Lagosuchus* and the Triassic pterosaur *Eudimorphodon*), basal dinosaurs (the herrerasaurids *Herrerasaurus* and *Eoraptor,* the coelophysid *Coelophysis,* and the basal ornithischian *Lesothosaurus*). Data (table 7.1) were either taken from direct measurements of specimens or compiled from monographs.

To look at the influences of phylogeny, body size, and mode of neck elongation, the data were graphed in several ways. First, to assess their allometric relationship, a plot of log neck lengths and log dorsal vertebral column lengths (fig. 7.4A) was prepared.

Because some sauropod taxa increased their cervical vertebral count through "cervicalization" of the most cranial dorsals, the possibility exists that the degree of neck elongation could be overemphasized when simply plotting neck length versus dorsal length, because each vertebra added to the neck results in the deletion of one from the trunk. Therefore, a second graph (fig. 7.4B) was created plotting log neck length versus the log of the sum of neck and dorsal column lengths, thus using total presacral column length as a linear estimator of body size.

Finally, femoral diameter was chosen as a third estimator of body size, with the assumption that it would be a more accurate representation of body mass, as opposed to body length, than would presacral vertebral column length. Because sauropod femora are generally not circular in cross section, this measurement was computed as the average of the anteroposterior and mediolateral widths of the femur at midshaft (fig. 7.4C).

To consider the effect of increased vertebral count on relative neck elongation, neck length was plotted against cervical count (fig. 7.4D). Finally, to assess the likelihood that sauropodomorphs were plesiomorphically bipedal or quadrupedal, the ratios of forelimb to hind limb lengths among sauropodomorphs and other dinosaurs are compared (table 7.2).

Results

Distribution of Browsing Heights among Terrestrial Tetrapods throughout the Paleozoic and Mesozoic

Vertebrate herbivory began in the latest Pennsylvanian with the edaphosaurid synapsids and diadectomorphs (Carroll & Sues, 2000). These squat, short-necked quadrupeds seem unlikely to have been capable of reaching browse more than a half meter above the ground

Table 7.1. Estimates of neck length and body length for Sauropodomorpha and selected outgroups known from relatively complete material

Taxon	Family	NeckL	Log NeckL	Cerv #	DorsalL	Log DorsL	Ratio (%)	TotL	Log TotL	NL/TotL	LNL/LTL	Source
Eudimorphodon	Pt	3.5	0.544068	9	6	0.778151	58	9.5	0.977724	0.368421	0.556464	Wild, 1978
Euparkeria	Ep	5.51	0.741152	9	13	1.113943	42	18.51	1.267406	0.297677	0.584778	Ewer, 1965
Lagosuchus	La	2.5	0.39794	9	4.3	0.633468	58	6.8	0.832509	0.367647	0.478001	Sereno and Arcucci, 1994
Lesothosaurus	Or	11.2	1.049218	9	25	1.39794	45	36.2	1.558709	0.309392	0.673133	Galton 1978
Herrerasaurus	Th	5	0.69897	9	12	1.079181	42	17	1.230449	0.294118	0.568061	Sereno and Novas 1992
Eoraptor	Th	11	1.041393	9	28	1.447158	39	39	1.591065	0.282051	0.654526	Sereno et al., 1993
Coelophysis	Tc	48.5	1.685742	10	42.5	1.628389	114	91	1.959041	0.532967	0.860493	Colbert, 1989
Massospondylus	Ma	84	1.924279	10	82	1.913814	102	166	2.220108	0.506024	0.86675	Cooper, 1984
Mussaurus	Pl	3.8	0.579784	10	7.4	0.869232	51	11.2	1.049218	0.339286	0.552586	Bonaparte and Vince, 1979
Thecodontosaurus	An	12	1.079181	10	16	1.20412	75	28	1.447158	0.428571	0.745725	Kermack, 1984
Riojasaurus	Me	189	2.276462	10	241	2.382017	78	430	2.633468	0.439535	0.864435	Bonaparte and Pumares, 1995
Anchisaurus	An	45	1.653213	10	57	1.755875	79	102	2.0086	0.441176	0.823067	Galton and Cluver, 1976
Saturnalia	Pr	41	1.612784		48	1.681241	85	89	1.94939	0.460674	0.827327	Langer et al., 1999

Taxon		NeckL			DorsalL			TotL		NL/ToTL	LNL/LTL	Reference
Plateosaurus	Pl	104	2.017033	10	140	2.146128	74	244	2.38739	0.42623	0.84487	Huene, 1926
Dicraeosaurus	Dc	227	2.356026	12	208	2.318063	109	435	2.638489	0.521839	0.892945	Janensch, 1929
Camarasaurus (juvenile)	Ca	103	2.012837	12	95	1.977724	108	198	2.296665	0.520202	0.876417	Gilmore, 1925
Camarasaurus	Ca	360	2.556303	12	240	2.380211	150	600	2.778151	0.6	0.920145	Osborn and Mook, 1921
Diplodocus	Di	643	2.808211	15	321	2.506505	200	964	2.984077	0.667012	0.941065	Hatcher, 1901
Apatosaurus	Di	574	2.758912	15	262	2.418301	219	836	2.922206	0.686603	0.944119	Gilmore, 1936
Omeisaurus	Eu	853	2.930949	17	225	2.352183	379	1078	3.032619	0.79128	0.966475	Young, 1939
Mamenchisaurus	Eu	946	2.975891	19	273	2.436163	347	1219	3.086004	0.776046	0.964319	Young and Zhao, 1972
Euhelopus	Eu	800	2.90309	17	300	2.477121	267					Wiman, 1929
Shunosaurus	Sh	267	2.426511	13	307	2.487138	87	574	2.758912	0.465157	0.879517	Zhang, 1988
Brachiosaurus	Br	868	2.93852	14	386	2.586587	225	1254	3.098298	0.692185	0.94843	Janensch, 1950
Amargasaurus	Di	239	2.378398	12	167	2.222716	143	406	2.608526	0.58867	122	Salgado and Bonaparte, 1991

Note: NeckL, total cervical column length (cm); DorsalL, total length of dorsal vertebral column; TotL, cervical + dorsal column lengths; NL/ToTL, NeckL/TotalL.; LNL/LTL, log neck length/log total length. An, Anchisauridae; Br, Brachiosauridae; Ca, Camarasauridae; Dc, Diplodocidae (Dicraeosavrinae); Di, Diplodocidae (Diplodocinae); Eu, Euhelopodidae; Ep, Euparkeridae; La, Lagosuchidae; Ma, Massospondylidae; Me, Melanorasauridae; O, basal Ornithischia; Pt, Pterosauria (Eudimorphodontidae); Pl, Plateosauridae; Pr, Prosauropoda (family indet.); Sh, *Shunosaurus* (familial associations disputed); Tc, Theropoda (Coelophysidae); Th, Theropoda or basal Dinosauria (Herrerasauridae).

Table 7.2. Forelimb and hind limb lengths and ratios for prosauropods and selected outgroups

Taxon	Family	H + R	Log FL	F + T	Log HL	Limb Ratio	Source
Eudimorphodon	Pt	7.6	0.880814	5.3	0.724276	1.433962	Wild, 1978
Euparkeria	Ep	7	0.845098	10.4	1.017033	0.673077	Ewer, 1965
Lagosuchus	La	4.6	0.662758	9.2	0.963788	0.5	Sereno & Arcucci, 1994
Lesothosaurus	Or	11	1.041393	25	1.39794	0.44	Galton, 1978
Herrerasaurus	Th	30	1.477121	66	1.819544	0.454545	Sereno & Novas, 1992
Eoraptor	Th	15	1.176091	29	1.462398	0.517241	Sereno et al., 1993
Coelophysis	Tc	18.5	1.267172	43.3	1.636488	0.427252	Colbert, 1989
Massospondylus	Ma	40	1.60206	74	1.869232	0.540541	Cooper, 1984
Vulcanodon	Me	39	1.591065	50	1.69897	0.78	Cooper, 1981
Mussaurus	Pl	4.8	0.681241	5	0.69897	0.96	Bonaparte & Vince, 1979
Riojasaurus	Me	63	1.799341	79	1.897627	0.797468	Bonaparte & Pumares, 1995
Melanorosaurus	Me	100	2	136	2.133539	0.735294	Van Heerden & Galton, 1997
Efraasia	Pl	72	1.857332	113	2.053078	0.637168	Galton & Cluver, 1976
Saturnalia	Pr	22	1.342423	34	1.531479	0.647059	Langer et al., 1999
Plateosaurus	Pl	67	1.826075	117	2.068186	0.57265	Huene, 1926

Note: Abbreviations as in Table 7.1; H + R, humerus length + radius length (cm); Log FL, log (humerus + radius lengths); F + T, femur length + tibia length (cm); Log HL, log (femur + tibia lengths).

(fig. 7.1). In the Permian, the diversity of plant-eating vertebrates and the diversity of their feeding mechanisms expanded (Sues & Reisz, 2001). Nonetheless the chief synapsid and diapsid herbivores still had squat, quadrupedal body plans, although the larger body sizes of groups such as the caseid synapsids and stahleckeriid dicynodonts resulted in the expansion of their potential vertical feeding zone to perhaps 1.5 m. With the appearance of rhynchosaurs in the Middle Triassic, herbivory emerged for the first time within the Archosauromorpha, and the great abundance of rhynchosaurs in Middle and Late Triassic sediments attests to their successful exploitation of the medium-sized quadrupedal herbivore niche.

With the appearance of the first undoubted prosauropods, including large forms such as *Euskelosaurus* and *Melanorosaurus,* in the Late Triassic (Late Carnian), the maximum browsing height for herbivores more than doubled. Not only did prosauropods have elongate necks, but their probable facultatively bipedal habits placed the dorsal vertebral column in a subvertical position, with the result that the base of the neck was elevated relative to the position in previous herbivorous vertebrates.

Thus, the elongation of the neck in the prosauropods, most notably in the plateosaurids and lufengosaurids, appears to have allowed them to exploit a large and hitherto untapped portion of the terrestrial plant biota (Weishampel, 1984; Galton, 1985). Taphonomic census data must be used with caution, but the relative abundance of prosauropod specimens in Carnian through Toarcian terrestrial deposits is striking, with at least seventy-five percent of the tetrapod specimens known from the Knollen-mergel (Norian) of Germany attributable to the single prosauropod genus *Plateosaurus* (Benton, 1983) and eighty-two percent of individuals from the upper part of the Lower Lufeng Series of China represented by *Lufengosaurus* (Galton, 1990).

Similar quantitative studies have not been done for Middle and Late Jurassic faunas, but sauropods are abundant in dinosaur-bearing sediments of these ages (Weishampel, et al., 2004). Sauropods were abundant worldwide during the Early Cretaceous but were absent from North America for most of the Late Cretaceous, although they remained the dominant large herbivores in the southern hemisphere (Wilson & Sereno, 1998). The Titanosauroidea, formerly thought to be a small radiation of morphologically conservative sauropods, now appear to have constituted a taxonomically and morphologically diverse clade that dominated the herbivore faunas of Gondwana in the Cretaceous (Curry Rogers & Forster, 2001; Jacobs et al., 1993; Salgado et al., 1997; Wilson, 2002).

Trends in Relative Neck Length among Archosaurs and within Sauropodomorpha

Relative neck length is variable among basal dinosauromorphs (*sensu* Padian, 1999). Both *Lagosuchus (*Sereno & Arcucci, 1994) and basal pterosaurs (e.g., Wild, 1978) have necks that are close to 60 percent of the length of the dorsal vertebral column (table 7.1, fig. 7.2). However the basal dinosaurs or basal theropods of the Family Herrerasauridae have necks roughly subequal in length to the dorsal series, in contrast to the longer necks in both coelophysoids (e.g., Colbert, 1989) and sauropodomorphs. In the basal ornithischian *Lesothosaurus,* the neck is 44 percent the length of the dorsal column, shorter than those in other basal dinosaurs. In adult sauropodomorphs, the ratio of neck length to dorsal column length ranges from 74 percent in *Plateosaurus* to 379 percent in *Omeisaurus* (table 7.1).

The sauropodomorph dinosaurs are among the most immediately recognizable clades of vertebrates by virtue of their relatively uniform and highly distinctive body form. The feature that most immediately

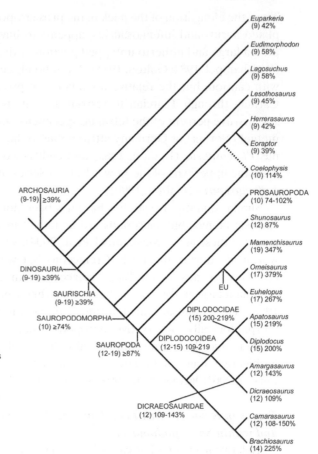

Figure 7.2. Cervical vertebral count (in parentheses) and ratios of neck length to presacral vertebral column length for Sauropodomorpha and selected outgroups graphed onto a cladogram based, in part, on Upchurch (1998) and Currie and Padian (1997).

characterizes sauropodomorphs is the elongation of the neck and tail through elongation of individual vertebrae and addition of extra vertebrae (through both creation of new elements and "cervicalization" of the most cranial dorsals). Cervical counts in sauropodomorph necks range from ten to nineteen in contrast to the nine cervicals found in most other dinosaurs (fig. 7.2), and the caudal count ranges from forty-five up to eighty or more in contrast to values in the low forties for basal theropods and ones ranging from forty to seventy in ornithischians (Mackovicky, 1999; McIntosh, 1999).

Even more striking than the increase in cervical count in sauropodomorphs is the relative elongation of the individual vertebrae. In basal archosaurs like *Euparkeria* (Ewer, 1965), the length of a midcervical is subequal to centrum height, whereas, in an adult *Plateosaurus* (Huene,

1926), the length of a comparable vertebra is nearly three times centrum height, and centrum length can reach more than four times centrum height in taxa such as *Omeisaurus* (Young, 1939).

Elongation of the tail is equally pronounced, most commonly by attenuation of individual elements rather than through a marked increase in vertebral count. In prosauropods, the number of caudal centra is generally close to the estimated plesiomorphic dinosaur count of fifty (Mackovicky, 1999). However, the number of caudals is much more variable in sauropods, ranging from thirty-four in *Opisthocoelicaudia* to well over eighty in diplodocids such as *Apatosaurus* (McIntosh, 1990).

Overview of Sauropodomorph Phylogeny

Sauropodomorph dinosaurs comprise two distinct groups, the Prosauropoda and Sauropoda, which most recent phylogenetic studies (fig. 7.3; Upchurch, 1995, 1998; Wilson & Sereno, 1998; Wilson, 2002) treat as separate monophyletic lineages. Historically, the Prosauropoda, which range from the late Carnian through the end of the Early Jurassic, were considered to be the temporal as well as phylogenetic precursors of the Sauropoda. However, the recent discoveries Late Triassic sauropods from Thailand (Buffetaut et al., 2000; 2002), and South Africa (Yates and Kitching, 2003) makes the first appearances in the fossil record of the two clades more nearly contemporaneous, as has been predicted by contemporary phylogenies (e.g., Wilson, 2002; Upchurch, 1998; Yates and Kitching, 2003).

Nonetheless, the Prosauropoda are by far the more abundantly represented sauropodomorph clade during the Carnian-Toarcian interval; hundreds of specimens are known from this interval (Galton, 1990), as contrasted with a handful of sauropod specimens, all but *Isanosaurus* coming from the Early Jurassic (McIntosh, 1990). Another striking difference is in both overall size and relative neck elongation, with adult prosauropods usually being smaller and overlapping the size range of sauropods in only a few instances within the Family Melanorosauridae (table 7.1). Another striking difference is observed between limb proportions in prosauropods and sauropods. In most prosauropods, the forelimbs are markedly shorter than the hind limbs, whereas even the earliest completely known sauropods have forelimbs nearly equal in length to their hind limbs. In the largest prosauropods, melanorosaurids such as *Riojasaurus,* the hind limb/forelimb proportions (range sixty-four to eighty percent) more nearly approximate those of sauropods. Using hind limb/trunk ratios, Galton (1990) hypothesized that the melanorosaurids

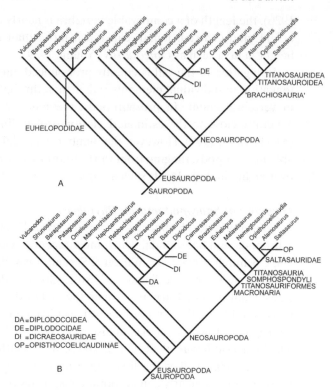

Figure 7.3. Phylogenetic relationships of Sauropoda. (A) Upchurch (1998). (B) Wilson (2002).

were obligate quadrupeds, whereas the remaining prosauropods for which complete postcrania are known appear to have had the capability of being at least facultative bipeds, and these ratios are consistent with this hypothesis.

Although many recent authors have hypothesized that sauropods may have had the capability of rearing tripodally (e.g., Bakker, 1986) their locomotion generally has been considered obligatorily quadrupedal (but, see Paul, 1998). Upchurch and Barrett (2000) list the elongation of the forelimbs relative to hind limbs as an apomorphy of the Sauropoda.

The recent phylogenetic analyses of the Sauropoda (fig. 7.3; Upchurch 1995, 1998; Wilson, 2002; Wilson & Sereno, 1998) all make a distinction between basal sauropods, including *Vulcanodon* and *Barapasaurus,* and the Eusauropoda, a group comprising most of the Middle Jurassic-Late Cretaceous sauropod lineages. Wilson and Sereno (1998) diverged from Upchurch (1998) in their placement of the sauropods from the Middle Jurassic of China. Upchurch (1998) placed the taxa *Euhelopus, Shunosaurus, Omeisaurus,* and *Mamenchisaurus* together as a monophyletic

Euhelopodidae that splits off from other Eusauropoda at the base of the clade. In Wilson's (2002) analysis, the Chinese sauropods are paraphyletic, with the short-necked *Shunosaurus* branching off at the base of the Eusauropoda, an *Omeisaurus–Mamenchisaurus* clade emerging as the sister-taxon of the Neosauropoda (a clade apparent in both studies comprising the common ancestor of *Diplodocus, Saltasaurus,* and all its descendants) and *Euhelopus* nested high in the cladogram as the sister-taxon of the Titanosauria. Otherwise, the two phylogenies are in substantial agreement, with a monophyletic Diplodocidae (Upchurch, 1998) or Diplodocoidea (Wilson, 2002) composing two distinct clades. The first, represented by *Diplodocus, Apatosaurus,* and *Barosaurus,* and is called Diplodocinae by Upchurch (1998) and Diplodocidae by Wilson (2002). The second, consisting of short-necked sauropods with elongate neural spines, and comprising *Amargasaurus* and *Dicraeosaurus,* is called Dicreaosaurinae by Upchurch (1998) and Dicraeosauridae by Wilson (2002). In both studies, the *Diplodocus*/*Dicraeosaurus* cluster is the sister-taxon of a second clade comprising *Camarasaurus,* Brachiosauridae, and Titanosauroidea, within which *Camarasaurus* is the sister-taxon of the other two groups.

Wilson and Sereno (1998) discussed sauropod neck structure within their phylogenetic construct and argued for thirteen being the primitive cervical number for Eusauropoda, resulting in plasticity of cervical number from twelve to seventeen within that group, entailing multiple reductions and increases in vertebral count, a conclusion that is also supported by the Upchurch (1998) cladogram. Therefore, although an increase or decrease in cervical count may characterize particular lineages (such as Euhelopodidae *sensu* Upchurch, 1998; or the Diplodocidae *sensu* Wilson, 2002), no prevailing trend toward an increase or reduction in vertebral count is apparent within eusauropods as a whole.

Previous Hypotheses involving Neck Structure, Function, and Feeding in Sauropodomorpha

Most previous studies of sauropodomorph neck structure and function have employed adaptationist scenarios, including the idea that the long neck served as a snorkel to permit aquatic sauropods to continue breathing while their trunks were under water (e.g., Cope, 1878; Hatcher, 1901), that greater elongation of the neck and/or greater elevation of the anterior trunk (in *Brachiosaurus*) permitted sauropods to graze higher in the forest canopy (e.g., Riggs, 1904; Bakker, 1971; Upchurch, 1994; Upchurch & Barrett, 2000), or that cervical elongation served primarily to permit a greater lateral sweep of the neck from one standing position (Martin,

1987; Barrett & Upchurch, 1994; Upchurch & Barrett, 2000; Stevens & Parrish, 1999). A prevailing idea within these studies is the concept that the elongate necks of sauropods were the product of direct selection for some functional trait that the neck facilitated. In this study, the goal is to correlate phylogenetic and temporal patterns of neck elongation in sauropodomorphs with body size to test the hypothesis that the greater neck elongation in some sauropods may be primarily a scaling effect.

The mode of feeding in sauropodomorphs has received much attention in recent years through studies focusing on general feeding and paleoecology (e.g., Bakker, 1971; Barrett, 2000; Barrett & Upchurch, 1994, 1995; Coombs, 1975; Cooper, 1981; Dodson, 1990; Farlow, 1987; Galton, 1985, 1986, 1990; Upchurch & Barrett, 2000), dental microwear (Fiorillo, 1991, 1998; Upchurch & Barrett, 2000), jaw mechanics (Calvo, 1994; Barrett & Upchurch, 1994, 1995), and overall body proportions (e.g., Martin, 1987; Stevens & Parrish, 1999). Although a few authors have concluded that prosauropods were carnivorous (e.g., Swinton, 1934; Cooper, 1981), most studies have concluded that they were either wholly herbivorous (e.g., Galton, 1985, 1990) or omnivorous (Barrett, 2000). Although Barrett (2000) compared the tooth forms of prosauropods with those of the omnivorous extant iguanine lizards, microwear studies of prosauropod teeth comparable to those done on sauropods have not been performed.

The large number of sympatric genera of sauropods, particularly in the Late Jurassic of North America, has led several workers to hypothesize that these taxa utilized different feeding strategies to partition plant resources (e.g., Stevens & Parrish, 1999; Fiorillo, 1991, 1998; Upchurch & Barrett, 2000). Fiorillo's (1998) work concluded that *Camarasaurus* had a stronger bite and consumed harder food than *Diplodocus*. Barrett and Upchurch (1994) and Upchurch and Barrett (2000) concluded that diplodocids had a unique feeding mechanism relative to those of other sauropods, involving wholesale stripping of vegetation from branches, and suggested, based primarily on dental and cranial evidence, that sauropods as a group apomorphically developed the ability to crop and consume low-quality forage.

Stevens and Parrish (1999, 2005) have analyzed neck flexibility among Late Jurassic sauropods and have concluded that, although cervical flexibility among this group is variable, all the taxa studied thus far (*Apatosaurus, Diplodocus, Camarasaurus, Euhelopus,* and *Brachiosaurus*) overlap broadly in the feeding envelopes determined by establishing the limits of flexibility defined by their articulated necks (Parrish & Stevens, 1998; Stevens & Parrish, 1999, 2001, 2005). Another counterintuitive result of this study was the discovery that upward mobility of the cervical columns of sauropods is limited by the mechanics of the joint capsules connecting the pre- and post-zygapophyses of the cervical vertebrae, with the apparent result that dorsiflexion of the neck was restricted to the point that the maximum height a sauropod head could reach in a quadrupedal stance was little, if any, higher than those of prosauropods.

Relationship between Neck Length and Body Size among Sauropodomorpha

When log neck length is plotted against the log of the length of the dorsal vertebral column, a tightly constrained monotonic trend for two parameters, significant at the 99 percent level ($r^2 = 0.94$), indicates that neck length has an allometric coefficient of 1.35 relative to dorsal length. This means that not only do the largest sauropodomorphs tend to have the longest necks relative to presacral length, but that the smallest (notably the juvenile specimen of the plateosaurid *Mussaurus;* Bonaparte and Vince, 1979) has the lowest neck-to-trunk ratio (34 percent) of any member of the clade.

When neck length is plotted against total presacral column length, the correlations between both the raw data and the log transformed data are highly significant ($r^2 = 0.99$). The allometric coefficient for neck length relative to total presacral length is smaller than the relationship to dorsal length alone (1.15 versus 1.35), and this decrease appears to reflect the buffering effect introduced by having increased neck length incorporated into the overall presacral length on the x axis.

Plotting log neck length versus the log of femoral midshaft diameter yields a similar positive allometry, with an allometric coefficient, also significant at the 99 percent level ($r^2 = 0.93$) of 1.26. Therefore, the presence of a consistent relationship between increased body size and greater neck length among sauropodomorphs is robust regardless of which estimator of body size is used.

Phylogenetic Distribution of Neck Patterns among Sauropodomorphs

Although a consistent, positively allometric relationship is maintained between presacral column length and neck length in sauropodomorphs, clade-specific trends are also apparent within this larger group. Taxa with the longest necks relative to dorsal column length are *Brachiosaurus,* the euhelopodids, and the diplodocine diplodocids (table 7.1, fig. 7.4). When cervical count is plotted against the ratio of neck length to presacral length (fig.7.4D), it is apparent that the highest ratios of neck length to presacral length are observed among taxa that have the greatest number of cervical vertebrae, although significant variation is observed in ratios among taxa

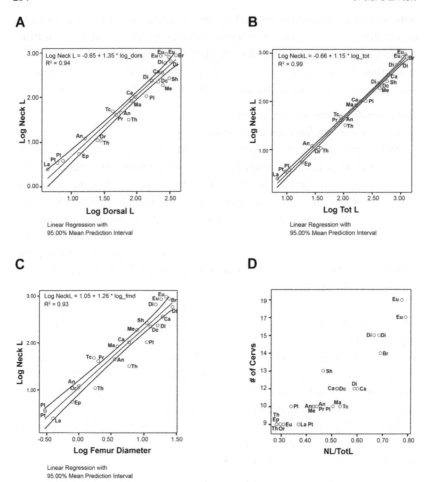

Figure 7.4. (A) Log neck length versus log dorsal vertebral column length. (B) Log neck length versus log total presacral column length. (C) Log neck length versus log femoral diameter. (D) Neck length/dorsal column length versus vertebral count. Abbreviations as in table 7.1.

with intermediate numbers (ten to twelve) of cervical elements. The most significant outliers in terms of both cervical count and neck length relative to presacral length are the Euhelopodidae *sensu* Upchurch (1998), raising the possibility that this group, which thus far is known only from the Jurassic of China, may have been subject to selection pressures and/or phylogenetic constraints unique among the sauropods. Certainly the anatomical variety of dental and cervical arrangements among sauropods suggests that clade-specific modifications occurred in cervical structure, and some of these modifications may well be related to feeding. However, our studies (Stevens and Parrish, 1999, 2005) do not indicate any notable variation in vertical partitioning of feeding envelopes among neosauropods (but, see Barrett & Upchurch, 1995; Christiansen, 2000).

Implications for Browsing in Sauropods in Light of the Record of Terrestrial Fossil Plants

Speculations of the potential correlation between clades of herbivorous dinosaurs and their preferred plant food sources have been widespread, particularly during the last two decades (e.g., Bakker, 1978; Barrett & Upchurch, 1994; Barrett & Willis, 2001; Farlow, 1987; Fiorillo, 1991, 1998; Sues, 2000; Tiffney, 1992; Weishampel, 1984; Weishampel & Norman, 1989; Wing & Tiffney, 1987; Ziegler et al., 1993; Rees et al., 2004). A frequently posed hypothesis is that differential neck forms among sauropod genera allowed niche partitioning (e.g., Dodson, 1990; Upchurch & Barrett, 2000), particularly when several sauropod taxa occurred sympatrically, as was the case in the Late Jurassic of North America (e.g., Weishampel, 1990). Several recent studies (Barrett & Upchurch, 1994; Calvo, 1994; Fiorillo, 1991, 1998; Upchurch & Barrett, 1995, 2000) have shown differences in cranial and/or dental structure in Jurassic to Cretaceous sauropods that are consistent with differential diets among brachiosaurids, camarasaurids, and diplodocids. Other studies (Stevens & Parrish, 1999; Parrish et al., 2004) have pointed out that the most abundant foliage that appears appropriate for sauropod feeding, in terms of both nutritional value and rapid resource renewal, occurred between ground level and 6 m and thus overlapped the feeding range observed among some of the larger prosauropods. This factor, taken in conjunction with recent mechanical studies (Martin, 1987; Martin et al., 1998; Stevens & Parrish, 1999, 2005) that demonstrate limited dorsal mobility within sauropod cervical columns, suggests that niche partitioning among taxa was not a direct result of differential feeding heights as determined by simple differences in the absolute length of the neck in different sauropod lineages.

Long Necks as a Phylogenetic Constraint and Alternative Functional Scenarios for Sauropodomorph Browsing

The data in table 7.1 suggest that the plesiomorphic pattern for Ornithodira is for a neck that is roughly 60 percent of the length of the dorsal vertebral column. When log neck length is plotted against the log of cervical plus dorsal columns (fig. 7.4B), the strong correlation between these two variables ($r = 0.99$) suggests that the greater neck length of large sauropods represents the expression of a positively allometric relationship between neck length and body length that is apparent among sauropodomorphs. Therefore, it appears that the greater elongation of the neck in sauropods may be largely, if not entirely, a reflection of positive allometry of neck length relative to body length.

One paradox this relationship raises involves the increase in vertebral count within the sauropodomorphs. Were this a simple developmentally mediated relationship between body size and neck length, the observed increase in cervical count within the Neosauropoda (Upchurch & Barrett, 2000) would not have been predicted. In all the long-necked sauropods (Euhelopodidae, Brachiosauridae, Diplodocidae), significant elongation of individual vertebrae co-occurs with an increase in vertebral count, so increased vertebral count may be simply an alternative solution to maintaining some mobility within the neck while preserving this scaling relationship with increasing size.

Clearly a similar scaling relationship is not observed in the theropods or within any of the ornithischian lineages, although the proportions of basal theropods, ornithischians, and nondinosaurian ornithodirans converge at and below a total cervical plus dorsal length of 1 m. Therefore, the pattern seen within the Sauropodomorpha appears to represent not only a scaling effect but also a phylogenetic constraint. In most theropods, the neck is subequal in length to the dorsal vertebral column, although the neck is slightly shorter than the trunk in large theropods such as *Tyrannosaurus* (56 percent) and *Carnotaurus* (43 percent), suggesting either an overall negative allometry or clade-specific phylogenetic constraints. By contrast, the neck remains notably shorter than the dorsal trunk in ornithischians, approaching the length of the trunk only in some large ornithopods such as *Ouranosaurus* (Taquet, 1976). Stegosaurs, notably *Kentrosaurus* (Galton, 1982; ratio = 0.24), have some of the shortest necks relative to trunk length among dinosaurs. Thus, the positively allometric trend observed in sauropodomorphs appears to be unique to that clade among dinosaurian lineages.

The Plesiomorphic Locomotor Patterns for Sauropodomorpha and Dinosauria

The question of whether the most basal dinosaurs were quadrupeds or bipeds has not been fully resolved, largely because of the phylogenetic controversies involving relative placement of taxa like Herrerasauridae and Prosauropoda. However, a consideration of the most basal members of the various dinosaurian clades and the putative dinosaur outgroups (table 7.2) offers some insights into this question. The most basal (and one of the earliest) well-known ornithischian, *Lesothosaurus,* had a fore-limb to hind limb ratio of 0.44, comparable to the ratios in *Coelophysis* (0.43) and *Herrerasaurus* (0.44). Among the Thyreophora, *Scutellosaurus* and *Scelidosaurus* are contemporaneous, and the former appears to have been a biped (Colbert, 1981). All ornithopods have been interpreted traditionally as bipeds with the possible exception of some hadrosaurs, and the basal members of both marginocephalian clades [Pachycephalosauria (Maryanska, 1990) and Psittacosauridae (Sereno, 1990)] are also generally considered to have been bipeds, so it appears most parsimonious to assume that bipedalism was also the plesiomorphic condition for Ornithischia.

Given that the basal condition for Theropoda is also a bipedal stance, as are those of the two most generally recognized dinosaurian outgroups, Lagosuchidae and Pterosauria, the case for the plesiomorphic condition for Sauropodomorpha also being bipedal seems very strong, particularly given the limb length disparity observed in prosauropod taxa such as *Anchisaurus* and *Plateosaurus.* The sole specimen of the earliest sauropod, *Isanosaurus,* is too incomplete to form the basis of a postural hypothesis, but the much better known basal sauropods, *Barapasaurus* and *Vulcanodon* (Cooper, 1981), both exhibit limb ratios that strongly suggest quadrupedality. Thus, it is unclear whether quadrupedalism was plesiomorphic for Sauropodomorpha or whether, as in the Melanorosauridae, an obligate quadrupedal stance was a morphological correlate of increased size within the two sauropodomorph clades. Yates and Kitching (2003) recently suggested that several taxa previously grouped as prosauropods, including the clearly bipedal *Anchisaurus,* were instead basal sauropods, which would also support the hypothesis that bipedalism was plesiomorphic for Sauropoda.

Conclusions

What appears to be the most plausible scenario is that all dinosaur lineages began with bipedal taxa, with all clades except for Theropoda moving toward a facultative and sometimes obligatory quadrupedal pose with increasing size and/or ecological specialization. Within the Sauropodomorpha, the initial elongation of the neck with size increase within the Prosauropoda appears to have made available to them a large amount of forage located between 2 and 6 m off the ground. As size increase resulted in various sauropodomorph lineages becoming facultative quadrupeds, the positively allometric relationship between neck length and presacral column length was maintained, resulting in increasingly elongate necks that nonetheless appear to have been limited in their capability of dorsal flexion (e.g., Stevens & Parrish, 1999, 2005), and at least the North American Jurassic sauropods overlapped widely in the geometry of their feeding envelopes (Stevens & Parrish, 2001, 2005). Nonetheless, the differential lengths and mechanical properties of sauropodomorph necks, taken in conjunction with demonstrated differences in cranial and dental structure, indicate that some niche partitioning did occur in feeding modes among Late Jurassic sauropods.

Acknowledgments

This study has been improved by conversations with Matt Bonnan, Dan Chure, Peter Dodson, Phil Senter, and Kent Stevens. Thoughtful reviews by Paul Barrett, Matt Carrano, Rick Blob, and an anonymous reviewer significantly improved the manuscript. Carl von Ende and Jim Robins helped with the statistical analyses, and Barb Ball and David Allen were instrumental in figure preparation. I am also grateful to the curatorial staffs at the American Museum, the Carnegie Museum, the Field Museum of Natural History, the Humboldt Museum, and the Museum am Lowentor for access to specimens. This project was funded in part by NSF Grant IBN-009329.

I am immensely grateful to Jim Hopson for taking me on as a fledgling graduate student back in 1977. Jim introduced me and my cohort of graduate students to the values of studying vertebrate evolution in a phylogenetic context and to the importance of generating testable hypotheses in paleontological inference. Since my own stint as a graduate student at the University of Chicago, I have watched Jim similarly mentor crop after crop of what have proved to be some of the most eminent workers in vertebrate paleontology, most of whom are also contributors to this book. His contributions to the field, through his own research and that of his students, have represented a major part of the advances in vertebrate paleontology that occurred during the last part of the twentieth century and beyond.

Literature Cited

Bakker, R. T. 1971. The ecology of the brontosaurs. *Nature* 229:172–174.
———. 1978. Dinosaur feeding behavior and the origin of flowering plants. *Nature* 274:661–663.
———. 1986. *Dinosaur Heresies.* New York: William Morrow.

Barrett, P. M. 2000. Speculations on prosauropod diets; pp. 42–78 *in* H.-D. Sues (ed.), *Evolution of Herbivory in Terrestrial Vertebrates: Perspectives from the Fossil Record.* Cambridge: Cambridge University Press.

Barrett, P. M. and P. Upchurch. 1994. Feeding mechanisms of *Diplodocus. GAIA* 10:195–204.

———. 1995. Sauropod feeding mechanisms: their bearing on palaeoecology. Pp. 107–110 in A. Sun and Y-Q. Wang (eds.). *Sixth Symposium on Mesozoic Terrestrial Ecosystems and Biotas, Short Papers.* Beijing: China Ocean Press.

Barrett, P. M. and K. J. Willis. 2001. Did angiosperms invent flowers? Dinosaur-andiosperm coevolution revisited. *Biological Reviews* 76:411–447.

Benton, M. J. 1983. Dinosaur success in the Triassic: a noncompetitive ecological model. *Quarterly Review of Biology* 58:29–55.

Bonaparte, J. F. and R. A. Coria. 1993. Un nuevo y gigantesco sauropodo Titanosaurio de la Formacio Rio Lemay (Albanio-Cenomaniano) de la provincia del Neuquen, Argentina. *Ameghiniana* 30:271–282.

Bonaparte, J. F. and J. A. Pumares. 1995. Notas sobre primer craneo de *Riojasaurus incertus* (Dinosauria: Prosauropoda, Melanorosauridae) del Triasico Superior de La Rioja, Argentina. *Ameghiniana* 32:341–349.

Bonaparte, J. F. and M. Vince. 1979. El hallazgo del primer nido de dinosaurios Triasicos (Saurischia: Prosauropoda), Triasico Superior de Patagonia, Argentina. *Ameghiniana* 16:173–182.

Buffetaut, E., V. Suteethorn, G. Cuny, H. Tong, J. Le Loeuff, S. Khansubha, and S. Jongautchariyakul. 2000. The earliest known sauropod dinosaur. *Nature* 407:72–74.

———, J. Le Loeuff, G. Cuny, H. Tong, and S. Khanshuba. 2002. The first giant dinosaurs: a large sauropod from the Late Triassic of Thailand. *Comptes Rendus Palevol* 1:103–109.

Calvo, J. O. 1994. Jaw mechanics in sauropod dinosaurs. *GAIA* 10:183–194.

Carroll, R. L. and H.-D. Sues. 2000. Herbivory in late Paleozoic and Triassic terrestrial vertebrates; pp. 9–41 *in* H.-D. Sues and N. C. Fraser (eds.), *Evolution of Herbivory in Terrestrial Vertebrates: Perspectives from the Fossil Record.* Cambridge: Cambridge University Press.

Christiansen, P. 2000. Feeding mechanisms of the sauropod dinosaurs *Brachiosaurus, Camarasaurus, Diplodocus,* and *Dicraeosaurus. Historical Biology* 14:137–152.

Colbert, E. H. 1981. A primitive ornithischian dinosaur from the Kayenta Formation of Arizona. *Bulletin of the Museum of Northern Arizona* 53:1–61.

———. 1989. The Triassic dinosaur *Coelophysis. Bulletin of the Museum of Northern Arizona* 57:1–174.

Coombs, W. P. 1975. Sauropod habits and habitats. *Palaeogeography, Palaeoclimatology, Palaeoecology* 17:1–33.

Cooper, M. A. 1981. The prosauropod dinosaur *Massospondylus carinatus* Owen from Zimbabwe: its biology, mode of life, and phylogenetic significance. *Occasional Papers of the National Museum and Monuments of Rhodesia, Series B* 6:689–840.

———. 1984. A reassessment of *Vulcanodon karibaensis* Raath (Dinosauria: Saurischia) and the origin of the Sauropoda. *Paleontologia Africana* 25:203–231.

Cope, E. D. 1878. On the saurians of the Dakota Cretaceous of Colorado. *Nature* 18:476.

Currie, P. J. and K. Padian. 1997. *Encyclopedia of Dinosaurs.* New York: Academic Press.

Curry Rogers, K. A. and C. A. Forster. 2001. The last of the dinosaur titans: a new sauropod from Madagascar. *Nature* 412:530–534

Dodson, P. M. 1990. Sauropod paleoecology; pp. 402–407 *in* D. B. Weishampel, P. Dodson, and H. Osmólska (eds.), *The Dinosauria.* Berkeley: University of California Press.

Ewer, R. F. 1965. The anatomy of the thecodont reptile *Euparkeria capensis* Broom. *Philosophical Transactions of the Royal Society of London B* 248: 379–435.

Farlow, J. O. 1987. Speculations about the diet and digestive physiology of herbivorous dinosaurs. *Paleobiology* 13:60–72.

Fiorillo, A. R. 1991. Dental microwear on the teeth of *Camarasaurus* and *Diplodocus:* implications for sauropod paleoecology. *Contributions from the Paleontological Museum, University of Oslo* 364:23–24.

———. 1998. Dental microwear patterns of sauropod dinosaurs *Camarasaurus* and *Diplodocus:* evidence for resource partitioning in the Late Jurassic of North America. *Historical Biology* 13:1–16.

Galton, P. M. 1978. Fabrosauridae, the basal family of ornithischian dinosaurs (Reptilia: Ornithopoda). *Paläontologische Zeitschrift* 52:138–159.

———. 1982. The postcranial anatomy of the stegosaurian dinosaur *Kentrosaurus* from the Upper Jurassic of Tanzania. *Geologica et Paleontologica* 15:139–160.

———. 1985. Diet of prosauropod dinosaurs from the Late Triassic and Early Jurassic. *Lethaia* 18:105–123.

———. 1986. Herbivorous adaptations of Late Triassic and Early Jurassic dinosaurs; pp. 203–221 *in* K. Padian (ed.), *The Beginning of the Age of Dinosaurs.* Cambridge: Cambridge University Press.

———. 1990. Basal Sauropodomorpha—Prosauropoda; pp. 320–344 *in* D. B. Weishampel, P. Dodson, and H. Osmólska (eds.), *The Dinosauria.* Berkeley: University of California Press.

Galton, P. M. and M. J. Cluver. 1976. *Anchisaurus capensis* and a revision of the Anchisauridae (Reptilia: Saurischia). *Annals of the South African Museum* 69:121–159.

Gilmore, C. W. 1925. A nearly complete articulated skeleton of *Camarasaurus,* a saurischian dinosaur from the Dinosaur National Monument. *Memoirs of the Carnegie Museum* 10:347–384.

————. 1936. Osteology of *Apatosaurus,* with special reference to specimens in the Carnegie Museum. *Memoirs of the Carnegie Museum* 11:175–300.

Hatcher, J. B. 1901. *Diplodocus* (Marsh): its osteology, taxonomy, and probable habits, with a reconstruction of the skeleton. *Memoirs of the Carnegie Museum* 1:1–63.

Huene, F. von. 1926. Vollstandige osteologie eines plateosauriden aud dem Schwabische Keuper. *Geologische und Paläontologische Abhandlungen* 15:129–179.

Jacobs, L. L., D. A. Winkler, W. R. Downs, and E. M. Gomani. 1993. New material of an Early Cretaceous sauropod dinosaur from Africa. *Palaeontology* 36:523–534.

Janensch, W. 1929. Die Wirbelsaule der gattung *Dicraeosaurus. Paleontographica (Supplement 7)* 3:39–133.

Janensch, W. 1950. Die skelettrekonstruction von *Brachiosaurus brancai. Palaeontographica (Supplement 7)* 3:97–103.

Kermack, D. 1984. New prosauropod dinosaur from South Wales. *Zoological Journal of the Linnean Society* 82:101–117.

Krassilov, V. A. 1981. Changes of Mesozoic vegetation and the extinction of the dinosaurs. *Palaeogeography, Palaeoclimatology, Palaeoecology* 34:207–224.

Langer, M., F. Abdala, M. Richter, and M. J. Benton. 1999. A sauropodomorph dinosaur from the Upper Triassic (Carnian) of southern Brazil. *Comptes Rendus de l'Académie des Sciences à Paris, Sciences de la Terre et des Planètes* 329:511–517.

Mackovicky, P. 1999. Postcranial axial skeleton; pp. 579–590 *in* P. J. Currie and K. Padian (eds.), *Encyclopedia of Dinosaurs.* San Diego: Academic Press.

Marsh, O. C. 1878. Principal characters of American sauropod dinosaurs Part 1. *American Journal of Science, Series 3* 14:411–416.

Martin, J. 1987. Mobility and feeding of *Cetiosaurus:* why the long neck? *Occasional Paper of the Tyrell Museum of Paleontology* 3:150–155.

————, V. Martin-Rolland, and E. Frey. 1998. Not cranes or masts, but beams: the biomechanics of sauropod necks. *Oryctos* 1:113–120.

Maryanska, T. 1990. Pachycephalosauria; pp. 564–577 *in* D. B. Weishampel, P. Dodson, and H. Osmólska (eds.), *The Dinosauria.* Berkeley: University of California Press.

McIntosh, J. S. 1990. Sauropoda; pp. 345–401 *in* D. B. Weishampel, P. Dodson, and H. Osmólska (eds.), *The Dinosauria.* Berkeley: University of California Press.

————. 1999. Sauropoda; pp. 654–658 *in* P. J. Currie and K. Padian (eds.), *Encyclopedia of Dinosaurs.* San Diego: Academic Press.

Osborn, H. F. and C. C. Mook. 1921. *Camarasaurus, Amphicoelias,* and other sauropods of Cope. *Memoirs of the American Museum of Natural History, new series,* 3:249–386.

Padian, K. 1999. Phylogeny of dinosaurs; pp. 546–551 *in* P. J. Currie and K. Padian (eds.), *Encyclopedia of Dinosaurs.* San Diego: Academic Press.

Parrish, J. M. and K. A. Stevens. 1998. Undoing the death pose: using computer imaging to restore the posture of articulated dinosaur skeletons. *Journal of Vertebrate Paleontology* 18:69A.

Parrish, J. T., F. Peterson, and C. Turner. 2004. Jurassic "savannah"—plant taphonomy and climate of the Morrison Formation (Jurassic, Western USA). *Sedimentary Geology* 167:137–162.

Paul, G. S. 1998. Differing bipedal and tripodal feeding modes in sauropods. *Journal of Vertebrate Paleontology* 18:69A.

Rees, P. M., C. R. Noto, J. M. Parrish, and J. T. Parrish. 2004. Late Jurassic climates, vegetation, and dinosaur distributions. Journal of Geology 112: 643–653.

Riggs, E. S. 1904. Structure and relationship of the opisthocoelian dinosaurs. Pt. II. Brachiosauridae. *Publications of the Field Columbian Museum, Geological Series* 2:229–248.

Salgado, L. and J. F. Bonaparte. 1991. Un nuevo sauropodo Dicraeosauridae, *Amargasaurus cazaui,* gen. et. sp. nov., de la Formación la Amarga, Neocomiano de la Province Neuquén, Argentina. *Ameghiniana* 28:333–346.

Salgado, L., R. A. Coria and J. O. Calvo. 1997. Evolution of titanosaurid sauropods. I. Phylogenetic analysis based on the postcranial evidence. *Ameghiniana* 34:3–32.

Sereno, P. C. 1990. Psittacosauridae; pp. 579–592 *in* D. B. Weishampel, P. Dodson, and H. Osmólska (eds.), *The Dinosauria.* Berkeley: University of California Press.

Sereno, P. C. and A. B. Arcucci. 1994. Dinosaurian precursors from the Middle Triassic of Argentina: *Marasuchus lilloensis* gen. nov. *Journal of Vertebrate Paleontology* 14:53–73.

Sereno, P. C. and F. E. Novas. 1992. The complete skull and skeleton of an early dinosaur. *Science* 258:1137–1140.

Sereno, P. C., C. A. Forster, R. R. Rogers and A. M. Monetta. 1993. Primitive dinosaur skeleton from Argentina and the early evolution of the Dinosauria. *Nature* 361:64–66.

Stevens, K. A. and J. M. Parrish. 1999. Neck posture and feeding habits of two Jurassic sauropod dinosaurs. *Science* 284:798–800.

———. 2001. Biological implications of digital reconstructions of the whole body of sauropod dinosaurs. *Journal of Vertebrate Paeontology* 22:104A.

———. 2005. Digital reconstructions of sauropod dinosaurs and implications for feeding; pp. 178–200 *in* K. A. Curry Rogers and J. A. Wilson (eds.), *The Sauropods: Evolution and Paleobiology.* Berkeley: University of California Press.

Sues, H.-D. (ed.). 2000. *Evolution of Herbivory in Terrestrial Vertebrates: Perspectives from the Fossil Record.* Cambridge: Cambridge University Press.

Sues, H.-D. and R. R. Reisz. 2001. Origins and early evolution of herbivory in tetrapods. *Trends in Ecology and Evolution* 13:141–145.

Swinton, E. D. 1934. *Dinosaurs: a Short History of a Great Group of Extinct Reptiles.* London: Thomas Mubry and Company.

Taquet, P. 1976. Geologie et paléontologie du gisement de Gadoufauna (Aptien du Niger). *Cahiers du Paleontologie, CNRS Paris* 1–191.

Tiffney, B. H. 1992. The role of vertebrate herbivory in the evolution of land plants. *Paleobotanist* 41:87–97.

Upchurch, P. 1994. Sauropod phylogeny and paleoecology. Gaia 10:249–260.

———. 1995. The evolutionary history of sauropod dinosaurs. *Philosophical Transactions of the Royal Society of London, Series B* 349:365–390.

———. 1998. The phylogenetic relationships of sauropod dinosaurs. *Zoological Journal of the Linnean Society* 124:43–103.

Upchurch, P. and P. M. Barrett. 2000. The evolution of sauropod feeding mechanisms; pp. 79–122 *in* H.-D. Sues (ed.), *Evolution of Herbivory in Terrestrial Vertebrates: Perspectives from the Fossil Record.* Cambridge: Cambridge University Press.

Van Heerden, J. and P. M. Galton. 1997. The affinities of *Melanorosaurus*—a Late Triassic prosauropod dinosaur from South Africa. *Neues Jahrbuch für Geologie und Paläontologie, Monatshefte* 1997:39–55.

Weishampel, D. B. 1984. Interactions between Mesozoic plants and vertebrates: fructifications and seed predation. *Neues Jahrbuch für Geologie und Paläontologie, Abhandlungen* 167:224–250.

———. 2004. Dinosaurian distributions; pp. 517–606 *in* D. B. Weishampel, P. Dodson, and H. Osmólska (eds.), *The Dinosauria* (Second Edition). Berkeley: University of California Press.

Weishampel, D. B., P. M. Barrett, R. A. Coria, J. Le Loeuff, X. Xing, Z. Sijin, A. Sahni, E. M. P. Gomani, and C. R. Noro. 2004. Dinosaur Distribution. Pp. 517–606 in D. B. Weishampel, P. Dodson, and H. Osmólska (eds.) *The Dinosauria* (Second Edition). Berkeley: University of California Press.

Weishampel, D. B. and D. B. Norman. 1989. Vertebrate herbivory in the Mesozoic: jaws, plants, and evolutionary metrics. *Geological Society of America Special Paper* 238:87–100.

Wild, R. 1978. Die flugsaurier (Reptilia: Pterosauria) aus der oberen Trias von Cene bei Bergamo. *Bolletino della Scienza Paleontologica Italia* 17:176–256.

Wilson, J. A. 2002. Sauropod dinosaur phylogeny: critique and cladistic analysis. *Zoological Journal of the Linnean Society* 136:217–276.

Wilson, J. A. and P. C. Sereno. 1998. Early evolution and higher-level phylogeny of sauropod dinosaurs. *Society of Vertebrate Paleontology Memoir* 5:1–68.

Wiman, C. 1929. Die Kriede-dinosaurier aus Shantung. *Paleontologica Sinica (Series C)* 6:1–67.

Wing, S. L. and B. H. Tiffney. 1987. The reciprocal interaction of angiosperm evolution and tetrapod herbivory. *Review of Paleobotany and Palynology* 50:179–210.

Yates, A. and J. W. Kitching. 2003. The earliest known sauropod dinosaur and the first steps towards sauropod locomotion. *Proceedings of the Royal Society of London, Series 8* 270:1753–1758.

Young, C. C. 1939. On a new Sauropoda, with notes on other fragmentary rep-
   tiles from Szechuan. *Bulletin of the Geological Society of China* 19:279–315.
Young, C. C. and X-J. Zhao. 1972. [*Mamenchisaurus hochuanensis,* sp. nov.].
   Academica Sinica 8:1–30.
Zhang, Y. 1988. The Middle Jurassic dinosaur fauna from Dashanpu, Zigong,
   Sichuan. *Journal of the Chengdu College of Geology* 3:1–87.
Ziegler, A. M., J. M. Parrish, Y. Jiping, E. D. Gyllenhaal, D. B. Rowley, J. T.
   Parrish, N. Shangyou, A. Bekker, and M. L. Hulver. 1993. Early Mesozoic
   phytogeography and climate. *Philosophical Transactions of the Royal Society
   of London, Series B* 341:297–305.

Body-Size Evolution in the Dinosauria

Matthew T. Carrano

Introduction

The evolution of body size and its influence on organismal biology have received scientific attention since the earliest decades of evolutionary study (e.g., Cope, 1887, 1896; Thompson, 1917). Both paleontologists and neontologists have attempted to determine correlations between body size and numerous aspects of life history, with the ultimate goal of documenting both the predictive and causal connections involved (LaBarbera, 1986, 1989). These studies have generated an appreciation for the thoroughgoing interrelationships between body size and nearly every significant facet of organismal biology, including metabolism (Lindstedt & Calder, 1981; Schmidt-Nielsen, 1984; McNab, 1989), population ecology (Damuth, 1981; Juanes, 1986; Gittleman & Purvis, 1998), locomotion (McMahon, 1975; Biewener, 1989; Alexander, 1996), and reproduction (Alexander, 1996).

An enduring focus of these studies has been Cope's Rule, the notion that body size tends to increase over time within lineages (Kurtén, 1953; Stanley, 1973; Polly, 1998). Such an observation has been made regarding many different clades but has been examined specifically in only a few (MacFadden, 1986; Arnold et al., 1995; Jablonski, 1996, 1997; Trammer & Kaim, 1997, 1999; Alroy, 1998). The discordant results of such analyses have underscored two points: (1) Cope's Rule does not apply universally to all groups; and (2) even when present, size increases in different clades may reflect very different underlying processes. Thus, the question, "does Cope's Rule exist?" is better parsed into two questions: "to which groups does Cope's Rule apply?" and "what process is responsible for it in each?"

Several recent works (McShea, 1994, 2000; Jablonski, 1997; Alroy, 1998, 2000a, 2000b) have begun to address these more specific questions, attempting to quantify patterns of body-size evolution in a phylogenetic (rather than strictly temporal) context, as well as developing methods for interpreting the resultant patterns. Perhaps surprisingly, none of these studies has focused on body-size evolution in nonavian dinosaurs (hereafter referred to as "dinosaurs"), a group for which body size increases are

axiomatic. Although dinosaurs are commonly perceived to have undergone dramatic size increases (and certainly the thousandfold size difference between outgroup "lagosuchians" and sauropods is remarkable), few
studies (Sereno, 1997) have attempted to quantify or analyze this pattern.

In this paper, I present the results of the first such study. Using
measurement data and a composite phylogeny of dinosaurs, I reconstruct patterns of body-size evolution in this group. Ultimately, dinosaurs
are brought into the context of Cope's Rule as the resulting patterns
are assessed and interpreted in light of several potential underlying
mechanisms.

Materials and Methods

Body-Size Estimation and Dinosaur Phylogeny

Estimating body size for any extinct organism is a difficult prospect,
particularly taxa that differ significantly from extant forms in body size
and shape. Dinosaurs have proven frustrating subjects for body-size estimation for this reason, and as a consequence different studies have generated widely varying results (Colbert, 1962; Bakker, 1975; Paul, 1988;
Alexander, 1985; Anderson et al., 1985; Peczkis, 1994; Henderson, 1999;
Seebacher, 2001). Much of this variation is tied to methodological differences (Alexander, 1985), the inherent subjectivity involved in creating
full-body reconstructions of extinct animals (Paul, 1988), and the uncer-
tainty surrounding predictions generated from the scaling relationships of
extant taxa (Carrano, 2001). Thus, estimates of body masses for specific
dinosaur taxa remain a subject of persistent debate.

However, it is not necessary to reconstruct absolute body masses to
analyze patterns of body-size evolution; only *relative* body sizes need to
be reconstructed. Therefore, it is possible to substitute proxies (or correlates) for body size in place of actual estimates, provided such proxies
have a consistent, linear relationship to body size. This relationship need
not even be specified, but the "fit" of the correlation and its linearity must
be demonstrated. Such a linearly correlated variable would then reflect
some multiple (or fraction) of body mass, allowing the relative sizes of
taxa to be compared on a single scale. Changes between one taxon (ancestor) and the next (descendant) can then be measured while maintaining the same relationship (differing only in some multiple or fraction)
that would have been present had actual masses been used.

In this study, I use femoral length (FL), anteroposterior diameter
(FAP), and mediolateral diameter (FML) as separate proxies for body
mass. These variables have been shown to be tightly linearly correlated

with body mass in many extant terrestrial taxa, particularly mammals (Alexander et al., 1979; Bou et al., 1987; Jungers et al., 1998; Christiansen, 1999) and birds (Maloiy et al., 1979; Cubo & Casinos, 1997). The linearity of this relationship is probably tied to the role of the femur (specifically, its cross-sectional area) in supporting body mass against gravity. Femoral measures have the added advantage of being relatively easily to obtain (even from photographs, when specimens are not directly accessible), and the femur is frequently preserved in dinosaur specimens.

I measured femoral length and diameters in 1,640 nonavian dinosaur specimens representing all major ingroup clades and nearly every taxon for which limb material is known ($N = 251$; appendix). I used the largest representative when multiple specimens were available, and excluded taxa when limbs were known only from juvenile specimens (e.g., *Lophorhothon, Pleurocoelus, Bellusaurus, Brachyceratops, Avaceratops, Maleevosaurus, Shanshanosaurus*). These measurements were mapped onto a composite phylogeny derived from several published sources (figs. 8.1, 8.2, 8.3, 8.4) and were analyzed by the methods described below. I incorporated several taxa that were not represented in published analyses into the phylogeny based on personal communications and observations. Although I attempted to include as many taxa as possible in this composite phylogeny, I omitted several taxa whose relationships were too uncertain to allow their placement in this context (e.g., *Saltopus, Kaijiangosaurus, Betasuchus, Tarascosaurus, Chuandongocoelurus, Podokesaurus, Tugulusaurus, Nanosaurus, Klamelisaurus, "Cetiosaurus" mogrebiensis, Lourinhasaurus*).

Controversy surrounds the details of several regions of this phylogeny and often extends to large collections of taxa. These groups include "ceratosaurian" theropods and "hypsilophodontid" ornithopods. In both cases, recent studies (Scheetz, 1999; Carrano et al., 2002) have favored rendering both formerly monophyletic groups as paraphyletic. For such examples, I rearranged the phylogeny to reflect these previous hypotheses and compared the reconstructed ancestral states. Similarly, I compared the effects of moving individual controversial taxa (e.g., *Euhelopus, Eoraptor, Heterodontosaurus*). In none of these instances were significant effects observed.

Identifying Evolutionary Patterns

Ideally, the identification and analysis of evolutionary trends are based on direct examination of actual ancestor-descendant pairs (Alroy, 1998), provided such forms and relationships could be identified. As few candidate ancestor-descendant pairs have been suggested among

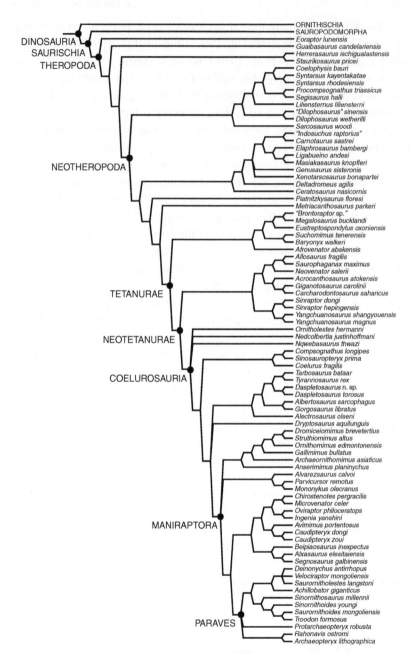

Figure 8.1. Phylogeny of Theropoda used in this study, after Sereno (1999), Holtz (2000), and Carrano et al. (2002). Additional taxa are included based on personal observations.

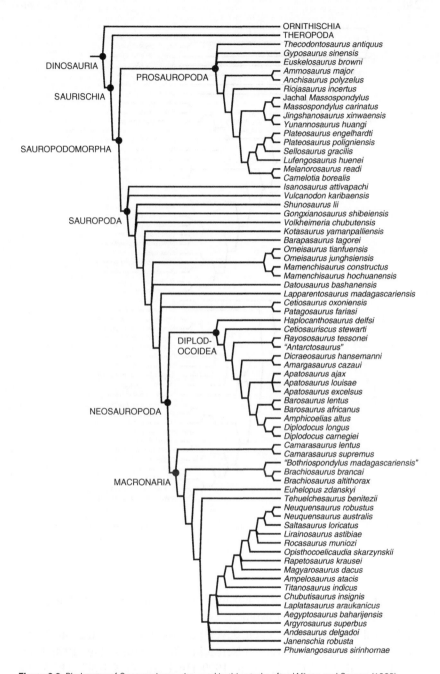

Figure 8.2. Phylogeny of Sauropodomorpha used in this study, after Wilson and Sereno (1998), Sereno (1999), Curry Rogers and Forster (2001), Wilson (2002), and K. A. Curry Rogers (personal communication). Additional taxa are included based on personal observations.

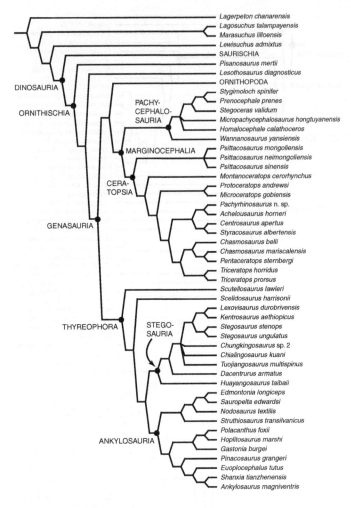

Figure 8.3. Phylogeny of Thyreophora and Marginocephalia used in this study, after Sereno (1999), Dodson et al. (2004), and R. V. Hill (personal communication.). Additional taxa are included based on personal observations.

dinosaurs, this option is not promising here. Instead, ancestral states must be either reconstructed or avoided.

Several methods have been developed for reconstructing ancestral states on a phylogenetic tree (Felsenstein, 1985; Schultz et al., 1996; Cunningham et al., 1998; Cunningham, 1999; Huelsenbeck & Bollback, 2001), perhaps the most straightforward being optimization of discrete characters directly onto a cladogram (Maddison et al., 1984). Continuous characters are problematic to reconstruct in this manner, largely because of the

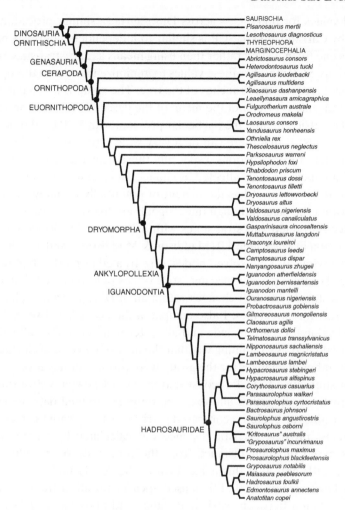

Figure 8.4. Phylogeny of Cerapoda used in this study, after Sereno (1999), Scheetz (1999), and Horner et al. (2004). Additional taxa are included based on personal observations.

difficulties associated with incorporating them into the discrete context of a cladistic analysis. Independent contrasts (Felsenstein, 1985) allows ancestral-state reconstructions of continuous characters, essentially by applying the mean of the two immediate daughter taxa to a given node. Unfortunately, this very procedure hampers its usefulness here, because averaging the changes contained within each set of ancestor-descendant pairs ultimately eliminates trends in the original data.

In this study, I used squared-change parsimony (SCP) to reconstruct nodal values for body mass across Dinosauria (a similar procedure was used by Carrano, 2000). SCP is similar to independent contrasts but includes one further step: rather than using means as ancestral-state reconstructions, SCP modifies these values to minimize the sum of the squared changes across all the branches of the tree. The result is that a particular ancestral state is often not the mean of the descendant values and could even lie outside them. One particular problem with SCP (and other similar methods) is that there can be a very wide error associated with ancestral-state reconstructions, especially those near the base of the tree, which are most strongly affected by changes in terminal taxon values and positions. However, it has the benefit of potentially retaining trend signals within the data, although these signals are likely to be weak.

I reconstructed ancestral states with SCP using the "Trace Continuous" option in MacClade 4.0 (Maddison & Maddison, 2000) and the composite phylogeny. Unresolved nodes were treated as hard polytomies, as required of this option. Once ancestral values were obtained, I analyzed the ancestor-descendant changes within all of Dinosauria as well as several large ingroup clades. I also compared the changes between the ancestral value for a given clade and all its terminal taxa. Overall changes were evaluated by examining whether the mean change, sum change, and total number of changes for the group were positive or negative.

An alternative to SCP reconstruction is to eschew ancestral-state reconstructions altogether and examine only the original data associated with terminal taxa. Here I compared patristic distance with the measured body-mass proxies. Patristic distance was calculated by numbering all nodes based on their distance from the root node of the phylogeny (Sidor, 2001; note that this is identical to the "clade rank" method described by Carrano, 2000). I then used Spearman-rank correlation to test for correlation between patristic distance and body mass. If taxa with higher patristic distance values tend to have larger (or smaller) body masses, this would be evident as a significant positive (or negative) correlation. In this manner, body mass-clade rank correlation allows both identification and evaluation of trends within the data. Again, I examined Dinosauria as a whole and several less-inclusive clades.

Analyzing Evolutionary Trends

Evolutionary radiations (or trends) in morphology can often be described as changes in morphospace occupation over time for a given clade. Such changes can comprise expansion of the morphological range, resulting in a greater amount of morphospace occupied by the clade (Fisher, 1986). Expansion into a morphospace may resemble simple diffusion, wherein the range of variation increases without bound through time from some ancestral condition. If, however, some maximum or minimum (or both) limits the variation expressed by the radiation, it can be described as diffusion with one (or more) bounds. Both situations fulfill McShea's (1994) definition of "passive" trends (see also McShea, 1998).

Change in morphospace occupation can also include displacement from one morphology to another, in which the location of a group or taxon shifts within the morphospace. In these cases, no range expansion is necessary, merely a change in the location of morphospace occupation; they conform to McShea's "driven" trends (1994) (here called "active"). (It should be noted that diffusion may also be overprinted on this pattern, so that range expansion may accompany a shift in location).

I employed McShea's (1994) three tests to determine whether trends in dinosaur body-size evolution could best be described as "passive" or "active":

1) The *minimum test* examines the behavior of the minimum bound of the size distribution through time. In a passive trend, this bound remains stable while the maximum bound increases, reflecting diffusive expansion of the group into morphospace. In contrast, the minimum and maximum bounds both increase in an active trend, reflecting the wholesale shift in morphospace occupation.
2) The *ancestor-descendant test* examines each change from an ancestor to its descendant and tallies the number of increases and decreases. Passive trends have near-equal numbers; active trends have a preponderance of one or the other.
3) The *subclade test* compares the frequency distributions of the state variable for ingroup clades with its distribution in the whole clade. Passively driven groups are expected to have distributions that deviate from the general pattern, whereas actively driven groups are expected to have patterns that mirror the general pattern.

The latter two tests are conducted on groups sufficiently far from the minimum bound so as to reduce the chance that the results are biased by it (McShea, 1994). For these two analyses, I examined taxa that were larger than the mean log size for each group (MacFadden, 1986).

Recently, Alroy (2000a) suggested that these tests did not adequately examine the subtleties present in many data sets but rather obscured them with the coarse designations "active" and "passive." He suggested that time-slice analyses were unlikely to reveal meaningful patterns, instead favoring examination of ancestor-descendant pairs. To illustrate this, Alroy presented twelve possible trend patterns based on plots of descendant-ancestor differences versus ancestral states. I present similar plots here to investigate whether the pattern of body-size evolution in dinosaurs is likely to be the result of random or nonrandom changes.

Alroy's (2000a) objections to time-slice analyses have merit, and one possible alternative to avoiding them altogether is to replace temporal data with phylogenetic data. Superficially, this will cause little change in the overall pattern because most vertebrate clades (including dinosaurs) show some correlation between age rank and clade rank (Benton & Hitchin, 1997). However, by replacing time data with patristic distance, the comparisons become explicitly phylogenetic even when specific ancestor-descendant comparisons are not made. In dinosaurs, where the fossil record is extremely variable and long ghost lineages are inferred for several major clades (Sereno, 1999), this difference can interfere with resulting perceptions of evolutionary patterns. In particular, late first appearances of basal taxa (due to an incomplete record) may still be in proper sequence but can alter the overall pattern, especially if these taxa are located at the edges of the distribution. Thus, I also perform a *modified minimum test* in which the minimum bound is tracked on plots of body size versus patristic distance. For these analyses, patristic distance is rescaled to 1.0 for each of the major ingroup clades.

Results

Evolutionary Patterns

SCP reconstructions of ancestral (nodal) states produce a general pattern consistent with an overall size increase throughout Dinosauria. This is evident in the positive mean, sum, and median changes as well as the left-skewed distribution of changes (tables 8.1–8.3). The results from comparisons between each ancestor and descendant are similar to those from comparisons between the single reconstructed ancestral value for Dinosauria and all its terminal taxa (table 8.4). This pattern is consistent regardless of which measured variable is examined.

The pattern is robust but complex, largely due to the overlap of numerous internal patterns associated with less-inclusive dinosaurian clades. When performed on these clades, the SCP analysis reveals size increases in most, but not all, dinosaur groups (tables 8.1–8.3). Again, the patterns are consistent for most measured variables, with a tendency for some clades to show weaker trends with diameter measures. The most notable exception occurs in coelurosaurian theropods, which occasionally show negative mean and sum changes, as well as a negative median change for FML, that suggest overall size decreases. The coelurosaur pat-

Table 8.1. Body-size statistics for Dinosauria and ingroup clades, using squared-change parsimony reconstructions based on measurements of femoral length

Group	Mean	Sum	Skew	Median	N	+	−	χ^2
Dinosauria	0.012	5.773	−0.442	0.018	466	**279**	187	18.163*
Saurischia	0.008	2.267	−0.482	0.016	273	116	**157**	6.158*
Theropoda	0.004	0.576	−0.694	0.018	154	**92**	62	5.844*
Coelurosauria	−0.001	−0.050	−0.465	0.006	78	**43**	35	0.821
Sauropodomorpha	0.013	1.523	0.493	0.009	118	**52**	43	1.661
Prosauropoda	0.003	0.094	0.132	0.019	28	**16**	12	0.571
Sauropoda	0.014	1.203	0.771	0.006	89	**47**	42	0.281
Macronaria	0.006	0.233	0.349	0.006	41	**21**	20	0.024
Ornithischia	0.019	3.491	−0.354	0.021	185	**118**	67	14.059*
Thyreophora	0.034	1.376	−0.197	0.041	41	**27**	14	4.122*
Stegosauria	0.037	0.559	0.390	0.041	15	**9**	6	1.389
Ankylosauria	0.021	0.438	−0.679	0.033	21	**14**	7	2.333
Marginocephalia	0.027	1.060	−0.480	0.041	38	**27**	11	6.737*
Pachycephalo	0.022	0.218	−0.159	0.057	10	**6**	4	0.400
Ceratopsia	0.029	0.823	−0.637	0.044	28	**21**	7	7.000
Ornithopoda	0.012	1.175	−0.407	0.016	99	**61**	38	5.343*

Notes: Statistics summarize differences between each reconstructed ancestral node and each descendant taxon. Skew, skewness; +, number of positive ancestor-descendant changes; −, number of negative ancestor-descendant changes; χ^2, chi-square results. Asterisks indicate χ^2 values that are significant to at least $P < 0.05$. Pachycephalo = Pachycephalosauria.

tern is also manifest at a higher level, within Theropoda as a whole. Sauropoda and Pachycephalosauria also show evidence of size decreases, although these are more weakly evident (usually as a right-skewed distribution of changes). The small sample size for Pachycephalosauria hampers further investigation, but the sauropod pattern seems to be influenced by size decreases concentrated within Macronaria.

Patristic-distance correlations clarify these trends, albeit at the expense of the increased number of data points afforded by SCP reconstructions (fig. 8.5; tables 8.5–8.7). These results are very similar to those produced by SCP, revealing size increases in nearly all dinosaur clades as well as in Dinosauria, and are consistent among the three measured variables. Spearman-rank correlations reveal positive trends in most groups, although these are not significant in Ankylosauria and Stegosauria for FAP, or Pachycephalosauria for FML. Negative correlations are also present—significant in Macronaria, Sauropoda, and Theropoda but nonsignificant in Saurischia and Coelurosauria—indicating trends toward size decreases in these groups.

Table 8.2. Body-size statistics for Dinosauria and ingroup clades, using squared-change parsimony reconstructions based on measurements of femoral anteroposterior diameter

Group	Mean	Sum	Skew	Median	N	+	−	χ^2
Dinosauria	0.006	2.400	−0.331	0.015	385	**216**	169	5.738*
Saurischia	0.002	0.545	−0.310	0.018	226	**127**	99	3.469
Theropoda	−0.005	−0.644	−0.428	0.020	130	**74**	56	2.492
Coelurosauria	−0.028	−1.850	−0.124	0.003	66	**34**	32	0.061
Sauropodomorpha	0.010	0.960	0.257	0.015	118	**64**	54	0.847
Prosauropoda	0.003	0.063	−0.206	0.011	24	**13**	11	0.167
Sauropoda	0.009	0.641	0.331	0.011	70	**38**	32	0.514
Macronaria	−0.013	−0.400	0.583	−0.003	32	15	**17**	0.125
Ornithischia	0.013	1.913	−0.265	0.014	151	**85**	66	2.391
Thyreophora	0.031	0.866	−0.177	0.004	28	**15**	13	0.143
Stegosauria	0.043	0.391	0.989	0.008	9	**6**	3	1.000
Ankylosauria	0.007	0.098	−0.267	−0.005	15	6	**9**	0.600
Marginocephalia	0.034	1.023	−0.060	0.049	30	**20**	10	3.333
Pachycephalo	0.023	0.162	0.488	−0.018	7	3	**4**	0.143
Ceratopsia	0.037	0.860	−0.316	0.059	23	**17**	6	5.260*
Ornithopoda	0.002	0.165	−0.463	0.013	87	**48**	39	0.931

Note: Statistics summarize differences between each reconstructed ancestral node and each descendant taxon. Abbreviations are as in Table 8.1.

Evolutionary Trends

Minimum Test. When plotted against time (i.e., age rank), body-size distribution in Dinosauria shows a rapid expansion in range throughout the Mesozoic, most of which occurs during the Late Triassic (age ranks 1–4) (fig. 8.6). However, this range expansion is almost entirely confined to the right of the distribution, representing size increases. Few taxa decrease in size from the ancestral dinosaurian condition, and as a result the distribution shows a relatively stable minimum bound, suggesting that the pattern is passive.

Within Dinosauria, the patterns are more complex: most groups mirror this overarching passive pattern, but a few do not. These exceptions — Sauropodomorpha, Thyreophora, and (although inconsistently among different variables) Stegosauria and Ceratopsia — instead show a loss of taxa at the minimum bound while the maximum bound increases. Thus, the entire distribution shifts toward the right, although an increase in range may also be present. Interestingly, macronarians show the reverse pattern: loss of larger taxa as smaller taxa appear, thus shifting

Table 8.3. Body-size statistics for Dinosauria and ingroup clades, using squared-change parsimony reconstructions based on measurements of femoral mediolateral diameter

Group	Mean	Sum	Skew	Median	N	+	−	χ^2
Dinosauria	0.014	5.110	−0.364	0.030	375	**225**	150	15.000*
Saurischia	0.006	1.352	−0.472	0.019	215	**123**	92	4.470*
Theropoda	−0.002	−0.225	−0.593	0.034	120	**67**	53	1.633
Coelurosauria	−0.009	−0.508	−0.356	−0.007	56	27	**29**	0.071
Sauropodomorpha	0.014	1.308	0.181	0.009	94	**55**	39	2.723
Prosauropoda	0.002	0.053	−0.293	0.038	23	**13**	10	0.391
Sauropoda	0.013	0.911	0.239	0.008	70	**41**	29	2.057
Macronaria	−0.001	−0.050	0.079	0.005	35	**20**	15	0.714
Ornithischia	0.024	3.758	−0.135	0.036	154	**99**	55	12.570*
Thyreophora	0.041	1.463	−0.337	0.046	36	**25**	11	5.444*
Stegosauria	0.038	0.573	0.521	0.027	15	**10**	5	1.667
Ankylosauria	0.028	0.471	−0.784	0.058	17	**12**	5	2.882
Marginocephalia	0.035	1.041	−0.539	0.044	30	**23**	7	19.200*
Pachycephalo	0.035	0.244	0.413	0.021	7	**5**	2	1.286
Ceratopsia	0.105	0.797	−0.914	0.048	23	**18**	5	7.348*
Ornithopoda	0.016	1.324	0.086	0.027	82	**48**	34	2.390

Note: Statistics summarize differences between each reconstructed ancestral node and each descendant taxon. Abbreviations as in Table 8.1.

the distribution to the left. Both types of exceptions may be described as active.

Ancestor-Descendant Test. When taxa greater than the mean log-size are considered, most dinosaur clades (including Coelurosauria) show a greater number of increases between ancestors and descendants than decreases (table 8.8). This active pattern is also seen in Dinosauria as a whole. There is only one weak instance of the reverse pattern (Macronaria, FAP), although a few groups (Macronaria, Sauropoda, Sauropodomorpha) show near-equal values for increases and decreases.

Subclade Test. The body-size distribution for Dinosauria is strongly right skewed, as is typical for most animal groups (fig.8.7; Stanley, 1973). When subclades whose means are larger than the mean log size are analyzed, their distributions are quite variable (table 8.9). This variation ranges from positively skewed distributions (very similar to that for Dinosauria) to near-normal and negatively skewed distributions, often differing for the same group depending on the measured variable.

Table 8.4. Body-size statistics for Dinosauria and ingroup clades, using squared-change parsimony reconstructions based on measurements of femoral length

Group	Mean	Sum	Skew	Median	N	+	−	χ^2
Dinosauria	0.826	208.036	−0.664	0.937	252	**246**	6	228.571*
Saurischia	0.404	60.159	−0.701	0.519	149	**118**	31	50.799*
Theropoda	0.277	22.419	−0.334	0.357	81	**60**	21	20.753*
Coelurosauria	0.068	2.870	0.060	0.015	42	**22**	20	0.009
Sauropodomorpha	0.399	27.151	−1.112	0.466	68	**62**	6	46.118*
Prosauropoda	0.133	2.128	−0.270	0.221	16	**10**	6	1.000
Sauropoda	0.209	10.879	−0.588	0.239	52	**46**	8	30.769*
Macronaria	−0.084	−2.108	−0.333	−0.035	25	10	**15**	1.000
Ornithischia	1.571	155.577	−0.635	1.675	99	**99**	0	99.000*
Thyreophora	0.528	11.609	−1.027	0.566	22	**21**	1	18.182*
Stegosauria	1.002	9.018	−0.343	1.004	9	**9**	0	9.000*
Ankylosauria	0.098	1.077	−0.933	0.138	11	**9**	2	4.455*
Marginocephalia	1.269	26.645	−0.092	1.272	21	**21**	0	21.000*
Pachycephalo	−0.154	−0.925	0.035	−0.101	6	1	**5**	2.667
Ceratopsia	0.392	5.880	−0.556	0.587	15	**11**	4	3.267
Ornithopoda	0.590	31.839	−0.747	0.725	54	**50**	4	39.185*

Note: Statistics summarize differences between the basal reconstructed ancestral node for each clade and all its descendant terminal taxa. Abbreviations as in Table 8.1.

Change Versus Ancestor Plots. When descendant-ancestor differences are plotted against ancestral states for Dinosauria, the resulting pattern suggests that body-size evolution is an active pattern, rather than due to simple random diffusion (fig. 8.8). Although the mean change is nearly zero, there is a weak trend within the data: the regression slope is small but significantly positive ($y = 0.72x − 0.193$; $r^2 = 0.062$; $P < 0.001$). Given the negative autocorrelation between the x and y variables, the corrected positive correlation would be even stronger (Alroy, 1998). Indeed, this pattern is remarkably similar to that for Cenozoic mammals in that it can also be described by a cubic equation favoring moderate-to-large body sizes.

The positive regression pattern is repeated at many more inclusive levels within Dinosauria. The best-sampled ingroups (Saurischia, Ornithischia, Theropoda, Thyreophora, Ornithopoda, Coelurosauria) have significant patterns that are very similar to that of Dinosauria. They can also be described by similar cubic equations. Other groups also have positive regression slopes but not significantly so. In addition, the reduced sample sizes of these other clades make it difficult to determine whether a cubic equation could also be appropriately fit to those data.

Modified Minimum Test. Several differences from the minimum test are detected in these patterns when body size is plotted against patristic distance instead of time (fig. 8.5). Dinosauria, Saurischia, Theropoda,

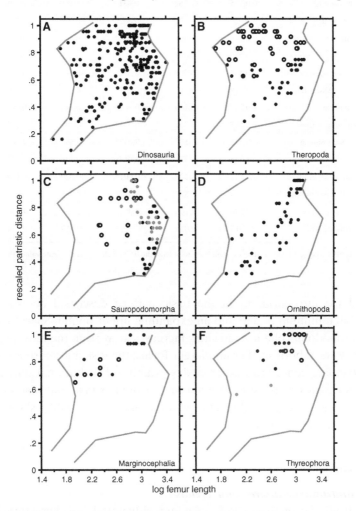

Figure 8.5. Patristic distance analyses results. Patristic distance (rescaled to 1.0 for major ingroup clades) is plotted against log femur length for Dinosauria and several representative ingroup clades. (A) Dinosauria. (B) Theropoda: open circles, Coelurosauria; closed circles, all other theropods. (C) Sauropodomorpha: open circles, Prosauropoda; gray circles, Macronaria; closed circles, other sauropods. (D) Ornithopoda. (E) Marginocephalia: open circles, Pachycephalosauria; closed circles, Ceratopsia. (F) Thyreophora: gray circles, basal thyreophorans; open circles, Stegosauria; closed circles, Ankylosauria.

Table 8.5. Spearman-rank correlations of body size (based on femoral length) and patristic distance for Dinosauria and ingroup clades

Group	rho	Z	P	rho[†]	Z[†]	P[†]
Dinosauria	0.285	6.142	<0.001*	0.283	6.118	<0.001*
Saurischia	−0.071	−1.169	0.243	−0.073	−1.206	0.228
Theropoda	−0.175	−2.162	0.031*	−0.178	−2.206	0.027*
Coelurosauria	−0.020	−0.174	0.862	−0.030	−0.265	0.791
Sauropodomorpha	0.340	3.681	<0.001*	0.339	3.664	<0.001*
Prosauropoda	0.556	2.891	0.004*	0.550	2.855	0.004*
Sauropoda	−0.104	−0.973	0.331	−0.106	−0.998	0.318
Macronaria	−0.718	−3.515	<0.001*	−0.727	−3.526	<0.001*
Ornithischia	0.745	10.100	<0.001*	0.744	10.093	<0.001*
Thyreophora	0.593	3.749	<0.001*	0.587	3.711	<0.001*
Stegosauria	0.571	2.135	0.033*	0.562	2.102	0.036*
Ankylosauria	0.548	2.450	0.014*	0.535	2.394	0.017*
Marginocephalia	0.831	5.120	<0.001*	0.829	5.107	<0.001*
Pachycephalo	0.858	2.573	0.010*	0.855	2.564	0.010*
Ceratopsia	0.802	4.166	<0.001*	0.798	4.148	<0.001*
Ornithopoda	0.913	9.035	<0.001*	0.913	9.034	<0.001*

Note: Asterisks highlight *p*-values that are significant to at least the 0.05 level; daggers indicate values that are corrected for ties.

and Coelurosauria retain the basic pattern described above, with a stable minimum bound and an expanding upper bound. Sauropoda and Sauropodomorpha instead show increases away from the most primitive size toward both smaller and larger forms; the distribution resembles an expanding cone. Most other clades exhibit a distributional shift toward larger forms with a concomitant loss of smaller taxa, but macronarians again show the unusual reverse pattern. The three variables show consistent overall patterns.

Discussion

Evolutionary Patterns and Trends

Body-size increases have been implicitly noted as a characteristic feature of dinosaur evolution since the early days of paleontological study. This was partly based on a tacit understanding of these immense animals as having necessarily descended from some smaller-sized member(s) of the Paleozoic fauna. Subsequent discoveries bolstered this opinion by documenting small early dinosaurs (Cope, 1889; Talbot, 1911; Huene,

Table 8.6. Spearman-rank correlations of body size (based on femoral anteroposterior diameter) and patristic distance for Dinosauria and ingroup clades

Group	rho	Z	P	rho[†]	Z[†]	P[†]
Dinosauria	0.296	5.827	<0.001*	0.295	5.806	<0.001*
Saurischia	−0.067	−0.998	0.318	−0.069	−1.030	0.303
Theropoda	−0.202	−2.294	0.022*	−0.205	−2.331	0.020*
Coelurosauria	−0.030	−0.242	0.809	−0.039	−0.315	0.753
Sauropodomorpha	0.292	2.828	0.005*	0.290	2.813	0.005*
Prosauropoda	0.476	2.281	0.023*	0.471	2.258	0.024*
Sauropoda	−0.201	−1.671	0.095*	−0.204	−1.697	0.090*
Macronaria	−0.205	−0.869	0.385	−0.211	−0.894	0.371
Ornithischia	0.746	9.233	<0.001*	0.746	9.228	<0.001*
Thyreophora	0.524	2.772	0.006*	0.518	2.743	<0.006*
Stegosauria	−0.212	−0.601	0.548	−0.271	−0.767	0.443
Ankylosauria	0.338	1.266	0.206	0.322	1.205	0.228
Marginocephalia	0.857	4.692	<0.001*	0.855	4.682	<0.001*
Pachycephalo	0.777	1.903	0.057*	0.771	1.888	0.059*
Ceratopsia	0.842	4.166	<0.001*	0.798	4.148	<0.001*
Ornithopoda	0.878	8.178	<0.001*	0.878	8.186	<0.001*

Note: Symbols as in Table 8.5.

1914) along with increasingly larger Jurassic and Cretaceous forms. This general notion became ensconced in scientific opinion, even coming to fulfill a perceived role in contributing both to their success and extinction (Benton, 1990).

From the many descriptive analyses presented here, it is clear that dinosaur evolution is indeed characterized by a marked, pervasive pattern of body-size increase. This is evident in most of the major ingroup clades as well, indicating that the overall pattern is not merely an artifact of overlapping—and potentially discordant—internal patterns. This perhaps belabors the rather obvious point that dinosaurs did, in fact, get bigger as time proceeded in the Mesozoic. However, the specificity of these analyses also allows a more complex pattern to be determined. For example, at least two less-inclusive clades (Macronaria and Coelurosauria) are typified by size *decreases*.

These two groups are interesting in their own right as the most diverse and morphologically divergent components of their parent clades. Macronarians (including "titanosaurs") display a host of unusual synapomorphies among sauropods that are likely tied to unique locomotor and

Table 8.7. Spearman-rank correlations of body size (based on femoral mediolateral diameter) and patristic distance for Dinosauria and ingroup clades

Group	rho	Z	P	rho[†]	Z[†]	P[†]
Dinosauria	0.243	4.717	<0.001*	0.241	4.691	<0.001*
Saurischia	−0.053	−0.783	0.435	−0.055	−0.815	0.415
Theropoda	−0.167	−1.817	0.070*	−0.170	−1.853	0.064*
Coelurosauria	−0.058	−0.432	0.666	−0.072	−0.531	0.596
Sauropodomorpha	0.287	2.796	0.005*	0.285	2.777	0.006*
Prosauropoda	0.386	1.853	0.064*	0.376	1.803	0.071*
Sauropoda	−0.219	−1.833	0.067*	−0.223	−1.863	0.063*
Macronaria	−0.552	−2.467	0.014*	−0.563	−2.519	0.012*
Ornithischia	0.592	7.347	<0.001*	0.591	7.361	<0.001*
Thyreophora	0.683	4.095	<0.001*	0.679	4.073	<0.001*
Stegosauria	0.553	2.068	0.039*	0.543	2.033	0.042*
Ankylosauria	0.631	2.525	0.012*	0.624	2.495	0.013*
Marginocephalia	0.782	4.282	<0.001*	0.779	4.267	<0.001*
Pachycephalo	0.634	1.553	0.121	0.624	1.529	0.126
Ceratopsia	0.762	3.576	<0.001*	0.758	3.555	<0.001*
Ornithopoda	0.845	7.604	<0.001*	0.845	7.602	<0.001*

Note: Symbols as in Table 8.5.

postural specializations in this group (Wilson & Sereno, 1998; Wilson & Carrano, 1999). Appearing during the Middle Jurassic (Wilson & Sereno, 1998), macronarians attained very large body sizes (e.g., *Brachiosaurus, Chubutisaurus, Argentinosaurus*), but eventually produced taxa as "small" as elephants (saltasaurines). It has been suggested that at least one taxon (*Magyarosaurus*) was the product of dwarfing within Macronaria (Jianu & Weishampel, 1999).

Coelurosaurs are most noteworthy as the clade including birds (although avians are excluded from this study), and marked size decrease has been specifically implicated in the origin of the latter group (Sereno, 1997; Carrano, 1998). However, this pattern extends well into the more basal nodes of Coelurosauria, with the largest taxa (tyrannosaurids) also representing the most basal major clade in the group. Only therizinosaurs—sister taxa to oviraptorosaurs—show a significant reversal of the size-decrease trend. Furthermore, coelurosaurs include the most dramatic size decrease in all of Dinosauria: five orders of magnitude from 1,000-kg tyrannosaurs to 0.1-kg basal avians. (If Neornithes are included, this size decrease spans *seven* orders of magnitude, down to 0.001-kg hummingbirds.)

Trend analyses do not produce consistent results for all groups or for all methods. Table 8.10 shows that several clades show characteristics of either active or passive trends depending on the test employed. Specifically, the ancestor-descendant test suggests that size increases in nearly all groups are the result of active trends, whereas many saurischian trends are characterized as passive under the minimum and modified minimum tests. A few of the passive trends in the minimum test results are identified

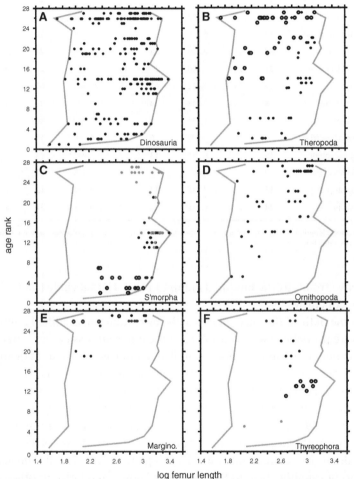

log femur length

Figure 8.6. Minimum test results. Age rank is plotted against log femur length for Dinosauria and several representative ingroup clades. The gray lines indicate the total distribution for Dinosauria for successive ingroup graphs. (A) Dinosauria. (B) Theropoda: open circles, Coelurosauria; closed circles, all other theropods. (C) Sauropodomorpha: open circles, Prosauropoda; gray circles, Macronaria; closed circles, other sauropods.(D) Ornithopoda. (E) Marginocephalia: open circles, Pachycephalosauria; closed circles, Ceratopsia. (F) Thyreophora: gray circles, basal thyreophorans; open circles, Stegosauria; closed circles, Ankylosauria.

Table 8.8. Ancestor-descendant test results, using taxa larger than the mean size for each variable

	FL			FAP			FML		
Group	+	−	χ^2	+	−	χ^2	+	−	χ^2
Dinosauria	193	77	49.837*	157	69	34.265*	158	66	39.786*
Saurischia	116	53	23.485*	94	44	18.116*	91	43	17.194*
Theropoda	56	9	33.985*	45	10	22.273*	40	9	19.612*
Coelurosauria	19	0	19.000*	14	2	9.000*	12	2	7.143*
Sauropodomorpha	60	44	2.462	49	34	2.711	51	34	3.400
Prosauropoda	13	3	6.250*	11	3	4.571*	10	5	1.667
Sauropoda	47	41	0.409	38	31	0.710	41	29	2.057
Macronaria	21	19	0.100	15	16	0.032	20	15	0.714
Ornithischia	77	24	27.812*	63	25	16.409*	67	23	21.511*
Thyreophora	18	5	7.348*	14	3	2.882	22	7	7.759*
Stegosauria	9	5	1.143	6	3	1.000	10	5	1.667
Ankylosauria	9	0	9.000*	6	0	6.000*	11	2	6.231*
Marginocephalia	15	2	9.941*	13	2	8.067*	14	2	9.000*
Pachycephalo	0	0	N/A	0	0	N/A	1	0	1.000
Ceratopsia	15	2	9.941*	13	2	8.067*	13	2	8.067*
Ornithopoda	44	17	11.951*	36	20	4.571*	31	14	6.422*

Note: FL, femoral length; FAP, femoral anteroposterior diameter; FML, femoral mediolateral diameter. Other abbreviations as in Table 8.1.

as active trends by the modified minimum test. The subclade test is generally inconclusive.

Some of the differences between the minimum and modified minimum tests are probably due to the nature of the dinosaur fossil record. Because many time intervals are poorly sampled, numerous dinosaur taxa appear later in time than their phylogenetic relationships suggest—i.e., later than the first appearance of their sister taxon. Although the *order* of appearance is not strongly affected (as demonstrated by strong age rank-clade rank correlations), the specific pattern is. As basal taxa are drawn into later time intervals, the size distribution of these primitive forms is drawn with them. This effect is mitigated by using the modified minimum test: by restoring primitive taxa to their "proper" position relative to other taxa, the size distribution is modified.

This discrepancy suggests that the active results may be inaccurate. It is largely due to the presence of small-bodied, derived taxa late in the Mesozoic, because these late-surviving taxa retain low patristic-distance values. Here the incompleteness of the fossil record is probably

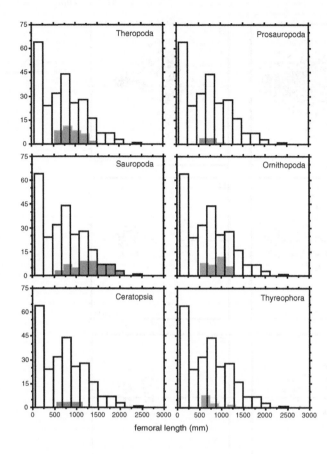

Figure 8.7. Subclade test results for representative ingroup clades. In each graph, the overall distribution for Dinosauria is shown by the open bars; each clade is represented by the black bars within it.

Table 8.9. Subclade test results, using raw data from subclades with means larger than (or close to) that for Dinosauria

Group	FL		FAP		FML	
	Mean	Skew	Mean	Skew	Mean	Skew
Dinosauria	717.71	0.618	75.45	1.025	98.31	0.930
Tetanurae	943.01	−0.086	100.20	1.212	114.13	1.154
Sauropoda	1324.73	0.188	138.66	0.905	201.02	0.033
Macronaria	1240.32	0.094	115.44	1.532	193.34	0.102
Iguanodontia	892.73	−0.477	100.38	−0.115	110.26	0.131
Stegosauria	911.89	0.109	96.90	1.043	133.50	−0.218

Note: FL, femoral length; FAP, femoral anteroposterior diameter; FML, femoral mediolateral diameter.

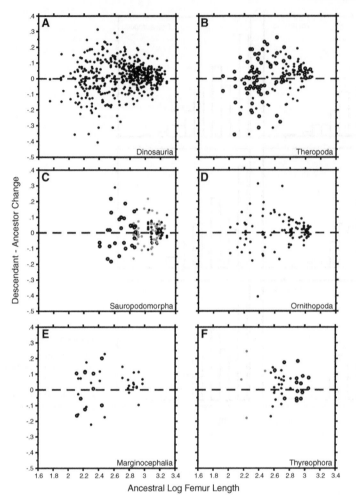

Figure. 8.8. Difference-versus-ancestor plots. The difference between each descendant and its reconstructed ancestor is plotted against the ancestral log femur length for that pair. The dashed line indicates zero change. (A) Dinosauria. (B) Theropoda: open circles, Coelurosauria; closed circles, all other theropods. (C) Sauropodomorpha: open circles, Prosauropoda; gray circles, Macronaria; closed circles, other sauropods. (D) Ornithopoda. (E) Marginocephalia: open circles, Pachycephalosauria; closed circles, Ceratopsia. (F) Thyreophora: gray circles, basal thyreophorans; open circles, Stegosauria; closed circles, Ankylosauria.

interfering with the underlying pattern. Because small-bodied forms are likely to be more poorly sampled, the forms that are sampled will have patristic distance values that are too low relative to those of large-bodied forms. For example, it is highly improbable that the ornithopod *Thescelosaurus* actually represents a single surviving form that originated in the

Table 8.10. Summary of trend analyses results

Group	min	mod min	anc-desc
Dinosauria	+ + +	+ + +	+ + +
Saurischia	– – –	– – –	+ + +
Theropoda	– – –	– – –	+ + +
Coelurosauria	– – –	– – –	+ + +
Sauropodomorpha	+ + +	+ + +	+ + +
Prosauropoda	– – –	– – –	+ + +
Sauropoda	– – –	– – –	+ + +
Macronaria	– + +	+ + +	+ – +
Ornithischia	+ + +	+ + +	+ + +
Thyreophora	+ + +	+ + +	+ + +
Stegosauria	+ – +	+ + +	+ + +
Ankylosauria	– – –	+ + +	+ + +
Marginocephalia	+ + +	+ + +	+ + +
Pachycephalo	+ + +	+ + +	00+
Ceratopsia	+ + +	+ + +	+ + +
Ornithopoda	+ + +	+ + +	+ + +

Note: min, minimum test; mod min, modified minimum test; anc-desc, ancestor-descendant test; +, active; –, passive; 0, insufficient sample. Each symbol refers to one variable illustrating that pattern, in the order FL, FAP, FML. Symbols in boldface represent significant trends for the ancestor-descendant test.

Early Jurassic and survived into the Maastrichtian without any other sister taxa along its lineage. Instead, its patristic distance is made artificially low by the absence of an intervening record. This problem is most apparent at the low end of the body-size range and affects most of the ornithischian lineages. The comparatively (and disproportionately) well-studied theropods do not exhibit this problem to the same degree, nor do the large-bodied sauropodomorphs. In this light, it is particularly interesting that the size decreases in Coelurosauria and Macronaria appear to be active trends.

The presence of upper and lower bounds is more difficult to discern. If the passive pattern most accurately describes the size increase in dinosaurs, then a lower size bound likely exists. Circumstantial evidence — specifically, the total lack of adult dinosaurs below FL ~ 45 mm and nonoverlap of dinosaurian and Mesozoic mammalian size ranges — supports such an inference. Similarly, size reduction in macronarians may be a reflective response to an upper size bound. This upper bound is very similar for most nonsauropod dinosaurs (in the 10-ton range), with sauropods alone achieving sizes a full order of magnitude larger. Why sauropods were uniquely free of the size constraint evident in other groups remains a mystery.

This evident complexity is manifest across several hierarchical levels, highlighting one problem fundamental to macroevolutionary trend analysis: scale. The pattern described for Dinosauria does not hold for all its constituent clades; is it an artifact? Could the pattern for each clade be broken down further, possibly revealing ingroup patterns that are equally discordant with the larger one? At some point these clades will have been atomized to their furthest level (in the case of fossil taxa, that of specimens), but long before this point we will have ceased to focus on "macroevolutionary" patterns.

Pattern Biases and Robustness

Certainly the patterns described here are potentially sensitive to analytical biases. For example, SCP attempts to minimize the sum of squared changes across all branches of the tree but in doing so effectively minimizes (although does not eliminate) trends within the data. Thus, it is not surprising that the values for overall mean and median changes are very close to zero. In light of this, it is considered significant that SCP fails to reduce these values to zero, and this "failure" is interpreted as support for the presence of a general size trend in the data.

The data sample itself is certainly not an unbiased representation of true dinosaur diversity, but it is difficult to assess the specific effects of various potential factors. Certain time periods are poorly sampled, and others exhibit strikingly dense sampling. Yet these variations can affect taxa of all sizes alike, especially when particular times and places have no known dinosaur record whatsoever. In these cases, there is no clear bias against any specific body size.

Smaller taxa face a number of preservational biases in the fossil record, the result being that smaller taxa should be relatively less common overall. This tendency should become more pronounced in older strata, as overall preservation quality (and rock outcrop area) declines. The expected result would be a record that lacks smaller taxa in older sediments. In fact, the actual record finds that early dinosaurs are predominantly small-bodied forms. The predicted taphonomic bias should be opposite this pattern, but instead small forms are most commonly found basally in their respective clades (and therefore earlier in time).

Interestingly, larger taxa also face a sampling bias—one involving collection. Large dinosaurs are more difficult to collect and harder to prepare and curate than smaller forms. As a result, collectors often sample — and museums curate—larger dinosaurs less frequently and less thoroughly than smaller forms (see Dodson 1996, 297, endnote 36 for one such story). As a result, although many small dinosaur taxa are certainly entirely missing from the current fossil record, it is equally likely that some larger taxa have been discovered but deliberately left uncollected.

Finally, the fundamental link between these patterns and the phylogenetic history of Dinosauria means that subsequent changes and refinements in dinosaur systematics will result in alterations to these patterns. These changes are unavoidable as new taxa are discovered and placed into the existing phylogenetic context, but also as that context shifts with future study. Nevertheless, the most basic aspects of the patterns described here are likely to remain relatively robust to such changes. For the overall pattern of size increase to be reversed, the currently sampled dinosaur record would have to entirely misrepresent the actual pattern — a large number of derived, smaller taxa would have to be missing along with a large number of primitive, larger taxa.

Underlying Mechanisms

In quantitatively documenting the patterns of body-size evolution in dinosaurs, this paper provides a structure within which hypotheses of underlying causal mechanisms, scenarios of competition and purported structural limits may be framed and, ultimately, tested. Although these are not exhaustively treated here, a few points warrant discussion.

The dinosaurian pattern of body-size evolution appears to be consistent whether within-lineage or among-lineage patterns are compared, implying that it is not an artifact of the latter (e.g., caused by preferential extinction of small-bodied lineages). The predominant absence of derived small-bodied taxa in numerous clades instead suggests that there is a significant within-lineage trend for size increase. However, it is also likely that this trend is the result of several competing (and coordinating) influences rather than one single underlying mechanism, as was recently suggested for similar patterns in North American Cenozoic mammals (Alroy, 1998). The resolution of the dinosaurian record makes investigation of these potential mechanisms difficult.

Without identifying any underlying mechanism per se, it can still be noted that size changes occur passively during the evolution of dinosaurs (and many other vertebrates). However, if these trend analyses are accurate, then dinosaurs are not characterized by a uniform pattern of body-size increase (or even of body-size change). Instead, numerous groups show patterns suggestive of active trends, including many ornithischians but few saurischian clades. The strongest active signal is present in Ornithopoda, which exhibits a steady progression from small to large body sizes through the Mesozoic. A similar trend is evident in Ceratopsia and probably also in Pachycephalosauria. For the reasons discussed above, these active signals are viewed with suspicion.

Like many animal groups, dinosaurs (and most of their constituent clades) originated at small body sizes relative to their later size diversity. Possible reasons for the commonness of this pattern among animals have been discussed elsewhere (Cope, 1896; Simpson, 1953; Stanley, 1973). One consequence is that the larger populations and higher turnover rates among smaller taxa would favor the more frequent generation of descendant taxa with different body sizes. With the presence of a lower size bound (structural, ecological, or competitive), the result would be a tendency to expand the size range at its upper end until some upper size limit were reached. This general pattern is observed in dinosaurs, having originated at sizes near the minimum of their range, as well as in most less-inclusive clades.

More recently, Burness et al. (2001) noted a strong correlation between the maximum body size in endotherms and ectotherms and the land area occupied by these taxa, which they linked to metabolic requirements. However, the authors noted that dinosaurs seem to achieve significantly larger sizes than would be predicted by any of the metabolic equations, based on estimated Mesozoic land areas. This suggests that whereas land area may impose one constraint on maximum body size, it may not represent the only such constraint. An additional factor (or factors) may act to restrict maximum body sizes for mammals below that seen in dinosaurs. For example, Farlow (1993) suggested that theropods were able to achieve significantly larger body sizes than carnivorous terrestrial mammals due to a combination of factors, potentially including higher population densities, lower metabolic rates, and higher turnover rates. Similar differences between dinosaurs and mammals may have extended to other dinosaur groups as well (Farlow et al., 1995).

Janis and Carrano (1992) suggested that mammals are constrained below a certain maximum size due to the positively allometric relationship between gestation period and body size. Increasingly large mammals require disproportionately long gestation times, which in turn limit reproductive turnover rates and increase susceptibility to population perturbations. Dinosaurs, as oviparous animals, were likely free from this constraint and able to achieve very large body sizes with little or no impact on reproductive turnover. This assumes that both groups (or, indeed, any group) eventually would have evolved both smaller and larger taxa unless and until some constraint prevented them from further doing so. However, birds do not appear to have benefited from oviparity in the same manner. Perhaps birds (even flightless forms) were constrained from reaching very large body sizes by the numerous biological modifications that occurred during the origin of flight, much as they appear to be constrained from evolving viviparity (Blackburn & Evans, 1986).

Conclusions

This study clearly demonstrates a significant, consistent pattern of body size increase in nonavian dinosaurs through the Mesozoic. This pattern is evident in nearly every ingroup clade, although it is difficult to detect in poorly sampled ones (e.g., Pachycephalosauria). Only two clades—Coelurosauria and Macronaria—are characterized by size decreases. These patterns are detected when either phylogenetic or temporal data are used, but the former allows more precision and alleviates problems associated with long unsampled lineages.

Trend analyses produce some conflicting interpretations of these patterns, but most size increases are ascribed to active patterns in ancestor-descendant and some minimum and modified minimum tests. Subclade test results are inconsistent. The overall pattern for Dinosauria appears to be due to passive trends, and there is some evidence of the existence of both lower and upper bounds. Size increases in ornithischians tend to be more strongly described as active compared with those in saurischians, but this is likely skewed by the poor record of small-bodied forms. Sauropods appear to have reached the upper bound on size, and macronarians may have decreased in size preferentially partially in response to this.

These patterns are unlikely to be the result of various sampling or methodological biases, despite the obvious incompleteness of the dinosaur fossil record and the dependence of this study on the underlying phylogenetic framework. In addition, the complexity of these patterns suggests that no single explanation is appropriate to describe body-size evolution in all dinosaurs. Instead, future studies should focus on less-inclusive dinosaur clades as taxa within them become more densely sampled. Parallel comparisons between dinosaur and terrestrial mammal clades are also likely to be fruitful in illuminating similarities and differences in body-size evolution in these two phylogenetically independent groups.

Acknowledgments

This paper is dedicated to Jim Hopson in thanks for years of instruction and guidance, both deliberate and unintentional, direct and through example. I am also grateful to several colleagues for their advice and helpful discussions, including John Alroy, Robin O'Keefe, David Polly, and Chris Sidor. The data used in this study were gathered at many different institutions throughout the world, thanks to the hospitality of many curators, collections managers, and other colleagues. Finally, this manuscript was improved thanks to the comments of two anonymous reviewers.

Literature Cited

Alexander, R. M. 1985. Mechanics of posture and gait of some large dinosaurs. *Zoological Journal of the Linnean Society* 83:1–25.

———. 1996. Biophysical problems of small size in vertebrates; pp. 3–14 *in* P. J. Miller (ed.), *Miniature Vertebrates: The Implications of Small Body Size.* London: Academic Press.

Alexander, R. M., A. S. Jayes, G. M. O. Maloiy, and E. M. Wathuta. 1979. Allometry of limb bones of mammals from shrews (*Sorex*) to elephant (*Loxodonta*). *Journal of Zoology, London* 189:305–314.

Alroy, J. 1998. Cope's Rule and the evolution of body mass in North American fossil mammals. *Science* 280:731–734.

———. 2000a. Understanding the dynamics of trends within evolving lineages. *Paleobiology* 26(3):319–329.

———. 2000b. New methods for quantifying macroevolutionary patterns and processes. *Paleobiology* 26(4):707–733.

Anderson, J. F., A. Hall-Martin, and D. A. Russell. 1985. Long-bone circumference and weight in mammals, birds and dinosaurs. *Journal of Zoology, London (A)* 207:53–61.

Arnold, A. J., D. C. Kelly, and W. C. Parker. 1995. Causality and Cope's Rule: evidence from the planktonic foraminifera. *Journal of Paleontology* 69(2): 203–210.

Bakker, R. T. 1975. Experimental and fossil evidence for the evolution of tetrapod bioenergetics; pp. 365–399 *in* D. M. Gates and R. B. Schmerl (eds.), *Perspectives of Biophysical Ecology.* New York: Springer Verlag.

Benton, M. J. 1990. Scientific methodologies in collision: the history of the study of the extinction of the dinosaurs. *Evolutionary Biology* 24:371–400.

Benton, M. J. and R. Hitchin. 1997. Congruence between phylogenetic and stratigraphic data on the history of life. *Philosophical Transactions of the Royal Society of London B* 264:885–890.

Biewener, A. A. 1989. Scaling body support in mammals: Limb posture and muscle mechanics. *Science* 245:45–48.

Blackburn, D. G. and H. E. Evans. 1986. Why are there no viviparous birds? *The American Naturalist* 128:165–190.

Bou, J., A. Casinos, and J. Ocaña. 1987. Allometry of the limb long bones of insectivores and rodents. *Journal of Morphology* 192:113–123.

Burness, G. P., J. Diamond, and T. Flannery. 2001. Dinosaurs, dragons, and dwarfs: the evolution of maximal body size. *Proceedings of the National Academy of Sciences, USA* 98(25):14518–14523.

Carrano, M. T. 1998. Locomotion in non-avian dinosaurs: integrating data from hindlimb kinematics, in vivo strains, and bone morphology. *Paleobiology* 24(4):450–469.

———. 2000. Homoplasy and the evolution of dinosaur locomotion. *Paleobiology* 26(3):489–512.

———. 2001. Implications of limb bone scaling, curvature and eccentricity in mammals and non-avian dinosaurs. *Journal of Zoology* 254:41–55.

Carrano, M. T., S. D. Sampson, and C. A. Forster. 2002. The osteology of *Masiakasaurus knopfleri,* a small abelisauroid (Dinosauria: Theropoda) from the Late Cretaceous of Madagascar. *Journal of Vertebrate Paleontology* 22(3): 510–534.

Christiansen, P. 1999. Scaling of the limb long bones to body mass in terrestrial mammals. *Journal of Morphology* 239:167–190.

Colbert, E. H. 1962. The weights of dinosaurs. *American Museum Novitates* 2076:1–16.

Cope, E. D. 1887. *The Origin of the Fittest.* New York: Appleton and Company.

———. 1889. On a new genus of Triassic Dinosauria. *The American Naturalist* 23:626.

———. 1896. *The Primary Factors of Organic Evolution.* Chicago: Open Court Publishing Company.

Cubo, J. and A. Casinos. 1997. Flightlessness and long bone allometry in Palaeognathiformes and Sphenisciformes. *Netherlands Journal of Zoology* 47(2): 209–226.

Cunningham, C. W. 1999. Some limitations of ancestral character-state reconstruction when testing evolutionary hypotheses. *Systematic Biology* 48: 665–674.

Cunningham, C. W., K. E. Omland, and T. H. Oakley. 1998. Reconstructing ancestral character states: a critical reappraisal. *Trends in Ecology and Evolution* 13:361–366.

Curry Rogers, K. A. and C. A. Forster. 2001. The last of the dinosaur titans: a new sauropod from Madagascar. *Nature* 412:530–534.

Damuth, J. 1981. Population density and body size in mammals. *Nature* 290: 699–700.

Dodson, P. 1996. *The Horned Dinosaurs: A Natural History.* Princeton, NJ: Princeton University Press.

Dodson, P., C. A. Forster, and S. D. Sampson. 2004. The Ceratopsidae; pp. 494–513 *in* D. B. Weishampel, P. Dodson, and H. Osmólska (eds.), *The Dinosauria* (2nd ed.). Berkeley: University of California Press.

Farlow, J. O. 1993. On the rareness of big, fierce animals: speculations about the body sizes, population densities, and geographic ranges of predatory mammals and large carnivorous dinosaurs. *American Journal of Science* 293-A:167–199.

Farlow, J. O., P. Dodson, and A. Chinsamy. 1995. Dinosaur biology. *Annual Review of Ecology and Systematics* 26:445–471.

Felsenstein, J. 1985. Phylogenies and the comparative method. *The American Naturalist* 125(1):1–15.

Fisher, D. C. 1986. Dynamics of diversification in state space; pp. 91–108 *in* M. L. McKinney and J. A. Drake (eds.), *Biodiversity Dynamics: Turnover of Populations, Taxa, and Communities.* New York: Columbia University Press.

Gittleman, J. L. and A. Purvis. 1998. Body size and species-richness in carnivores and primates. *Proceedings of the Royal Society of London B* 265: 113–119.

Henderson, D. M. 1999. Estimating the masses and centers of mass of extinct animals by 3-D mathematical slicing. *Paleobiology* 25(1):88–106.

Holtz, T. R., Jr. 2000. A new phylogeny of the carnivorous dinosaurs. *GAIA* 15:5–61.

Horner, J. R., D. B. Weishampel, and C. A. Forster. 2004. The Hadrosauridae; pp. 438–463 *in* D. B. Weishampel, P. Dodson, and H. Osmólska (eds.), *The Dinosauria* (2nd ed.). Berkeley: University of California Press.

Huelsenbeck, J. P., and J. P. Bollback. 2001. Empirical and hierarchical Bayesian estimation of ancestral states. *Systematic Biology* 50:351–366.

Huene, F. v. 1914. Das natürliche System der Saurischia. *Centralblatt für Mineralogie, Geologie und Paläontologie* 1914:154–158.

Jablonski, D. 1996. Body size and macroevolution; pp. 256–289 *in* D. Jablonski, D. H. Erwin, and J. H. Lipps (eds.), *Evolutionary Paleobiology.* Chicago: University of Chicago Press.

———. 1997. Body-size evolution in Cretaceous molluscs and the status of Cope's Rule. *Nature* 385:250–252.

Janis, C. M. and M. Carrano. 1992. Scaling of reproductive turnover in archosaurs and mammals: why are large terrestrial mammals so rare? *Annales Zoologici Fennici* 28:201–216.

Jianu, C.-M. and D. B. Weishampel. 1999. The smallest of the largest: a new look at possible dwarfing in sauropod dinosaurs. *Geologie en Mijnbouw* 78: 335–343.

Juanes, F. 1986. Population density and body size in birds. *The American Naturalist* 128:921–929.

Jungers, W. L., D. B. Burr, and M. S. Cole. 1998. Body size and scaling of long bone geometry, bone strength, and positional behavior in cercopithecoid primates; pp. 309–330 *in* E. Strasser, J. Fleagle, A. Rosenberger, and H. McHenry (eds.), *Primate Locomotion: Recent Advances.* New York: Plenum Press.

Kurtén, B. 1953. On the variation and population dynamics of fossil and Recent mammal populations. *Acta Zoologica Fennica* 76:1–122.

LaBarbera, M. 1986. The evolution and ecology of body size; pp. 69–98 *in* D. M. Raup and D. Jablonski (eds.), *Patterns and Processes in the History of Life.* Berlin: Springer-Verlag.

———. 1989. Analyzing body size as a factor in ecology and evolution. *Annual Review of Ecology and Systematics* 20:97–117.

Lindstedt, S. L. and W. A. Calder III. 1981. Body size, physiological time, and longevity of homeothermic animals. *The Quarterly Review of Biology* 56(1): 1–16.

MacFadden, B. J. 1986. Fossil horses from "Eohippus" (*Hyracotherium*) to *Equus:* scaling, Cope's law, and the evolution of body size. *Paleobiology* 12(4):355–369.

Maddison, W. P. and D. R. Maddison. 2000. MacClade 4.0. Sunderland, MA: Sinauer Associates.

Maddison, W. P., M. J. Donoghue, and D. R. Maddison. 1984. Outgroup analysis and parsimony. *Systematic Zoology* 33:83–103.

Maloiy, G. M. O., R. M. Alexander, R. Njau, and A. S. Jayes. 1979. Allometry of the legs of running birds. *Journal of Zoology, London* 187:161–167.

McMahon, T. A. 1975. Using body size to understand the structural design of animals: quadrupedal locomotion. *Journal of Applied Physiology* 39(4): 619–627.

McNab, B. K. 1989. Basal rate of metabolism, body size, and food habits in the Order Carnivora; pp. 335–354 *in* J. L. Gittleman (ed.), *Carnivore Behavior, Ecology, and Evolution.* Ithaca, NY: Cornell University Press.

McShea, D. W. 1994. Mechanisms of large-scale evolutionary trends. *Evolution* 48(6):1747–1763.

———. 1998. Progress in organismal design; pp. 99–117 *in* D. M. Raup and D. Jablonski (eds.), *Patterns and Processes in the History of Life.* Berlin: Springer Verlag.

———. 2000. Trends, tools, and terminology. *Paleobiology* 26(3):330–333.

Paul, G. S. 1988. *Predatory Dinosaurs of the World: A Complete Illustrated Guide.* New York: Simon & Schuster.

Peczkis, J. 1994. Implications of body-mass estimates for dinosaurs. *Journal of Vertebrate Paleontology* 14(4):520–533.

Polly, P. D. 1998. Cope's Rule. *Science* 282:50–51.

Scheetz, R. 1999. Osteology of *Orodromeus makelai* and the phylogeny of basal ornithopod dinosaurs. Unpublished Ph.D. thesis, Montana State University, Bozeman.

Schmidt-Nielsen, K. 1984. *Scaling: Why Is Animal Size So Important?* New York: Cambridge University Press.

Schultz, T. R., R. B. Cocroft, and G. A. Churchill. 1996. The reconstruction of ancestral character states. *Evolution* 50:504–511.

Seebacher, F. 2001. A new method to calculate allometric length-mass relationships of dinosaurs. *Journal of Vertebrate Paleontology* 21(1):51–60.

Sereno, P. C. 1997. The origin and evolution of dinosaurs. *Annual Reviews of Earth and Planetary Sciences* 25:435–489.

———. 1999. The evolution of dinosaurs. *Science* 284:2137–2147.

Sidor, C. A. 2001. Simplification as a trend in synapsid cranial evolution. *Evolution* 55(7):1419–1442.

Simpson, G. G. 1953. *The Major Features of Evolution.* New York: Columbia University Press.

Stanley, S. M. 1973. An explanation for Cope's Rule. *Evolution* 27(1):1–26.

Talbot, M. 1911. *Podokesaurus holyokensis,* a new dinosaur from the Triassic of the Connecticut Valley. *American Journal of Science* 31(186):469–479.

Thompson, D. W. 1917. *On Growth and Form.* Cambridge: Cambridge University Press.

Trammer, J. and A. Kaim. 1997. Body size and diversity exemplified by three trilobite clades. *Acta Palaeontologica Polonica* 42(1):1–12.

———. 1999. Active trends, passive trends, Cope's Rule and temporal scaling: new categorization of cladogenetic changes in body size. *Historical Biology* 13:113–125.

Wilson, J. A. 2002. Sauropod dinosaur phylogeny: critique and cladistic analysis. *Zoological Journal of the Linnean Society* 136:217–276.

Wilson, J. A. and M. T. Carrano. 1999. Titanosaurs and the origin of "wide-gauge" trackways: a biomechanical and systematic perspective on sauropod locomotion. *Paleobiology* 25(2):252–267.

Wilson, J. A. and P. C. Sereno. 1998. Early evolution and higher-level phylogeny of sauropod dinosaurs. *Journal of Vertebrate Paleontology Memoir 5,* 18(2, suppl.):1–68.

Appendix 8.1.

Body-size measurements used in this study: FL, femoral length; FAP, femoral anteroposterior diameter; FML, femoral mediolateral diameter. Original measurements in millimeters. Data were log-transformed before analysis.

Institutional abbreviations: AM, Albany Museum, Grahamstown, South Africa; AMNH, American Museum of Natural History, New York; ANSP, Academy of Natural Sciences, Philadelphia; BHI, Black Hills Institute, Hill City, South Dakota; BMNH, The Natural History Museum, London; BSP, Bayerische Staatssammlung für Paläontologie, München; BYU, Brigham Young University, Provo, Utah; CEUM, College of Eastern Utah Prehistoric Museum, Price; CH/P.W., Department of Mineral Resources, Bangkok; CM, Carnegie Museum of Natural History, Pittsburgh; CMNH, Cleveland Museum of Natural History, Ohio; CV, Municipal Museum of Chungqing, China; FMNH, Field Museum of Natural History, Chicago; GMNH, Gunma Museum of Natural History, Japan; GMV, China Geological Museum, Beijing; HMN, Humboldt Museum für Naturkunde, Berlin, Germany; IGM/GIN, Institute of Geology, Ulan Baatar, Mongolia; IRSNB, Institute Royale des Sciences Naturelles de Belgique, Brussels; ISI, Indian Statistical Institute, Kolkata; IVPP, Institute of Palaeoanthropology and Palaeontology, Beijing; JM, Jura-Museum, Eichstatt, Germany; KMV, Kunming Museum of Vertebrate Palaeontology, China; LV, Laboratory of Vertebrate Palaeontology, Geological Survey of China, Beijing; MACN, Museo Argentino de Ciencias Naturales "Bernardino Rivadavia," Buenos Aires; MAFI, Geological Institute of Hungary, Budapest; MCN, Museu de Ciências Naturais, Rio Grande do Sul, Brazil; MCNA, Museo de Ciencias Naturales de Alava, Vitoria-Gasteiz, Spain; MCZ, Museum of Comparative Zoology, Harvard University, Cambridge; MDE, Musée des Dinosaures, Esperaza, France; MIWG, Museum of Isle of Wight Geology, Sandown, U.K.; ML, Museu da Lourinha, Portugal; MLP, Museo de La Plata, Argentina; MNA, Museum of Northern Arizona, Flagstaff; MNHN, Muséum National d'Histoire Naturelle, Paris; MNN, Musée National d'Histoire Naturelle, Niamey, Niger; MNUFR, Mongolian National University, Ulan Baatar; MOR, Museum of the Rockies, Bozeman, Montana; MPCA, Museo Provincial "Carlos Ameghino," Cipoletti, Argentina; MPEF, Museo Paleontolòlogico "Egidio Feruglio," Trelew, Argentina; MUCP, Museo de Ciencias Naturales de la Universidad Nacional del Comahue, Neuquén, Argentina; NCSM, North Carolina State Museum of Natural Sciences, Raleigh; NGMC, National Geological Museum, Beijing; NIGP, Nanjing Institute of Geology and Palaeontology, China; NMC, Canadian Museum of Nature, Ottawa; OMNH, Sam Noble Oklahoma Museum of Natural History, University of Oklahoma, Norman; OUM, Oxford University Museum, UK; PIN, Paleontological Institute, Moscow; PMU, Palaeontological Museum, University of Uppsala, Sweden; POL, Musée de Poligny, Jura, France; PVSJ, Museo Provincial de San Juan, Argentina; PVL, Instituto Miguel Lillo, Tucumán, Argentina; PVPH, Paleontología de Vertebrados, Museo "Carmen Funes," Plaza

Huincul, Argentina; QG, Queen Victoria Museum, Harare, Zimbabwe; ROM, Royal Ontario Museum, Toronto; RTMP, Royal Tyrrell Museum of Palaeontology, Drumheller, AB; SAM, South African Museum, Cape Town; SGM, Société Geologique de Morocco, Rabat; SMNS, Staatliches Museum für Naturkunde, Stuttgart; SMU, Shuler Museum of Paleontology, Southern Methodist University, Dallas; TATE, Tate Geological Museum, Casper College, Wyoming; UA, Université d'Antananarivo, Madagascar; UALB, University of Alberta, Edmonton, Canada; UCMP, University of California Museum of Paleontology, Berkeley; UC OBA, Department of Organismal Biology and Anatomy, University of Chicago; UHR, Hokkaido University, Sapporo, Japan; UMNH, Utah Museum of Natural History, Salt Lake City; UNPSJB, Universidad Nacional de la Patagonia "San Juan Bosco," Comodoro Rivadavia, Argentina; UPLR, Universidad Provincial de La Rioja, Argentina; USNM, National Museum of Natural History, Smithsonian Institution, Washington, DC; YPM, Peabody Museum of Natural History, Yale University, New Haven, Connecticut; ZDM, Zigong Dinosaur Museum, Dashanpu, China; ZPAL, Muzeum Ziemi Polska Akademia Nauk, Warsaw.

Table A8.1. List of taxa, specimens, and measurements used in this study

Taxon	Specimen	FL	FAP	FML
Dinosauriformes				
Lagerpeton chanarensis	PVL 06	76.8	7.7	
Lagosuchus talampayensis	UPLR 09	40.6	2.5	122.2
Lewisuchus admixtus	PVL 4629	1115.0		
	PVL 3456		112.0	
	PVL 3454			7.7
Marasuchus lilloensis	PVL 3871	56.3		5.1
	PVL 3870		3.7	
Dinosauria				
Saurischia				
Theropoda				
Eoraptor lunensis	PVSJ 512	154.0	19.6	12.7
Guaibasaurus candelariensis	MCN-PV 2355	214.0	25.0	19.0
Herrerasaurus ischigualastensis	PVL 2566	482.0	54.4	55.1
Staurikosaurus pricei	MCZ 1669	220.0	24.7	
Coelophysoidea				
Coelophysis bauri	UCMP 129618	245.0	22.0	19.0
"Dilophosaurus" sinensis	KMV 8701	587.0		
Dilophosaurus wetherilli	UCMP 37302	552.0	35.9	67.8
Liliensternus liliensterni	HMN R.1291	424.0	37.8	31.1

Table A8.1. *(continued)*

Taxon	Specimen	FL	FAP	FML
Procompsognathus triassicus	SMNS 12951	96.0	9.4	7.1
Sarcosaurus woodi	BMNH 4840	321.0	31.4	35.9
Segisaurus halli	UCMP 32101	142.9		
Syntarsus kayentakatae	MNA V2623	272.0	21.6	23.1
Syntarsus rhodesiensis	QG 1	203.0	16.5	17.0
Neoceratosauria				
Carnotaurus sastrei	MACN-CH 894	1018.0	95.6	94.6
Ceratosaurus nasicornis	UMNH VP 5278	759.0	71.7	90.4
Deltadromeus agilis	BSP 1912 VIII 70	1230.0	138.0	147.0
Elaphrosaurus bambergi	HMN Gr. S. 38–44	529.0	59.0	45.0
Genusaurus sisteronis	MNHN Bev-1	390.0	34.1	33.7
"Indosuchus raptorius"	ISI R 401–454	872.0	77.0	
Ligabueino andesi	MACN-N 42	81.4		7.0
Masiakasaurus knopfleri	UA 8681	202.5		
	UA 8684		21.5	19.8
Xenotarsosaurus bonapartei	UNPSJB-Pv 184	611.0	77.0	69.0
Tetanurae				
Metriacanthosaurus parkeri	OUM J.12114	849.0	85.0	96.1
Piatnitzkysaurus floresi	PVL 4073	548.0	67.3	91.3
Spinosauroidea				
Baryonyx walkeri	BMNH R.9951		92.0	100.0
"Brontoraptor sp.*"*	TATE 1012	832.6	110.9	123.6
Afrovenator abakensis	UC OBA 1	761.0	65.0	83.0
Eustreptospondylus oxoniensis	OUM J.13558	510.0	56.1	56.2
Megalosaurus bucklandi	OUM mount	830.0	102.7	111.3
Suchomimus tenerensis	MNN GDF 500	1080.0	85.7	145.2
Allosauroidea				
Acrocanthosaurus atokensis	NCSM 14345	1180.0		
	SMU 74646		189.7	214.2
Allosaurus fragilis	AMNH 630	1001.0		133.6
	AMNH 680		115.7	
Carcharodontosaurus saharicus	BSP 1922 X46	1260.0		126.0
Giganotosaurus carolinii	MUCPv-CH-1	1350.0	151.0	
Neovenator salerii	MIWG 6348	780.0		90.0
Saurophaganax maximus	OMNH 01123	1135.0	107.0	150.0
Sinraptor dongi	IVPP 87001	869.5	78.7	91.4
Sinraptor hepingensis	ZDM 0024	995.0		100.0

(continued)

Table A8.1. (*continued*)

Taxon	Specimen	FL	FAP	FML
Yangchuanosaurus magnus	CV 00216	1200.0		
Yangchuanosaurus shangyouensis	CV 00215	850.0	96.0	
Coelurosauria				
Coelurus fragilis	YPM 1991	220.0	17.0	
Compsognathus longipes	MNHN CNJ 79	109.7	9.2	
Dryptosaurus aquilunguis	ANSP 9995/10006	778.0	83.5	88.0
Nedcolbertia justinhoffmani	CEUM 5071	144.8		18.2
Nqwebasaurus thwazi	AM 6040		11.0	
Ornitholestes hermanni	AMNH 619	215.0	17.0	24.1
Sinosauropteryx prima	NIGP 127587	86.4		
	NIGP 127586		6.0	
Tyrannosauroidea				
Albertosaurus sarcophagus	ROM 807	1066.0	85.2	123.6
Alectrosaurus olseni	AMNH 6554	661.0	72.3	
Daspletosaurus n. sp.	OMNH 10131	1033.0	118.0	132.0
Daspletosaurus torosus	AMNH 5438	1030.0		108.3
	FMNH 5336		125.3	
Gorgosaurus libratus	NMC 2120	1040.0		
	NMC 530		133.0	103.0
Tarbosaurus bataar	PIN 551-1	1200.0	110.0	
	PIN 551-2			110.0
Tyrannosaurus rex	FMNH PR 2081	1342.5		188.1
	BHI 3033		168.0	
Ornithomimosauria				
Anserimimus planinychus	GIN AN MPR 100/300	433.0		
Archaeornithomimus asiaticus	AMNH 6570	402.0	46.0	
Dromiceiomimus brevitertius	NMC 12228	468.0		
	ROM 852		41.0	33.5
Gallimimus bullatus	ZPAL MgD-I/1	680.0	63.1	59.2
Ornithomimus edmontonensis	NMC 12441	500.0		
	ROM 851		29.2	25.3
Struthiomimus altus	AMNH 5375	501.0	39.8	
	AMNH 5339			44.0
Alvarezsauria				
Alvarezsaurus calvoi	MUCPv-54		9.7	10.3
Mononykus olecranus	GIN 107/6	132.5	11.5	10.3
Parvicursor remotus	PIN 4487/25	52.6	3.3	3.1

Table A8.1. *(continued)*

Taxon	Specimen	FL	FAP	FML
Oviraptorosauria				
Alxasaurus elesitaiensis	IVPP 88402	555.0	70.0	70.0
Avimimus portentosus	PIN 3907/1	185.9	16.3	14.3
Beipiaosaurus inexpectus	IVPP V.11559	260.0		38.0
Caudipteryx dongi	IVPP 12344	149.0	16.0	17.0
Caudipteryx zoui	NGMC 97-9-A	149.0		
Chirostenotes pergracilis	RTMP 79.30.1	311.0	16.6	30.7
Ingenia yanshini	GIN 100/30	228.1	22.2	22.8
Microvenator celer	AMNH 3041	122.8	11.2	10.7
Oviraptor philoceratops	AMNH 6517	262.0		
Segnosaurus galbinensis	GIN 100/82	840.0		
Deinonychosauria				
Achillobator giganticus	MNUFR-15	505.0		67.3
Deinonychus antirrhopus	MCZ 4371	335.0	32.1	37.8
Saurornithoides mongoliensis	AMNH 6516	198.0		
	IVPP V.10597		6.5	6.5
Saurornitholestes langstoni	RTMP 88.121.39	212.0	22.1	15.6
Sinornithoides youngi	IVPP V.9612	140.0	11.0	
Sinornithosaurus millenii	IVPP V.12811	148.0	9.5	
Troodon formosus	MOR 748	317.0		
	MOR 553s		32.8	32.9
Velociraptor mongoliensis	IGM 100/988	238.0	25.2	26.1
Protarchaeopteryx robusta	GMV 2125	120.0		16.0
Archaeopteryx lithographica	Solnhofen specimen	70.0		
	JM SoS 2257		2.9	
Rahonavis ostromi	UA 8656	86.9	6.0	5.5
Prosauropoda				
Ammosaurus major	YPM 208	221.0	33.2	42.3
Anchisaurus polyzelus	YPM 1883/2128	210.0	20.0	24.3
Camelotia borealis	BMNH R.2870	985.7	114.2	132.8
Euskelosaurus brownii	SAM 3349/02	590.0	70.0	88.0
Gyposaurus sinensis	IVPP V.27	235.5		
	IVPP V.26		28.0	
	IVPP V.43			27.7
Jingshanosaurus xinwaensis	LV003	845.0		
Lufengosaurus huenei	IVPP V.98	780.0		
	IVPP V.82			115.0

(continued)

Table A8.1. (*continued*)

Taxon	Specimen	FL	FAP	FML
Massospondylus carinatus	QG 1159	335.0	40.6	42.1
Jachal *Massospondylus*	PVSJ uncat.	220.0	25.6	24.9
Melanorosaurus readi	SAM 3450	583.3	59.1	81.6
Plateosaurus engelhardti	SMNS uncat.	750.0	82.3	90.8
Plateosaurus poligniensis	POL 75	821.0		138.6
	POL 76		96.0	
Riojasaurus incertus	PVL 3808	608.0		106.2
	PVL 3669		67.2	
Sellosaurus gracilis	SMNS 12843	551.0		
	SMNS 17928		65.1	
	SMNS 5715			66.9
Thecodontosaurus antiquus	SMNS uncat.	280.0	43.4	
Yunnanosaurus huangi	IVPP AS V20	435.0	54.6	60.0
Sauropoda				
Gongxianosaurus shibeiensis	holotype	1164.0		270.0
Isanosaurus attavipachi	CH4-1	760.0	71.5	121.6
Vulcanodon karibaensis	QG 24	1100.0	140.0	174.0
Eusauropoda				
Barapasaurus tagorei	ISI R.50	1365.0	131.0	187.0
Cetiosaurus oxoniensis	OUM J.13899	1626.0		305.0
Datousaurus bashanensis	IVPP V.7262	1057.0	147.0	
Kotasaurus yamanpalliensis	111/S1Y/76	1130.0	80.0	160.0
Lapparentosaurus madagascariensis	"individu taille max"	1590.0	240.0	
Mamenchisaurus constructus	IVPP V.948	1280.0	207.0	
Mamenchisaurus hochuanensis	IVPP holotype	860.0		
Omeisaurus junghsiensis	IVPP AS holotype		103.0	
Omeisaurus tianfuensis	ZDM T5701	1310.0		206.0
Patagosaurus fariasi	PVL 4076	1542.0	135.5	
	PVL 4170			255.0
Shunosaurus lii	IVPP V.9065	1250.0		188.0
Volkheimeria chubutensis	PVL 4077	1156.0	148.0	75.1
Neosauropoda				
Diplodocoidea				
Amargasaurus cazaui	MACN-N 15	1050.0	128.8	180.0
Amphicoelias altus	AMNH 5764	1770.0	210.0	216.0
Apatosaurus ajax	YPM 1860	2500.0		
Apatosaurus excelsus	FMNH 7163	1830.0	310.0	310.0

Table A8.1. *(continued)*

Taxon	Specimen	FL	FAP	FML
Apatosaurus louisae	CM 3018	1785.0	174.0	332.3
Barosaurus africanus	HMN NW 4	1361.0	150.6	204.2
Barosaurus lentus	AMNH 6341	1440.0	120.2	204.3
Cetiosauriscus stewarti	BMNH R.3078	1360.0	190.0	195.0
Dicraeosaurus hansemanni	HMN m	1220.0	142.5	
	HMN dd 3032			192.3
Diplodocus carnegii	CM 84	1542.0	174.0	
	CM 94			186.0
Diplodocus longus	YPM 1920	1645.0		
	AMNH 223		143.0	
Haplocanthosaurus delfsi	CMNH 10380	1745.0		
	CM 572		207.0	
Rayososaurus tessonei	MUCPv-205	1440.0		220.0
Macronaria				
"Bothriospondylus madagascariensis"	MNHN uncat.	1460.0	110.0	
Brachiosaurus altithorax	FMNH P25107	2000.0		365.0
Brachiosaurus brancai	HMN St	1913.0	151.7	299.0
Camarasaurus lentus	DINO 4514	1470.0		252.0
	CM 11338		86.5	
Camarasaurus supremus	AMNH 5761a	1800.0	255.0	
	GMNH-PV 101			228.0
Euhelopus zdanskyi	PMU R234	955.0	100.0	142.0
Titanosauriformes				
Aegyptosaurus baharijensis	BSP 1912 VIII 61	1290.0	75.0	223.0
Ampelosaurus atacis	MDE uncat. 1	802.0		157.5
	MDE uncat. 2		66.0	
Andesaurus delgadoi	MUCPv-132	1550.0		226.0
Argyrosaurus superbus	PVL 4628	1910.0	160.0	300.0
Chubutisaurus insignis	MACN 18222	1715.0	265.0	
Janenschia robusta	HMN IX	1330.0		
	HMN P		131.5	188.8
Laplatasaurus araukanicus	MLP-Av 1047/1128	1000.0		
Lirainosaurus astibiae	MCNA 7468	686.0		97.0
Magyarosaurus dacus	BMNH R.3856	488.0	43.7	66.8
Neuquensaurus australis	MLP-Cs 1121/1103	700.0	110.0	
Neuquensaurus robustus	MLP-Cs 1094	799.0		134.5
	MLP-Cs 1480		120.0	

(continued)

Table A8.1. (continued)

Taxon	Specimen	FL	FAP	FML
Opisthocoelicaudia skarzynskii	ZPAL MgD-I/48	1395.0	108.0	280.0
Phuwiangosaurus sirinhornae	P.W. 1-1/1-21	1250.0	85.0	215.0
Rapetosaurus krausei	SUNY uncat.	687.0	91.2	63.1
Rocasaurus muniozi	MPCA-Pv 56	768.0		117.0
Saltasaurus loricatus	PVL 4017-80	875.0		164.6
	PVL 4017-79		90.0	
Tehuelchesaurus benitezii	MPEF-PV 1125	1530.0		243.0
Titanosaurus indicus	BMNH R.5934	865.0	67.3	80.9
Ornithischia				
Lesothosaurus diagnosticus	BMNH RU B.17	102.0	10.5	12.3
Pisanosaurus mertii	PVL 2577	170.3	9.8	6.6
Marginocephalia				
Ceratopsia				
Psittacosauridae				
Psittacosaurus mongoliensis	AMNH 6254	162.0		
	AMNH 6541			22.9
Psittacosaurus neimongoliensis	IVPP 12-0888-2	129.5	15.0	
Psittacosaurus sinensis	IVPP V.740-741	96.0	11.8	
Neoceratopsia				
Microceratops gobiensis	holotype	123.0		12.0
Montanoceratops cerorhynchus	MOR 300	346.0	33.7	42.6
Protoceratops andrewsi	AMNH 6416	226.0		
	AMNH 6424		28.3	29.0
Ceratopsidae				
Achelousaurus horneri	MOR 591	612.0	68.2	85.9
Centrosaurus apertus	AMNH 5427	800.0		
	YPM 2015		99.3	111.6
Chasmosaurus belli	ROM 839	825.0		
	BMNH R.4948		71.1	130.8
Chasmosaurus mariscalensis	UTEP P.37.3.031	676.5		97.5
Pachyrhinosaurus n. sp.	RTMP 87.55.57	676.0	99.6	106.5
Pentaceratops sternbergi	OMNH 10165	1096.0		226.5
	PMU R286		114.5	
Styracosaurus albertensis	AMNH 5372	749.0	112.0	
Triceratops horridus	AMNH 5033	1033.5	105.2	175.2
Triceratops prorsus	USNM 4842	1104.0	120.0	190.0

Table A8.1. *(continued)*

Taxon	Specimen	FL	FAP	FML
Pachycephalosauria				
Homalocephale calathoceros	GI SPS 100/1201	218.0		25.2
Micropachycephalosaurus hongtuyanensis	IVPP V.5542	125.8	14.2	
Prenocephale prenes	ZPAL MgD-I/104	221.5	22.0	26.0
Stegoceras validum	UALB 2	221.0	19.5	17.9
Stygimoloch spinifer	RTMP 97.27.4	441.0	46.7	69.1
Wannanosaurus yansiensis	IVPP V.447.1	90.0		
Thyreophora				
Scelidosaurus harrisonii	BMNH R.1111	403.0	56.5	52.4
Scutellosaurus lawleri	MNA.Pl.1752	114.0		
	MCZ 8797		19.2	12.2
Stegosauria				
Chialingosaurus kuani	IVPP V.2300	690.0		101.0
Chungkingosaurus sp. 2	CV 00205	670.0		110.0
Dacentrurus armatus	BMNH R.46013	1232.0		176.0
				135.0
Huayangosaurus taibaii	ZDM T7001	475.0		75.5
Kentrosaurus aethiopicus	HMN bbl	784.0	88.0	120.0
Lexovisaurus durobrivensis	BMNH R.1989	979.0	82.0	138.0
Stegosaurus stenops	YPM 1387	1157.0		
	AMNH 650		78.5	172.0
Stegosaurus ungulatus	YPM 1853	1348.0		
	YPM 1858		101.0	168.0
Tuojiangosaurus multispinus	CV 00209	872.0		141.0
Ankylosauria				
Ankylosauridae				
Ankylosaurus magniventris	AMNH 5214	660.0	76.4	136.8
Euoplocephalus tutus	AMNH 5404	547.5	85.6	
	ROM 784			103.2
Pinacosaurus grangeri	ZPAL MgD-II/1	243.0	23.0	42.0
Shanxia tianzhenensis	IVPP V.11276	402.6		67.7
Nodosauridae				
Edmontonia longiceps	NMC 8531	660.0		
Gastonia burgei	CEUM 1307	574.4		103.8
Hoplitosaurus marshi	USNM 4572	476.0	41.0	90.0
Nodosaurus textilis	YPM 1815	609.0	63.4	101.1

(continued)

Table A8.1. *(continued)*

Taxon	Specimen	FL	FAP	FML
Polacanthus foxii	BMNH R.175	548.0	41.0	
Sauropelta edwardsi	AMNH 3032	764.0	88.0	104.0
Struthiosaurus transylvanicus	MCNA 6531/6540	300.0		
	BMNH R.11011		26.5	33.5
Ornithopoda				
Heterodontosauridae				
Abrictosaurus consors	BMNH RU B.54	77.1	8.4	6.2
Heterodontosaurus tucki	SAM K337	113.2	12.8	8.9
Euornithopoda				
Agilisaurus louderbacki	ZDM 6011	198.5	24.5	
Agilisaurus multidens	ZDM T6001	153.4	18.8	11.9
Fulgurotherium australe	NMV P208186	186.5	24.5	
	NMV P186326			17.4
Hypsilophodon foxii	BMNH R.5829	200.0		
	BMNH R.192a		22.2	21.2
Laosaurus consors	YPM 1882	246.0	29.5	27.1
Leaellynasaura amicagraphica	NMV P186047	135.0	17.0	
	NMV P186333			12.7
Orodromeus makelai	MOR 473	166.1		
	PU 23443		116.0	15.0
Othnielia rex	BYU ESM-163R	139.1	15.8	15.4
Parksosaurus warreni	ROM 804	270.0		
Thescelosaurus neglectus	AMNH 5891	448.0	56.3	61.2
Xiaosaurus dashanpensis	IVPP V.6730A	110.0		11.6
Yandusaurus honheensis	holotype			77.0
Ankylopollexia				
Camptosaurus dispar	YPM 1877	591.0		
	YPM 1880		80.0	100.3
Camptosaurus leedsi	BMNH R.1993	311.0	40.0	
Draconyx loureiroi	ML 434		64.0	66.0
Dryosaurus altus	CM 1949	470.0		
	YPM 1876		40.3	42.1
Dryosaurus lettowvorbecki	BMNH R.12777	343.0		45.0
	BMNH R.12278		42.0	
Gasparinisaura cincosaltensis	MUCPc-208	94.7	11.7	10.3
Rhabdodon priscum	MNHN uncat.	550.0	92.0	
	BMNH 3814			43.0

Table A8.1. (*continued*)

Taxon	Specimen	FL	FAP	FML
Tenontosaurus dossi	FWMSH 93B1	557.0		84.6
Tenontosaurus tilletti	UALB 22	700.0		
	YPM 5535		70.8	81.8
Valdosaurus canaliculatus	BMNH R.184/185	135.4	15.9	14.8
Valdosaurus nigeriensis	MNHN GDF 332	224.0	26.8	26.3
Iguanodontia				
Claosaurus agilis	YPM 1190	673.0		
Gilmoreosaurus mongoliensis	AMNH 6551	704.0	84.3	
Iguanodon atherfieldensis	BMNH R.3741	840.0		
	BMNH R.5764		79.0	
	IRSNB 1551			95.4
Iguanodon bernissartensis	BMNH 2649	1090.0	121.0	185.1
Iguanodon mantelli	BMNH 2650	822.0	84.8	131.4
Muttaburrasaurus langdoni	BMNH R.9604	990.0	108.9	168.8
Nipponosaurus sachaliensis	UHR 6590	544.0	57.0	59.0
Nanyangosaurus zhugeii	IVPP V.11821	517.0	62.0	
Orthomerus dolloi	BMNH 42955	495.0	57.0	57.0
Ouranosaurus nigeriensis	MNHN GDF 381	830.7	86.2	99.7
Probactrosaurus gobiensis	PIN AN SSR 2232/1	565.0		
Telmatosaurus transsylvanicus	MAFI v.10338	731.0		104.0
Hadrosauridae				
Anatotitan copei	AMNH 5730	1150.0	140.0	
Bactrosaurus johnsoni	AMNH 6553	781.0	103.0	
Corythosaurus casuarius	AMNH 5240	1080.0	130.0	
	RTMP 80.40.1			143.7
Edmontosaurus annectens	ROM 801	1278.5		
	NMC 8399		140.8	
	ROM 867			126.0
"Gryposaurus" incurvimanus	ROM 764	1053.0	122.8	
	RTMP 80.22.1			119.6
Gryposaurus notabilis	AMNH 5465	1215.0	137.8	
Hadrosaurus foulkii	ANSP 10005	1055.0		
Hypacrosaurus altispinus	NMC 8501	1074.0		
	AMNH 5217		106.8	
	RTMP 81.10.1			100.9
Hypacrosaurus stebingeri	MOR 773	1190.0	100.1	142.4
"Kritosaurus" australis	MACN-RN 02	790.0	80.7	109.2

(*continued*)

Taxon	Specimen	FL	FAP	FML
Lambeosaurus lambei	NMC 351	1117.0		
	ROM 6474		104.7	
	RTMP 82.38.1			155.2
Lambeosaurus magnicristatus	RTMP 66.4.1	1000.0	102.8	
Maiasaura peeblesorum	MOR 005 1989	960.0	138.3	
	MOR 005 7-12-91-45			93.3
Parasaurolophus cyrtocristatus	FMNH P27393	1041.5	124.5	142.4
Parasaurolophus walkeri	ROM 768	1059.0	146.8	98.7
Prosaurolophus blackfeetensis	MOR 454	855.0	62.7	104.8
Prosaurolophus maximus	RTMP 84.1.1	1050.0		
	ROM 7871		107.9	190.1
Saurolophus angustirostris	PIN 551-8	1200.0		
Saurolophus osborni	AMNH 5220	1150.0		
	AMNH 5271		145.5	

Major Changes in the Ear Region and Basicranium of Early Mammals

Guillermo W. Rougier
and John R. Wible

Introduction

The mammalian ear region is a complex, tight space, influenced by several intricate anatomical systems, including the auditory apparatus and various cranial nerves, arteries, and veins (see reviews by Moore, 1981; Novacek, 1993). The distinctive complexity of the ear region and the arguably independent variation of the multiple characters provided by the array of vessels, nerves, and osteological features have resulted in a wealth of information for systematics and functional morphology (Kampen, 1905; Gregory, 1910; Klaauw, 1931; Simpson, 1945; McDowell, 1958; MacPhee, 1981).

The basicranium and ear region are ideally suited for paleontological study, because their numerous soft tissues leave osseous impressions in the form of grooves, canals, and foramina, allowing the reconstruction of the missing structures with various degrees of confidence (MacIntyre, 1972; Wible, 1983, 1986, 1987; Rougier et al., 1992; Wible & Hopson, 1993, 1995). Furthermore, the bone preserving most of these soft-tissue impressions, the petrosal, is one of the densest of the mammalian skeleton. This sturdy bone can resist the rigors of the fossilization process and occasionally is found as an isolated element, frequently amid the remains recovered by screen-washing techniques. The petrosal is one of few skeletal remains that in most cases allows the attribution of isolated elements to a high-level taxonomic group (Clemens, 1966; MacIntyre, 1972; Archibald, 1979; Prothero, 1983; Wible et al., 1995, 2001; Rougier et al., 1996a; Ekdale et al., 2004; Ladevèze, 2004).

The study of the mammalian basicranium and ear region has a long tradition, especially the early recognition of the different developmental patterns of the chondrocranium among mammals and its consequence on the overall architecture of the skull (Gaupp, 1913; Goodrich, 1930; De Beer, 1937; Starck, 1967; Presley, 1981; Kuhn & Zeller, 1987; Novacek, 1993). Perhaps the most influential papers dealing with the development and/or morphological diversity of the basicranium and ear region among generalized mammals are the seminal contributions of Kampen (1905),

Klaauw (1931), and MacPhee (1981). These studies provide a detailed reference point for therians that can be used as a basis for comparison and extrapolation of soft tissues in extinct Mesozoic groups. The comprehensive monographs by Kuhn (1971) on the echidna and by Zeller (1989) on the platypus serve a similar purpose for monotremes.

Skulls of Mesozoic mammals or Cenozoic representatives of Mesozoic lineages were extremely rare until the discovery of the rich deposits of the Nemegt valley in the Gobi Desert, Mongolia (Kielan-Jaworowska, 1971, 1981, 1984). There had been early attempts to include the few known basicrania in a global hypothesis of the major transformations of the mammalian skull base (Simpson, 1937; Kermack, 1963, 1967), but it was not until the description of well-preserved therians and multituberculates from Mongolia that a fundamental dichotomy in early mammalian history based on cranial features in addition to dental characters gained wider acceptance (Hopson, 1970; Kermack & Kielan-Jaworowska, 1971; Crompton & Jenkins, 1979). The late 1960s and the 1970s were dominated by this idea of a basal split of Mammalia into two high-level taxa usually recognized as subclasses: Prototheria (or non-therians) and Theria. The basic tenet of this precladistic hypothesis was that Prototheria is characterized by teeth with cusps in line and a broad anterior lamina forming a great portion of the lateral wall of the braincase; in contrast, therians have triangular molariforms and lack the anterior lamina (Kermack, 1967; Hopson, 1970; Kermack & Kielan-Jaworowska, 1971; Crompton & Jenkins, 1979). Evidence on monotreme development (Kuhn, 1971; Presley, 1981; Kemp, 1983; Kuhn & Zeller, 1987) was mustered to support basicranial similarities between multituberculates and monotremes (Kermack & Kielan-Jaworowska, 1971; Meng & Wyss, 1995, 1996). However, the study of the lateral wall of the skull in marsupials (Maier, 1987, 1993) and in fossils (Hopson & Rougier, 1993) convincingly rejected Prototheria as a monophyletic group.

More recently, isolated petrosals have been successfully integrated into phylogenetic attempts to resolve Mesozoic mammal interrelationships based on basicranial characters alone (Wible, 1990; Wible et al., 1995; Rougier et al., 1996a, 1996b) or integrated into more comprehensive analyses including other areas of the anatomy (Luo, 1994; Hu et al., 1997; Rougier et al., 1998; Ji et al., 1999, 2002; Luo et al., 2001a, 2001b, 2002, 2003; Wang et al., 2001; Meng et al., 2003). Although the number of basicranial characters can be relatively high (more than fifty), only a handful of them have been discussed in some detail in these studies, which attempt to resolve high-level relationships among the major groups of mammals and their fossil relatives. We take this opportunity here to review the major transformations recognized in the basicranium and ear region of mammaliaforms as a unified problem rather than individual characters and to discuss other topics that have been excluded in contributions by us where phylogenetic resolution was the primary goal of the character analysis.

Materials and Methods

As a general reference point to discuss the sequence of events, we follow the cladogram in figure 9.1, which is a simplification of previous studies (Wible et al., 1995, 2001; Rougier et al., 1996a, 1998, 2001a; Hu et al., 1997; Ji et al., 1999, 2002; Luo et al., 2001a, 2001b, 2002, 2003; Wang et al.,

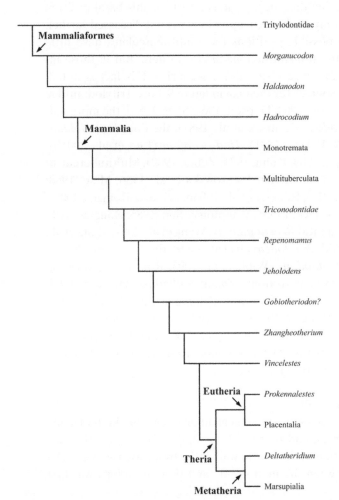

Figure 9.1 Cladogram of relationships among the major Mesozoic and Recent taxa treated here. This tree is a simplification of previous studies, including Wible et al. (1995, 2001), Rougier et al. (1996a, 1996b, 1998, 2001a), Hu et al. (1997), Ji et al. (1999, 2002), Luo et al. (2001a, 2001b, 2002, 2003), and Wang et al. (2001).

2001; Meng et al., 2003). Perhaps the most disputed point concerns the position of Multituberculata, which has been posited variously as the sister-group to mammals (Kielan-Jaworowska & Hurum, 2001; Luo et al., 2002), to triconodontids (Rougier et al., 1996a, 1996b), and to monotremes (Meng and Wyss, 1995; Wang et al., 2001). We believe the argument for a very basal position for Multituberculata (as the sister-group of Mammalia) is not well supported, having not been recovered in any of the most recent phylogenetic analyses using parsimony (Rougier et al., 1996a, 1996b; Hu et al., 1997; Ji et al., 1999, 2002; Luo et al., 2001a, 2001b, 2002, 2003; Wang et al., 2001; Meng et al., 2003), although it is defended as an ad hoc hypothesis statistically not significantly different from more parsimonious arrangements by Luo et al. (2002) that place them with Mammalia. Considering that there is no support for this basal position in any recent parsimony study, such a possibility is not discussed further. The two remaining possible positions for multituberculates have little impact on most of the transformations discussed here, but in those instances in which they suggest alternative scenarios, this fact is noted. Primary anatomical sources for the taxa in figure 9.1 are Tritylodontidae (Kühne, 1956; Crompton, 1964; Hopson, 1964; Sues, 1986), the morganucodontid *Morganucodon* (Kermack et al., 1981), the docodont *Haldanodon* (Lillegraven & Krusat, 1991), *Hadrocodium* (Luo et al., 2001b), Monotremata (Gaupp, 1908; Kuhn, 1971; Zeller, 1989), Multituberculata (Kielan-Jaworowska, 1971; Kielan-Jaworowska et al., 1986; Miao, 1988; Lillegraven & Hahn, 1993; Rougier et al., 1996b; Wible & Rougier, 2000), Triconodontidae (Kermack, 1963; Crompton & Sun, 1985; Rougier et al., 1996a), the gobiconodontid *Repenomamus* (Wang et al., 2001; Meng et al., 2003), the "amphilestid" *Jeholodens* (Ji et al., 1999); the "symmetrodont"? *Gobiotheriodon* (=*Gobiodon*) (Wible et al., 1995; Rougier et al., 1996a; Trofimov, 1997), the "symmetrodont" *Zhangheotherium* (Hu et al., 1997, 1998), the prototribosphenidan *Vincelestes* (Bonaparte and Rougier, 1987; Rougier et al., 1992; Rougier, 1993), the basal eutherian *Prokennalestes* (Wible et al., 2001), Placentalia (Kampen, 1905; Klaauw, 1931; MacPhee, 1981), the basal metatherian *Deltatheridium* (Rougier et al., 1998), and Marsupialia (Archer, 1976; Wible, 2003).

Lateral Wall of the Skull

Perhaps the most dramatic osteological difference in the skulls of living mammals, monotremes, and therians (marsupials and placentals) pertains to the composition of the lateral wall of the braincase (fig. 9.2). The basic distinction between the monotreme and therian pattern was first

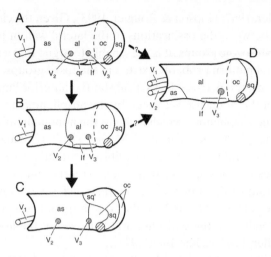

Figure 9.2. Diagrammatic representation of the major features of the lateral wall of the braincase among mammals and immediate fossil forerunners. Arrows denote possible evolutionary pathways; parallel lines on the squamosal represent the cut zygomatic process; gray denotes foramen. (A) Condition in nonmammalian mammaliaforms, such as *Morganucodon, Megazostrodon,* and docodonts, which still retain a sizable quadrate ramus of the alisphenoid (epipterygoid). (B) Condition in therian stem taxa, such as *Vincelestes,* in which the alisphenoid is somewhat enlarged and the quadrate ramus strongly reduced. However, the overall plesiomorphic pattern is still retained. (C) Therian condition in which the anterior lamina is lost, the glenoid fossa is displaced forward, and the squama of the squamosal appears. Most of these changes are likely to be related to the acquisition of a more modern, forwardly placed jaw articulation. (D) Multituberculate-monotreme (and probably triconodontid) condition in which the alisphenoid is strongly reduced, restricted to a small portion forming the lateral wall of the cavum epiptericum, and the anterior lamina is an extensive element forming most of the lateral wall. This derived condition could have arisen from a morphology such as that in A and B; we are unable to decide between these options at present. Abbreviations: al, anterior lamina; as, alisphenoid = epipterygoid; lf, lateral flange; oc, otic capsule; qr, quadrate ramus of alisphenoid; sq, squamosal; sq', squama of squamosal; V_1, ophthalmic division of trigeminal nerve; V_2, maxillary division of trigeminal nerve; V_3, mandibular division of trigeminal nerve.

noted by Watson (1916) and subsequently elaborated and extended to fossils during the 1960s and 1970s (Kermack 1963, 1967; Hopson, 1970; Kermack & Kielan-Jaworowska, 1971; Crompton & Jenkins, 1979). In monotremes (fig. 9.2D) and most of the Mesozoic groups (fig. 9.2A), the anterior lamina, a broad laminar bone that is continuous or fused with the petrosal, forms the side of the braincase posterior to the opening of the maxillary branch of the trigeminal nerve (V_2) (Hopson, 1970; Kermack & Kielan-Jaworowska, 1971; Griffiths, 1978). In monotremes, the anterior lamina (lamina obturans) forms as an intramembranous ossification in the sphenoobturator membrane (Watson, 1916; Kuhn, 1971; Presley, 1981;

Kuhn & Zeller, 1987; Hopson & Rougier, 1993). Given the close similarities existing between the ossifications of the lateral wall of monotremes and several Mesozoic groups, it is reasonable to argue that a comparable developmental pattern probably existed in the fossil groups. In contrast, among therians, the petrosal lacks a substantial anterior lamina and the lateral wall of the braincase is formed by the alisphenoid and squamosal (fig. 9.2C). The alisphenoid, which includes the nonmammalian epipterygoid (Goodrich, 1930; De Beer, 1937; Moore, 1981), is an ancestral component of the lateral braincase that exhibits substantial variation among the major Mesozoic groups. It can have an extensive laminar process developed in front of the anterior lamina that encloses the endochondral processus ascendens, or it can be reduced to a small basal portion probably corresponding to the endochondral "anlage" of the ala temporalis (Kuhn & Zeller, 1987; Zeller, 1989, 1993).

The constitution of the lateral wall of the braincase can be seen as the interplay between the anterior lamina, the processus ascendens, and the squama of the squamosal, a neomorphic formation present among therians. In derived cladotheres, such as *Vincelestes,* the glenoid fossa is displaced forward with regard to the petrosal, effectively reducing the length of the lateral wall of the braincase in front of the glenoid and, therefore, necessitating a rostral shift of the squamosal in general. This process is likely to be related to the reduction of the anterior lamina and the preeminence gained by the squamosal through the development of the squama. The squamosal, however, has a minimal contribution, if any, to the formation of the inner surface of the braincase in basal therians, where the petrosal, alisphenoid, and parietal are the dominant elements.

The anterior lamina is absent or very small in tritylodontids and tritheledontids (Sues, 1986; Hopson, 1994). *Sinoconodon* (Patterson & Olson, 1961; Crompton & Luo, 1993), *Morganucodon* (Kermack et al., 1981), and allies are the first to exhibit a substantial anterior lamina (fig. 9.2A). Although not dramatic, there is a proportional enlargement in the cranial volume of these early mammaliaforms compared with more generalized forms, such as the tritheledontid *Pachygenelus,* that might explain the need for a proportional increase in the size of the elements forming the braincase. The anterior lamina is ubiquitously present in all Mesozoic groups for which the braincase is known up to, and including, *Vincelestes.* The loss of an extensive anterior lamina and its replacement by the squama of the squamosal and alisphenoid is a derived feature of Theria (Kemp, 1983; Maier, 1987; Hopson & Rougier, 1993). This simple picture is complicated by the remnants of a rudimentary anterior lamina in the petrosal attributed to the basal eutherian *Prokennalestes* (Wible et al., 2001) and in isolated "zhelestid" petrosals (labeled as lateral flange by Ekdale et al., 2004). A reduced lateral flange (see below) has been reported in *Pucadelphys* and other basal metatherians (Marshall & Muizon, 1995; Ladevèze, 2004), but so far there is no evidence of an anterior lamina exposed on the lateral braincase in any metatherian. The presence of a rudimentary anterior lamina is likely to be plesiomorphic for Eutheria in light of its presence in *Prokennalestes* and "zhelestids," two taxa that are unlikely to form a monophyletic group (Archibald et al., 2001), suggesting this is indeed a widespread primitive feature and not an autapomorphic condition of a clade of basal eutherians.

The alisphenoid is tall and anteroposteriorly narrow among the non-mammalian cynodonts basal to morganucodontids (Rowe, 1988, 1993; Hopson, 1994) and probably represents little else than the ossified processus ascendens. In basal mammaliaforms (fig. 9.2A), such as *Morganucodon* (Kermack et al., 1981), *Megazostrodon* (Gow, 1986), *Sinoconodon* (Patterson & Olson, 1961; Crompton & Luo, 1993), and *Adelobasileus* (Lucas and Luo, 1993), the alisphenoid is still a tall element bridging the basicranium and the skull roof, but it is proportionately expanded anteroposteriorly. Anterior and medial to the alisphenoid, the orbitosphenoid forms the ventral boundary of the frontal lobes and, therefore, constitutes the anteroventral extent of the braincase (as, for example, in *Probainognathus* MCZ 98743). The alisphenoid increases in size among more derived mammals, such as *Vincelestes* (fig. 9.2B; Hopson & Rougier, 1993), achieving proportions similar to that present in therians. In contrast, multituberculates, monotremes, and likely triconodontids reduce the alisphenoid to the ossified remnant of the ala temporalis (fig. 9.2D; Hopson, 1970; Kermack & Kielan-Jaworowska, 1971; Crompton & Jenkins, 1979; Hopson & Rougier, 1993). This condition is a derived feature that has been used repeatedly to unite these groups and constitutes perhaps the most convincing character linking them (Kermack & Kielan-Jaworowska, 1971; Wible, 1991; Meng & Wyss, 1995; Wible & Hopson, 1993, 1995; Wang et al., 2001; Meng et al., 2003), which, however, results as a convergent acquisition in most recent phylogenies because multituberculates and monotremes seem to be only distantly related (Rowe, 1988; Wible, 1991; Wible et al., 1995; Rougier et al., 1996a, 1996b; Hu et al., 1997; Ji et al., 1999, 2002; Luo et al., 2001a, 2001b).

The reptilian epipterygoid derives from the first branchial arch (Goodrich, 1930; De Beer, 1937; Romer, 1956; Moore, 1981), as does the quadrate. Because of this developmental link, the epipterygoid has a posteriorly directed process, the quadrate ramus, that without reaching the quadrate (or incus) extends ventral to the lateral flange of the anterior lamina. This process is present among basal mammaliaforms (fig. 9.2A; Kermack et al., 1981), is retained in triconodontids (Crompton and Sun, 1985) and gobiconodontids (Wang et al., 2001; Meng et al., 2003), and is reduced in *Vincelestes* (fig. 9.2C; Rougier et al., 1992). Monotremes and multituberculates lack this process, as do therians, and this loss appears to be convergent among the three. The progressive reduction of the quadrate ramus seems to be logically correlated with the general reduction and loss of mechanical function of the quadrate (incus) in its transition from a component of the jaw articulation into an auditory ossicle.

The divisions of the trigeminal nerve (V_{1-3}) can perforate either the anterior lamina or the alisphenoid, depending on the general size of these elements (fig. 9.2). If a large anterior lamina is present, then most or all of V_3 leaves through that bone and V_2 is frequently found in, or close by, the suture located between the ascending process of the alisphenoid and the anterior lamina. When the anterior lamina is small or absent (therians), and the alisphenoid replaces it (fig. 9.2C), V_3 leaves through it (wholly or in part), and V_2 alternatively can perforate the alisphenoid or come out of the braincase through the sphenorbital fissure between the alisphenoid and orbitosphenoid.

Another feature of the ear region that is probably related to the degree of development of the anterior lamina is the presence and extent of a lateral flange (fig. 9.2). The lateral flange is the thickened ventral edge of the anterior lamina and probably the lateralmost portion of the petrosal proper. The lateral flange forms the lateral margin of the lateral trough (Wible & Hopson, 1993), which includes the facial sulcus (Kielan-Jaworowska et al., 1986; Rougier et al., 1992). With the sole exception of multituberculates, the lateral flange is ventrally directed and forms a sharp edge between the plane determined by the anterior lamina and the ventral aspect of the ear region (Kermack et al., 1981; Crompton & Sun, 1985; Crompton & Luo, 1993; Luo, 1994; Rougier et al., 1996a, 1996b). In multituberculates, the lateral flange is inflected medially (fig. 9.3E; Rougier et al., 1992; Wible & Rougier, 2000), contacting the promontorium and forming a canal that likely transmitted the ramus inferior of the stapedial artery and the posttrigeminal vein (see fig. 9.5E; Kielan-Jaworowska et al., 1986; Rougier et al., 1992; Wible & Hopson, 1993, 1995). In all mammaliaforms with a large anterior lamina (fig. 9.2A, B, D), the lateral flange runs the length of the promontorium/anterior lamina. This is the condition in *Sinoconodon, Adelobasileus, Morganucodon,* and allies (Kermack et al., 1981; Gow, 1986; Crompton & Luo, 1993), triconodontids (Kermack, 1963; Crompton & Sun, 1985; Rougier et al., 1996a), probable "symmetrodonts" (Wible et al., 1995; Rougier et al., 1996a), zhangheotheriids (Hu et al., 1997, 1998; Rougier et al., 2003), and *Vincelestes* (Bonaparte & Rougier, 1987; Rougier et al., 1992). Also, predictably, the lateral flange is absent in most therians (fig. 9.2C). There are a few instances in which a small lateral flange has been identified in therians. Among eutherians, a small lateral flange has been described only for isolated petrosals referred to *Prokennalestes* and "zhelestids," being present only from the level of the prootic canal back (Wible et al., 2001; Ekdale et al., 2004). This has been interpreted by Wible et al. (2001) as an intermediate condition between the extensive lateral flange of most Mesozoic taxa and the complete absence of this structure in more derived eutherians. Among metatherians, an anterior lamina has been described in *Pucadelphys* (Marshall & Muizon, 1995) and isolated Paleocene petrosals (Ladevèze, 2004). However, the structure identified as the anterior lamina has no exposure on the lateral wall of the braincase and projects anteriorly on the side of the promontorium. The amount of bone present in *Pucadelphys* and the isolated petrosals is not very different from that in didelphids or other basal metatherians, but if that structure is the remnant of a primitive morphology lost among later metatherians, it is better considered as part of a lateral flange than an anterior lamina. However, despite the somewhat thicker bone on the side of the petrosal, there is no suture or other indication of a separate origin from the pars cochlearis of the petrosal. It is true that even in forms with well-developed lateral flanges, sutural demarcations between the contributions of different elements in this area are rare; perhaps the only noteworthy exceptions are juvenile platypus (Wible & Hopson, 1993) and *Vincelestes* (Rougier et al., 1992), which have a clear demarcation between the lateral flange and promontorium indicated by a crease in the lateral trough. Nevertheless, we consider this portion of bone in *Pucadelphys* as a part of the petrosal rather than an ossification of the lamina obturans, because of its clear continuity with the area of the facial foramen, which almost undoubtedly is part of the petrosal proper. Therefore, this portion of bone would be unlikely to be homologous with the intramembranous anterior lamina and lateral flange.

Middle Ear

As mentioned above, the middle ear is affected by changes in the veins, arteries, and nerves traversing the region in addition to the transformations affecting the auditory apparatus itself. Both middle and inner ear have a profound impact on the morphology of the basicranium. The external ear is less constrained by osteological features, and its reconstruction in fossil cynodonts is challenging (Allin, 1975, 1986; Allin & Hopson, 1991). The likely presence of a mandibular tympanum in basal mammaliaforms and most forerunners, for which no modern analog exists, makes the study of the external ear a fairly intangible matter that we do not address here. For clarity, we treat middle ear osteology and vascular systems separately.

Middle Ear Osteology

The middle ear is the space between the tympanic membrane and the fenestra vestibuli (Sisson, 1910). In modern mammals, this space is occupied by three ear ossicles (malleus, incus, and stapes) surrounded by an air sac (cavum tympani); in turn, this apparatus is frequently enclosed in an ossified auditory bulla in which several elements in various proportions can contribute (Kampen, 1905; Klaauw, 1931; Novacek, 1977). The major contributing elements include the alisphenoid, basisphenoid, petrosal, squamosal, entotympanic(s), and ectotympanic. A complete osseous auditory bulla is not known among Mesozoic mammals. However, a somewhat expanded ectotympanic is known for some Cretaceous eutherians (asioryctitheres, Kielan-Jaworowska, 1975, 1981; Novacek et al., 1997; *Daulestes,* McKenna et al., 2000; and *Zalambdalestes,* Wible et al., 2004), and a strong alisphenoid tympanic process is known for some Cretaceous metatherians (*Didelphodon,* Clemens, 1966; the undescribed Gurlin Tsav skull, Kielan-Jaworowska & Nessov, 1990). The middle ear ossicles are probably the poorest known elements in fossil mammals, because their fragile nature and tenuous connections between themselves and with the skull makes them extremely vulnerable to preservational biases. Rougier et al. (1996b) have reviewed the scanty evidence of middle ear ossicles in Mesozoic mammals, building on earlier reports by Archibald (1979), Novacek and Wyss (1986), Miao and Lillegraven (1986), Meng (1992), Meng and Wyss (1995), and Hurum et al. (1995, 1996). Allin (1975, 1986) and Allin and Hopson (1991) presented detailed scenarios for the origin of the ear ossicles in mammals from nonmammalian cynodonts.

Recently, Wang et al. (2001), Meng et al. (2003), and Li et al. (2003) described a bony rod articulating with the medial side of the dentary in several Early Cretaceous taxa and interpreted it as the ossified Meckel's cartilage. These remarkable finds help explain some problematic facets and pits on the back of the jaw of gobiconodontids (Rougier et al., 2001a) and, in turn, suggest a possible explanation for other facets found on the back of the jaw of more derived forms, such as *Peramus* and *Amphitherium* (Meng et al., 2003). Undisputable articulated bony rods articulating with the dentary are known for *Repenomamus robustus* and *Gobiconodon zofiae* (Wang et al. 2001; Meng et al., 2003; Li et al., 2003). Based on the extremely close morphological similarities in jaw morphology that exist among all gobiconodontids, we believe it likely that an ossified Meckel's cartilage is ubiquitous in this monophyletic clade. Extension of this interpretation to other Mesozoic groups is more problematic. Meng et al. (2003) cite two other examples of putative ossified Meckel's cartilage, the "symmetrodont" *Zhangheotherium* and an Early Cretaceous triconodont from Japan. The latter has been studied by Rougier et al. (1999, in preparation), and it has no trace of an ossified Meckel's cartilage; what Meng et al. (2003) identify as such is the inturned, broken ventral edge of the jaw. The case for *Zhangheotherium* rests on the identification of a bony rod, underlying the posteroventral edge of the left dentary. We agree that this might be an ossified Meckel's cartilage, but we consider it equally likely to be part of the hyoid arch, resembling the hyoid element preserved in the multituberculate *Kryptobaatar* (Wible & Rougier, 2000, fig. 14). The extension of the gobiconodontid paradigm to other fossil taxa with grooves on the medial side of the dentary for which no ossified rods are known (e.g., *Peramus* and *Amphitherium*) is possible, but more hypothetical. Admittedly, the gobiconodontid example opens new avenues to investigate how the definitive mammalian middle ear evolved, but at present we are unsure whether this is an autapomorphy of this distinctive group or represents a transitional state en route to a therian arrangement.

The facets in *Peramus* and *Amphitherium* that Meng et al. (2003) interpreted as for the ossified Meckel's cartilage have been said (Allin & Hopson, 1991) to contain some of the postdentary bones still attached to the lower jaw (for a review, see Allin, 1975) and have been used to argue for the independent acquisition of middle ear ossicles suspended from the skull base among some precladothere mammals on the one hand and therians on the other. The basic similarity and close correspondence of detailed morphological features between the ear ossicles of monotremes, multituberculates, and therians make a primary statement of homology possible that when optimized on the cladogram presented here and in most other recent cladograms argues against the independent origin of the mammalian middle ear.

Another challenge to the single acquisition of the typical mammalian middle ear has been raised recently, although in a covert form, by the suggestion that monotremes are related to a poorly known clade of mammals from the southern continents, the australosphenidans (Luo et al., 2001b, 2002, 2003; Rauhut et al., 2002). Monotremes would be an end branch of this clade, preceded by several stem groups represented only by isolated lower jaws with well-developed grooves on their medial aspect that would have lodged postdentary bones. Thus, the monotreme and therian middle ear would have been achieved independently. This idea, however, depends wholly on the putative relationship of monotremes with these poorly understood taxa. The support for such affiliation is tantalizing but far from solid (for an alternative view, see Woodburne et al., 2003; Woodburne, 2003), particularly because of the fragmentary nature (lower teeth and jaws) of the australosphenidan stem groups. The monotreme-australosphenidan hypothesis and its consequences are of great interest, but, given that characters of the middle ear that could link monotremes with the therian stem group are not included in the analysis, the challenge to a monophyletic origin of the mammalian middle ear remains inconclusive. Another challenge to the single origin is *Hadrocodium* from the Early Jurassic of the Lower Lufeng Formation of Yunnan, China, which has middle ear ossicles separated from the mandible, as predicted by Luo et al. (2001a), because grooves on the medial aspect of its mandible apparently are lacking. In the phylogenies of Luo et al. (2002, 2003) and Ji et al. (2002), *Hadrocodium* is the sister of crown group Mammalia. Accepting these morphological and phylogenetic interpretations, *Hadrocodium* represents another homoplastic acquisition of the typical mammalian middle ear.

The overall aspect of the petrosal in ventral view and, therefore, of the ear region is dominated by the presence/absence of the lateral flange and the development of the promontorium (fig. 9.3). The promontorium houses the pars cochlearis of the petrosal (Klaauw, 1931), which includes the organ of hearing, the cochlea. The promontorium is a rounded ventral projection of the petrosal that generally correlates well with the shape and length of the enclosed cochlea, being more prominent among therians with an elongated cochlea with at least one full turn (Doran, 1878; West, 1985, 1986; Wible et al., 2001).

Luo et al. (1995) have surveyed in detail the origin of the promontorium, with special emphasis on the morphology of basal mammaliaforms. Bonaparte and Crompton (1994) provided further evidence of an incipient promontorium in a juvenile probainognathian and postulated that the promontorium in mammals was the expression of a retained neotenic feature. The promontorium is ventrally covered by a wing of the

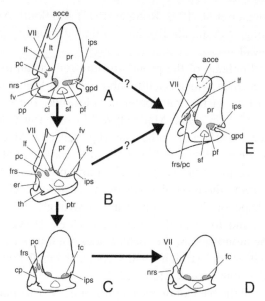

Figure 9.3. Diagrammatic representation of right petrosal bones in ventral view. Gray denotes apertures. (A) Condition in derived nonmammalian mammaliaforms, such as *Morganucodon*. (B) Condition in the therian stem taxa, such as *Vincelestes*. (C) Presumed ancestral morphology for therians, such as the basal eutherian *Prokennalestes*. (D) Basal placental condition. (E) Condition in multituberculates. Abbreviations: aoce, anterior opening of cavum epiptericum; ci, crista interfenestralis; cp, crista parotica; er, epitympanic recess; fc, fenestra cochleae; frs, foramen for ramus superior; frs/pc, foramen for ramus superior and prootic canal; fv, fenestra vestibuli; gpd, groove for perilymphatic duct; ips, foramen for inferior petrosal sinus; lf, lateral flange; lt, lateral trough; nrs, notch for ramus superior; pc, prootic canal; pf, perilymphatic foramen; pp, paroccipital process; pr, promontorium; ptr, post-promontorial tympanic recess; sf, stapedius fossa; th, tympanohyal; VII, foramen for facial nerve.

sphenoid among basal forms, such as *Adelobasileus* (Luo et al., 1995), and it is barely inflated. *Morganucodon* has a more prominent promontorium that completely lacks a basisphenoid wing and has a very steep lateral wall, suggesting the site of attachment of the tensor tympani muscle (fig. 9.3A). The long and narrow, steep sided, and mostly unadorned promontorium of *Morganucodon* reflects a long, rather straight cochlea (Kermack et al., 1981; Graybeal et al., 1989). A similar morphology is retained by pretherian Mesozoic groups, including docodonts (Lillegraven & Krusat, 1991), multituberculates (fig. 9.3E; Kielan-Jaworowska et al., 1986; Luo & Ketten, 1991; Wible & Hopson, 1993, 1995; Hurum, 1998a, 1998b), triconodontids (Kermack, 1963; Rougier et al., 1996a), gobicon-odontids (Wang et al., 2001), and "symmetrodonts" (Wible et al., 1995; Hu et al., 1997, 1998). More derived therian stem taxa, such as *Vincelestes* (fig. 9.3B; Rougier et al., 1992; Rougier, 1993) and South American dryolestoids (Rougier et al., 2001b), have rounded promontoria without a steep lateral wall.

The ventral bulging of the promontorium reflects an increase in the length of the cochlea, which is a little above 270° in *Vincelestes* and more than one full turn in the Patagonian dryolestoids. Early metatherians (Wible, 1990; Muizon, 1994; Marshall & Muizon, 1995; Muizon et al., 1997; Ladevèze, 2004) and eutherians (fig. 9.3C, D; MacIntyre, 1972; Meng & Fox, 1995a, 1995b; Wible et al., 2001; Ekdale et al., 2004) show bulging, rounded promontoria generally more prominent laterally than medially, following the modiolar axis of the enclosed cochlea. As noted above, the shape of the promontorium is strongly influenced by the shape, length, and form of the enclosed cochlea. However, at least in the case of the monotremes, the cochlea does not follow the shape of the cochlear canal, as the straight, wide, cochlear canal houses a curved cochlea (Zeller, 1989; Fox & Meng, 1997).

In therians, at the back of the promontorium are two openings, the laterally placed fenestra vestibuli (fenestra ovalis) and the medial fenestra cochleae (fenestra rotunda). The fenestra vestibuli, which tightly conforms to the shape of the stapedial footplate, poses no homological problem and is thought to be homologous among all amniotes (Goodrich, 1930; Starck, 1995). The fenestra cochleae has a more complex origin as discussed below.

The fenestra vestibuli is circular or slightly elliptical among most basal mammaliaforms and preplacental Mesozoic mammals. The stapedial ratio of Segall (1970) (the ratio of the longer axis/shorter axis of the ellipse) is less than 2 in these forms. In many groups of placentals, the fenestra vestibuli lies within a shallow recessed area, the fossula fenestra vestibuli (MacPhee, 1981). As a general rule, the fenestra vestibuli becomes progressively more oval in more derived groups. The shape of the stapes, however, might have some influence on the final appearance of the fenestra. Collumelliform stapes (Novacek & Wyss, 1986; Meng, 1992) have a fairly circular footplate and thus a stapedial ratio close to 1. Multituberculates show some heterogeneity in this feature, with the Mongolian Late Cretaceous taxa showing fairly elliptical fenestrae (Wible & Rougier, 2000) but with an almost circular footplate in the derived taeniolabidoid *Lambdopsalis* from the Chinese Paleocene (Meng, 1992).

The fenestra cochleae of therians (fig. 9.3C, D) has a complex origin, being a derived feature present in those taxa from "symmetrodonts" up (Wible et al., 1995). Among more basal forms (fig. 9.3A, E), the medial opening in the back of the promontorium is not the fenestra cochleae but the perilymphatic foramen transmitting the perilymphatic duct from the inner ear to the inside of the braincase through the adjacent jugular foramen. The difference between a perilymphatic foramen and a fenestra cochleae is the presence of an osseous processus recessus (De Beer, 1929, 1937; Bast, 1946; Allen, 1964; Zeller, 1985a, 1985b), which walls off the perilymphatic duct from the middle ear cavity into a canal called the cochlear canaliculus (cochlear aqueduct). The cochlear canaliculus usually opens endocranially at the edge of the jugular foramen or in its vicinity.

The presence of an exposed perilymphatic duct usually can be inferred from the presence of a groove leading from the dorsomedial aspect of the perilymphatic foramen to the jugular foramen (fig. 9.3A, E). Among living mammals, monotremes are the only ones with an exposed perilymphatic duct (Kuhn, 1971; Zeller, 1989, 1991). Both tachyglossids, but especially *Zaglossus,* have bony lappets that enclose parts of the perilymphatic duct in a bony channel. These lappets become more extensive in specimens of older age and are absent in neonates and juveniles. A true cochlear canaliculus is present in the probable symmetrodont petrosals from Khoobur (represented by ?*Gobiotheriodon* in fig. 9.1; Wible et al., 1995; Rougier et al., 1996a), *Vincelestes* (Rougier et al., 1992; Rougier, 1993), petrosals of dryolestoids (Rougier et al., 2001b), and therians. An exposed perilymphatic duct is a plesiomorphic feature for more basal taxa.

Multituberculates show some variation in the degree of enclosure of the perilymphatic duct. Jurassic multituberculates (Lillegraven & Hahn, 1993) and djadochtatherians (Kielan-Jaworowska et al., 1986; Hurum, 1998a, 1998b; Wible & Rougier, 2000) have the duct fully exposed in the middle ear. Some isolated petrosals from the Late Cretaceous of North America (Luo, 1989) considered taeniolabidoids ("*Catopsalis*" *joyneri* AMNH 119445; *Mesodma thompsoni* FMNH PM 53904; Wible & Hopson, 1995), however, show partial enclosure of the perilymphatic duct by bony lappets that extend both ventrally and dorsally without actually contacting each other. These lappets in multituberculates therefore are not likely to be homologous with the processus recessus of therians and stem group relatives.

Along the medial edge of the petrosal and between the suture of this bone and the exoccipital, the jugular foramen is universally found in mammals. This foramen transmits cranial nerves IX–XI and often one or more components of the dural sinus system—namely, the sigmoid and inferior petrosal sinuses, which either singly or jointly continue extracranially as the internal jugular vein. The jugular foramen is not fully differentiated from the perilymphatic foramen and the two are usually coalescent in the forms basal to mammaliamorphs. The separation of the jugular and perilymphatic foramina is well established among *Morganucodon* and allies (Kermack et al., 1981; Gow, 1986; Lucas & Luo, 1993). The only important variation of the jugular foramen apparent in Mesozoic mammals is its relative size and the presence or absence of a cochlear canaliculus. Late Cretaceous djadochtatherian multituberculates develop a very conspicuous and spacious depression posteromedial to the promontorium that Kielan-Jaworowska et al. (1986) called the jugular fossa. This fossa is an expansion of the middle ear cavity that reaches an extreme development among large multituberculates, such as *Catopsbaatar* and *Tombaatar* (Kielan-Jaworowska et al., 1986; Rougier et al., 1997; Wible & Rougier, 2000; Spurlin et al., 2000), where this fossa is so large and sunken into the skull base that the final effect is that of a middle ear space surrounded by steep bony walls that likely acts as an analog of the therian auditory bulla.

In therians, immediately behind the promontorium is a space, the postpromontorial tympanic recess, which extends from the jugular foramen to the level of the fenestra vestibuli (fig. 9.3B). This space is also present in *Vincelestes,* in the probable symmetrodont petrosals from Khoobur (Wible et al., 1995; Rougier et al., 1996a), and in zhangheotheriids (Hu et al., 1997; Rougier et al., 2003). More basal forms, with the exception of some multituberculates (Rougier et al., 1996b), lack this space, which is interrupted by a crest that separates the fenestra vestibuli from the perilymphatic foramen, the crista interfenestralis (fig. 9.3A). This crest extends posteriorly to reach the variously developed caudal tympanic process of the petrosal (Wible et al., 1995). Lateral to the crista interfenestralis is a depression for the stapedius muscle (absent in monotremes though; Wible, 1991). Predictably, given the proportionately larger size of the postdentary elements and ear ossicles among basal mammaliaforms, the depression and by inference the stapedius muscle are larger than they are among therians. The size of the fossa for the stapedius muscle has been gauged against the size of the fenestra vestibuli (Wible et al., 1995, 2001; Rougier et al., 1996b), but the significance of the differences is unclear and may be artificial.

Even more laterally on the petrosal, between the promontorium and the lateral flange (in the taxa that have one), there is a broad concave space, the lateral trough (fig. 9.3A). The lateral trough is continuous with the postpromontorial tympanic recess, if the latter is present, and provides a floor for the cavum epiptericum, the endocranial space that lodges the trigeminal ganglion. The floor for the cavum epiptericum can be wide open ventrally, as in the basal members of Mammaliamorpha, including tritylodontids (Crompton, 1964; Sues, 1986), morganucodontids (fig. 9.3A; Kermack et al., 1981; Luo et al., 2001a), docodonts (Lille-graven & Krusat, 1991), *Dinnetherium* (Crompton & Luo, 1993), and tri-conodontids (Crompton & Sun, 1985). The opening is reduced in more derived forms, such as *Vincelestes,* to an opening probably just a little larger than the artery it probably transmitted, the ramus inferior of the stapedial artery (figs. 9.3B, 9.5B; Rougier et al., 1992). Multituberculates have a floored cavum epiptericum (Kielan-Jaworowska et al., 1986; Lille-graven & Hahn, 1993; Hurum et al., 1996, Hurum 1998a, 1998b; Wible & Rougier, 2000), a condition masked by the medially inflected lateral flange, which covers the lateral trough and contacts the promontorium (fig. 9.3E). Metatherians lack any remnant of the opening of the cavum epiptericum, but, among eutherians, the piriform fenestra, which occurs in some Re-cent and Cretaceous taxa (MacPhee, 1981; Kielan-Jaworowska, 1981; Wible et al., 2004), could be homologous. The skeletal elements flooring the cavum epiptericum vary dramatically among the taxa considered here. In basal mammaliaforms and nontherian mammals, with or without a ventrally open cavum epiptericum, the petrosal is the principal or sole flooring element. In therians, where the lateral flange and/or anterior lamina are reduced or absent, the petrosal plays a minor role in providing a floor for the trigeminal ganglion; other neighboring bones, chief among them the alisphenoid, become the main flooring elements.

The paroccipital process is a prominent process that projects ventrally from the posterolateral corner of the petrosal among Mesozoic mam-maliaforms (fig. 9.3A; Rougier et al., 1992, 1996a, 1996b; Wible & Hopson, 1993; Luo, 1994). This process is homologous with the portion of the petrosal providing support for the quadrate among basal nonmammalian cynodonts in which the mammalian temporomandibular joint has not yet been established (Crompton, 1958, 1964; Crompton & Hylander, 1986; Luo & Crompton, 1994). A variously developed, anteriorly directed crest extends from the paroccipital process and is straddled by the quadrate. This crest, the crista parotica (fig. 9.3C), marks the medial edge of the fossa incudis in mammals, which lodges the crus breve (or short process) of the incus = quadrate (MacPhee, 1981; Kielan-Jaworowska et al., 1986). The epitympanic recess is the rostral continuation of the fossa incudis and accommodates the body of the malleus and incus (fig. 9.3B; Klaauw, 1931). The quadrate (incus) is supported by the squamosal in nonmam-malian cynodonts, such as *Probainognathus, Thrinaxodon* (Crompton & Hylander, 1986), tritylodontids (Crompton, 1964; Sues, 1986), and *Pachy-genelus* (Crompton & Hylander, 1986). In morganucodontids, the con-dition is the same, with the quadrate supported medially only by the paroccipital process of the petrosal, but its bulk still articulating in deep fossae on the medial aspect of the squamosal (Kermack et al., 1981, although their reconstruction of the position of the quadrate was corrected by Crompton & Sun, 1985). The same basic morphology is present in *Sinoconodon* (Crompton & Sun, 1985) and *Megazostrodon* (Gow, 1986). Among more derived taxa, such as docodonts, (Lillegraven & Krusat, 1991), triconodontids (Kermack, 1963; Rougier et al., 1996a), multituberculates (Kielan-Jaworowska et al., 1986; Hurum et al., 1996; Wible & Rougier, 2000), therian stem groups (Hu et al., 1997, 1998; Rougier et al., 1992, 2001a), and therians, the quadrate (incus) is supported mostly or solely by the petrosal. The fossa incudis is limited to the petrosal, lateral to the crista parotica, and the epitympanic recess extends anterolaterally. However, among basal metatherians (*Didelphodon, Andinodelphis, ?Turgidodon*) and eutherians (*Prokennalestes,* zalambdalestids, and asioryctitheres), the lateral wall of the epitympanic recess is bounded laterally by the squamosal (Wible, 1990; Rougier et al., 1996a, 1998; Wible et al., 2001, 2004).

With the progressive reduction of the quadrate-incus, the paroccipital process loses its function as a main anchor for the primary jaw articulation and gains in importance as a site of muscular attachment (Crompton & Hylander, 1986). The paroccipital process is very prominent in zhangheotheriids (Hu et al., 1997, 1998; Rougier et al., 2003), probable "symmetrodonts" (Wible et al., 1995; Rougier et al., 1996a), and *Vincelestes* (Rougier et al., 1992). The paroccipital process of the petrosal becomes progressively more medial along the caudal tympanic process of the petrosal in more derived taxa, and it is not very conspicuous in derived therians, where the name paroccipital process is commonly applied to a presumably nonhomologous process of the exoccipital (Wible et al., 2004). In contrast, based on the similarity of muscular attachments to the paroccipital process of the petrosal and the paracondylar process of the exoccipital, Voss and Jansa (2003) argue in favor of their homology. A likely nonhomologous process is present roughly halfway along the caudal tympanic process of the petrosals in asioryctitheres and zalambdalestids, the tympanic process which in these taxa is connected to a strong crista interfenestralis (Kielan-Jaworowska, 1981; Novacek et al., 1997; Rougier et al, 1998; Wible et al., 2004).

Vasculature of the Basicranial Region

Veins. The venous pattern is traditionally considered very variable among mammals and, therefore, with few exceptions (Gelderen, 1924, 1926; Butler, 1967; Archer, 1976; Presley, 1993; Wible & Hopson, 1995), it has received less attention than arteries and nerves in dealing with the systematic implications of the changes in soft-tissue anatomy of the ear region. An outline of vascular evolution among Mesozoic mammals has been provided by Rougier et al. (1992) and the discussion pertaining to the prootic canal was subsequently refined and augmented by Wible and Hopson (1995). Wible et al. (1995, 2001) and Rougier et al. (1996a) further discussed some vascular transformations among Mesozoic mammals.

Among vertebrates, a fairly conservative embryonic pattern is present (Goodrich, 1930; Hafferl, 1933) in which the primary head vein is the main vein running longitudinally near the developing skull base. A portion of the primary head vein lies medial to the dorsal root ganglia of cranial nerves V–X; this is the medial head vein (vena capitis medialis). Later in ontogeny, the medial head vein is joined behind the trigeminal ganglion by another vein that develops lateral to the cranial nerve ganglia, the lateral head vein (vena capitis lateralis). The lateral head vein connects with the medial head vein behind the dorsal root ganglion of cranial nerve IX in nonmammalian tetrapods, but in mammals this connection is lost. The lateral head vein ultimately becomes more prominent in later stages and the portion of the medial head vein medial to the developing otic capsule and ganglia of cranial nerves VII–IX disappears (Gelderen, 1924, 1926; Goodrich, 1930). Therefore, the main longitudinal drainage of the developing skull has a composite origin: the medial head vein anterior to the trigeminal ganglion and the lateral head vein posteriorly. Several major veins, mostly dorsoventrally directed, drain the developing brain. A large vein, the prootic sinus (middle cerebral vein), joins the main longitudinal drainage between the ganglia for cranial nerves V and VII (fig. 9.4A). This connection is used as a landmark to somewhat arbitrarily determine the boundary between the contributions of the medial and lateral head veins to the new composite drainage of the skull. The segment of the medial head vein between the trigeminal ganglion and the prootic sinus receives a new name, the posttrigeminal vein (fig. 9.4A). The ultimate fate of the different portions of this system varies among living mammals (fig. 9.4; see Rougier et al., 1992; Wible & Hopson, 1995). The medial head vein is universally retained as the cavernous sinus around the pituitary gland and serves as a major connection among several cranial sinuses draining the anterior portion of the skull (Butler, 1967). The retention of the lateral head vein is related to the presence of a prootic sinus, both of which are large in monotremes (Gaupp, 1908; Kuhn, 1971; Zeller, 1989), but a portion of this system is very small or relictual among several groups of basal marsupials, including didelphids, some caenolestids, peramelinans, and dasyuromorphians (fig. 9.4D; Wible, 1990; Sánchez-Villagra & Wible,

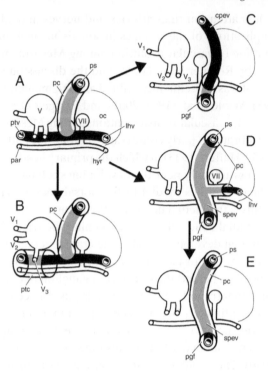

Figure 9.4. Diagrammatic representation of the major veins in the vicinity of cranial nerves V and VII. Black denotes veins; gray is for venous segments within osseous canals. (A) Primitive mammalian venous pattern present in juvenile monotremes (post-trigeminal vein lacking in adult platypus and enclosed within a canal in adult echidna) and inferred to be ancestral for mammaliaforms. (B) Condition in adult echidna and inferred for multituberculates. (C) Condition in placentals. (D) Ancestral pattern inferred for metatherians and present in basal marsupials. (E) Derived marsupial pattern. Abbreviations: cpev, capsuloparietal emissary vein; hyr, hyomandibular ramus of facial nerve; lhv, lateral head vein; oc, otic capsule; par, palatine ramus of facial nerve; pc, prootic canal; pgf, postglenoid foramen; ps, prootic sinus; ptc, posttrigeminal canal; ptv, post-trigeminal vein; spev, sphenoparietal emissary vein; V, trigeminal nerve ganglion; V_1, ophthalmic division of trigeminal nerve; V_2, maxillary division of trigeminal nerve; V_3, mandibular division of trigeminal nerve; VII, facial nerve ganglion.

2002). A distinct posttrigeminal vein is known only in monotremes among living mammals.

In fossils, the variations can be viewed as losses of different portions of this generalized pattern among the members of Mammalia and the inclusion of the several canals by different bones and/or inclusion of other structures in common openings for the prootic canal-lateral head vein system. Among the mammaliaforms *Morganucodon* (Kermack et al., 1981), *Sinoconodon* (Crompton & Sun, 1985; Crompton & Luo, 1993),

Adelobasileus (Lucas & Luo, 1993), docodonts (Lillegraven & Krusat, 1991), and *Hadrocodium* (Luo et al., 2001a), all the major elements described above seem to be present (fig. 9.4A). The prootic canal is conspicuous and generally large, with a tympanic opening directly lateral to the secondary facial foramen for the hyomandibular branch of cranial nerve VII. The endocranial opening is generally on the anterodorsal rim of the subarcuate fossa, posterior to the internal acoustic meatus. However, in *Sinoconodon,* it is in the rear of the cavum epiptericum (Crompton & Luo, 1993), which is likely to be an intermediate condition between that seen in more derived mammaliaforms and more basal nonmammalian cynodonts, such as chiniquodontids where the prootic sinus is only partially included in an osseous canal of the prootic bone (Rougier et al., 1992; Wible & Hopson, 1995). The bony elements forming the prootic canal also differ between the two main clades of monotremes; in the platypus, the prootic canal is formed between the anterior lamina and the petrosal proper, whereas it is formed wholly within the petrosal in the echidna (Wible & Hopson, 1995). However, in both monotremes, the canal is large, opens on the tympanic surface lateral to the secondary facial foramen, and is connected to a well-developed lateral head vein (Wible & Hopson, 1995). The same pattern is inferred for triconodontids (Kermack, 1963; Crompton & Sun, 1985; Rougier et al., 1996a), *Dinnetherium* (Crompton & Luo, 1993), a probable symmetrodont (Wible et al., 1995), and *Vincelestes* (Rougier et al., 1992). Zhangheotheriids (Hu et al., 1997, 1998; Rougier et al., 2003) and the "amphilestid" *Jeholodens* (Ji et al., 1999) are not known well enough to be sure of the presence or absence of a prootic canal, but on the basis of their phylogenetic position, bracketed between forms with a prootic canal, we believe a prootic canal was likely present in these taxa. Most multituberculates show a modification of the primitive pattern by having a common opening on the tympanic surface of the petrosal for the prootic sinus and the ramus superior of the stapedial system (fig. 9.3E; Wible & Rougier, 2000), a feature also present in the platypus but not in the echidna (Wible & Hopson, 1995).

Among eutherians, a prootic canal has been described only for an isolated Albian/Aptian? petrosal attributed to *Prokennalestes* (fig. 9.3C; Wible et al., 2001) and isolated Coniacian petrosals attributed to zhelestids (Ekdale et al., 2004). The absence of a prootic canal is a traditional eutherian synapomorphy, but if the attributions by Wible et al. (2001) and Ekdale et al. (2004) are correct, then the loss of the canal and sinus would be synapomorphic somewhere higher up the eutherian line. Asioryctitheres and zalambdalestids, likely members of a basal eutherian radiation as evidenced by the retention of plesiomorphic traits in their skull (Novacek et al., 1997; Wible et al., 2004) and postcranium (Novacek et al., 1997; Horovitz, 2000, 2003), lack a prootic canal. The morphology in *Prokennalestes* and "zhelestids" can be interpreted as suggesting a reduction of the size of the prootic sinus and certainly a shortening of the length of the prootic canal (Wible et al., 2001; Ekdale et al., 2004); alternatively, but unlikely, a *de novo* origin(s) of the prootic canal in *Prokennalestes* and "zhelestids" can be postulated.

The morphology and relationships of the prootic canal are further modified among metatherians, where the osseous canal interpreted as the prootic canal is of small diameter and runs mostly horizontally (Wible, 1990; Sánchez-Villagra & Wible, 2002). In therians, there is no distinct remnant of the posttrigeminal vein (fig. 9.3C–E), so that there is no real maker to determine the boundary between the prootic sinus and lateral head vein. This is not an issue among living placentals, because they also lack a prootic sinus-lateral head vein. In those marsupials that retain the prootic canal (didelphids, some caenolestids, peramelinans, and dasyuromorphians), there is no objective way to determine whether the osseous canal includes a prootic sinus or a segment of the lateral head vein. Ultimately, this maybe a small difference from a developmental point of view, and regardless, of the ultimate nature of the vein in the "prootic canal" of metatherians, we believe it is homologous with the same structure of other Mesozoic mammals. The morphology of the prootic canal in *Prokennalestes,* with its short, dorsoventrally directed prootic canal, does not seem to be an intermediate condition en route to the metatherian morphology. In the "zhelestid" eutherians, the prootic canal has been described as short and horizontal as occurs in metatherians (Ekdale et al., 2004). This raises the possibility that the "zhelestid"/metatherian morphology is primitive for Theria and the vertical prootic canal of *Prokennalestes* is a reversal or an autapomorphic condition.

Intimately associated with the prootic canal is the fate of the posttrigeminal vein. As mentioned above, this vein extends from behind the trigeminal ganglion to in front of the connection of the prootic sinus and the lateral head vein (fig. 9.4A). Among living mammals, the posttrigeminal vein is present only in monotremes and has never been reported for therians (Wible & Hopson, 1995). However, Wible et al. (2001) suggested that a shallow notch on the anterior edge of the tympanic aperture of the prootic canal indicated the possibility that a posttrigeminal vein might have been present in *Prokennalestes*. Similar notches are known in the front of the prootic canal in *Morganucodon* (Kermack et al., 1981; personal observation), *Haldanodon* (Lillegraven & Krusat, 1991: "third foramen of the anterior lamina"), *Priacodon* (Rougier et al., 1996a), a mesungulatid dryolestoid (Rougier et al., 2001b), *Vincelestes* (Rougier et al., 1992), and a probable "symmetrodont" from the Early Cretaceous of Mongolia (Wible et al, 1995). The evidence in favor of a late retention of a posttrigeminal vein is attractive, but at the same time inconclusive, because the posttrigeminal vein would run, if present, on the roof of the middle ear in the deep portion of the lateral trough and thus open ventrally. The osteological evidence is limited to the small rostral emargination of the prootic canal and the grooves that continue rostrally from it. In a few instances, the position of the posttrigeminal vein is more clearly demarcated by a complete osseous channel. In multituberculates and in the echidna, the posttrigeminal vein is wholly enclosed in a canal (fig. 9.4B; Wible & Hopson, 1995; Wible & Rougier, 2000). This canal, the posttrigeminal canal (also called the canal for the ramus inferior), is formed in multituberculates by the medially inflected lateral flange, which contacts the lateral wall of the promontorium (fig. 9.3E; Kielan-Jaworowska et al., 1986; Rougier et al., 1992).

Major drainage of venous blood from the cranial cavity is accomplished through the system of dural sinuses, which follows a very conservative pattern among mammals in general (Gelderen, 1924, 1926; Butler, 1967; Wible & Hopson, 1993, 1995). The superior (dorsal) sagittal sinus drains blood posteriorly from the dorsal portions of the brain into the transverse sinus, which runs fairly horizontally, embedded in the tentorium cerebelli between the cerebellum and the cerebrum. The transverse sinus splits into anterior and posterior distributaries, the prootic sinus and the sigmoid sinus. Among living mammals, the sigmoid sinus drains through the foramen magnum in monotremes and marsupials (and all Mesozoic groups for which there is evidence from the grooves on the endocranial surface). In most eutherians, however, the jugular foramen is the major drainage route for both the sigmoid and the inferior petrosal sinuses (Wible, 1984). Possible exceptions to this include some basal eutherians, such as *Prokennalestes* (Wible et al., 2001) and *Zalambdalestes* (Wible et al., 2004). Endocranial grooves for the sigmoid sinus are the most reliable indicator of this vein's pathway in fossils; the size of the jugular foramen can be deceptive. For instance, the jugular foramen in monotremes is proportionally large, several times larger than that of other foramina in the ear region; yet, the major drainage of the sigmoid and inferior petrosal sinus is through the foramen magnum (Wible & Hopson, 1995).

Another venous sinus that shows variation among Mesozoic mammaliaforms and Recent mammals is the inferior petrosal sinus. This sinus connects the cavernous sinus anteriorly with the internal jugular vein posteriorly, running along the base of the skull, on or about the medial edge of the petrosal. A canal on the medial aspect of isolated Late Cretaceous therian petrosals interpreted as for the inferior petrosal sinus was described by MacIntyre (1972). This canal is intramural—that is, between the petrosal and basioccipital in most metatherians; it is also probably the primitive condition for placentals, although an endocranial course is found in several members of Placentalia (Wible, 1983; Rougier et al., 1996a). In most nontherian mammaliaforms, such as *Morganucodon* (Kermack et al., 1981), a probable "symmetrodont" (Wible et al., 1995), triconodontids (Rougier et al., 1996a), multituberculates (Wible & Rougier, 2000), undetermined cladotherian petrosals (Rougier et al., 2001b), *Vincelestes* (Rougier et al., 1992), and the metatherian *Didelphodon* (Rougier et al., 1998), the inferior petrosal sinus is enclosed in a canal within the substance of the petrosal (fig. 9.4A–C, E), a condition referred by Wible et al. (1995) as intrapetrosal. In most of these forms with an intrapetrosal inferior petrosal sinus, the cochlea is not fully coiled and a substantive proportion of the promontorium is present rostral to the anteriormost extent of the cochlea, a primitive mammaliaform condition (Luo et al., 1995). In the taxa with an intrapetrosal inferior petrosal sinus, this venous channel generally curves laterally at the anterior end of the promontorium. The inferior petrosal sinus also drains blood from the spongy tissues that form most of the areas of the promontorium not occupied by the cochlea. This system of cancellous bone includes a circumpromontorial venous plexus that has variable connections with the middle ear region and the canal of the inferior petrosal sinus in *Morganucodon* (Kermack et al., 1981), docodonts (Lillegraven & Krusat (1991), triconodontids (Kermack, 1963; Rougier et al., 1996a), a probable "symmetrodont" (Wible et al., 1995), and *Vincelestes* (Rougier et al., 1992). Among living mammals, monotremes serve as the best model for reconstruction of such a sinus in the petrosal (Kuhn, 1971; Zeller, 1989; Rougier et al., 1992). The presence of a well-developed circumpromontorial plexus is probably the result of the opportunistic development of venous sinuses in an area of bone devoid of function other than providing housing for the cochlea, which in many instances occupies only a small portion of the promontorium. Rougier et al. (1992) documented the presence of a circumpromontorial plexus in a variety of nonmammalian cynodonts.

Arteries. Perhaps the most significant contributions to our understanding of the evolution of the blood supply to the mammalian head were by the Austrian anatomist Tandler (1899, 1901, 1902). Tandler

(1899) constructed a Bauplan of the mammalian cranial arterial pattern based on dissections of two species of monotremes, five marsupials, and forty-eight placentals including xenarthrans, carnivorans, lipotyphlans, artiodactyls, perissodactyls, rodents, chiropterans, and primates. The primary components of Tandler's Bauplan were the internal carotid and vertebral arteries supplying the brain and the stapedial artery supplying the tissues in the distribution area of the trigeminal nerve. The external carotid artery had a limited primary role; it secondarily captured the area of supply of the stapedial artery to various degrees in marsupials and in many placentals.

Subsequent embryological and neontological research for the most part has supported Tandler's major findings; in fact, the amniote Bauplan resembles in general the mammalian except that the brain supply is through the internal carotid alone (Hafferl, 1933). Among the noteworthy contributions since Tandler are those of the Danish anatomist Bugge, who between 1967 and 1979 published a series of studies of acrylic-resin corrosion casts of various placentals (Bugge, 1974). The most significant addition to Tandler's Bauplan has been the arteria diploëtica magna, first described in the Tasmanian echidna by Hyrtl (1853). Wible (1984, 1987) found that this vessel occurs embryologically in both the echidna and platypus as well as in various marsupials and placentals, suggesting it is a component of the primitive mammalian and therian pattern.

Among mammals, the vessel that has received the most attention from paleontologists and neontologists is the internal carotid artery (fig. 9.5B–E). This vessel's course across the skull base en route to its foramen of entrance into the cranial cavity varies considerably between two extremes in Recent mammals; the extreme medial position is just off the midline along the basioccipital and basisphenoid (e.g., monotremes) and the extreme lateral position is on the promontorium of the petrosal (e.g., microchiropterans). In addition, this vessel often occupies a groove and/or canal in Recent mammals, which means its course can be reconstructed in fossils. Because of the presence of two sets of grooves on or near the petrosal in some fossil eutherians, Matthew (1909) proposed that two internal carotid arteries (medial and lateral) were present primitively, with one or the other being retained in all Recent placentals. This was the generally accepted view until Presley (1979) showed that all internal carotid arteries in Recent mammals are derived from the embryonic dorsal aorta and that the various adult pathways result from relative growth of the different basicranial components. A vascular model accounting for the two supposed carotid grooves identified by Matthew (1909) was

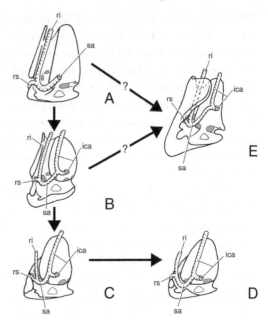

Figure 9.5. Diagrammatic representation of right petrosal bones in ventral view with major arteries. Gray denotes apertures. (A) Condition in derived nonmammalian mammaliaforms, such as *Morganucodon;* internal carotid artery probably ran medial to the promontorium in an extrabullar position (see text). (B) Condition in therian stem taxa, such as *Vincelestes.* (C) Presumed ancestral morphology for therians, such as the basal eutherian *Prokennalestes.* (D) Basal placental condition. (E) Condition in multituberculates. Abbreviations: ica, internal carotid artery; ri, ramus inferior; rs, ramus superior; sa, stapedial artery.

offered by Wible (1983), who reconstructed the inferior petrosal sinus in the medial groove and the internal carotid artery in the lateral one. Finally, Wible (1986) categorized the various courses for the internal carotid artery in relationship to the membranous or osseous auditory bulla: extrabullar or medial to the auditory bulla, perbullar or within a canal in the medial wall of the auditory bulla, and transpromontorial or on the cochlear housing enclosed within the middle ear cavity by the auditory bulla.

The extracranial course of the internal carotid artery is not always indicated by grooves or canals in extinct or extant mammaliamorphs. For isolated fossil petrosals lacking such grooves (e.g., probable "symmetrodonts," Wible et al., 1995; Rougier et al., 1996a), it is not possible to reconstruct the course of the internal carotid. However, if more of the basicranium is preserved, the artery's course may be estimated. For example, because the carotid foramina are within the basisphenoid so close to the midline in *Pachygenelus* (Wible & Hopson, 1993), *Sinoconodon* (Crompton & Luo, 1993), and *Morganucodon* (Kermack et al., 1981), and because of the general layout of the basicranium in these forms, it seems likely that the internal carotid artery had a medial course similar to that in monotremes. Therefore, Presley's (1979) prediction that a medial course was primitive for mammals is supported. The metatherian internal carotid generally follows the same pathway as in monotremes, but some extinct taxa (e.g., *Pucadelphys,* Marshall & Muizon, 1995) have a deep groove for the internal carotid at the anterior pole of the promontorium, in front of the cochlear housing. This is not the equivalent of a transpromontorial course, which is within the middle ear, but a variant on the extrabullar pathway in that the vessel was likely anteromedial to rather than medial to the middle ear.

Until recently, the transpromontorial course has been considered to be a derived feature within Eutheria. However, appropriate grooves on the promontorium now indicate that a transpromontorial course occurs in some multituberculates (fig. 9.5E; Wible & Rougier, 2000), some dryolestoids (Rougier et al., 2001b), and *Vincelestes* (fig. 9.5B; Rougier et al., 1992). Moreover, a transpromontorial course is not universally present among Cretaceous eutherians. A transpromontorial groove is reported for *Prokennalestes* (fig. 9.5C; Wible et al., 2001), *Daulestes* (McKenna et al., 2000), and *Protungulatum* (Archibald et al., 2001), but is lacking in asioryctitheres (Kielan-Jaworowska, 1981; Novacek et al., 1997), zalambdalestids (Kielan-Jaworowska & Trofimov, 1980; Kielan-Jaworowska, 1984; Archibald et al., 2001; Wible et al., 2004), and "zhelestids" (Archibald et al., 2001; Ekdale et al., 2004). For the asioryctitheres and zalambdalestids, enough of the basicranium is preserved to reconstruct a medial extrabullar course for the internal carotid.

As reconstructed by Wible (1987) and Rougier et al. (1992), the primitive mammalian stapedial artery branched off the internal carotid, ran through the middle ear, and divided into two primary rami lateral to the fenestra vestibuli (fig. 9.5). The ramus superior moved dorsally and divided into anterior and posterior divisions; the former ran forward to the orbit as the ramus supraorbitalis to be distributed with the ophthalmic division of the trigeminal nerve and the latter ran posteriorly to the occiput as the arteria diploëtica magna. The ramus inferior ran forward and divided into the ramus infraorbitalis and ramus mandibularis, which were distributed with the maxillary and mandibular divisions of the trigeminal nerve, respectively. Variations from this primitive pattern found in extinct and extant mammaliamorphs include the courses of various components (e.g., extracranial, intramural, or endocranial) and the presence/absence of various components (with various degrees of annexation to the external carotid system). The role of the external carotid system in fossils is very hard to reconstruct, because the usual points of annexation are either extracranial and leave no osseous impression (e.g., the maxillary artery's union with the distal ramus inferior) or are via foramina that are not exclusively arterial in nature (e.g., cranial nerve foramina).

Most of the components of the stapedial system of extant mammals leave osseous impressions in the form of grooves, canals, and foramina. The most problematic component to reconstruct is the main stem of the stapedial artery. Among Recent mammals, this vessel occurs in the platypus and many placentals but is wholly lacking in the echidna, marsupials, and other placentals (Tandler, 1899, 1901; Wible, 1984, 1987). When this vessel is present, it sometimes runs in a groove on the promontorium directed at the fenestra vestibuli, but this is not the case, for example, in the platypus and microchiropterans. Also, when this vessel is present in placentals, it invariably runs through a foramen in the stapes. However, the platypus does not have a foramen in its stapes; moreover, a foramen is found in the stapes of many taxa that do not have a stapedial artery. Actual evidence for the artery in the form of a groove on the promontorium is fairly limited in fossils, occurring among Mesozoic taxa in some multituberculates (Wible & Rougier, 2000), *Vincelestes* (Rougier et al., 1992), *Prokennalestes* (Wible et al., 2001), *Daulestes* (McKenna et al., 2000), asioryctitheres (Rougier et al., 1998), and zalambdalestids (Archibald et al., 2001; Wible et al., 2004). We also deem it likely that *Morganucodon* had a stapedial artery because of a shallow notch on the rim of the fenestra vestibuli visible in some specimens (Kermack et al., 1981, fig. 71A). Are these the only taxa that had an intact stapedial artery? Elsewhere, we have reconstructed this vessel more widely. For example, in the isolated petrosal of the symmetrodont ?*Gobiotheriodon*, Wible et al. (1995) reconstructed the stapedial artery based on the size of the grooves, indicating the presence of both the rami superior and inferior. This is one possible reconstruction, but we cannot exclude the absence of the stapedial artery with the annexation of the rami superior and inferior to alternative vessels (e.g., arteria diploëtica magna, external carotid artery). In fact, as suggested by Rougier et al. (1992), given the large size of the posttemporal canal between the squamosal and petrosal in most Mesozoic mammaliamorphs, within which the arteria diploëtica magna is reconstructed, it is possible that this vessel primitively was the major supplier of the stapedial system. Under this scenario, the stapedial system is a distributary of the arteria diploëtica magna, which in turn is derived from the occipital artery. This is unlike the condition in Recent mammals when both the arteria diploëtica magna and stapedial artery are present.

In recent mammals, body size can be a reliable predictor of the importance of the stapedial system (Diamond, 1989). Above a certain body size (Conroy [1982] suggested 7 kg), the stapedial artery cannot widen to supply its original territory without impinging on the walls of the intercrural foramen. Diamond (1989) predicted that the proportionally large am- philestid from the Early Cretaceous Cloverly Formation, later described as *Gobiconodon ostromi* (Jenkins & Schaff, 1988), was of a size (about 40 cm) that would have affected the ability of the stapedial artery to irri- gate its original domain. The stapedial system also would be limited by size constraints in other large Mesozoic taxa, such as *Repenomamus* and some mesungulatid dryolestoids. Diamond (1989) suggested coarctation as a mechanism to ameliorate the auditory perturbance produced by an artery impinging on the stapes. An alternative solution to the size con- straint imposed by a bicrurate stapes is the one present in the platypus, the only example among recent mammals in which the stapedial artery is present with a columelliform stapes. Obviously, the size of the arteria diploëtica magna was an important factor in the architecture of the stape- dial system, which adds to the flexibility of the stapedial system to supply it territory.

The component of the stapedial system that has preserved the most detail in Mesozoic mammaliamorphs for hypothesizing an evolutionary transformation is the ramus superior and its branches, including the ar- teria diploëtica magna. From this record, Rougier et al. (1992) traced the ramus superior from a pathway largely in extracranial sulci in nonmam- malian cynodonts; to a partial intramural or endocranial pathway in *Sinoconodon, Morganucodon,* monotremes, and paulchoffatiid multitu- berculates; to a complete intramural or endocranial pathway in other multituberculates and *Vincelestes;* to an essentially complete endocranial pathway in eutherians and metatherians.

Conclusions

Mammalian basicrania provide an extremely fertile ground for char- acters to be used in systematic studies. Numerous transformations affect- ing the middle and inner ear, the arterial and venous patterns, and cranial nerves V, VII–XII make the petrosal and immediate surrounding bones perhaps the most densely character-packed area of mammalian anatomy.

In therians, the lateral wall of the skull has suffered one of the most dramatic transformations of the mammalian skeleton. The primitively present anterior lamina, an intramembranous ossification widely distrib- uted among Mesozoic mammals and retained by extant monotremes, is lost in therians and replaced mostly by the alisphenoid (Kemp, 1983; Hopson & Rougier, 1993). The forward shift of the glenoid fossa and the development of a broad squama of the squamosal in therians may be cor- related with the acquisition of a more modern jaw articulation where the dentary is the dominant or sole mandibular element. The lateral flange, the thickened ventral ridge lateral to the promontorium, is largely the thickened ventral edge of the anterior lamina and, therefore, correlated with the size and relationships of the former element.

The osteology of the ear region reflects the transformations of the middle and inner ear. The promontorium, the most distinctive portion of the pars cochlearis of the petrosal, reflects the shape of the enclosed cochlea; in taxa with straight cochlea, the petrosal is long and narrow, with a steep lateral wall, whereas it is rounded and shorter in taxa with a curved or spiral cochlea.

A plesiomorphic trait of Mammalia is the exposure of the perilymphatic duct to the middle ear space. Partial or total enclosure of the duct in a canal has been independently achieved by monotremes, multituberculates, and the therian lineage. The middle ear space is ancestrally small and divided by a crista interfenestralis that extends posteriorly between the fenestra vestibuli and the fenestra cochleae-perilymphatic foramen. The middle ear space is increased in volume in the therian lineage by formation of the postpromontorial tympanic recess as a result of the more vertical orientation of the crista interfenestralis. Asioryctitheres and zalambdalestids have a strong crista interfenestralis connected to the caudal tympanic process of the petrosal, as well as a postpromontorial tympanic recess. Djadochtatherian multituberculates increase further the size of the middle ear cavity by developing a relatively enormous jugular fossa that results in an analog of the therian auditory bulla.

The acquisition of a dentary-squamosal articulation with the concomitant reduction in size and attachment of the ossicular chain and postdentary elements transforms the paroccipital process from an articulation site for the first arch elements involved in the jaw articulation into mostly a muscular process. Several processes of the therian basicranium have been called a paroccipital process, including some on the exoccipital; it is unlikely they are homologous with the same feature on the petrosal of basal mammals.

Derived members of Mammalia have simplified the primitive venous pattern consisting of a posttrigeminal vein, a prootic sinus, and a well-developed lateral head vein. The prootic sinus and its osseous counterpart, the prootic canal, have a complex history (Wible & Hopson, 1995), and a homologous prootic canal and sinus likely were present in the common ancestor of Theria. The prootic canal in basal metatherians and eutherians, however, is modified and ultimately disappears. It is totally lost in post-*Prokennalestes* and post-"zhelestid" eutherians and only a reduced remnant survives in didelphids, some caenolestids, peramelinans, and dasyuromorphians among metatherians. Both eutherians and metatherians develop neomorphic drainage of venous blood through the postglenoid foramen, which may transmit nonhomologous vessels in these groups.

The arterial pattern also shows a progressive simplification in more derived members of Mammalia. Of the putative three main components of the primitive ancestral cranial arterial pattern, the stapedial, internal carotid, and vertebral arteries, the stapedial is associated with the most osteological features, allowing a reasonable reconstruction in fossils. The stapedial artery is likely universally present among Mesozoic mammals and in most cases connects with a well-developed transpromontorial internal carotid and a large ramus superior. A ramus inferior is present in the lateral trough and, in the case of multituberculates, is enclosed in an osseous canal that likely also included the posttrigeminal vein (Kielan-Jaworowska et al., 1986; Wible & Rougier, 2000). The stapedial system is drastically reduced in metatherians, but the ancestral pattern is mostly retained in basal eutherians.

The study of the mammalian basicranium and ear region has gained substantial momentum in the last twenty years. Isolated petrosals and dentitions can in most cases can be attributed to major groups with some degree of confidence, and basicranial characters do not seem to be better or worse than other traditional character systems (i.e., the dentition). We summarized above what we believe to be a consensual view of the current status of many issues dealing with a wide array of morphological problems ranging from basic bone homology to relatively minor details of cranial circulation. Today we have skulls, or skull fragments, of most of the major groups of Mesozoic mammals, a radical development brought about mostly by the spectaculars finds in the Yixian Formation, China, and by notable fossils from Mongolia and South America. Recent students of the mammalian ear region and basicranium address a broad array of questions and include specialists from many areas of the world. These circumstances, a rich material base and an awakened academic interest in mammalian anatomy and systematics, point to a bright future for cranial characters and their phylogenetic use. However, our excitement about our general understanding of what we believe are the major events of the evolving mammalian skull has to be tempered by the déjà vu feel of some of the more crucial questions we are facing today. Has the mammalian middle ear appeared more than once during mammalian history? What is the interpretation of the sulci on the medial side of the mammalian jaw? Are monotremes tribosphenic mammals? These questions, or variations of them, have stimulated paleontological research for decades and are still crucial today. The detailed study of the ear region and basicrania has provided new information for tackling these problems. The resolution of some of these questions, however, has the potential to deeply alter the currently accepted general hypotheses of homology and the sequence of events outlined in this chapter. We have learned much, but it is likely that a review of the mammalian basicranium twenty years from now will be facing similar overarching questions under a substantially different systematic consensus and basic assumptions of homology.

Acknowledgments

Both authors are deeply indebted to Jim Hopson, who has served as mentor and friend since our introductions to the discipline of vertebrate paleontology. Jim has had considerable influence on our current research agendas and whatever successes we have achieved collectively or individually. He remains our "spiritual coauthor" on nearly every paper we write, including this one!

For access to comparative material and for providing important additional information we thank Dr. José F. Bonaparte, Museo Argentino de Ciencias Naturales, Buenos Aires; Dr. Demberelyin Dashzeveg, the Geological Institute of the Mongolian Academy of Sciences, Ulaan Baatar; Dr. John Flynn, Field Museum of Natural History (FMNH), Chicago; Dr. James A. Hopson, University of Chicago, Chicago; Dr. Zofia Kielan-Jaworowska, Institute of Paleobiology, Polish Academy of Sciences, Warsaw; Dr. Farish A. Jenkins, Jr., Museum of Comparative Zoology (MCZ), Harvard University, Cambridge; Dr. Michael J. Novacek, American Museum of Natural History (AMNH), New York, and Dr. Zhe-Xi Luo, Carnegie Museum of Natural History, Pittsburgh. The figures were drawn by Gina Scanlon, Carnegie Museum of Natural History. Earlier versions of the manuscript benefited from comments by Drs. Zofia Kielan-Jaworowska, Mike Novacek, and Steve Nettleton. This research has been supported by NSF grants DEB 94-07999, DEB 95-27811, DEB 96-25431, DEB 99-96051, DEB 99-96172, DEB 0129061, DEB 0129127, and by a Ralph. E. Powe Junior Faculty Enhancement Award from The Oak Ridge Foundation and a Research Incentive Grant from the University of Louisville to G.W.R. A grant from the Antorchas Foundation to the AMNH and to G.W.R. has also provided support for study trips and presentations at meetings.

Literature Cited

Allen, G. W. 1964. Endolymphatic sac and cochlear aqueduct. *Archives of Oto-laryngology* 79:322–327.

Allin, E. F. 1975. Evolution of the mammalian middle ear. *Journal of Morphology* 147:403–438.

———. 1986. The auditory apparatus of advanced mammal-like reptiles and early mammals; pp. 283–294 *in* N. Hotton III, P. D. MacLean, J. J. Roth, and E. C. Roth (eds.), *The Ecology and Biology of Mammal-like Reptiles.* Washington, DC: Smithsonian Institution Press.

Allin, E. F. and J. A. Hopson. 1991. Evolution of the auditory system in Synapsida ("mammal-like reptiles" and primitive mammals) as seen in the fossil record; pp. 587–614 *in* D. B. Webster, R. R. Fay, and A. N. Popper (eds.), *The Evolutionary Biology of Hearing.* New York: Springer-Verlag.

Archer, M. 1976. The basicranial region of marsupicarnivores (Marsupialia), interrelationships of carnivorous marsupials, and affinities of the insectivorous peramelids. *Zoological Journal of the Linnean Society* 59:217–322.

Archibald, J. D. 1979. Oldest known eutherian stapes and a marsupial petrosal bone from the Late Cretaceous of North America. *Nature* 281:669–670.

Archibald, J. D., A. O. Averianov, and E. G. Ekdale. 2001. Late Cretaceous relatives of rabbits, rodents and other extant eutherian mammals. *Nature* 414: 62–65.

Bast, T. H. 1946. Development of the aquaeductus cochleae and its contained periotic duct and cochlear vein in human embryos. *Annals of Otology, Rhinology and Laryngology* 55:278–297.

Bonaparte, J. F. and A. W. Crompton. 1994. A juvenile probainognathid cynodont skull from the Ischigualasto Formation and the origin of mammals. *Revista del Museo Argentino de Ciencias Naturales Bernardino Rivadavia, Paleontología* 5:931–936.

Bonaparte, J. F. and G. W. Rougier. 1987. *Mamíferos del Cretácico Inferior de Patagonia. IV Congreso Latinoamericano de Paleontología de Vertebrados* I:343–359.

Bugge, J. 1974. The cephalic arterial system in insectivores, primates, rodents and lagomorphs, with special reference to the systematic classification. *Acta Anatomica* 87 (supplement 62): 1–159.

Butler, H. 1967. The development of mammalian dural venous sinuses with special reference to the post-glenoid vein. *Journal of Anatomy* 102: 33–56.

Clemens, W. A. 1966. Fossils mammals of the Type Lance Formation Wyoming. Part II. Marsupialia. *University of California Publications, Geological Sciences* 62:1–122.

Conroy, G. C. 1982. A study of cerebral vascular evolution in primates; pp. 247–261 *in* E. Armstrong and D. Falk (eds.), *Primate Brain Evolution: Methods and Concepts.* New York: Plenum Press.

Crompton, A. W. 1958. The cranial morphology of a new genus and species of ictidosaurian. *Proceedings, Zoological Society of London* 130:183–216.

———. 1964. On the skull of *Oligokyphus. Bulletin of the British Museum (Natural History)* 9:69–82.

Crompton, A. W. and W. L. Hylander. 1986. Changes in mandibular function following the acquisition of a dentary-squamosal jaw articulation; pp. 263–282 *in* N. Hotton III, P. D. MacLean, J. J. Roth, and E. C. Roth (eds.), *The Ecology and Biology of Mammal-like Reptiles.* Washington, DC: Smithsonian Institution Press.

Crompton, A. W. and F. A. Jenkins, Jr. 1979. Origin of mammals; pp. 59–73 *in* J. A. Lillegraven, Z. Kielan-Jaworowska, and W. A. Clemens (eds.), *Mesozoic Mammals: the First Two-thirds of Mammalian History.* Berkeley: University of California Press.

Crompton, A. W. and Z. Luo. 1993. Relationships of the Liassic mammals *Sinoconodon, Morganucodon oehleri,* and *Dinnetherium;* pp. 30–44 *in* F. S. Szalay, M. J. Novacek, and M. C. McKenna (eds.), *Mammal Phylogeny, Mesozoic Differentiation, Multituberculates, Monotremes, Early Therians and Marsupials.* New York: Springer-Verlag.

Crompton, A. W. and A. L. Sun. 1985. Cranial structure and relationships of the Liassic mammal *Sinoconodon. Zoological Journal of the Linnean Society* 85:99–119.

De Beer, G. R. 1929. The development of the skull of the shrew. *Philosophical Transactions of the Royal Society* B217:411–480.

———. 1937. *The Development of the Vertebrate Skull.* Oxford: Clarendon Press.

Diamond, M. K. 1989. Coarctation of the stapedial artery: an unusual adaptive response to competing functional demands in the middle ear of some eutherians. *Journal of Morphology* 200:71–86.

Doran, A. H. G. 1878. Morphology of the mammalian ossicula auditûs. *Transactions of the Linnean Society of London, 2nd Series, Zoology* 1:391–497.

Ekdale, E. G, J. D. Archibald, and A. O. Averianov. 2004. Petrosal bones of placental mammals from the Late Cretaceous of Uzbekistan. *Acta Palaeontologica Polonica* 49:161–176.

Fox, R. C. and J. Meng. 1997. An X-radiographic and SEM study of the osseous inner ear of multituberculates and monotremes (Mammalia): implications for mammalian phylogeny and evolution of hearing. *Zoological Journal of the Linnean Society* 121:249–291.

Gaupp, E. 1908. Zur Entwicklungsgeschichte und vergleichenden Morphologie des Schädels von *Echidna aculeata* var. *typica. Semon's Zoologische Forschungsreisen in Australien, Denkschriften der medicinisch-naturwissenschaftliche Gesellschaft zu Jena* 6:539–788.

———. 1913. Die Reichertsche Theorie (Hammer-, Amboss- und Kieferfrage). *Archiv für Anatomie und Enwicklungsgeschichte, supplement* 1912:1–416.

Gelderen, C. van. 1924. Die Morphologie der Sinus durae matris. Zweiter Teil. Die vergleichende Ontogenie der neurokraniellen Venen der Vögel und

Säugetiere. *Zeitschrift für Anatomie und Entwickelungsgeschichte* 74: 432–508.

———. 1926. *Die Vergleichende Ontogenie der Hirnhäute. Mit besonderer berücksigung der Lage der Neurokraniellen Venen.* München-Berlin: J. F. Bergmann and J. Springer.

Goodrich, E. S. 1930. *Studies on the Structure and Development of Vertebrates.* London: Macmillan.

Gow, C. E. 1986. A new skull of *Megazostrodon* (Mammalia, Triconodonta) from the Elliot Formation (Lower Jurassic) of southern Africa. *Palaeontologia Africana* 26:13–23.

Graybeal, A., J. J. Rosowski, D. R. Ketten, and A. W. Crompton. 1989. Inner-ear structure in *Morganucodon*, an Early Jurassic mammal. *Zoological Journal of the Linnean Society* 96:107–117.

Gregory, W. K. 1910. The order of mammals. *Bulletin of the American Museum Natural History* 27:1–427.

Griffiths, M. 1978. *The Biology of the Monotremes.* New York: Academic Press.

Hafferl, A. 1933. Das Arteriensystem; pp. 563–684 *in* L. Bolk, E. Göppert, E. Kallius, and W. Lubosch (eds.), *Handbuch der vergleichenden Anatomie der Wirbeltiere, vol. 6.* Reimpression 1967. Amsterdam: Ascher.

Hopson, J. A. 1964. The braincase of the advanced mammal-like reptile *Bienotherium. Postilla* 87:1–30.

———. 1970. The classification of non-therian mammals. *Journal of Mammalogy* 51:1–9.

———. 1994. Synapsid evolution and the radiation of non-eutherian mammals; pp. 190–219 *in* R. S. Spencer (ed.), *Major Features of Vertebrate Evolution.* Paleontological Society Short Courses in Paleontology 7. Knoxville: University of Tennessee Press.

Hopson, J. A. and G. W. Rougier. 1993. Braincase structure in the oldest known skull of a therian mammal: implications for mammalian systematics and cranial evolution. *American Journal of Science* 293-A:268–299.

Horovitz, I. 2000. The tarsus of *Ukhaatherium nessovi* (Eutheria, Mammalia) from the Late Cretaceous of Mongolia: an appraisal of the evolution of the ankle in basal therians. *Journal of Vertebrate Paleontology* 20: 547–560.

———. 2003. Postcranial skeleton of *Ukhaatherium nessovi* (Eutheria, Mammalia) from the Late Cretaceous of Mongolia. *Journal of Vertebrate Paleontology* 23:857–868.

Hu, Y.-M., Y.-Q. Wang, C.-K. Li, and Z.-L. Luo. 1998. Morphology of dentition and forelimb of *Zhangheotherium. Vertebrata Palasiatica* 36:102–125.

Hu, Y.-M., Y.-Q. Wang, Z.-X. Luo, and C.-K. Li. 1997. A new symmetrodont mammal from China and its implications for mammalian evolution. *Nature* 390:137–142.

Hurum, J. H. 1998a. The braincase of two Late Cretaceous Asian multituberculates studied by serial sections. *Acta Palaeontologica Polonica* 43:21–52.

————. 1998b. The inner ear in two Late Cretaceous multituberculate mammals, and its implications for multituberculate hearing. *Journal of Mammalian Evolution* 5:65–93.

Hurum, J. H., R. Presley, and Z. Kielan-Jaworowska. 1995. Multituberculate ear ossicles; pp. 243–246 *in* A. Sun and Y. Wang (eds.), *Sixth Symposium on Mesozoic Terrestrial Ecosystems and Biota, Short Papers.* Beijing: China Ocean Press.

Hurum, J. H., R. Presley, and Z. Kielan-Jaworowska. 1996. The middle ear in multituberculate mammals. *Acta Palaeontologica Polonica* 41:253–275.

Hyrtl, J. 1853. Beiträge zur vergleichenden Angiologie. IV. Das arterielle Gefässsystem der Monotremen. *Denkschriften Akademie der Wissenschaften, Wien, mathematisch-naturwissenschaftliche Klasse* 5:1–20.

Jenkins, F. A., Jr. and C. A. Schaff. 1988. The Early Cretaceous mammal *Gobiconodon* (Mammalia, Triconodonta) from the Cloverly Formation in Montana. *Journal of Vertebrate Paleontology* 8:1–24.

Ji, Q., Z.-X. Luo, and S.-A. Ji. 1999. A Chinese triconodont mammal and mosaic evolution of the mammalian skeleton. *Nature* 398:326–330.

Ji, Q., Z.-X. Luo, C.-X. Yuan, J. R. Wible, J.-P. Zhang, and J. A. Georgi. 2002. The earliest known eutherian mammal. *Nature* 416:816–822.

Kampen, P. N. van. 1905. Die Tympanalgegend des Säugetierschädels. *Gegenbaurs Morphologisches Jahrbuch* 34:321–722.

Kemp, T. S. 1983. The relationships of mammals. *Zoological Journal of the Linnean Society* 77:353–384.

Kermack, K. A. 1963. The cranial structure of the triconodonts. *Philosophical Transactions of the Royal Society of London* B246:83–102.

————. 1967. The interrelations of early mammals. *Zoological Journal of the Linnean Society* 47:241–249.

Kermack, K. A. and Z. Kielan-Jaworowska. 1971. Therian and non-therian mammals; pp. 103–115 *in* D. M. Kermack and K. A. Kermack (eds.), *Early Mammals. Zoological Journal of the Linnean Society, 50, supplement 1.*

Kermack, K. A., F. Musset, and H. W. Rigney. 1981. The skull of *Morganucodon. Zoological Journal of the Linnean Society* 71:1–158.

Kielan-Jaworowska, Z. 1971. Skull structure and affinities of the Multituberculata. *Palaeontologia Polonica* 25:4–41.

————. 1975. Evolution of the therian mammals in the Late Cretaceous of Asia. Part I. Deltatheriidae. *Palaeontologia Polonica* 33:103–131.

————. 1981. Evolution of the therian mammals in the Late Cretaceous of Asia. Part IV. Skull structure in *Kennalestes* and *Asioryctes. Palaeontologia Polonica* 42:25–78.

————. 1984. Evolution of the therian mammals in the Late Cretaceous of Asia. Part VII. Synopsis. *Palaeontologia Polonica* 46:173–183.

Kielan-Jaworowska, Z. and J. H. Hurum. 2001. Phylogeny and systematics of multituberculate mammals. *Palaeontology* 44:389–429.

Kielan-Jaworowska, Z. and L. A. Nessov. 1990. On the metatherian nature of the Deltatheroida, a sister-group of Marsupialia. *Lethaia* 23:1–10.

Kielan-Jaworowska, Z. and B. A. Trofimov. 1980. Cranial morphology of the Cretaceous eutherian mammal *Barunlestes. Acta Palaeontologica Polonica* 25:167–185.

Kielan-Jaworowska, Z., R. Presley, and C. Poplin. 1986. The cranial vascular system in taeniolabidoid multituberculate mammals. *Philosophical Transactions of the Royal Society of London* B313:525–602.

Klaauw, C. J. van der. 1931. The auditory bulla in some fossil mammals, with a general introduction to this region of the skull. *Bulletin of the American Museum of Natural History* 62:1–352.

Kuhn, H.-J. 1971. Die Entwicklung und Morphologie des Schädels von *Tachyglossus aculeatus. Abhandlungen der senckenbergischen naturforschenden Gesellschaft* 528:1–192.

Kuhn, H.-J. and U. Zeller. 1987. The cavum epiptericum in monotremes and therian mammals; pp. 50–70 *in* H.-J. Kuhn and U. Zeller (eds.), *Morphogenesis of the Mammalian Skull. Mammalia Depicta 13.* Hamburg: Paul Parey.

Kühne, W. G. 1956. *The Liassic Therapsid* Oligokyphus. London: Trustees of the British Museum.

Ladevèze, S. 2004. Metatherian petrosals from the Late Paleocene of Itaboraí, Brazil and their phylogenetic implications. *Journal of Vertebrate Paleontology* 24:202–213.

Li, C.-K., Y.-Q. Wang, Y.-M. Hu, and J. Meng. 2003. A new specimen of *Gobiconodon* from the Jehoh Biota and its implication to the age of the fauna. *Chinese Science Bulletin* 48:177–182.

Lillegraven, J. A. and G. Hahn. 1993. Evolutionary analysis of the middle and inner ear of Late Jurassic multituberculates. *Journal of Mammalian Evolution* 1:47–74.

Lillegraven, J. A. and G. Krusat. 1991. Cranio-mandibular anatomy of *Haldanodon exspectatus* (Docodonta; Mammalia) from the Late Jurassic of Portugal and its implications to the evolution of mammalian characters. *Contributions to Geology, University of Wyoming* 28:39–138.

Lucas, S. G. and Z. Luo. 1993. *Adelobasileus* from the Upper Triassic of west Texas: the oldest mammal. *Journal of Vertebrate Paleontology* 13:309–334.

Luo, Z. 1989. Structure of the petrosals of Multituberculata (Mammalia) and morphology of the molars of early arctocyonids (Condylarthra, Mammalia). Ph.D. dissertation, University of California, Berkeley.

———. 1994. Sister-group relationships of mammals and transformations of diagnostic mammalian characters; pp. 98–128 *in* N. C. Fraser and H.-D. Sues (eds.), *In the Shadow of the Dinosaurs: Early Mesozoic Tetrapods.* Cambridge: Cambridge University Press.

Luo, Z. and A. W. Crompton. 1994. Transformation of the quadrate (incus) through the transition from non-mammalian cynodonts to mammals. *Journal of Vertebrate Paleontology* 14:341–374.

Luo, Z. and D. R. Ketten. 1991. CT scanning and computerized reconstructions of the inner ear of multituberculate mammals. *Journal of Vertebrate Paleontology* 11:220–228.

Luo, Z., R. L. Cifelli, and Z. Kielan-Jaworowska. 2001b. Dual origin of tribosphenic mammals. *Nature* 409:53–57.

Luo, Z., R. L. Cifelli, and Z. Kielan-Jaworowska. 2002. In quest for a phylogeny of Mesozoic mammals. *Acta Palaeontologica Polonica* 47:1–78.

Luo, Z., A. W. Crompton, and S. G. Lucas. 1995. Evolutionary origins of the mammalian promontorium. *Journal of Vertebrate Paleontology* 15: 113–121.

Luo, Z., A. W. Crompton, and A.-L. Sun. 2001a. A new mammaliaform from the Early Jurassic and evolution of mammalian characters. *Science* 292:1535–1540.

Luo, Z., Q. Ji, J. R. Wible, and C.- X. Yuan. 2003. An Early Cretaceous tribosphenic mammal and metatherian evolution. *Science* 302:1934–1940.

MacIntyre, G. T. 1972. The trisulcate petrosal pattern of mammals; pp. 275–303 *in* T. Dobzhansky, M. K. Hecht, and W. C. Steere (eds.), *Evolutionary Biology 6.* New York: Appleton-Century-Crofts.

MacPhee, R. D. E. 1981. Auditory regions of primates and eutherian insectivores. *Contributions to Primatology* 18:1–282.

Maier, W. 1987. The ontogenetic development of the orbitotemporal region in the skull of *Monodelphis domestica* (Didelphidae, Marsupialia), and the problem of the mammalian alisphenoid; pp. 71–90 *in* H.-J. Kuhn and U. Zeller (eds.), *Morphogenesis of the Mammalian Skull. Mammalia Depicta 13.* Hamburg: Paul Parey.

———. 1993. Cranial morphology of the therian common ancestor, as suggested by the adaptations of neonate marsupials; pp. 165–181 *in* F. S. Szalay, M. J. Novacek, and M. C. McKenna (eds.), *Mammal Phylogeny, Mesozoic Differentiation, Multituberculates, Monotremes, Early Therians and Marsupials.* New York: Springer-Verlag.

Marshall, L. G. and C. de Muizon. 1995. Part II: the skull. *Mémoires du Muséum National d'Histoire Naturelle* 165:21–90.

Matthew, W. D. 1909. The Carnivora and Insectivora of the Bridger Basin, Middle Eocene. *Memoirs of the American Museum of Natural History* 9: 289–567.

McDowell, S. B., Jr. 1958. The Greater Antillean insectivores. *Bulletin of American Museum of Natural History* 115:113–214.

McKenna, M. C., Z. Kielan-Jaworowska, and J. Meng. 2000. Earliest eutherian mammal skull, from the Late Cretaceous (Coniacian) of Uzbekistan. *Acta Palaeontologica Polonica* 45:1–54.

Meng, J. 1992. The stapes of *Lambdopsalis bulla* (Multituberculata) and transformational analyses on some stapedial features in Mammaliaformes. *Journal of Vertebrate Paleontology* 12:459–471.

Meng, J. and R. C. Fox. 1995a. Therian petrosals from the Oldman and Milk River Formations (Late Cretaceous), Alberta. *Journal of Vertebrate Paleontology* 15:122–130.

Meng, J. and R. C. Fox. 1995b. Osseous inner ear structures and hearing in early marsupials and placentals. *Zoological Journal of Linnean Society* 115:47–71.

Meng, J. and A. R. Wyss. 1995. Monotreme affinities and low frequency hearing suggested by multituberculate ear. *Nature* 377:141–144.

Meng, J. and A. R. Wyss. 1996. Multituberculate phylogeny. *Nature* 379:407.

Meng, J., Y.-M. Hu, Y.-Q. Wang, and C.-K. Li. 2003. The ossified Meckel's cartilage and internal groove in Mesozoic mammaliaforms: implications to origin of the definitive mammalian middle ear. *Zoological Journal of the Linnean Society* 138:431–448.

Miao, D. 1988. Skull morphology of *Lambdopsalis bulla* (Mammalia, Multituberculata) and its implications to mammalian evolution. *Contributions to Geology, University of Wyoming, Special Paper* 4:1–104.

Miao, D. and J. A. Lillegraven. 1986. Discovery of three ear ossicles in a multituberculate mammal. *National Geographic Research* 2:500–507.

Moore, W. J. 1981. *The Mammalian Skull.* Cambridge: Cambridge University Press.

Muizon, C. de. 1994. A new carnivorous marsupial from the Palaeocene of Bolivia and the problem of marsupial monophyly. *Nature* 370:208–211.

Muizon, C. de, R. L. Cifelli, and R. Céspedes Paz. 1997. The origin of the doglike borhyaenoid marsupials of South America. *Nature* 389:486–489.

Novacek, M. J. 1977. Aspects of the problem of variation, origin and evolution of the eutherian auditory bulla. *Mammal Review* 7:131–149.

———. 1993. Patterns of skull diversity in the mammalian skull; pp. 438–545 *in* J. Hanken and B. K. Hall (eds.), *The Skull, Volume 2, Patterns of Structural and Systematic Diversity.* Chicago: University of Chicago Press.

Novacek, M. J. and A. Wyss. 1986. Origin and transformation of the mammalian stapes; pp. 35–53 *in* K. M. Flanagan and J. A. Lillegraven (eds.), *Vertebrates, Phylogeny, and Philosophy. Contributions to Geology, University of Wyoming, Special Paper 3.*

Novacek, M. J., G. W. Rougier, J. R. Wible, M. C. McKenna, D. Dashzeveg, and I. Horowitz. 1997. Epipubic bones in eutherian mammals from the Late Cretaceous of Mongolia. *Nature* 389:483–486.

Patterson, B. and E. C. Olson. 1961. A triconodontid mammal from the Triassic of Yunnan; pp. 129–191 *in* G. Vandebroek (ed.), *International Colloquium on the Evolution of Lower and Specialized Mammals.* Brussels: Koninklijke Vlaamse Academe voor Wetenschappen, Letteren en Schone Kunsten van België.

Presley, R. 1979. The primitive course of the internal carotid artery in mammals. *Acta Anatomica* 103:238–244.

———. 1981. Alisphenoid equivalents in placentals, marsupials, monotremes and fossils. *Nature* 294:668–670.

———. 1993. Development and the phylogenetic features of the middle ear region; pp. 21–29 *in* F. S. Szalay, M. J. Novacek, and M. C. McKenna (eds.), *Mammal Phylogeny, Mesozoic Differentiation, Multituberculates, Monotremes, Early Therians and Marsupials.* New York: Springer-Verlag.

Prothero, D. R. 1983. The oldest mammalian petrosal from North America. *Journal of Paleontology* 57:1040–1046.

Rauhut, O. W. M., T. Martin, E. Ortiz-Jaureguizar, and P. Puerta. 2002. A Jurassic mammal from South America. *Nature* 416:165–168.

Romer, A. S. 1956. *Osteology of Reptiles.* Chicago: University of Chicago Press.

Rougier, G. W. 1993. *Vincelestes neuquenianus* (Mammalia, Theria), un primitivo mamífero del Cretácico inferior de la Cuenca Neuquina. Thesis, Buenos Aires University.

Rougier, G. W., S. Isaji, and M. Manabe. 1999. An Early Cretaceous Japanese triconodont and a revision of triconodont phylogeny. *Journal of Vertebrate Paleontology* 19, supplement to 3:72A.

Rougier, G. W., Q. Ji, and M. J. Novacek. 2003. A new symmetrodont mammal with fur impressions from the Mesozoic of China. *Acta Geologica Sinica* 77:2–16.

Rougier, G. W., M. J. Novacek, and D. Dashzeveg. 1997. A new multituberculate from the Late Cretaceous locality Ukhaa Tolgod, Mongolia. Considerations on multituberculate interrelationships. *American Museum Novitates* 3191: 1–26.

Rougier, G. W., M. J. Novacek, M. C. McKenna, and J. R. Wible. 2001a. Gobiconodontids from the Early Cretaceous of Oshih (Ashile), Mongolia. *American Museum Novitates* 3348:1–30.

Rougier, G. W., M. J. Novacek, R. Pascual, R. V. Hill, and E. Ortiz-Jaureguizar. 2001b. Mammalian petrosals from the Late Cretaceous of South America: implications for the evolution of the mammalian ear region. *Journal of Vertebrate Paleontology* 21, supplement to 3:94A–95A.

Rougier, G. W., J. R. Wible, and J. A. Hopson. 1992. Reconstruction of the cranial vessels in the Early Cretaceous mammal *Vincelestes neuquenianus:* implications for the evolution of the mammalian cranial vascular system. *Journal of Vertebrate Paleontology* 12:188–216.

Rougier, G. W., J. R. Wible, and J. A. Hopson. 1996a. Basicranial anatomy of *Priacodon fruitaensis* (Triconodontidae, Mammalia) from the Late Jurassic of Colorado and a reappraisal of mammaliaform interrelationships. *American Museum Novitates* 3183:1–37.

Rougier, G. W., J. R. Wible, and M. J. Novacek. 1996b. Middle-ear ossicles of the multituberculate *Kryptobaatar* from the Mongolian Late Cretaceous: implications for mammaliamorph relationships and the evolution of the auditory apparatus. *American Museum Novitates* 3187:1–43.

Rougier, G. W., J. R. Wible, and M. J. Novacek. 1998. Implications of *Deltatheridium* specimens for early marsupial history. *Nature* 396:460–463.

Rowe, T. 1988. Definition, diagnosis, and origin of Mammalia. *Journal of Vertebrate Paleontology* 8:241–264.

———. 1993. Phylogenetic systematics and the early history of mammals; pp. 129–145 *in* F. S. Szalay, M. J. Novacek, and M. C. McKenna (eds.), *Mammal Phylogeny: Mesozoic Differentiation, Multituberculates, Monotremes, Early Therians, and Marsupials.* New York: Springer-Verlag.

Sánchez-Villagra, M. R. and J. R. Wible. 2002. Patterns of evolutionary transformation in the petrosal bone and some basicranial features in marsupial mammals, with special reference to didelphids. *Journal of Zoological Systematics and Evolutionary Research* 40:26–45.

Segall, W. 1970. Morphological parallelisms of the bulla and auditory ossicles in some insectivores and marsupials. *Fieldiana, Zoology* 51:169–205.

Simpson, G. G. 1937. Skull structure of the Multituberculata. *Bulletin of the American Museum of Natural History* 73:727–763.

———. 1945. The principles of classification and a classification of mammals. *Bulletin of the American Museum of Natural History* 85:1–350.

Sisson, S. 1910. *A Text-Book of Veterinary Anatomy.* Philadelphia: W. B. Saunders Company.

Spurlin, B. K., M. J. Novacek, and G. W. Rougier. 2000. Ear region evolution in Multituberculata: a new taxon from the Late Cretaceous locality Udan Sayr, Mongolia. *Journal of Vertebrate Paleontology* 20(3, supplement):70A–71A.

Starck, D. 1967. Le crâne des mammifères; pp. 405–549, 1095–1102 *in* P.-P. Grassé (ed.), *Traité de Zoologie, Volume 16.* Paris: Masson.

———. 1995. *Lehrbuch der speziellen Zoologie. Band II: Wirbeltiere. Teil 5/1: Säugetiere.* Jena: Gustav Fischer.

Sues, H.-D. 1986. The skull and dentition of two tritylodontid synapsids from the Lower Jurassic of western North America. *Bulletin, Museum of Comparative Zoology* 151:217–268.

Tandler, J. 1899. Zur vergleichenden Anatomie der Kopfarterien bei den Mammalia. *Denkschriften Akademie der Wissenschaften, Wien, mathematisch-naturwissenschaftliche Klasse* 67:677–784.

———. 1901. Zur vergleichenden Anatomie der Kopfarterien bei den Mammalia. *Anatomishe Hefte* 18:327–368.

———. 1902. Zur Entwicklungsgeschichte der Kopfarterien bei den Mammalia. *Gegenbaurs Morphologisches Jahrbuch* 30:275–373.

Trofimov, B. 1997. A new generic name *Gobiotheriodon* for a symmetrodont mammal *Gobiodon* Trofimov, 1980. *Acta Palaeontologica Polonica* 42:496.

Voss, R. S. and S. A. Jansa. 2003. Phylogenetic studies on didelphid marsupials. 2, nonmolecular data and new IRBP sequences: separate and combined analyses of didelphine relationships with denser taxon samplings. *Bulletin of the American Museum of Natural History* 276:1–86.

Wang, Y.-Q., Y.-M. Hu, J. Meng, and C.-K. Li. 2001. An ossified Meckel's cartilage in two Cretaceous mammals and the origin of the mammalian middle ear. *Science* 294:357–361.

Watson, D. M. S. 1916. The monotreme skull: a contribution to mammalian morphogenesis. *Philosophical Transactions of the Royal Society of London* B207:311–374.

West, C. D. 1985. The relationship of the spiral turns of the cochlea and the length of the basilar membrane to the range of audible frequencies in ground dwelling mammals. *Journal of the Acoustic Society of America* 77:1091–1101.

———. 1986. Cochlear length, spiral turns and hearing. *12th International Congress of Acoustics, Canada* 1986:123–124.

Wible, J. R. 1983. The internal carotid artery in early eutherians. *Acta Palaeontologica Polonica* 28:281–293.

———. 1984. The ontogeny and phylogeny of the mammalian cranial arterial pattern. Ph.D. dissertation, Duke University, Durham, NC.

———. 1986. Transformations in the extracranial course of the internal carotid artery in mammalian phylogeny. *Journal of Vertebrate Paleontology* 6: 313–325.

———. 1987. The eutherian stapedial artery: character analysis and implications for superordinal relationships. *Zoological Journal of Linnean Society* 91:107–135.

———. 1990. Petrosals of Late Cretaceous marsupials from North America, and a cladistic analysis of the petrosal in therian mammals. *Journal of Vertebrate Paleontology* 10:183–205.

———. 1991. Origin of Mammalia: the craniodental evidence reexamined. *Journal of Vertebrate Paleontology* 11:1–28.

———. 2003. On the cranial osteology of the short-tailed opossum *Monodelphis brevicaudata* (Didelphidae, Marsupialia). *Annals of Carnegie Museum* 72:137–202.

Wible, J. R. and J. A. Hopson. 1993. Basicranial evidence for early mammal phylogeny; pp. 45–62 *in* F. S. Szalay, M. J. Novacek, and M. C. McKenna (eds.), *Mammal Phylogeny: Mesozoic Differentiation, Multituberculates, Monotremes, Early Therians, and Marsupials.* New York: Springer-Verlag.

———. 1995. Homologies of the prootic canal in mammals and non-mammalian cynodonts. *Journal of Vertebrate Paleontology* 15:331–356.

Wible, J. R. and G. W. Rougier. 2000. Cranial anatomy of *Kryptobaatar dashzevegi* (Mammalia, Multituberculata), and its bearing on the evolution of mammalian characters. *Bulletin of the American Museum of Natural History* 247:1–124.

Wible, J. R., M. J. Novacek, and G. W. Rougier. 2004. New data on the skull and dentition of the Mongolian Cretaceous eutherian mammal *Zalambdalestes. Bulletin of the American Museum Natural History* 281:1–144.

Wible, J. R., G. W. Rougier, M. J. Novacek, and M. C. McKenna. 2001. Earliest eutherian ear region: a petrosal referred to *Prokennalestes* from the Early Cretaceous of Mongolia. *American Museum Novitates* 3322:1–44.

Wible, J. R., G. W. Rougier, M. J. Novacek, M. C. McKenna, and D. Dashzeveg. 1995. A mammalian petrosal from the Early Cretaceous of Mongolia: implications for the evolution of the ear region and mammaliamorph interrelationships. *American Museum Novitates* 3149:1–19.

Woodburne, M. O. 2003. Monotremes as pretribosphenic mammals. *Journal of Mammalian Evolution* 10:195–248.

Woodburne, M. O., T. H. Rich, and M. S. Springer. 2003. The evolution of tribospheny and the antiquity of mammalian clades. *Molecular Phylogenetics and Evolution* 28:360–385.

Zeller, U. 1985a. Die Ontogenese und Morphologie der Fenestra rotunda und des Aquaeductus cochleae von *Tupaia* und anderen Säugern. *Gegenbaurs Morphologishes Jahrbuch* 131:179–204.

———. 1985b. The morphogenesis of the fenestra rotunda in mammals; pp. 153–157 *in* H.-R. Duncker and G. Fleischer (eds.), *Functional Morphology of Vertebrates.* Stuttgart: Fischer.

———. 1989. Die Entwicklung und Morphologie des Schädels von *Ornithorhynchus anatinus* (Mammalia: Prototheria: Monotremata). *Abhandlungen der senckenbergischen naturforschenden Gesellschaft* 545:1–188.

———. 1991. Foramen perilymphaticum und Recessus scalae tympani von *Ornithorhynchus anatinus* (Monotremata) und anderen Säugern. *Verhandlungen der Anatomischen Gesellschaft* 84:441–443.

———. 1993. Ontogenetic evidence for cranial homologies in monotremes and therians; pp. 95–128 *in* F. S. Szalay, M. J. Novacek, and M. C. McKenna (eds.), *Mammal Phylogeny: Mesozoic Differentiation, Multituberculates, Monotremes, Early Therians, and Marsupials.* New York: Springer-Verlag.

PART THREE   Functional Morphology

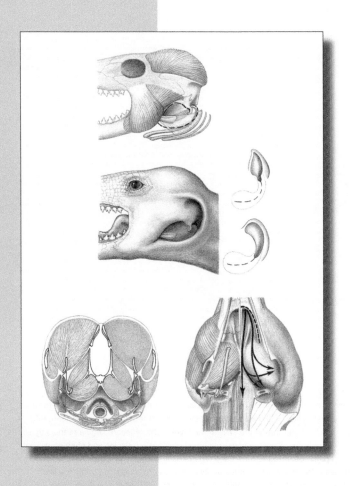

Plate 3. Recontruction of the jaw musculature and ear region in the therapsid *Thrinaxodon*. Top: jaw musculature, ear, and hyobranchial apparatus, showing location of the tympanic membrane (dashed line). Center: flesh reconstruction, with two possible restorations of the external ear. Bottom left: posterior view of a cross section through the skull at the level of the reflected lamina of the angular bone. Bottom right: ventral view of flesh reconstruction, showing jaw muscles and pharyngeal roof, with arrows indicating paths of air to the throat and typanum. Illustrations by Edgar Allin, from: E. F. Allin & J. A. Hopson (1992), Evolution of the auditory system in Synapsida ("mammal-like reptiles" and primitive mammals) as seen in the fossil record; pp. 587–614 *in* D. B. Webster, R. R. Fay and A. N. Popper (eds.), *The Evolutionary Biology of Hearing* (New York: Springer-Verlag).

10

Shoulder Girdle and Forelimb in Multituberculates: Evolution of Parasagittal Forelimb Posture in Mammals

Paul C. Sereno

Introduction

During the early evolution of mammals, the shoulder girdle and fore-limb underwent a profound functional transformation: the rigid shoulder girdle and laterally divergent forelimb posture of the earliest mammals were transformed into the mobile shoulder girdle and more parasagittal forelimb posture that characterizes nearly all living therian (marsupial and placental) mammals (Gregory, 1912; Romer, 1922; Jenkins, 1970a, 1971a, 1973; Jenkins & Weijs, 1979). The fossil record has begun to yield decisive evidence of how and when this musculoskeletal reorganization occurred.

In 1984, I discovered a skull and partial skeleton of a cimolodontan multituberculate that preserve in articulation most of the ribcage and sternum, all elements of the shoulder girdle, and the major long bones of the forelimb. Sereno and McKenna (1995) briefly described this specimen, now referred to as *Kryptobaatar dashzevegi* (Sereno, in review) and suggested that its shoulder girdle and elbow joint provide key evidence for three related hypotheses:

1. Multituberculates, like therians, are characterized by a mobile shoulder girdle and a more parasagittal forelimb posture than that in monotremes and nonmammalian cynodonts.
2. A mobile shoulder girdle and more parasagittal forelimb posture arose once within Mammalia sometime before the Late Jurassic.
3. The shoulder girdle provides key character evidence placing multi-tuberculates within crown mammals closer to therians than to monotremes.

Some authors have questioned these hypotheses, arguing that (1) the proposed postural interpretation is incorrect (Gambaryan & Kielan-Jaworowska, 1997; Kielan-Jaworowska, 1997, 1998); (2) the postural interpretation, if correct, arose more than once within Mammalia (Ji et al., 1999; Luo et al., 2002); and (3) the character evidence from the shoulder girdle is not decisive or compelling with regard to the alliance of multituberculates and therians (Rougier et al., 1996a; Gambaryan & Kielan-Jaworowska, 1997; Luo et al., 2002).

In this report, the phylogenetic significance of the shoulder girdle and elbow for early mammalian relationships is reevaluated, a task more easily accomplished with the recent publication of a data set for basal mammals that includes postcranial characters (Luo et al., 2002). Then, the posture and probable function of the shoulder and elbow joints in multituberculates are reconsidered, with special reference to cineradiographic data from recent mammals (Jenkins, 1970b; 1971a; Pridmore, 1985).

Taxonomic Framework

A phylogenetic taxonomic scheme is utilized here for basal mammalian clades that recognizes node-stem triplets with integrative phylogenetic definitions (for more discussion, see Sereno, 2005). At present paleontologists are split over whether to adhere to a more inclusive, traditional definition of "Mammalia" (Hopson & Barghusen, 1986; Crompton & Jenkins, 1979; Kemp, 1982, 1983; Cifelli, 2001; Luo et al., 2002) or to adopt a less inclusive, crown group definition and erect new taxa at more inclusive levels (Rowe, 1988, 1993; Rowe & Gauthier, 1992). For clarity of meaning in the following paper, the rationale behind the choice of taxonomic definitions is briefly considered.

Traditional Mammalia

What may be termed the "traditional" concept of the taxon Mammalia was formulated by paleontologists. The definition included extinct genera, such as *Sinoconodon* and *Morganucodon,* with an advanced jaw joint that were regarded as basal, if not ancestral, to later mammals. These genera, however, are now widely accepted to lie outside crown mammals (i.e., the clade bounded by living monotremes, marsupials and placentals). Maintenance of a definition of Mammalia that extends beyond crown mammals to include stem genera has been based in good measure on (1) preference for an apomorphy-based definition featuring the jaw joint and/or (2) inclusion of basal taxa to mirror as much as possible traditional taxonomic content.

Kermack and Mussett (1958) identified the dentary-squamosal jaw joint as the necessary and sufficient key derived character (apomorphy) for Mammalia. As this joint was eventually discovered in tritheledontids, the concavoconvex form of the joint became the key apomorphy. Key apomorphies, however, inevitably require continued modification and, therefore, are of questionable utility in phylogenetic taxonomy (for a review, see Sereno, 1999). The jaw joint, in this particular case, is poorly described and/or preserved in critical immediate outgroups and basal ingroups, and homoplasy or further variation in the form of the jaw joint is possible, if not probable. Perhaps in recognition of these shortcomings, some authors who endorse the form of the jaw joint as a criterion for delineating Mammalia have chosen to *define* Mammalia on the basis of reference taxa (Luo et al., 2002, 19).

Because the phylogenetic meaning and taxonomic content of Mammalia has varied over the years (Rowe & Gauthier, 1992), it is difficult to isolate a single tradition. Some authors have viewed mammals as a grade or polyphyletic assemblage (Simpson, 1959; Kielan-Jaworowska, 1992) and others view them as a clade with monotremes in loose alliance with several extinct basal lineages known as Prototheria (fig. 10.1; Kermack, 1967; Crompton & Jenkins, 1979). Only in the last twenty years has cladistic work supported specific phylogenetic hypotheses positioning taxa, such as *Sinoconodon* and *Morganucodon,* outside a clade comprising crown mammals (fig. 10.2B; Rowe, 1988; Crompton & Luo, 1993; Wible

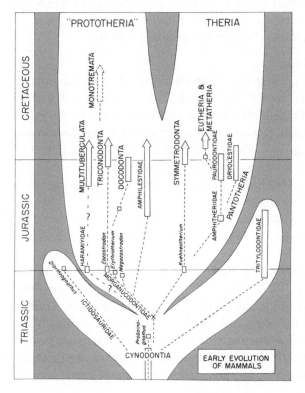

Figure 10.1. Phylogenetic diagram showing Mammalia as potentially monophyletic with a basal split along two lines, Prototheria and Theria (from Crompton & Jenkins, 1978, 1979).

& Hopson, 1993). That is why the important clade comprising crown mammals is left unnamed when Mammalia is used at a more inclusive level (Cifelli, 2001; Luo et al., 2002, 22).

The taxon Aves, in this regard, presents an instructive contrast to Mammalia. Traditionally, Aves was regarded as a monophyletic clade that comprised a recognized subclade of crown birds (Neornithes), outside of which were positioned several extinct stem taxa (e.g., *Archaeopteryx*). That century-long phylogenetic consensus provides the opportunity for phylogenetic definitions to maintain the traditional phylogenetic meaning and taxonomic content of both Aves and Neornithes (Sereno, 1998, 2005). There is no such historical precedence for the taxon Mammalia and, as a result, no widely used alternative name for crown mammals.

Crown "Mammalia"

Rowe (1987, 209) proposed a crown definition of Mammalia as "the most recent common ancestor of its two principal divisions, Monotremata and Theria, and all of its descendants." Later definitions presented slight variations on this theme (Rowe, 1988, 247, 1993, 138). The definition of Mammalia presented here (table 10.1) is very similar but employs species as reference taxa, such as the genus *Ornithorhynchus* in place of "monotremes," to increase clarity of meaning and stability.

Recommended Revision

Given the foregoing, a crown group definition of Mammalia is preferred, because there appears to be little historical basis for a more inclusive phylogenetic definition of Mammalia. Available crown-group definitions of Mammalia, however, use vulgarizations of high-level clades as reference taxa (e.g., monotremes or therians) and are not linked by definition to subordinate stem-based taxa in node-stem triplets (Sereno, 1999). Stability of taxonomic content, nonetheless, is greatest when definitions (1) utilize formal lower-level, reference taxa that are phylogenetically remote from the node of concern and (2) are constructed for mutual stability in the face of new taxa or altered relationships (Sereno, 1999).

One solution outlines four node-stem triplets at the base of the mammalian clade (fig. 10.2A, table 10.1, Sereno, in review). Two of these comprise crown-total taxa defined by survivorship (Mammalia, Theria), and two are diversity-based node-stem triplets (Mammaliamorpha, Theriiformes) that link particularly diverse stem taxa (Tritylodontidae, Allotheria, respectively). Both tritylodontids (i.e., *Tritylodon*) and tritheledontids (i.e., *Pachygenelus*) are used as ingroup reference taxa for

Table 10.1. Indented taxonomic hierarchy and proposed phylogenetic definitions for four node-stem triplets among basal mammalian clades (see fig. 10.2A)

Taxonomic hierarchy	Phylogenetic definition
Mammaliamorpha Rowe, 1988	The least inclusive clade containing *Tritylodon langaevus* Owen,1884, *Pachygenelus monus* Watson, 1913, and *Mus musculus* Linnaeus, 1758.
Tritylodontidae Kühne, 1956	The most inclusive clade containing *Tritylodon longaevus* Owen, 1884, but not *Pachygenelus monus* Watson, 1913, *Mus musculus* Linnaeus, 1758.
Tritheledontidae Broom, 1912	The most inclusive clade containing *Pachygenelus monus* Watson, 1913 but not *Tritylodon longaevus* Owen, 1884, *Mus musculus* Linnaeus, 1758.
Mammaliaformes Rowe, 1988	The most inclusive clade containing *Mus musculus* Linnaeus, 1758 but not *Tritylodon longaevus* Owen, 1884 or *Pachygenelus monus* Watson, 1913.
Mammalia Linnaeus, 1758	The least inclusive clade containing *Ornithorhynchus anatinus* (Shaw, 1799) and *Mus musculus* Linnaeus, 1758.
Prototheria Gill, 1872	The most inclusive clade containing *Ornithorhychus anatinus* (Shaw, 1799) but not *Mus musculus* Linnaeus, 1758.
Monotremata Bonaparte, 1837	The least inclusive clade containing *Ornithorhynchus anatinus* (Shaw, 1799) and *Tachyglossus aculeatus* (Shaw, 1792).
Theriimorpha Rowe, 1993	The most inclusive clade containing *Mus musculus* Linnaeus, 1758 but not *Ornithorhynchus anatinus* (Shaw, 1799).
Theriiformes Rowe, 1988	The least inclusive clade containing *Mus musculus* Linnaeus, 1758 and *Taeniolabis taoensis* (Cope, 1882).
Allotheria Marsh, 1880	The most inclusive clade containing *Taeniolabis taoensis* (Cope, 1882) but not *Mus musculus* Linnaeus, 1758 or *Ornithorhynchus anatinus* (Shaw, 1799).
Multituberculata Cope, 1884	The least inclusive clade containing *Taeniolabis taoensis* (Cope, 1882) and *Paulchofattia delgadoi* Kühne, 1961.
Trechnotheria McKenna, 1975	The most inclusive clade containing *Mus musculus* Linnaeus, 1758 but not *Taeniolabis taoensis* (Cope 1882) or *Ornithorhynchus anatinus* (Shaw, 1799).
Theria Parker & Haswell, 1897	The least inclusive clade containing *Mus musculus* Linnaeus, 1758 an *Didelphis marsupialis* Linnaeus, 1758.
Metatheria Huxley 1880	The most inclusive clade containing *Didelphis marsupialis* (Linnaeus, 1758) but not *Mus musculus* Linnaeus, 1758.

(continued)

Table 10.1. *(continued)*

Taxonomic hierarchy	Phylogenetic definition
Marsupialia Illiger, 1811	The least inclusive clade containing *Didelphis marsupialis* (Linnaeus 1758) and *Phalanger orientalis* (Pallas, 1766).
Eutheria Gill 1872	The most inclusive clade containing *Mus musculus* Linnaeus, 1758 but not *Didelphis marsupialis* Linnaeus 1758.
Placentalia Owen, 1837	The least inclusive clade containing *Dasypus novemcinctus* Linnaeus, 1758, *Elephas maximus* Linnaeus, 1758, *Erinaceus europaeus* (Linnaeus, 1758), *Mus musculus* Linnaeus 1758.

Note: Boldface indicates node-based definitions; regular typeface indicates stem-based definitions. Monotremata, Marsupialia, and Placentalia are best employed as node-based, crown taxa within more inclusive stem-based Allotheria, Metatheria, and Eutheria, respectively. Twelve deeply nested or type species are used as specifiers.

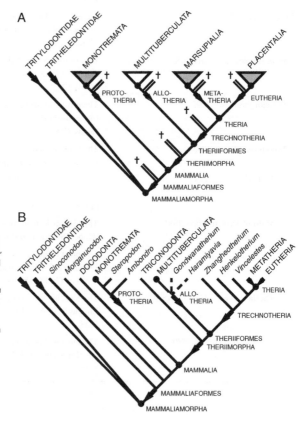

Figure 10.2. (A) Phylogenetic diagram at the base of Mammaliamorpha showing the arrangement of crown clades (shaded triangles), a diversity-based stem clade (empty triangle), extinct taxa (crosses), node-based taxa (dots), and stem-based taxa (arrows). (B) Phylogenetic diagram showing the arrangement of major groups at the base of Mammaliamorpha consistent with Luo et al. (2002) and others, utilizing phylogenetic definitions for higher-level taxa presented in this paper (table 10.1). Dashed lines indicate particularly weak support.

Mammaliamorpha, because opinion has varied as to which taxon is closer to Mammalia (Wible, 1991; Crompton & Luo, 1993; Wible & Hopson, 1993; Hopson & Kitching, 2001; Luo et al., 2002).

Two taxa in the proposed taxonomic framework have fallen from wide-spread use—Prototheria and Allotheria. Here they are revived as stem-based taxa that contain less inclusive node-based groups, Monotremata and Multituberculata, respectively (fig. 10.2A, table 10.1). Prototheria, so defined, specifically includes the extinct and often poorly known relatives of monotremes (e.g., *Steropodon* and *Obdurodon*), as the taxon was used in the earliest cladistic classifications of Mammalia (McKenna, 1974, 1975).

It should be noted that another taxon, Australosphenida, has been erected recently with taxonomic content similar to Prototheria as defined above: "the common ancestor of *Ambondro, Ausktribosphenos,* living monotremes and all its descendants" (Luo et al., 2002, 22). Employing poorly known stem taxa (e.g., *Ausktribosphenos*) as reference taxa in a node-based definition, however, is a recipe for instability, because the relationships of genera based on such fragmentary material is likely to remain controversial (e.g., Rich et al., 2002). Archer et al. (1993), as one solution to this problem, included these extinct stem taxa within a stem-based Monotremata (Archer et al., 1993). The solution outlined here, in contrast, restricts Monotremata to a node-based crown clade and employs a stem-based Prototheria to gather together extinct, and often poorly known, stem taxa (fig. 10.2A, table 10.1).

Allotheria, like Prototheria, may be defined as a stem-based taxon more inclusive than Multituberculata, which has already been given a node-based definition (Simmons, 1993). A stem-based Allotheria includes a node-based Multituberculata and functions like Prototheria and Monotremata, respectively. Allotheria, as defined here, has the capacity to incorporate various poorly known extinct relatives of multituberculates, such as gondwanatherians and haramiyids, without basing its definition on such enigmatic forms (fig. 10.2A).

Metatheria and Eutheria, likewise, are defined here as stem-based taxa, following comments of similar intent by Novacek et al. (1997, 483) regarding the definition of Eutheria. Each includes its respective crown clade, Marsupialia and Placentalia (Archibald & Deutschman, 2001, 108), here explicitly defined (table 10.1). In the case of the crown clade Placentalia, the configuration of basal taxa is controversial; candidate clades include xenarthrans, afrotherians, or erinaceomorphs or some combination of the three (Waddell et al., 1999; Liu & Miyamoto, 1999; Liu et al.,

2001; Cifelli, 2001). Using several representative genera as ingroup reference taxa (*Dasypus, Elephas, Erinaceus, Mus*) in the definition of Placentalia serves to identify the same common placental ancestor regardless of which hypothesis ultimately proves to be the most robust (table 10.1).

Thus, in this scheme, four familiar high-level taxa that comprise the bulk of mammalian diversity—Prototheria, Allotheria, Metatheria, and Eutheria—are given stem-based definitions. Nested within each of these clades is a widely used node-based taxon, three of which are crown groups (Monotremata, Marsupialia, Placentalia) and the last of which constitutes the most diverse mammalian clade of the Mesozoic (Multituberculata) (fig. 10.2A, table 10.1). Node-stem triplets stabilize the relationship between taxa; Theria, for example, equals Metatheria plus Eutheria in taxonomic content, regardless of the discovery of new basal therians or the repositioning of existing basal taxa. Crown groups, in addition, are associated with widely used names (Mammalia, Theria, Monotremata, Marsupialia, Placentalia).

The following institutional abbreviations are used: FMNH, Field Museum of Natural History, Chicago; IVPP, Institute of Vertebrate Paleontology and Paleoanthropology, Beijing; MCZ, Museum of Comparative Zoology, Cambridge; PSS-MAE, Paleontology and Stratigraphic Section of the Geologic Institute, Mongolian Academy of Sciences, Mongolian-American Museum Expedition, Ulaan Baatar/New York; UCPC, University of Chicago Paleontology Collection, Chicago; ZPAL, Institute of Paleobiology, Polish Academy of Sciences, Warsaw.

Previous Work

Forelimb Posture in Multituberculates

Simpson (1928a) regarded the multituberculate shoulder girdle and forelimb as similar in design to that in therians, basing his observations on the proximal ends of the scapulocoracoid and humerus in *Ptilodus* and *Djadochtatherium*. About the ventral orientation of the glenoid, he remarked that it was "exactly that of higher mammals and fundamentally unlike that in monotremes" (1928a, 11). "The importance of the scapula," he continued with remarkable prescience, is that it demonstrates that multituberculates "cannot possibly have been ancestral or closely related to the monotremes." Simpson, however, did not further characterize multituberculate forelimb posture.

By manipulating isolated elements of Late Cretaceous multituberculates, Sloan and Van Valen (1965, 3) concluded, similarly, that in the shoulder joint "movement of the humerus is apparently in the same plane as the blade of the scapula." This orientation presupposes a more parasagittal forelimb posture. In the elbow joint, they continued, "movement of the ulna . . . is restricted to a single plane" by the shapes of opposing surfaces.

On the basis of a well-preserved skeleton of the Paleocene multituberculate *Ptilodus* and isolated girdle and forelimb bones from the Late Cretaceous, Krause and Jenkins (1983, 235) came to similar conclusions, noting that "multituberculate shoulder girdle posture and mobility were comparable to those of modern therians."

In marked contrast to previous authors, Kielan-Jaworowska and Gambaryan (1994) concluded that the forelimbs in multituberculates are significantly abducted (i.e., sprawling) and that their gait, when moving fast, "was different from those occurring in modern mammals" (1994, 82). Their model was based in large part on Late Cretaceous cimolodontan multituberculates from Mongolia. In midstride, the humerus was shown extending from the glenoid nearly perpendicular to the skeletal axis (Gambaryan & Kielan-Jaworowska, 1997, fig. 10, fig. 10.12). In contrast to previous studies of the mammalian elbow joint (Jenkins, 1973), they proposed a causal link between the trochlear elbow joint and the parasagittal posture of the forelimb in therians; the condylar elbow joint in multituberculates, they suggested to the contrary, is functionally consistent with a sprawling forelimb posture and "does not support multituberculate-therian sister-group relationship" (Gambaryan & Kielan-Jaworowska, 1997, 40).

Locomotor Specialization in Multituberculates

On the basis of fragmentary postcranial remains, Gidley (1909, 621) suggested, "the relatively large proportions of the pelvis and hind limbs strongly suggest that *Ptilodus* was saltatorial." Simpson (1926) challenged this interpretation on the basis of comparative limb proportions of recent mammals. In his view, multituberculate postcranial remains then available were insufficient to determine conclusively whether multituberculates were primarily terrestrial or arboreal herbivores. He concluded (1926, 247), "it would be remarkable if many of them were not at least semi-arboreal."

Jenkins and Krause (1983) and Krause and Jenkins (1983) described tarsal, pedal, and caudal features in *Ptilodus* that allowed significant rotational capability of the pes, enhanced pedal grasping with a divergent hallux, and indicated the presence of a prehensile tail, all potentially assisting in headfirst descent within an arboreal habitat. They stated, nevertheless, that these "adaptations for climbing" do not imply that *Ptilodus* or other multituberculate genera were "exclusively arboreal" (Krause & Jenkins, 1983, 243).

More recently, Kielan-Jaworowska and Gambaryan (1994) argued, to the contrary, that these same osteological features are present in running and jumping mammals. On the basis of a more complete multituberculate pes in *Kryptobaatar*, they concluded that the hallux might not have been divergent, at least in Asian cimolodontans. Returning to Gidley's view of multituberculates as terrestrial saltators, they argued multituberculates used an asymmetrical jumping gait at fast speeds.

Forelimb Posture in Other Nontherian Mammals

Forelimb posture in other basal mammals that preserve at least part of the pectoral girdle and forelimb have been interpreted either as sprawling or more parasagittal. The triconodont *Gobiconodon* was reconstructed with the forelimb in a more parasagittal posture, based on the lower portion of the scapula and a partial forelimb (Jenkins & Schaff, 1988, fig. 1). A similar forelimb posture was regarded as probable based on the more complete remains of *Henkelotherium* (Krebs, 1991, fig. 12) and *Vincelestes* (Rougier, 1993, fig. 104), mammals usually placed closer to therians (fig. 10.2B).

Recent discoveries of nearly complete, albeit flattened, skeletons of the triconodont *Jeholodens* (Ji et al., 1999) and the more advanced mammal *Zhangheotherium* (Hu et al., 1997) have led to other interpretations. Regarding *Jeholodens,* Ji et al. (1999, 327, 330) proposed that the mobile shoulder girdle arose independently from that in therians and that, despite the advanced form of the girdle with its ventrally facing glenoid, it was associated with a "sprawling elbow." Similarly for *Zhangheotherium,* Hu et al. (1997, 141, fig. 5) argued that it possessed a mobile shoulder girdle, but that the humerus was "more abducted" than that in most living therians. The evidence for both of these interpretations of forelimb posture, however, is questioned below.

New Fossil Evidence

Kryptobaatar, a Multituberculate

The discovery of an articulated skeleton of *Kryptobaatar dashzevegi* (formerly tentatively referred to as *Bulganbaatar;* Sereno & McKenna, 1995) that preserves both shoulder girdles and forelimbs in articulation clarifies many details regarding the structure and function of the shoulder and elbow joints (figs. 10.2–10.4, 10.7; Sereno, in review).

Shoulder Girdle. The shoulder girdle is composed of scapulocoracoids, clavicles and an interclavicle. Unlike most mammals, the scapulocoracoid is longer than the humerus and approximately five times longer than its greatest width. The blade is narrower than in other mammals and mammalian outgroups. Although both dorsal and ventral edges project laterally, the former is usually regarded as the homolog of the therian scapular spine (figs. 10.3, 10.7). The long axis of the blade may have been oriented only about 60° above the horizontal, rather than vertical as in some mammals (fig. 10.8; Sereno, in review). The so-called "incipient supraspinous fossa" occurs as a subtle depression along the dorsal margin of the blade near the base of the acromial process (fig. 10.3). The large, thin, T-shaped acromial process is nearly completely preserved on the right scapulocoracoid. The large size of the acromial process effectively positions the acromioclavicular articulation lateral to the humeral head and glenoid (fig. 10.7). The ventrally facing glenoid cavity forms an elongate, arched articular surface (fig. 10.7). It is broadest under the blade, where the surface is concave, and canted slightly laterally (fig. 10.5). The coracoid is

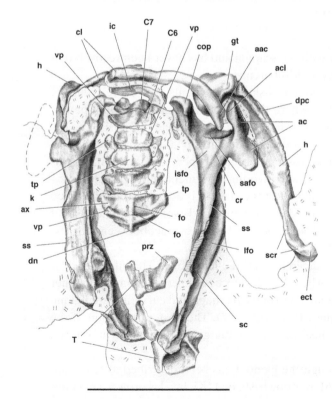

Figure 10.3. Partial skeleton of *Kryptobaatar dashzevegi* (PSS-MAE 103) in dorsal view. Scale bar equals 1 cm. Cross-hatching indicates broken bone surface. Abbreviations: 6,7, vertebral number; aac, articular surface for the acromion; ac, acromion; acl, articular surface for the clavicle; ax, axis; C, cervical; cl, clavicle; cop, coracoid process; cr, crest; dn, dens; dpc, deltopectoral crest; ect, ectepicondyle; fo, foramen; gt, greater tuberosity; h, humerus; ic, interclavicle; isfo, incipient supraspinous fossa; k, keel; lfo, lateral fossa; prz, prezygapophysis; safo, subacromial fossa; sc, scapula; scr, supinator crest; ss, scapular spine; T, thoracic vertebra; tp, transverse process; vp, ventral process.

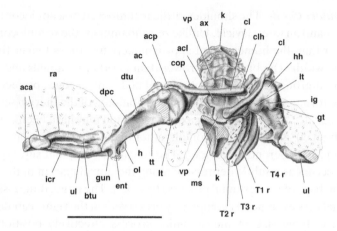

Figure 10.4. Partial skeleton of *Kryptobaatar dashzevegi* (PSS-MAE 103) in anterior view. Scale bar equals 1 cm. Cross-hatching indicates broken bone surface. Abbreviations: 1-4, rib number; ac, acromion; aca, articular surface for the carpus; acl, articular surface for the clavicle; acp, acromial process; ax, axis; btu, biceps tubercle; cl, clavicle; clh, clavicular head; cop, coracoid process; dpc, deltopectoral crest; dtu, deltoid tubercle; ent, entepicondyle; gt, greater tuberosity; gun, groove for the ulnar nerve; h, humerus; hh, humeral head; icr, interosseous crest; ig, intertubercular groove; k, keel; lt, lesser tuberosity; ms, manubrium sterni; ol, olecranon; r, rib; ra, radius; T, thoracic; tt, teres tuberosity; ul, ulna; vp, ventral process.

completely co-ossified with the scapula and has never been recovered as separate element in any multituberculate.

The clavicles are slender rod-shaped elements that curve from the acromion to the midline, where they presumably articulate in a pair of shallow fossae on the anterior face of the interclavicle (fig. 10.4). The left and right clavicle have been dislodged ventrally and dorsomedially, respectively (fig. 10.4). The fossae are broadly open, and the joint, for this reason, appears to have allowed pivoting movement of the clavicle in several directions. The lateral half of the clavicle is dorsoventrally flattened and curves posteriorly toward the acromion and over the bulbous head of the humerus. There does not appear to be any contact between the clavicle and humeral head.

The interclavicle is shaped like a three-leaf clover. The interclavicle is strongly bowed posteriorly, unlike the more flattened interclavicle in other mammals and mammalian outgroups. Its shape and curvature are unique to multituberculates (figs. 10.4, 10.5, 10.7). The posterior margin of the ventral flange rests against the anterior margin of the manubrium sterni.

Girdle Joints. Although the glenoid has been described as teardrop-shaped (McKenna, 1961, 6) or pear-shaped (Kielan-Jaworowska & Gam-

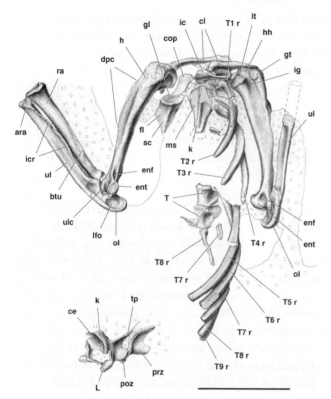

Figure 10.5. Partial skeleton of *Kryptobaatar dashzevegi* (PSS-MAE 103) in ventral view. Scale bar equals 1 cm. Cross-hatching indicates broken bone surface. Abbreviations: 1-9, rib number; ara, articular surface for the radius; btu, biceps tubercle; ce, centrum; cl, clavicle; cop, coracoid process; dpc, deltopectoral crest; enf, entepicondylar foramen; ent, entepicondyle; fl, flange; gl, glenoid; gt, greater tuberosity; h, humerus; hh, humeral head; ic, interclavicle; icr, interosseous crest; ig, intertubercular groove; k, keel; lfo, lateral fossa; lt, lesser tuberosity; L, lumbar vertebra; ms, manubrium sterni; ol, olecranon; poz, postzygapophysis; prz, prezygapophysis; r, rib; ra, radius; sc, scapula; T, thoracic vertebra; tp, transverse process; ul, ulna; ulc, ulnar condyle.

baryan, 1994, 22) in multituberculates, this may be based on incomplete specimens. In PSS-MAE 103, the glenoid is narrow and subrectangular; it is approximately four times longer than wide and is longer than the diameter of the humeral head (fig. 10.7F). The form of the glenoid appears very similar in other multituberculate species. The posterior (scapular) portion is transversely broader, more concave, and faces ventrally. The scapular portion of the glenoid is offset laterally (figs. 10.5, 10.7F). The anterior (coracoid) portion of the glenoid extends to the end of the coracoid process and is transversely flat. In lateral view, it arches anteroventrally (fig. 10.7A) and extends between the greater and lesser trochanters (fig. 10.7F). In this region, the glenoid is one-third the width of the humeral head, allowing considerable transverse mobility of the humeral head at the shoulder joint. The anteroposterior length of the glenoid, in contrast, exceeds that of the humeral head. During maximum extension, anteriormost and posteriormost portions of the glenoid extend beyond the humeral head. The longer length of the glenoid does not appear to limit anteroposterior flexion-extension at the shoulder joint (Jenkins, 1971b).

The right acromioclavicular joint is slightly disarticulated (fig. 10.3). The smooth, narrow joint surface suggests that the clavicular brace may have had a somewhat flexible attachment at the shoulder. The medial end of each clavicle is disarticulated from the interclavicle (fig. 10.4). The medial one-third of the clavicle has a subcylindrical shaft and a rounded medial head that appears to have articulated in a shallow fossa on the interclavicle (figs. 10.4, 10.7B). This joint doubtless allowed some movement, as there are no ligament striae, deeply contoured articular surfaces, or fusion, as occurs in monotremes or immediate mammalian outgroups (Jenkins & Parrington, 1976, fig. 3).

Forelimb. The humeral head has been described as hemispherical, but the medial portion of the head (closer to the lesser trochanter) is the most convex and has the greatest contact with the glenoid (fig. 10.7). Greater and lesser tuberosities are nearly identical in size. The lesser tuberosity is slightly lower in position than the greater tuberosity, and its perimeter is better demarcated from the articular surface of the head (fig. 10.4; Kielan-Jaworowska & Qi, 1990; Kielan-Jaworowska & Gambaryan, 1994), as in *Morganucodon* (Jenkins & Parrington, 1976). The deltopectoral crest develops from the ridge below the greater tuberosity and extends distally to an apex located near the midpoint of the humeral shaft (figs. 10.4, 10.7D). The distal end is only moderately expanded compared with humeral length. The transverse axis of the distal end of the humerus, as measured against an axis through the lesser and greater tuberosity, exhibits torsion of only about 15° (fig. 10.7G). The entepicondyle is more expanded transversely than the ectepicondyle, and the radial condyle is broader than the ulnar condyle, as seen in ventral view (figs. 10.6C, 10.11B; Kielan-Jaworowska & Qi, 1990; Kielan-Jaworowska & Gambaryan, 1994). The asymmetry of the ulnar condyle is of particular interest. In proximal or distal view, the ventral surface of the condyle is beveled strongly, so that the entire surface faces ventrolaterally toward the radial condyle (figs. 10.6, 10.11B). The anterior portion of the ulnar condyle, thus, might better be described as a "hemicondyle," because it lacks a median eminence and symmetrical (ventromedially facing) articular surface. This asymmetric articular surface resembles the ulnar portion of the therian trochlear joint (Sereno & McKenna, 1995). In dorsal view, the articular surface narrows to form a keel-shaped condyle (fig. 10.11B). The prominent, hemispherical radial condyle is surrounded by a nonarticular trough, the radial fossa (figs. 10.6C, 10.7D). The radial condyle flattens as it passes under the humerus and becomes transversely concave in dorsal

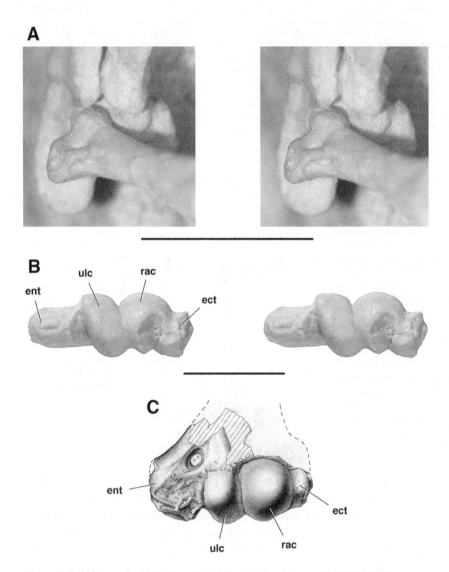

Figure 10.6. (A) Stereopair of right elbow joint of *Kryptobaatar dashzevegi* (PSS-MAE 103) in ventromedial view. (B) Stereopair in distal view and (C) drawing in ventral view of the distal end of left humerus of *Lambdopsalis bulla* (IVPP V7151.150). Scale bar in (A) equals 5 mm; scale bar in (B) and (C) equals 1 cm. Abbreviations: ect, ectepicondyle; ent, entepicondyle; rac, radial condyle; ulc, ulnar condyle.

Figure 10.7. Reconstruction of the pectoral girdle, manubrium sterni, and humerus of *Kryptobaatar dashzevegi* (PSS-MAE 103) (D–G enlarged by twenty percent relative to A–C). (A) Pectoral girdle and manubrium sterni in right lateral view with cross sections of the scapular blade and clavicle; (B, C) pectoral girdle and manubrium sterni in anterior and posterior views; (D) left humerus in ventral view; (E) left humerus in proximal view; (F) left humerus in proximal view and left glenoid (shaded); (G) left humerus in proximal view and left distal end (shaded). Abbreviations: ac, acromion; cl, clavicle; cop, coracoid process; dpc, deltopectoral crest; ect, ectepicondyle; ent, entepicondyle; enf, entepicondylar foramen; gl, glenoid; gt, greater tuberosity; hh, humeral head; ic, interclavicle; isfo, incipient supraspinous fossa; lfo, lateral fossa; lt, lesser tuberosity; ms, manubrium sterni; rac, radial condyle; rfo, radial fossa; sc, scapula; ss, scapular spine; ulc, ulnar condyle.

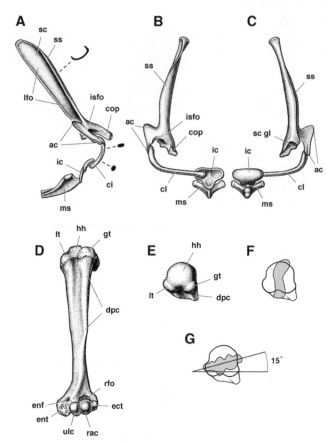

view (figs. 10.6B, 10.11B). The dorsal portion of the radial condyle, thus, also bears a strong resemblance to the therian trochlear joint.

The olecranon process of the ulna is well developed and the semilunar notch forms a deep articular trough with raised proximal and distal lips (fig. 10.11B). The semilunar notch is divided by an intercondylar crest into a larger medial socket for the ulnar condyle and a smaller, more strongly beveled, lateral socket for the radial condyle of the humerus. An oval concave articular facet—the radial notch—truncates the distal margin of the semilunar notch and accommodates the head of the radius. A crescentic interosseous crest builds from a ridge at midshaft to a well-developed flange on the distal one-third of the shaft (fig. 10.5). The small distal end of the ulna is nearly circular with a concave distal articular surface for the carpus (fig. 10.4).

The expanded head of the radius has a subquadrate contour in proximal view (figs. 10.4, 10.6A). Its concave proximal articular surface rotates against the hemispherical radial condyle on the humerus. During such long-axis rotation, the swollen rim on the medial aspect of the head of the radius slides against the concave radial notch on the ulna (fig. 10.11B; Krause & Jenkins, 1983, fig. 15). The shaft narrows in width distal to the head and then is slightly swollen again by the biceps tubercle (fig. 10.4). The distal two-thirds of the shaft is flattened transversely, and the distal end flares to twice the width of that of the ulna. In distal view, the long axis of the distal articular ends of the radius and ulna is oriented antero-posteriorly, at approximately 90° to the transverse axis of the elbow joint (fig. 10.5).

Elbow Joint. The right and left forearms are flexed to angles of approximately 20° and 60°, respectively, to the long axis of the humerus (fig. 10.5). Extension to an angle of approximately 180° appears to have been possible, given the dorsal continuation of the ulnar condyle and deep olecranon fossa on the humerus. During extension, the elbow operates as a transverse hinge joint that is slightly skewed from a transverse axis. A small amount of abduction of the manus occurs with extension at the elbow joint due to the spiral form of the ulnar condyle (figs. 10.4, 10.6B, 10.11B; Jenkins, 1973). The anterior portion of the ulnar condyle of the humerus forms a cam for the ulna that is flattened and beveled posteromedially (Jenkins, 1973). When the elbow is flexed (fig. 10.6A), the ulna articulates only on the flattened lateral aspect of the condyle. The bulbous radial condyle and prominent biceps tubercle suggest that considerable long-axis rotation of the radius was possible.

Jeholodens, a Triconodont

Recent discovery of an articulated triconodont skeleton from China, *Jeholodens jenkinsi* (Ji et al., 1999), supplements previous information on the triconodont shoulder girdle and forelimb, (*Gobiconodon ostromi;* Jenkins & Schaff, 1988). Ji et al. (1999) believed that the shoulder joint in *Jeholodens* was mobile but that the elbow joint was abducted into a sprawling posture. They described a small interclavicle with loose attachment to the clavicle and an "incipient ulnar trochlea" at the distal end of the humerus resembling that in therians.

Reexamination of the specimen, however, reveals that the medial ends of the clavicles, the interclavicle, and the manubrium sterni are completely hidden from view by other elements and matrix. It is not possible, likewise, to verify the presence of an incipient trochlea at the distal end

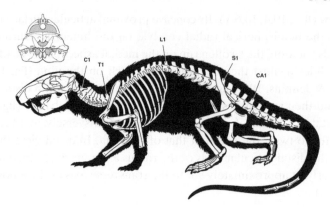

Figure 10.8. Skeletal reconstruction of *Kryptobaatar dashzevegi*. Skull, neck, ribcage, pectoral girdle, and forelimbs based on PSS-MAE 103; lumbar and proximal and midcaudal vertebrae, pelvic girdle, and hind limbs based on ZPAL MgM-I/41. Lumbar vertebrae also based on *Nemegtbaatar gobiensis* (ZPAL MgM-I/81). Top drawing shows the skull and the broadest rib in the ribcage. Abbreviations: C1, atlas; T1, first thoracic vertebra; L1, first lumbar vertebra; S1, first sacral vertebra; CA1, first caudal vertebra.

of the humerus, which has a somewhat convex distal articular margin. As this portion of the humerus is preserved and exposed in dorsal view, it resembles neither a trochlea nor well-formed distal condyles.

In *Jeholodens,* the broad supraspinous fossa, ventrally positioned acromion, and ventrally facing glenoid of the scapula are clearly derived, as established in *Gobiconodon* (Jenkins & Schaff, 1988). The clavicle in *Jeholodens* is strap-shaped but more robust than in *Kryptobaatar.* The elbow joint in triconodonts is best exposed in *Gobiconodon* (Jenkins & Schaff, 1988, figs. 14, 15). A distal trochlea on the humerus has been reported but not figured. The proximal end of the ulna has a deep semilunar notch with an intercondylar crest, similar to that in multituberculates and therians (fig. 10.11B, 10.C). The elbow joint appears to have been designed for considerable flexion-extension. Based on *Jeholodens* and *Gobiconodon,* therefore, triconodonts appear to have an advanced mobile shoulder girdle with a fully developed supraspinous fossa. The forelimb appears to have had a more parasagittal than sprawling posture (*contra* Ji et al., 1999) that emphasized hinge-like flexion-extension at the elbow, as previously shown in a skeletal reconstruction of *Gobiconodon* (Jenkins & Schaff, 1988, fig. 1).

Zhangheotherium, a Symmetrodont

Recent discovery of an articulated symmetrodont skeleton from China, *Zhangheotherium quinquecuspidens* (Hu et al., 1997), provided information on the symmetrodont shoulder girdle and forelimb. Like Ji et al. (1999), Hu et al. (1997) concluded that the shoulder girdle of *Zhangheotherium* was mobile, with a pivotal clavicle-interclavicle joint, but that the elbow joint is more primitive than that in therians, based on three features of the humerus: the "incipient trochlea for the ulna," "large lesser tubercle relative to the greater tubercle," and more pronounced torsion (30°) in the humeral shaft. Later in the same report, however, they observed the reverse about the size of the humeral tuberosities (tubercles), stating that the greater tuberosity is "slightly wider" than the lesser. Likewise, they noted a "weakly developed ulnar condyle" in addition to the trochlea at the distal end of the humerus (Hu et al., 1997, 140–141).

The holotypic skeleton of *Zhangheotherium* is preserved in ventral view, exposing the entire pectoral girdle (Hu et al., 1997, fig. 1), opposite the condition in the holotypic skeleton of *Jeholodens*. Reexamination of the specimen has revealed new information. Several derived features are present in the pectoral girdle including the supraspinous fossa, ventrally facing glenoid, ventrally positioned acromioclavicular articulation, and reduced size of the interclavicle. There is ample evidence for mobility in the pectoral girdle but no evidence that the humerus was held in a subhorizontal, or sprawling, posture, as proposed by Hu et al. (1997). Both forelimbs are preserved, although neither fully exposes either the proximal or distal articular ends of the humerus. The relative size of the humeral tuberosities, thus, is a moot point (*contra* Hu et al., 1997). The prominence and relative size of the tuberosities, in any case, have been shown to vary considerably within subgroups of extant mammals (e.g., Argot, 2001, fig. 8.1) and are not indicative of erect versus sprawling postures. The same is true regarding humeral torsion and forelimb posture. The torsion reported in *Zhangheotherium* (30°) is a rough approximation at best, because the proximal axis across the tuberosities cannot be precisely established. And the form of the elbow joint remains uncertain, because the lateral portion of the distal end of the humerus is not fully exposed.

Thus, on the basis of *Zhangheotherium,* symmetrodonts appear to have an advanced mobile shoulder girdle with a fully developed supraspinous fossa. The great capacity for hinge-like flexion-extension at the elbow joint is shown on the right side of the skeleton, where flexion at the elbow has brought the forearm against the humerus. The forelimb may well have had a more parasagittal than sprawling posture that emphasized hinge-like flexion-extension at the elbow, as in *Jeholodens*.

Phylogenetic Comparisons

Specializations in the Pectoral Girdle and Forelimb

Multituberculata. Six features in the pectoral girdle and forelimb of *Kryptobaatar* are unusual compared with other mammals and mammalian outgroups and appear to characterize all multituberculates, including both Late Cretaceous and Paleocene multituberculates (Krause & Jenkins, 1983; Kielan-Jaworowska & Gambaryan, 1994).

1. Scapular blade convex with C-shaped cross section. The transversely convex scapular blade in multituberculates contrasts with that in therians, monotremes, and mammalian outgroups, in which the central portion of the blade is relatively flat (figs. 10.9, 10.10).
2. Coracoid glenoid subrectangular (Sereno & McKenna, 1995). The elongate, subrectangular shape of the anterior portion of the glenoid in multituberculates (fig. 10.7E) is easily distinguished from the subtriangular shape that characterizes therians, monotremes, and mammalian outgroups (figs. 10.9, 10.10).
3. Coracoid posterior process absent (Sereno & McKenna, 1995). In multituberculates, there is no development of a discrete posterior process on the coracoid, which presumably forms the narrow anterior extension of the glenoid. The prominent anteromedial corner of the coracoid process is likely the much reduced homolog of the posterior process of therians, monotremes, and mammalian outgroups.
4. Interclavicle clover-shaped (Sereno & McKenna, 1995). The interclavicle in multituberculates has a unique three-leaf clover shape. The presence of a clover-shaped interclavicle in the fossorial Paleocene taeniolabidoid *Lambdopsalis* (IVPP, unnumbered, articulated clavicle and manubrium sterni) suggests that this bone, like much of the multituberculate skeleton, differs little in size and shape. In therians, the interclavicle is reduced in size and fused to the manubrium sterni (Klima, 1987). In monotremes and mammalian outgroups, the ventral process is usually longer than the lateral process and has a distinctive shape (Jenkins & Parrington, 1976; Kemp, 1982).
5. Interclavicle dorsoventrally arched (Sereno & McKenna, 1995). In multituberculates the interclavicle is dorsoventrally bowed, with the concave anterior surface articulating with the clavicle (figs. 10.7). The interclavicle in monotremes, in contrast, is transversely

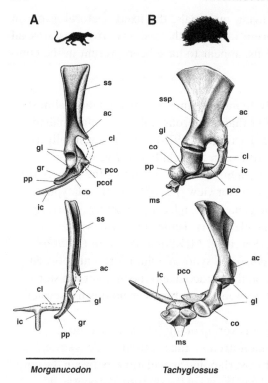

Figure 10.9. Pectoral girdle of (A) *Morganucodon watsoni*, a basal mammaliaform, and (B) *Tachyglossus aculeatus* (UCPC, recent collection), an extant monotreme in lateral (above) and posterior (below) views. The distal scapular blade of *Morganucodon* based on the closely related trithelodotid *Pachygenelus*. Scale bars equal 1 cm. Abbreviations: ac, acromion; cl, clavicle; co, coracoid; gl, glenoid; gr, groove; ic, interclavicle; ms, manubrium sterni; pco, procoracoid; pcof, procoracoid foramen; pp, posterior process; ss, scapular spine; ssp, secondary scapular spine.

arched (fig. 10.9B). In immediate mammalian outgroups, the interclavicle is flat (Jenkins & Parrington, 1976; Sun & Li, 1985).

6. Ulnar distal articular end approximately one-third the area of that of the radius. The distal end of the ulna tapers significantly in diameter. The radius, in contrast, expands in width so that the articular surface of its distal end is three times the area of that of the ulna (fig. 10.4). A marked size differential of the articular ends of the bones of the forearm may well occur in all multituberculates (e.g., Krause & Jenkins, 1983, fig. 14). In monotremes, the triconodonts *Jeholodens* and *Gobiconodon* (Ji et al, 1999; Jenkins & Schaff, 1988), *Henkelotherium* (Krebs, 1991), and the basal eutherian *Eomaia* (Ji et al., 2002), the distal ends of the radius and ulna appear to be subequal. In many other mammals and mammalian outgroups, however, the distal end of the radius is larger, although not to the degree observed in *Kryptobaatar* (e.g., *Morganucodon*; Jenkins & Parrington, 1976; *Kayentatherium*, MCZ 8812).

Monotremata. Among living mammals, the fixed pectoral girdle in monotremes is strikingly primitive. Nevertheless, six derived features, all possible fossorial adaptations, appear to have been present in the common monotreme ancestor.

1. Glenoid facing laterally. In monotremes, the glenoid faces almost directly laterally rather than posterolaterally as in mammalian outgroups (Jenkins & Parrington, 1976) or posteroventrally as in theriimorphs (fig. 10.10). In the echidna *Tachyglossus,* the glenoid is angled very slightly posteriorly, so that the edge of the glenoid is visible in posterior view (fig. 10.9). In the platypus *Ornithorhynchus,* the glenoid is angled very slightly anteriorly, so that the edge of the glenoid is visible in anterior view.

2. Coracosternal contact (Sereno & McKenna, 1995). In *Ornithorhynchus* and *Tachyglossus,* a synovial joint is present between a transverse process on the coracoid and the manubrium sterni (fig. 10.9B). Contact between these bones is unknown elsewhere among synapsids.

3. Coracoid and procoracoid overlap of interclavicle (Sereno & McKenna, 1995). In monotremes, the pectoral girdle is strengthened in the midline by overlapping articulations between the coracoid and procoracoid and between the opposing procoracoids (fig. 10.9B). In other mammals and mammalian outgroups, opposing pectoral girdles never contact, and contact between the pectoral girdle and axial skeleton is limited to the claviculosternal joint.

4. Clavicle-interclavicle fusion (Sereno & McKenna, 1995). These bones fuse only in monotremes. Complete fusion occurs late in development and characterizes fully grown adults.

5. Interclavicle lateral process long, contacting acromion (Sereno & McKenna, 1995). The lateral process of the interclavicle is long enough to contact the acromion only in monotremes among tetrapods (fig. 10.9B).

6. Interclavicle posterior process fan-shaped (Sereno & McKenna, 1995). In monotremes, the posterior process of the interclavicle has an unusual fan shape. It expands in width toward its posterior contact with the manubrium sterni. A unique low median keel is also present on the ventral aspect of the interclavicle in both *Ornithorhynchus* and *Tachyglossus.* In multituberculates (fig. 10.7), basal mammaliamorphs (fig. 10.9A; Jenkins & Parrington, 1976; Sun & Li, 1985), and more distant outgroups (Jenkins, 1971b),

the posterior process is tongue-shaped or parallel-sided and does not expand in width distally.

Synapomorphies in the Pectoral Girdle and Forelimb

This section describes synapomorphies in the pectoral girdle and forelimb among basal mammalian clades (fig. 10.15, tables 10.2, 10.3). The authors cited after a synapomorphy (also table 10.2) were the first to use the character in a phylogenetic analysis (for a more detailed discussion, see Sereno, in review).

Mammalia. Four synapomorphies from the pectoral girdle and forelimb characterize mammals.

1. Scapulocoracoid suture fused (Sereno & McKenna, 1995). With rare exception among fossil and recent mammals (McKenna, 1961; Klima, 1987; Jenkins & Schaff, 1988), the scapulocoracoid suture fuses at some point during development. In monotremes (*Ornithorhynchus, Tachyglossus*), fusion occurs only in older individuals (e.g., unfused in fig. 10.15B). In therians, fusion may occur in neonates or subadults, and, in a few cases, the suture remains open in adults (e.g., xenarthrans). Mammalian outgroups (*Morganucodon,* tritheledontids, tritylodontids) invariably show an open scapulocoracoid suture (fig. 10.9A; Jenkins & Parrington, 1976; Sun & Li, 1985).

2. Coracoid groove absent (Sereno & McKenna, 1995). The groove, or trough, on the posterior margin of the coracoid in mammalian outgroups (fig. 10.9A) is not present on the coracoid in monotremes (fig. 10.9B) or on the reduced, co-ossified coracoid in multituberculates and therians (fig. 10.10).

3. Coracoid posterior process shorter than glenoid diameter (modified from Sereno & McKenna, 1995). In monotremes, the robust coracoid has a posterior process that is shorter than the glenoid diameter (fig. 10.9B). In *Jeholodens* and *Zhangheotherium,* the posterior process is not fully exposed but appears to have been short as in *Henkelotherium.* This process has been completely reduced in multituberculates. In therians, the posterior process is usually reduced to a relatively small anteromedially projecting protuberance (fig. 10.10B). In mammalian outgroups, in contrast, the posterior process is equal to, or longer than, glenoid diameter (fig. 10.9A; Jenkins, 1971b; Sun & Li, 1985).

Table 10.2. Characters and character-states in the pectoral girdle and forelimb among basal mammals and mammalian outgroups, listed by clade (A) and anatomical region (B)

A. Listed by Clade

Mammalia

1. Scapulocoracoid suture: open (0); fused (1). (Sereno & McKenna, 1995)

2. Coracoid groove: present (0); absent (1). (Sereno & McKenna, 1995)

3. Coracoid posterior process, length: longer or subequal to (0), or shorter than (1), maximum glenoid diameter. (modified from Sereno & McKenna, 1995)

4. Procoracoid foramen: present (0); absent (1). (Sereno & McKenna, 1995)

Theriimorpha

5. Acromion position: dorsal (0), or lateral (1), to the glenoid. (Rowe, 1988)

6. Supraspinous fossa along anterior margin of scapular blade: absent or rudimentary (0); present (1). (Rowe, 1988)

7. Glenoid width: subequal to (0), or approximately 60% of (1), transverse width of the humeral head. (Sereno & McKenna, 1995)

8. Coracoid glenoid, orientation: posterolateral (0); lateral (1); posteroventral (2). (Sereno & McKenna, 1995)

9. Coracoid glenoid, form: convex (0); concave (1). (Sereno & McKenna, 1995)

10. Procoracoid: present (0); reduced/fused with manubrium sterni (1). (Rowe, 1988)

11. Clavicular head, form: tongue-shaped (0); rod-shaped (1).

12. Interclavicle size: larger (0), or smaller (1), than the sternum (anteroposterior and transverse dimensions). (modified from Rowe, 1988)

13. Humeral ulnar condyle, form: condylar (0); hemicondylar (1); trochlear (2). (Rowe, 1988)

14. Ulnar olecranon process length: less (0), or equal to or more (1), than the length of the semilunar notch.

15. Ulnar semilunar notch, intercondylar crest (anterior view): absent or rudimentary (0); present (1).

16. Ulnar anconeal process: absent (0); present (1).

Trechnotheria

17. Acromion, form: perpendicular (0), or recurved to lie parallel (1), to scapular blade. (Sereno & McKenna, 1995)

Vincelestes + Theria

18. Interclavicle-sternum coossification: absent (0); present (claviculosternal contact) (1). (Rowe, 1988)

B. Listed by anatomical region

Pectoral Girdle

1. Scapulocoracoid suture: open (0); fused (1).

5. Acromion position: dorsal (0), or lateral (1), to the glenoid.

7. Acromion form: perpendicular (0), or recurved to lie parallel (1), to scapular blade.

6. Supraspinous fossa along anterior margin of length of scapular blade: absent or rudimentary (0); present (1).

7. Glenoid width: subequal to (0), or approximately 60% of (1), transverse width of the humeral head.

3. Coracoid posterior process, length: longer or subequal to (0), or shorter than (1), the maximum diameter of the glenoid.

Table 10.2. *(continued)*

8. Coracoid glenoid, orientation: posterolateral (0); lateral (1); posteroventral (2).

9. Coracoid glenoid, form: convex (0); concave (1).

2. Coracoid groove: present (0); absent (1).

10. Procoracoid: present (0); reduced/fused with manubrium sterni (1).

4. Procoracoid foramen: present (0); absent (1).

11. Clavicular head, form: tongue-shaped (0); rod-shaped (1).

12. Interclavicle size: larger (0), or smaller (1), than the sternum (anteroposterior and transverse dimensions).

18. Interclavicle-sternum coossification: absent (0); present (claviculosternal contact) (1).

Humerus

13. Humeral ulnar condyle, form: condylar (0); hemicondylar (1); trochlear (2).

Radius and Ulna

14. Ulnar olecranon process length: less (0), or equal to or more (1), than the length of the semilunar notch.

15. Ulnar semilunar notch, form (anterior view): shallow, subrectangular (0); deep, hourglass-shaped (1).

16. Ulnar anconeal process: absent (0); present (1).

Note: 0, primitive state; 1, 2, derived states.

Table 10.3. Taxon-by-character matrix

	Character States			
Taxon	**5**	**10**	**15**	
Tritylodontidae	00000	00000	00000	000
Tritheledontidae	00000	00000	00000	000
Morganucodon	00000	00000	00000	000
Monotremata	11110	00000	00000	000
Gobiconodon	1???1	11??1	???11	10?
Jeholodens	1?1?1	11111	???1?	???
Multituberculata	1X111	01111	11111	100
Zhangheotherium	111?1	11111	11?0?	?10
Henkelotherium	11111	111?1	?1?11	?10
Vincelestes	11111	11111	?1211	111
Metatheria	11111	11111	11211	111
Eutheria	11111	11111	11211	111

Note: ?, not preserved or exposed; X, too transformed to score; multistate character 13 is ordered, because the derived states are cumulative. Embryological evidence is used to score relevant characters in terminal taxa with living representatives. Other abbreviations as in Table 10.2.

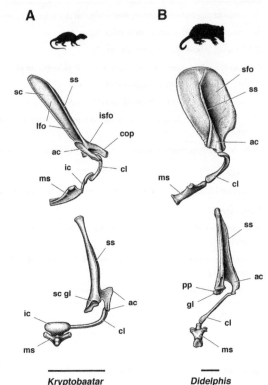

Figure 10.10. Pectoral girdle of (A) *Kryptobaatar dashzevegi* and (B) *Didelphis virginiana* (UCPC, recent collection) in lateral (above) and posterior (below) views. Scale bars equal 1 cm. Abbreviations: ac, acromion; cl, clavicle; cop, coracoid process; gl, glenoid; ic, interclavicle; isfo, incipient supraspinous fossa; lfo, lateral fossa; ms, manubrium sterni; pp, posterior process; sc, scapula; sfo, supraspinous fossa; ss, scapular spine.

4. Procoracoid foramen absent (Sereno & McKenna, 1995). The procoracoid foramen has not been recorded at any developmental stage in mammals (Klima, 1987). Mammalian outgroups (*Morganucodon,* tritheledontids, tritylodontids), in contrast, invariably retain the foramen for transit of the supracoracoideus nerve and associated vasculature (fig. 10.9A; Jenkins & Parrington, 1976; Sun & Li, 1985).

Theriimorpha. Twelve synapomorphies in the pectoral girdle and forelimb characterize theriimorphs (table 10.2).

5. Acromioclavicular joint positioned lateral to the glenoid (modified from Rowe, 1988). In theriimorphs, the acromion extends ventrally from the lateral aspect of the scapula, so that the joint with the clavicle is positioned lateral to the glenoid (fig. 10.10). This condition is clearly present in triconodonts (*Gobiconodon, Jeholodens;* Jenkins & Schaff, 1988; Ji et al., 1999), symmetrodonts (*Zhangheotherium;* Hu et al., 1997), and the basal trechnotherian

Vincelestes (Rougier, 1993). The complete multituberculate acromial process, seen for the first time in *Kryptobaatar* (figs. 10.3, 10.7), is particularly long and places the acromioclavicular joint immediately lateral to the midpoint of the arc of the glenoid. In this position, the acromion and scapular spine lie along an axis passing through the glenoid, so that muscles attaching to either side of the scapular spine may now be involved in the mobilization and/or stabilization of the shoulder joint. In monotremes and mammalian outgroups (fig. 10.9), the acromion projects from the anterior margin of the scapular blade and does not descend to the level of the glenoid (Jenkins, 1971b; Jenkins & Parrington, 1976; Sun & Li, 1985).

6. Supraspinous fossa along anterior margin of scapular blade (Rowe, 1988). In all theriimorphs except multituberculates, a well-marked supraspinous fossa extends along the anterior margin of the scapular blade, anterior to the scapular spine and base of the acromial process (fig. 10.10B; Jenkins & Schaff, 1988; Krebs, 1991; Rougier, 1993; Hu et al., 1997; Ji et al., 1999, 2002). Multituberculates, a notable exception among theriimorphs, are similar to outgroups, in which the protruding anterior edge of the blade has been interpreted as the homolog of the therian scapular spine (figs. 10.9, 10.10A).

7. Glenoid width only approximately 60 percent of transverse width of the humeral head (Sereno & McKenna, 1995). In theriimorphs, the maximum width of the glenoid (transverse width of scapular portion) is approximately one-half the maximum diameter of the opposing surface of the humeral head (fig. 10.7E). The marked differential width of the articular surfaces is present in multituberculates (fig. 10.11E) and *Vincelestes* (Rougier, 1993) and appears to be present in triconodonts, symmetrodonts, and basal trechnotherians (Jenkins & Schaff, 1988; Krebs, 1991; Hu et al., 1997; Ji et al., 1999). In theriimorph outgroups, in contrast, the scapular glenoid more closely approximates the size of the opposing surface of the humeral head, as in monotremes (Jenkins, 1971b; Sun & Li, 1985).

8. Coracoid glenoid with posteroventral orientation (Sereno & McKenna, 1995). In theriimorphs, the coracoid glenoid faces posteroventrally. In lateral view, little of the surface of the coracoid glenoid is exposed, because it is positioned parallel to the line of sight (fig. 10.10). In theriimorph outgroups, in contrast, the coracoid glenoid faces either laterally (monotremes) or posterolater-

ally (basal mammaliamorphs, basal synapsids). As a result, the
coracoid glenoid is broadly exposed in lateral view but largely
hidden in posterior view (fig. 10.9).

9. Coracoid glenoid concave (Sereno & McKenna, 1995). In theri-
    imorphs, the glenoid is anteroposteriorly concave and transversely
    flat (multituberculates) or very slightly concave (e.g., *Vincelestes;*
    Rougier, 1993, fig. 82). In triconodonts and symmetrodonts, the
    coracoid process is also anteroposteriorly concave (Hu et al., 1997:
    Ji et al., 1999). In theriimorph outgroups, in contrast, the coracoid
    glenoid is concavoconvex (fig. 9; Jenkins, 1971b; Jenkins &
    Parrington, 1976).

10. Procoracoid reduced/fused with the manubrium sterni (Rowe,
    1988). In theriimorphs, the procoracoid fails to ossify as a separate
    element. In the embryos of living marsupials, a procoracoid con-
    densation is occasionally retained as a small cartilaginous element
    (praeclavium) at the medial end of the clavicle (Klima, 1987). In
    placentals, the procoracoid anlagen of the procoracoid is incorpo-
    rated into the manubrium sterni (Klima, 1987). In theriimorph
    outgroups such as monotremes, *Morganucodon,* and tritylodon-
    tids, the procoracoid is an ossified element sutured to the coracoid
    and scapula (fig. 10.9).

11. Clavicular head rod-shaped (Sereno & McKenna, 1995). The medial
    end of the clavicle in theriimorphs is rod-shaped rather than flat-
    tened, presumably to allow more flexibility at the claviculointer-
    clavicle joint. A clavicle of this type is known in multituberculates,
    the triconodont *Jeholodens,* the symmetrodont *Zhangheotherium,*
    and *Henkelotherium.* In theriimorph outgroups, the clavicle has a
    flattened, fluted medial end that is firmly attached to the interclav-
    icle (Jenkins, 1971b).

12. Interclavicle smaller than the sternum (anteroposterior and trans-
    verse dimensions) (modified from Rowe, 1988). In theriimorphs,
    the interclavicle is reduced in size, either as a small ossification
    (fig. 10.7; Hu et al., 1997) or as embryological condensation
    (Klima, 1987). The interclavicle appears to be partially co-ossified
    to the manubrium sterni in *Vincelestes* (Rougier, 1993, fig. 83).
    In theriimorph outgroups, the interclavicle is larger than the
    manubrium sterni.

13. Humeral ulnar hemicondyle (modified from Rowe, 1988). In theri-
    imorphs, the anterior portion of the humeral ulnar condyle takes
    the form of a hemicondyle, in which only the lateral, intercondy-
    lar portion of the primitive hemispherical condyle is developed.

The hemicondyle, thus, has an asymmetrical cam shape (Jenkins, 1973, 290). In multituberculates the cam is developed only on the anterior half of the ulnar condyle adjacent to the bulbous, hemispherical portion of the radial condyle (fig. 10.11B). In therians, the ulnar condyle is cam-shaped posteriorly as well, which, together with the broadened intercondylar groove, forms the trochlea (fig. 10.11C). An ulnar condyle that is, at least in part, cam-shaped is transitional to a fully developed therian trochlea. In theriimorph outgroups, the ulnar condyle is more symmetrically developed, although somewhat variable in its convexity.

14. Ulnar olecranon exceeds the length of the semilunar notch. In theriimorphs, the long olecranon process increases the lever arm of the triceps musculature, reflecting the increased activity of the elbow joint in the stride in a forelimb with a more parasagittal posture (fig. 10.11B, C). In theriimorph outgroups, the olecranon is somewhat shorter than the semilunar notch (Jenkins, 1976; Jenkins & Parrington, 1976; Sun & Li, 1985). In monotremes, the olecranon is subequal to the length of the semilunar notch and may have gained this length secondarily as a fossorial adaptation.

15. Ulnar intercondylar crest present. In theriimorphs, the semilunar notch of the ulna is divided by an intercondylar crest separating articular surfaces for the radial and ulnar condyles of the humerus (fig. 10.11B). The intercondylar crest is present in multituberculates (fig. 10.11B), *Vincelestes* (Rougier, 1993), and therians (fig. 10.11C) but is poorly documented in other basal theriimorphs. There is no intercondylar crest and, consequently, no articular surface on the ulna for the radial condyle of the humerus in monotremes, tritylodontids (fig. 10.11A), or more distant theriimorph outgroups (Jenkins, 1971b).

16. Ulnar anconeal process present. In theriimorphs, a distinct anconeal process is present at the posteriormost end of the intercondylar crest on the ulna in multituberculates (fig. 10.11B), triconodonts (Jenkins & Schaff, 1988, fig. 15), *Vincelestes* (Rougier, 1993, fig. 87), and therians (fig. 11C). In theriimorph outgroups, the ulnar cotylus for the humerus is broad with little development of its posterior rim, as seen in monotremes, tritylodontids (fig. 10.11A; Sun & Li, 1985, fig. 10), and more basal synapsids (Jenkins, 1971b).

Trechnotheria. One synapomorphy in the pectoral girdle characterizes trechnotherians (table 10.2). This synapomorphy eventually may be shown to have a broader distribution among basal theriimorphs.

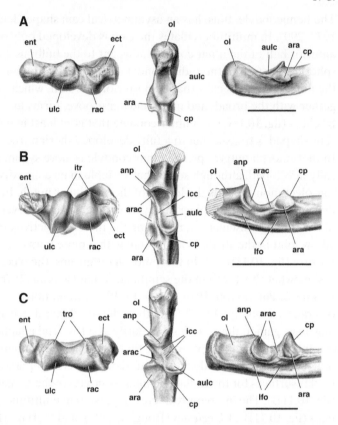

Figure 10.11. Distal right humerus in ventral view (anterior toward bottom) and proximal right ulna in anterior (left) and medial (right) views in (A) *Bienotherium yunnanense* (FMNH CUP 2286, 2291, 2300), (B) Cretaceous multituberculates from North America and Asia (*Nemegtbaatar gobiensis*), and (C) *Didelphis virginiana* (UCPC, recent collection). Distal right humerus and proximal right ulna in (B) drawn from stereophotographs in Jenkins (1973, fig. 19) and (Kielan-Jaworowska & Gambaryan, 1994, fig. 14D, E), respectively. Scale bars equal 2 cm, 5 mm, and 1 cm, respectively, for (A), (B), and (C). Abbreviations: anp, anconeal process; ara, articular surface for the radius; arac, articular surface for the radial condyle; aulc, articular surface for the ulnar condyle; cp, coronoid process; ect, ectepicondyle; ent, entepicondyle; icc, intercondylar crest; itr, incipient trochlea; lfo, lateral fossa; ol, olecranon; rac, radial condyle; tro, trochlea; ulc, ulnar condyle.

17. Acromion recurved to lie parallel to scapular blade (Sereno & Mc-
    Kenna, 1995). In *Vincelestes* (Rougier, 1993, fig. 82) and therians
    (fig. 10.10B), the acromial process curves posteriorly so that the
    expanded portion of the process lies in a plane parallel to that of
    the scapular blade. In *Henkelotherium,* the acromial process may
    be developed in a similar manner (Krebs, 1991, fig. 7). The condi-
    tion is poorly preserved in basal theriimorphs such as tricono-
    donts and symmetrodonts. In multituberculates (figs. 10.3, 10.7)

and *Morganucodon* (fig. 10.9A; Jenkins & Parrington, 1976, fig. 4), the acromial process curves laterally to a lesser degree, positioned perpendicular rather than parallel to the scapular blade. In monotremes (fig. 10.9B), tritylodontids (Sun & Li, 1985, fig. 7), and more basal synapsids (Jenkins, 1971b, fig. 17), the acromial process projects anteriorly from the scapular blade.

Vincelestes plus Theria.
One synapomorphy from the pectoral girdle characterizes *Vincelestes* plus Theria (table 10.2).

18. Interclavicle and sternum co-ossified, claviculosternal contact (Rowe, 1988). In *Vincelestes* (Rougier, 1993, fig. 83) and therians (fig. 10.10B), the interclavicle fails to ossify as a separate element, and, as a result, the clavicle articulates with the manubrium sterni. In therians the endochondral portion of the interclavicle forms the anteromedian portion of the manubrium sterni (Klima, 1987, fig. 29). Among outgroups, the presence of a separate interclavicle remains uncertain in *Henkelotherium* and triconodonts. In multituberculates (fig. 10.10A) and *Zhangheotherium* (Hu et al., 1977), the interclavicle is present as a separate element, albeit reduced in size.

Phylogenetic Significance

Multituberculates as Theriimorph Mammals

The dentition in multituberculates, monotremes, and several other basal mammalian taxa is so modified that dental characters have had limited impact in sorting their relationships at the root of Mammalia. Likewise, the composition of the sidewall of the braincase was once viewed as key to resolving multituberculate relationships (Broom, 1914; Kielan-Jaworowska, 1971). A recent review of basicranial evidence (51 characters), however, proved indecisive; multituberculates were joined with therians in the shortest trees but linked with prototherians with one additional step (Rougier et al., 1996b, 24).

Sereno and McKenna (1995), elaborating on earlier results by Rowe (1988), suggested that postcranial characters, especially those from the shoulder girdle and forelimb, decisively link multituberculates and therians within Mammalia. Rougier et al. (1996a; 406) countered that "evaluation of the competing phylogenetic hypotheses of multituberculate relationships must await the inclusion of additional taxa and characters," a reasonable caution. Now we are in a position to reevaluate.

Luo et al. (2002) presented the most comprehensive analysis for basal mammalian phylogeny (275 characters, forty-six taxa). Using *Probainog-nathus* as an outgroup, they generated forty-two equally parsimonious trees from 271 informative characters [935 steps, consistency index (CI) = 0.499, retention index (RI) = 0.762; Luo et al., 2002, 6]. Rerun of their data yielded similar results (273 informative characters, forty-two trees, 933 steps, CI = 0.498, RI = 0.762, multistate taxa scored as polymorphic) and the same strict consensus tree (Luo et al., 2002, fig. 1).

Despite significant character support in their data placing multi-tuberculates within crown mammals (Mammalia) as a sister-taxon to Theria (figs. 10.2A, B), Luo et al. (2002) regarded their position outside crown mammals as equally likely. Regarding postcranial support, Luo et al. (2002, 33) stated that only "four synapomorphies for multituberculates and trechnotherians are unambiguous." In their shortest trees, however, ten postcranial synapomorphies provide unambiguous support for this node. A total of twenty-one unambiguous characters (dental, cranial, postcranial) reside at Mammalia or nodes within that include multituberculates.

To test the significance of this result, Luo et al. (2002, fig. 2) con-strained multituberculates plus *Haramiyavia* to a position outside Mam-malia (crown mammals), generating trees seven steps longer (942 steps). Rerunning the analysis with similar constraints yielded similar results. With multituberculates and *Haramiyavia* as a clade outside Mammalia (crown mammals), length increases seven steps (933 to 940 steps). A sim-ilar result is obtained by constraining only multituberculates to a position outside crown mammals; in this case, *Haramiyavia* returns to the base of the tree (rather than as a sister-taxon to Multituberculata), reducing tree length by one step to 939 steps, or six steps beyond minimum length (933 steps). Regarding the most parsimonious and constrained positions for multituberculates, Luo et al. (2002, 33) reasoned, "this is a small differ-ence between these two contrasting tree topologies," and further, "non-parametric tests . . . show that the difference is not significant." Thus, they concluded, "we do not favor one interpretation over the other."

Why do multituberculate relationships deserve this special six-step moratorium from phylogenetic interpretation? Is this another expression of the traditional viewpoint that holds that apomorphic resemblances be-tween multituberculates and therians *must* be homoplastic? Luo et al. (2002, 32–33) regard the phylogenetic position of multituberculates es-pecially difficult to determine because they (1) exhibit a mosaic of prim-itive and derived character states, (2) exhibit many masticatory and den-tal autapomorphies, and (3) lack a good early fossil record. This is not clear in the data set at hand or in current stratigraphic ranges (fig. 10.16).

Multituberculates are not particularly unusual among Mesozoic mammaliaforms regarding missing data (via lack of preservation or transformation) or missing stratigraphic range.

Decay analysis shows that few nodes derived from the data presented by Luo et al. (2002) have as many unambiguous synapomorphies as those linking multituberculates and therians within Mammalia. Hundreds of trees exist at just one step beyond minimal length (934 steps), and thousands of trees exist at two steps beyond minimal length (935 steps). The strict or semistrict consensus of trees at 935 steps breaks down all nodes except two within Trechnotheria. If character support placing multituberculates within Mammalia (crown mammals) is deemed insignificant, then little significance resides elsewhere in this data set.

The decisiveness of characters from the shoulder girdle and forelimb regarding the affinities of multituberculates is real and documents a major transition in mammalian locomotor function and posture that occurred sometime during the Jurassic (fig. 10.16, node 2). In the data set of Luo et al. (2002), most of the twelve unambiguous characters linking multituberculates and trechnotherians are postcranial; fifty percent come from the pectoral girdle and forelimb. These data, furthermore, do not include six synapomorphies in the pectoral girdle and forelimb outlined above (7, 11, 12, 14–16, table 10.2). The structure of character data from the pectoral girdle and forelimb presented here is very simple (table 10.3). Several taxa are actual or potential taxonomic equivalents (Wilkinson, 1995) with identical, or potentially identical, character states, respectively (Tritylodontidae / Tritheledontidae / *Morganucodon;Gobiconodon/ Jeholodens;Vincelestes*/ Metatheria / Eutheria).

Problematic Characters in the Pectoral Girdle and Forelimb

Twenty characters from the pectoral girdle and forelimb from previous analyses (Rowe, 1988; Luo et al., 2002) are rejected or regarded as uninformative for nodes at the root of Mammalia (for more discussion, see Sereno, in review). Three are uninformative with the present terminal taxa (fig. 10.15) or after rescoring character states (Rowe, 1988: radial condyle form; radial styloid process; Luo et al., 2002: supinator ridge with rectangular extension). One is a functional interpretation (Luo et al., 2002: claviculosternal apparatus joint) and another is clearly redundant (Luo et al., 2002: interclavicle-manubrium sterni contact/relationship). Most of the rejected characters are ill-defined or quite variable and, thus, are difficult to interpret and score (humeral head shape/inflection; bicipital (intertubercular) groove; lesser tuberosity muscle insertions; greater and lesser tuberosities form pronounced ridges; radial/ulnar condyle size

(Rowe, 1988); cranial margin of interclavicle; curvature of the clavicle; acromioclaviclular joint; fossa for teres major muscle; medial surface of the scapula; lesser tuberosity size; deltopectoral crest length; humeral torsion; epicondylar robustness (Luo et al., 2002).

Functional Evolution

Multituberculates as Sprawling Saltators?

Unique Locomotor Model. There are no living quadrupedal mammalian saltators with a sprawling forelimb posture, nor have any extinct species been so interpreted previously. Thus, the functional model proposed for multituberculates by Gambaryan and Kielan-Jaworowska (1997) (fig. 10.12)—here termed the "sprawling saltator" model—is unique. The proposed stride sequence for multituberculates, furthermore, differs from locomotor patterns in living mammals—including the sprawling monotremes.

First, during the stride in living mammals, the elbow joint moves only

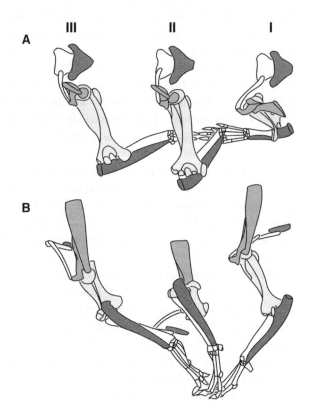

Figure 10.12. Reconstruction of phase I–III of the walking step of a multituberculate in (A) dorsal and (B) lateral views as proposed by Gambaryan and Kielan-Jaworowska (1997; after their fig. 10). Bones are differentiated by shading (not in the original).

slightly in a transverse direction (Jenkins, 1973) and is located either close to the thorax, as in therians (Jenkins, 1971a, 1979), or strongly abducted, as in monotremes (Pridmore, 1985)—but not transient in transverse position, moving from one to the other, as shown in the sprawling saltator model (fig. 10.12A, I, II). In this regard, Kielan-Jaworowska and Gambaryan (1997) stated, "the main difference between the abducted (sprawling) and parasagittal limbs does not concern the angle of humeral or femoral abduction from the sagittal plane, but the positions of hands and feet in the propulsive phase with respect to this plane." This is not true for the forelimb of living mammals as documented by cineradiography.

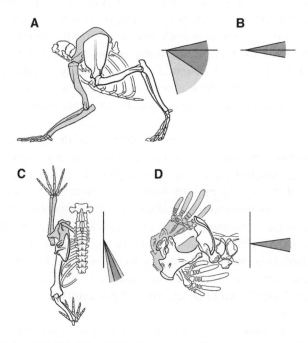

Figure 10.13. Lateral (A, B) and dorsal (C, D) views of the walking step in the eutherian *Didelphis virginiana* (A, C) and monotreme *Tachyglossus aculeatus* (B, D), showing the posture of the forelimb bones during the initiation (shaded, phase I) and completion (unshaded, phase III) of propulsion, as interpreted from cineradiography. Associated excursion arcs for the humerus show usual range of movement (dark shading) and occasional deviations beyond that range (light shading), with the apex of the arc located at the shoulder joint (after Jenkins 1970b, 1971b). (A) *Didelphis virginiana* in lateral view showing the posteroventral excursion of the shoulder joint, posture of the forelimb, and usual humeral arc of depression/elevation (about 50°; dark shading) during the walking step. (B) Smaller observed range of humeral depression/extension (about 20°) in the monotreme *Tachyglossus aculeatus* that is not synchronized with the step cycle. (C) *Didelphis virginiana* in dorsal view showing the posterior excursion of the shoulder joint, posture of the forelimb, and usual angle of humeral adduction from the sagittal plane (about 20°; dark shading) during the walking step. (D), *Tachyglossus aculeatus* in dorsal view showing the posture of the forelimb and usual angle of humeral adduction from the sagittal plane (about 90°; dark shading) during the walking step.

Therian mammals with a more erect posture show far less humeral abduction than sprawling monotremes (figs. 10.13C, D).

Second, movement of the shoulder joint relative to the trunk in therian mammals follows a short arc during propulsion, from a position anterodorsal to the manubrium sterni near the dorsal margin of the thoracic outlet to a position lateral to the manubrium near the ventral margin of the thoracic outlet (fig. 10.13A; Jenkins, 1979, fig. 5). In the sprawling saltator model, in contrast, the shoulder joint is shown originating in a position anteroventral to the manubrium, relocating to a position posterodorsal to the manubrium, and then returning far ventral to the manubrium and the associated thoracic outlet (fig. 10.12B, I–III).

Finally, the forearm assumes a vertical, or near vertical, posture toward the end of propulsion in both erect and sprawling mammals (Jenkins, 1970b, 1971a, 1979; Pridmore, 1985). In the sprawling saltator model, in contrast, the forelimb is oriented at approximately 60° from the vertical (fig. 10.12B, III). All these aspects of the proposed model for multituberculate locomotion are without parallel in extant mammals.

Other Discrepancies. Dorsal and lateral views of the sprawling saltator model differ in significant ways from each other and from preserved multituberculate skeletons. In dorsal view of phase I of the stride, the rotational axis of the elbow joint is shown as nearly transverse; in lateral view, the same axis is angled anterolaterally. In dorsal view of phase II, the clavicle is transversely oriented, and the interclavicle is located medial to the scapula; in lateral view, the clavicle is shown angling posterolaterally, and the interclavicle is positioned anterior to the scapula. In dorsal view of phase III, the interclavicle/manubrium and the olecranon process of the ulna are positioned farther anteriorly and posteriorly, respectively, than in lateral view.

Other proportions or articulations do not match preserved skeletal material. The scapula is shown as shorter, rather than longer, than the humerus; the interclavicle is shown as flat rather than strongly arched; the clavicle is shown articulating with the dorsal, rather than ventral, side of the interclavicle; and the claviculoacromial joint is positioned anterior, rather than lateral, to the glenoid (figs. 10.7A, B).

The form of the elbow joint in multituberculates clearly suggests that significant rotation occurred. In the sprawling saltator model for multituberculate locomotion, however, flexion-extension at the elbow is as limited as in monotremes (figs. 10.12, 10.13D). Gambaryan and Kielan-Jaworowska (1997, 38), nevertheless, stated that enhanced flexion-extension at the elbow joint also characterizes their model: "When during landing the forelimbs were stretched anteroventrally, the elbow joint extended and the olecranon fitted into a deep fossa olecrani on the dorsal side of the humerus." During this, or any other, phase of their model, however, the olecranon is not close to the olecranon fossa, and flexion-extension at the elbow is limited (fig. 10.12, II). Likewise, they reported "strong flexion of the elbow joint during the middle of the propulsive phase" that required "a deep coronoid fossa on the ventral side of the humerus, into which fitted the radial head." In their model, however, the elbow never flexes at an angle much less than 90°, which keeps the radial head far from the radial fossa. In *Kryptobaatar* (PSS-MAE 103), for example, it is clear that flexion of the elbow at an angle as tight as 50° still does not engage the radial fossa (fig. 10.5). The radial fossa comes into play under situations of much greater flexion, as preserved in the left forelimb of PSS-MAE 103 (fig. 10.5).

Evidence for Sprawling Forelimb Posture. In the sprawling saltator model (Gambaryan & Kielan-Jaworowska, 1997), movement of the humerus is closest to that observed in monotremes, which have a rigid pectoral girdle, laterally facing glenoid, strong muscles attached to hypertrophied tuberosities for humeral rotation and retraction, and limited flexion-extension at the elbow joint (figs. 10.13B, D; Pridmore, 1985). In multituberculates, in contrast, the shoulder joint is surely mobile during the stride, judging from its construction; the glenoid faces ventrally; the humeral epicondyles are much less expanded relative to humeral length, even in the presumed fossorial genus *Lambdopsalis* (less than 50 percent humeral length); and flexion-extension at the elbow joint is greatly enhanced, judging from the arc of the ulnar condyle of the humerus and other features. These attributes characterize therians with a more parasagittal forelimb posture.

Gambaryan and Kielan-Jaworowska (1997) largely sidestepped these fundamental, functionally relevant differences among extant mammals, basing their argument for sprawling posture instead on attributes of the humerus—namely, the (1) increased size and/or prominence of the lesser tuberosity compared with the greater tuberosity, (2) broad width of the intertubercular (bicipital) groove, (3) high degree of humeral torsion, and (4) condylar structure of the distal end. None of these characterize multituberculates, in particular, or provides convincing evidence for sprawling forelimb posture, as discussed below.

In multituberculates, the lesser tuberosity is subequal to the greater tuberosity in size and prominence (figs. 10.4, 10.7E); there is no significant size differential. The relative size of the tuberosities, furthermore, is not indicative of a particular forelimb posture. In *Tachyglossus* and some mammalian outgroups, in contrast, the lesser tuberosity can be as much as twice the dimensions of the greater tuberosity (Jenkins, 1971b, figs. 27, 28). In some extant therians, the reverse proportions obtain (MacPhee, 1994, fig. 26). Among therians, the lesser tuberosity can be smaller, subequal, or larger than the greater tuberosity, depending on the species examined. All three proportions, for example, occur among didelphimorphs (Argot, 2001, fig. 8.1). In fossorial therians, either tuberosity may be enlarged (MacPhee, 1994; *contra* Gambaryan & Kielan-Jaworowska, 1997, 30). In sum, the humeral tuberosities in multituberculates, which are subequal in size, do not provide evidence favoring a sprawling forelimb posture.

The depression between the humeral tuberosities (i.e., the intertubercular or bicipital groove) is developed as a broad, open trough in monotremes (Gambaryan & Kielan-Jaworowska, 1997, fig. 3C), basal mammaliamorphs such as *Morganucodon* (Jenkins & Parrington, 1976, fig. 5C), and mammaliamorph outgroups (Jenkins, 1971b, fig. 26B). In multituberculates and therians, the proximal end of the humerus is proportionately narrower, and the space between the tuberosities is much reduced. Gambaryan and Kielan-Jaworowska (1997, 30) suggested that the reduction in width—from a trough to a groove—is correlated with a reduction in musculature passing from the coracoid to the humerus, the groove occupied only by the tendon of the biceps brachii caput longum (in some therians, a slip of the coracobrachialis may also pass along the groove; Argot, 2001, 59).

Differentiating "broad" versus "narrow" regarding the space between the tuberosities, however, was not well specified. Kielan-Jaworowska and Gambaryan (1994, 61) stated, "the intertubercular groove is very wide in multituberculates; it is 32 percent of the width of the proximal epiphysis in *Nemegtbaatar* and 28–31 percent in ?*Lambdopsalis*, while in modern small rodents it is only 14–23 percent." Later they reported a groove width of 30 percent in *Nemegtbaatar*, 40 percent in *Chulsanbaatar*, and 40 percent in an unidentified Asian multituberculate (Gambaryan & Kielan-Jaworowska, 1997, 31). In *Kryptobaatar* (PSS-MAE 103), the groove constitutes approximately 25 percent of the width across the tuberosities and is narrower than the width of either tuberosity (figs. 10.4, 10.7E). In fact, a range of 20–30 percent appears to characterize all the Asian genera mentioned above including *Lambdopsalis* (see photographs in Kielan-Jaworowska & Gambaryan, 1994, and Gambaryan & Kielan-Jaworowska, 1997).

In the basal trechnotherian *Vincelestes,* the groove appears to be proportionately broader than in multituberculates (Rougier, 1993, 295, fig. 84). In the opossum *Didelphis,* the groove occupies 50 percent that of the width across the tuberosities and is subequal in width to the largest tuberosity (UCPC, recent collection). Among extant didelphimorphs, the width of the groove relative to that across the tuberosities varies by as much as a factor of two (Argot, 2001, fig. 8.1). Among extant fossorial therians with various forelimb postures, likewise, the width of the groove varies from extremely narrow or partially enclosed (*Myospalax,* Gambaryan & Kielan-Jaworowska, 1997, fig. 6B; *Euphractus,* MacPhee, 1994, fig. 26) to more broadly open (*Manis,* MacPhee, 1994, fig. 26).

In sum, the space between the tuberosities in multituberculates takes the form of a groove and more closely resembles the bicipital groove in therians (with various forelimb postures) than it does the broadly open trough that characterizes monotremes and mammalian outgroups. The groove in multituberculates, therefore, does not provide evidence favoring a sprawling forelimb posture.

Torsion in the humeral shaft—i.e., the angle between the axis across the tuberosities proximally and the axis through the distal condyles and epicondyles distally (Simpson, 1928b; Kielan-Jaworowska & Gambaryan, 1994, fig. 46)—varies among mammals and mammalian outgroups. Among mammalian outgroups, torsion is 20–30° in the tritylodontid *Bienotherium* (FMNH CUP2286, 2300) and more distant taxa (Jenkins, 1971b, figs. 27, 28). Jenkins and Schaff (1988, 15) reported similar torsion (about 33°) in the triconodont *Gobiconodon.* Hu et al. (1997, 140), likewise, reported torsion of about 30° in the symmetrodont *Zhangheotherium,* although the proximal end of the humerus is not fully exposed (see above). Humeral torsion in extant fossorial mammals is sometimes greater (45° in Tachyglossidae; 60° in Chrysochloridae), although considerable variation exists (MacPhee, 1994, fig. 26).

Humeral torsion in *Kryptobaatar* is approximately 15°, as measured in both humeri of PSS-MAE 103 (fig. 10.7G). Other estimates of torsion for this genus (30°) are based on separate proximal and distal ends (Gambaryan & Kielan-Jaworowska, 1997) or on a humerus that is not fully exposed proximally (Kielan-Jaworowska, 1998, fig. 2). In the latter case, the preserved landmarks on this humerus in ventral view closely match those in PSS-MAE 103 (fig. 10.7D), and thus it would be very surprising if its torsion is greater by a factor of two. In *Lambdopsalis,* torsions of 24° and 38° were measured from two specimens; fractures in the shaft of each specimen, however, may well have artificially increased the apparent range of torsion in this genus (Gambaryan & Kielan-Jaworowska, 1997, fig. 9). Humeral torsion in multituberculates probably falls within a range of 15–30°.

In the metatherian *Didelphis,* humeral torsion measures 20–25°, and this range is typical of many therians (UCP, recent collection). Rougier (1993, 298) reported stronger torsion (40°) in the basal trechnotherian *Vincelestes.* Humeral torsion may have decreased somewhat from 20–30° in theriimorph outgroups to 20° or less among theriimorphs. But it is clear, first, that *Kryptobaatar* and perhaps other slender-limbed multituberculates have less than 20° of humeral torsion and, second, that humeral torsion among living and fossil theriimorphs with a variety of forelimb postures is more variable than previously was assumed. For these reasons, humeral torsion in multituberculates does not substantiate a sprawling forelimb posture.

Detailed study of the elbow joint in extinct mammalian outgroups has shown that during propulsion the primitive spiral condylar form of the distal humerus "maintains the forearm in a sagittal plane as the humerus adducts, elevates and medially rotates" (Jenkins, 1973, 288), much as does the asymmetrical trochlea of basal extinct therians or the living opossum (fig. 10.13A, C). Contrary to statements by Gambaryan and Kielan-Jaworowska (1997), the spiral condylar structure of the multituberculate humeroulnar joint does not support a sprawling forelimb posture. Krause and Jenkins (1983, 235) concluded, "as similar excursions are accommodated by the spiral trochlear humeroulnar joint of primitive therians, the significance of the difference between the condylar and trochlear joint types is unclear." Sloan and Van Valen (1965, 222) stated that ulnar movements in multituberculates are "restricted to a single plane by the shapes of the trochlea and semilunar notch" (fig. 10.11B). This is strongly corroborated in this study and is preserved *in situ* by comparing the articulated positions of left and right forelimbs in PSS-MAE 103 (figs. 10.4, 10.5).

In multituberculates, the ulna was not free to slide transversely across a broadly convex ulnar condyle on the humerus, as shown by Gambaryan and Kielan-Jaworowska (1997, fig. 1A). Rather, the ventral portion of the ulnar condyle in multituberculates is cam-shaped (fig. 10.6), the transitional form of the ulnar condyle predicted to have occurred during evolution of the therian trochlea (Jenkins, 1973, 290). The narrow dorsal half of the ulnar condyle is fitted to an arcuate fossa in the semilunar notch and is bounded laterally and dorsally by an intercondylar crest and anconeal process, respectively (fig. 10.11B). In sum, there is no evidence from the multituberculate humeroulnar joint that supports a sprawling forelimb posture.

Evidence for Saltatory Locomotion. The argument for saltatory locomotion in multituberculates was based on a single feature—the reconstructed length of the spinous processes of the lumbar vertebrae in
Nemegtbaatar (Kielan-Jaworowska & Gambaryan, 1994, fig. 36B). The
length of the lumbar transverse processes, in turn, was cited as an indicator
of an asymmetrical gait, in which both forelimbs move before the hind
limbs (Kielan-Jaworowska & Gambaryan, 1994, 65, 72). Hypotheses regarding the steepness of the trajectory of the jump, the shock-absorbing
function of the forelimbs, and others stemmed from the initial conjecture
of saltatory habits in multituberculates based on lumbar neural spine
length.

The length of the lumbar neural spines in multituberculates, however,
is not greater than that in other basal mammals relative to either the
centrum or transverse processes. In lumbar vertebrae of *Eucosmodon*
(Granger & Simpson, 1929, fig. 26A; Krause & Jenkins, 1983, fig. 28A), a
complete spine measures 50–70 percent of the length of the centrum or
transverse processes and is approximately subequal to the height of the
centrum. The lumbar spines are erect, rather than anterodorsally inclined, and are either subquadrate or somewhat subrectangular in shape
(Krause & Jenkins, 1983, fig. 28).

In *Kryptobaatar* (PSS-MAE 103), an anterior lumbar vertebra is preserved with all processes intact. The spine is proportionately shorter,
measuring only 1 mm in height. The centrum of this vertebra has a height
of 1.8 mm and the transverse processes measure 2 mm in length. Like
Nemegtbaatar, the spine is narrower than in *Eucosmodon* and is anterodorsally inclined.

In *Nemegtbaatar,* all the spines in the single available lumbar series are
broken and have been reconstructed at twice their preserved length
(Kielan-Jaworowska & Gambaryan, 1994, fig. 36B). The evidence from
Kryptobaatar brings into question this reconstruction.

In the triconodont *Gobiconodon,* the lumbar neural spines and transverse processes are longer than centrum length and thus relatively longer
than in multituberculates (Jenkins & Schaff, 1988, fig. 12C, E, G). Relatively long lumbar transverse processes and spines, furthermore, are
present in many nonsaltatory mammals (MacPhee, 1994, fig. 18). In fact,
there are few skeletal correlates among the less modified vertebrate saltators that are not also present in fossorial or cursorial mammals (Emerson, 1985; Hildebrand & Goslow, 2001).

In sum, the evidence from the lumbar vertebrae of multituberculates
suggests that significant variation might exist, some species with subquadrate, erect spines and others with somewhat narrower, anterodorsally

inclined spines. There is no evidence at present to suggest that the lumbar spines of multituberculates are inordinately long or that increased length of the lumbar neural spines is correlated with saltatory habits alone. Two features common among habitual mammalian saltators, in fact, are absent in multituberculates—namely, significant lengthening and strengthening of the hind limb and a posterior, rather than anterior, centered body mass (Emerson, 1985; Sereno, in review). New evidence is needed if a plausible case is to be made for habitual saltatory locomotion in multituberculates.

Checklist for Forelimb Posture

Three major, biomechanically significant features in shoulder girdle and forelimb architecture in living mammals may best constrain the range of possible forelimb postures in extinct relatives: shoulder girdle orientation and shoulder and elbow joint mobility.

Shoulder Joint Orientation. In the sprawling monotremes and in extinct mammalian outgroups long held to have a semierect forelimb posture, the glenoid faces laterally or posterolaterally and the coracoid forms a significant portion of the articular socket (fig. 10.9). In sprawling or semierect postures, the forelimbs and position of the manus are farther from the body axis, which necessarily generates a significant transverse, or compressive, force acting against a laterally facing shoulder socket (figs. 10.13D, 10.14A; Gray, 1944; Jenkins, 1970b, 1971a, 1971b). In living therians and their extinct theriimorph relatives, in contrast, the forelimbs operate in a parasagittal plane closer to the body axis (Jenkins & Weijs, 1979). Consequently, most of the glenoid faces ventrally, the coracoid contribution is reduced, and the manus is positioned ventral to the glenoid (figs. 10.10, 10.14B, C). The glenoid is designed to oppose the humeral force vector, and its orientation, therefore, appears to be an excellent predictor of forelimb posture.

Shoulder Joint Mobility. In the sprawling monotremes and in extinct mammalian outgroups long held to have a semierect forelimb posture, opposing pectoral girdles are joined by rigid articulations of the clavicle, interclavicle, and (in monotremes) procoracoid to form a stable platform for the shoulder socket (figs. 10. 9, 10.14A). This stable shoulder joint is so designed to "sustain the compressive forces generated by sprawling limb posture" (Jenkins, 1971b, 135). In living therians and their extinct theriimorph relatives, in contrast, the shoulder joint participates in stride generation (fig. 10.13A, C), the clavicular articulations are reduced to al-

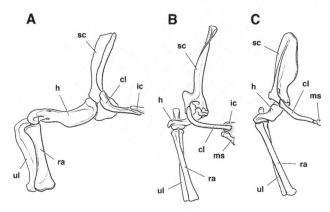

Figure 10.14. Anterior view of pectoral girdle and forelimb in a multituberculate compared with more basal and derived synapsids. (A) Cynodont based on *Cynognathus* and *Thrinaxodon* (Jenkins, 1971b). (B) *Kryptobaatar dashzevegi* based on PSS-MAE 103. (C) *Didelphis virginiana* (after Jenkins & Weijs, 1979). Abbreviations: cl, clavicle; ic, interclavicle; h, humerus, ms, manubrium sterni; ra, radius; sc, scapula; ul, ulna.

low such movement of the shoulder joint, and the clavicle and interclavicle are reduced or lost (figs. 10.10, 10.14B, C; Jenkins & Weijs, 1979). The stability of the union between opposing pectoral girdles depends on the humeral force vector and the presence of substantial transverse compressive stresses. Mobility of the shoulder joint, therefore, appears to be an excellent predictor of forelimb posture.

Elbow Joint Mobility. Flexion-extension at the elbow joint is minimal during terrestrial locomotion in the sprawling monotremes (fig. 10.13B) and in extinct mammalian outgroups (fig. 10.14A; Jenkins, 1971b, fig. 42). In monotremes, the ulnar portion of the distal humeral condyle is received in a shallow semilunar notch in the ulna. In tritylodontids, the ulnar condyle of the humerus is nearly flat and does not round onto either the anterior or posterior sides of the distal end (fig. 10.11A). In multituberculates and therians, in contrast, flexion-extension is clearly greatly enhanced. The bulbous ulnar condyle of the humerus has a dorsoventral articular arc of 180° or more (figs. 10.6B, 10.11B). The ulna, likewise, shows adaptations for increased flexion-extension, such as a distinct anconeal process and intercondylar crest, lengthened olecranon process, and marked radial and ulnar fossae. Enhanced flexion-extension at the elbow joint is linked with generating stride length in a parasagittal, or near parasagittal, plane (Jenkins 1971b, 1973) and thus appears to be an excellent predictor of forelimb posture.

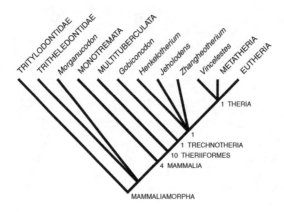

Figure 10.15. Semistrict consensus tree summarizing twenty-two minimum-length trees of twenty steps generated from eighteen characters in the pectoral girdle and forelimb across select mammaliamorph taxa (Tritylodontidae and Tritheledontidae as outgroups). Numbers show the distribution of unambiguous synapomorphies (consistencey index = 0.95, retention index = 0.98). Character support linking multituberculates and basal trechnotherians with Theria is strong and overwhelms available data from the skull and dentition regarding placement among crown mammal clades.

Evolution of Shoulder and Elbow Joints

From the available fossil record, it is possible to visualize three major stages in the evolution of mammalian shoulder and elbow joints. The skeletal changes associated with each stage appear to have evolved only once in mammalian history, an interpretation based on phylogenetic analysis (figs. 10.15, 10.16). The onus now is on the oft-repeated hypothesis of parallel evolution of these functional attributes in several mammalian lineages. Defending this hypothesis requires an alternative phylogenetic arrangement and/or a significant amount of new and contradictory character information to that summarized in this paper.

Girdle Simplification. Reduction and coossification of the coracoid with the scapula occurred at the base of Mammalia probably sometime during the Jurassic (fig. 10.16, node 1). In living mammals, fusion of the scapulocoracoid suture is nearly universal, but it often occurs quite late in posthatching/postnatal development. A second round of simplification occurred before the close of the Jurassic among basal therians involving fusion of the procoracoid and interclavicle to the manubrium sterni. The fossil record shows the loss of the procoracoid and interclavicle; the proposed fusions are deduced solely from embryological evidence in living therians (Klima, 1987).

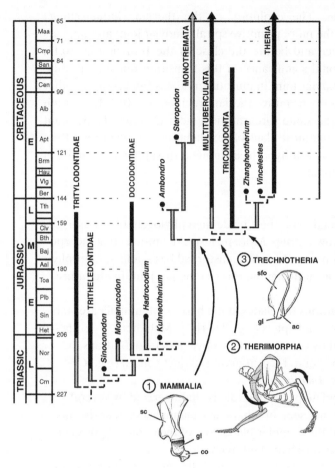

Figure 10.16. Temporally calibrated phylogeny for key early mammalian taxa showing major functional and skeletal innovations in the shoulder and elbow (1, girdle coossification; 2, enhanced shoulder/elbow joint mobility during locomotion; 3, girdle simplification). Known age estimates (dots), temporal durations (solid bars), and missing ranges (shaded bars) are shown. Time scale is based on Gradstein et al. (1999). Abbreviations: ac, acromion; co, coracoid; gl, glenoid; sc, scapula; sfo, supraspinous fossa.

Shoulder and Elbow Joint Mobility Mobility of the shoulder joint and enhanced flexion-extension of the elbow joint both appear to have evolved once, possibly in concert, among basal theriimorphs sometime before the close of the Jurassic (fig. 10.16, node 2). In living quadrupedal therians with a mobile shoulder joint, the elbow joint is located posterior to the shoulder joint near the thorax wall and, therefore, is subject to a greater degree of flexion and extension than in forms with sprawling or semierect forelimb postures. At present, phylogenetic evidence and the fossil record suggest that these two critical functional attributes of parasagittal forelimb posture in mammals evolved once over the same temporal interval before the close of the Jurassic.

Joint Stabilization. The evolution of the therian (humeroulnar) trochlear joint from the more primitive spiral condylar joint also appears to have occurred once and before the close of the Jurassic (fig. 10.16, node 3). The functional significance of this transition remains speculative, as the spiral condylar joint in multituberculates (and possibly other basal mammals) and the primitive therian trochlea (fig. 10.11B, C) both appear to generate the same forearm movement (Jenkins, 1973). Joint stabilization may be the most plausible explanation for evolution of the therian trochlea, but it is a hypothesis in need of supporting empirical data from living mammals.

Conclusions

1. The taxon Mammalia may best be defined phylogenetically as a node-based crown group. More inclusive definitions that incorporate extinct stem taxa have little historical basis, and apomorphy-based definitions utilizing the form of the jaw joint invite future ambivalence.

2. Taxonomic definitions for clades at the base of Mammalia are stabilized by node-stem triplets (e.g., Theria = Metatheria + Eutheria) delineated by complementary phylogenetic definitions that employ well-preserved recent and/or extinct genera (e.g., *Ornithorhynchus, Taeniolabis*).

3. The multituberculate pectoral girdle is characterized by several specializations not seen in other early mammals such as the unusual length of the scapular blade, loss of the coracoid posterior process, and clover-shaped interclavicle.

4. Character evidence from the pectoral girdle and forelimb plays a major role in establishing phylogenetic arrangements at the base of Mammalia, in particular positioning Multituberculata and several other Mesozoic clades within Mammalia in alliance with Theria.

5. Skeletal evidence from fossils and modern mammalian analogs does not support a sprawling forelimb posture or habitual saltatory locomotion in multituberculates. Evidence in favor of a more parasagittal forelimb posture in multituberculates includes the small ventrally facing glenoid, mobile shoulder joint, cam-shaped ulnar condyle of the humerus, and elbow joint with enhanced flexion-extension capability.

6. A mobile shoulder joint, an elbow joint with enhanced flexion-extension, and parasagittal forelimb posture appear to have evolved once, possibly in concert, among basal theriimorphs near the close of the Jurassic.

Acknowledgments

I thank Carol Abraczinskas for the finished illustrations and for the elegant drawings of bones under magnification. For permission to examine specimens in their care, I thank James Hopson, Farish Jenkins, Jr., Zofia Kielan-Jaworowska, Desui Miao, Yuanqin Wang, and Ji-Cai Yin. For their comments on earlier drafts of this manuscript, I thank Jack Conrad, David Krause, Farish Jenkins, Jr., Gregory Wilson, and Jeffrey Wilson. For planting the seed that sent me to Mongolia and collaborating on the initial paper on which this work was based, I thank Malcolm McKenna. For being an inspirational teacher and supportive colleague for many years at the University of Chicago, I am indebted to James Hopson. This research was supported by the American Museum of Natural History, National Geographic Society, The David and Lucile Packard Foundation, and the Pritzker Foundation.

Literature Cited

Archer, M., P. Murray, S. Hand and H. Godthelp. 1993. Reconsideration of monotreme relationships based on the skull and dentition of the Miocene *Obdurodon dicksoni;* pp. 75–94 *in* F. S. Szalay, M. J. Novacek and M. C. McKenna (eds.), *Mammal Phylogeny: Mesozoic Differentiation, Multituberculates, Monotremes, Early Therians, and Marsupials.* New York: Springer-Verlag.

Archibald, J. D. and D. H. Deutschman. 2001. Quantitative analysis of the timing of the origin and diversification of extant placental orders. *Journal of Mammalian Evolution* 8:107–124.

Argot, C. 2001. Functional-adaptive anatomy of the forelimb in the Didelphidae, and the paleobiology of the Paleocene marsupials *Mayulestes ferox* and *Pucadephys andinus. Journal of Morphology* 247:51–79.

Broom, R. 1912. On a new type of cynodont from the Stormberg. *Annals of the South African Museum* 7:334–336.

Cifelli, R. L. 2001. Early mammalian radiations. *Journal of Paleontology* 75: 1214–1226.

Crompton, A. W. and F. A. Jenkins, Jr. 1978. Mesozoic mammals; pp. 46–55 *in* V. J. Maglio and H. B. S. Cooke (eds.), *Evolution of African Mammals.* Cambridge, MA: Harvard University Press.

———. 1979. Origin of mammals, pp. 59–73 *in* J. A. Lillegraven, Z. Kielan-Jaworowska and W. A. Clemens (eds.), *Mesozoic Mammals: The First Two-Thirds of Mammalian History.* Berkeley: University of California Press.

Crompton, A. W. and Z. Luo. 1993. Relationships of the Liassic mammals *Sinoconodon, Morganucodon oehleri,* and *Dinnetherium;* pp. 30–44 *in* F. S. Szalay, M. J. Novacek and M. C. McKenna (eds.), *Mammal Phylogeny:*

Mesozoic Differentiation, Multituberculates, Monotremes, Early Therians, and Marsupials. New York: Springer-Verlag.

Emerson, S. B. 1985. Jumping and leaping; pp. 58–72 in M. Hildebrand, D. M. Bramble, K. F. Liem, and D. B. Wake (eds.), Functional Vertebrate Morphology. Cambridge, MA: Harvard University Press.

Gambaryan, P. and Z. Kielan-Jaworowska. 1997. Sprawling versus parasagittal stance in multituberculate mammals. Acta Palaeontologica Polonica 42:13–44.

Gidley, J. W. 1909. Notes on the fossil mammalian genus Ptilodus, with descriptions of new species. Proceedings of the United States National Museum 36:611–626.

Gill, T. 1872. Arrangement of the families of mammals with analytical tables. Smithsonian Miscellaneous Collections 11:1–98.

Gradstein, F. M., F. P. Agterberg, J. G. Ogg, J. Hardenbol and S. Blackstrom. 1999. On the Cretaceous time scale. Neues Jahrbuch für Geologie und Paläontologie Abhandlungen 212:3–14.

Granger, W. W. and G. G. Simpson. 1929. A revision of the Tertiary Multituberculata. Bulletin of the American Museum of Natural History 56:601–676.

Gray, J. 1944. Studies in the mechanics of the tetrapod skeleton. Journal of Experimental Biology 20:88–116.

Gregory, W. K. 1912. Notes on the principles of quadrupedal locomotion and on the mechanism of the limbs in hoofed mammals. Annals of the New York Academy of Science 22:267–294.

Hildebrand, M. and G. E. Goslow, Jr. 2001. Analysis of Vertebrate Function, 5th ed. New York: John Wiley & Sons.

Hopson, J. A. and H. R. Barghusen. 1986. An analysis of therapsid relationships; pp. 83–106 in N. Hotton, III, P. D. MacLean, J. J. Roth, and E. C. Roth (eds.), The Ecology and Biology of Mammal-like Reptiles. Washington, DC: Smithsonian Institution Press.

Hopson, J. A. and J. W. Kitching. 2001. A probainognathian cynodont from South Africa and the phylogeny of nonmammalian cynodonts. Bulletin of the Museum of Comparative Zoology 156:5–35.

Hu, Y., Y. Wang, Z. Luo and C. Li. 1997. A new symmetrodont mammal from China and its implications for mammalian evolution. Nature 390:137–142.

Huxley, T. H. 1880. On the application of the laws of evolution to the arrangement of the Vertebrata, and more particularly of the Mammalia. Proceedings of the Zoological Society of London 43:649–662.

Jenkins, F. A., Jr. 1970a. Cynodont postcranial anatomy and the "prototherian" level of mammalian organization. Evolution 24:230–252.

———. 1970b. Limb movements in a monotreme (Tachyglossus aculeatus): a cineradiographic analysis. Science 198:1473–1475.

————. 1971a. Limb posture and locomotion in the Virginia opossum (*Didelphis marsupialis*) and in other non-cursorial mammals. *Journal of Zoology* 165:303–315.

————. 1971b. The postcranial skeleton of African cynodonts. *Bulletin of the Peabody Museum of Natural History* 36:1–216.

————. 1973. The functional anatomy and evolution of the mammalian humero-ulnar articulation. *American Journal of Anatomy* 137:281–298.

Jenkins, F. A., Jr. and Krause, D. W. 1983. Adaptations for climbing in North American multituberculates (Mammalia). *Science* 220:712–715.

Jenkins, F. A., Jr. and F. R. Parrington. 1976. The postcranial skeletons of the Triassic mammals *Eozostrodon, Megazostrodon* and *Erythrotherium. Philosophical Transactions of the Royal Society of London, B, Biological Sciences* 273:387–431.

Jenkins, F. A., Jr. and C. R. Schaff. 1988. The Early Cretaceous mammal *Gobiconodon* (Mammalia, Triconodonta) from the Cloverly Formation in Montana. *Journal of Vertebrate Paleontology* 6:1–24.

Jenkins, F. A., Jr. and W. A. Weijs. 1979. The functional anatomy of the shoulder girdle in the Virginia opossum (*Didelphis virginiana*). *Journal of Zoology, London* 188:379–410.

Ji, Q., Z. Luo and S. Ji. 1999. A Chinese triconodont mammal and mosaic evolution of the mammalian skeleton. *Nature* 398:326–330.

Ji, Q., Z. Luo, C. Yuan, J. R. Wible, J. Zhang and J. A. Georgi. 2002. The earliest known eutherian mammal. *Nature* 416:816–822.

Kemp, T. S. 1982. *Mammal-like Reptile and the Origin of Mammals.* London: Academic Press.

————. 1983. The relationships of mammals. *Zoological Journal of the Linnean Society* 77:353–384.

Kermack, K. A. 1967. The interrelationships of early mammals. *Journal of the Linnean Society (Zoology)* 47:241–249.

Kermack, K. A. and F. Mussett. 1958. The jaw articulation of the Docodonta and the classification of Mesozoic mammals. *Proceedings of the Royal Society of London B,* 149:204.

Kielan-Jaworowska, Z. 1971. Results of the Polish-Mongolian Palaeontological Expeditions. Pt. III. Skull structures and affinities of the Multituberculata. *Acta Palaeontologia Polonica* 25:5–41.

————. 1992. Interrelationships of Mesozoic mammals. *Historical Biology* 6:185–202.

————. 1997. Characters of multituberculates neglected in phylogenetic analyses of early mammals. *Lethaia* 29:249–266.

————. 1998. Humeral torsion in multituberculate mammals. *Acta Palaeontologica Polonica* 43:131–134.

Kielan-Jaworowska, Z. and P. P. Gambaryan. 1994. Postcranial anatomy and habits of Asian multituberculate mammals. *Fossils and Strata* 36:1–92.

Kielan-Jaworowska, Z. and T. Qi. 1990. Fossorial adaptations of a taenio-labidoid multituberculate mammal from the Eocene of China. *Vertebrata PalAsiatica* 28:81–94.

Klima, M. 1987. Early development of the shoulder girdle and sternum in marsupials (Mammalia: Metatheria). *Advances in Anatomy, Embryology and Cell Biology* 109:1–91.

Krause, D. W. and F. A. Jenkins, Jr. 1983. The postcranial skeleton of North American multituberculates. *Bulletin of the Museum of Comparative Zoology* 150:199–246.

Krebs, B. 1991. Das Skelett von *Henkelotherium guimarotae* gen. et sp. nov. (Eupantotheria, Mammalia) aus dem Oberen Jura von Portugal. *Berliner geowissenschaften Abhandlungen A* 133:1–121.

Kühne, O. 1956. *The Liassic Therapsid* Oligokyphus. London: British Museum (Natural History).

Linnaeus (Linné), C. 1758. *Systema Naturae, sive Regna Tria Naturae Systematice Proposita per Ckasses, Ordines, Genera, & Species.* Leiden: Lugduni Batavorum, Theodorum Haak.

Liu, F. R. and M. M. Miyamoto. 1999. Phylogenetic assessment of molecular and morphological data for eutherian mammals. *Systematic Biology* 48:54–64.

Liu, F. R., M. M. Miyamoto, N. P. Freire, P. Q. Ong, M. R. Tennent, T. S. Young and K. F. Gugel. 2001. Molecular and morphological supertrees for eutherian (placental) mammals. *Science* 291:1786–1789.

Luo, Z.-X., Z. Kielan-Jaworowska and R. Cifelli. 2002. In quest for a phylogeny of Mesozoic mammals. *Acta Palaeontologica Polonica* 47:1–78.

MacPhee, R. D. E. 1994. Morphology, adaptations, and relationships of *Plesiorycteropus,* and a diagnosis of a new order of eutherian mammals. *Bulletin of the American Museum of Natural History* 220:1–214.

Marsh, O. C. 1880. Notice of Jurassic mammals representing two new orders. *American Journal of Science, Series 3* 20:235–239.

McKenna, M. C. 1961. On the shoulder girdle of the mammalian Subclass Allotheria. *American Museum Novitates* 2066:1–27.

———. 1974. The phylogenetic relationships of Eutheria. Unpublished manuscript prepared for the Burg Wartenstein Symposium No. 61, Phylogeny of the Primates: An Interdisciplinary Approach, July 6–14, 1974. Werner-Gren Foundation for Anthropological Research.

———. 1975. Toward a phylogenetic classification of the Mammalia, pp. 21–46 *in* W. P. Luckett and S. F. Szalay (eds.), *Phylogeny of the Primates.* New York: Plenum Press.

Novacek, M. J., G. W. Rougier, J. R. Wible, M. C. McKenna, D. Dashzeveg and I. Horovitz. 1997. Epipubic bones in eutherian mammals from the Late Cretaceous of Mongolia. *Nature* 389:483–486.

Parker, T. J. and W. A. Haswell. 1897. *A Text-book of Zoology, Volume 2, Chordata.* London: Macmillan and Company.

Pridmore, P. A. 1985. Terrestrial locomotion in monotremes (Mammalia: Monotremata). *Journal of the Zoological Society of London (A)* 205:53–73.

Rich, T. H., T. F. Flannery, P. Trusler, L. Kool, N. A. Van Klaveren and P. Vickers-Rich. 2002. Evidence that monotremes and ausktribosphenids are not sister-groups. *Journal of Vertebrate Paleontology* 22:466–469.

Romer, A. S. 1922. The locomotor apparatus of certain primitive and mammal-like reptiles. *Bulletin of the American Museum of Natural History* 46: 517–606.

Rougier, G. W. 1993. *Vincelestes neuquenianus* Bonaparte (Mammalia, Theria) un primitivo mamifero del Cretacico Inferior de la cuenca Neuquina. Ph. D. dissertation, University of Buenos Aires.

Rougier, G. W., J. R. Wible and M. J. Novacek. 1996a. Multituberculate phylogeny. *Nature* 379:406–407.

Rougier, G. W., J. R. Wible, and J. A. Hopson. 1996b. Basicranial anatomy of *Priacodon fruitaensis* (Triconodontidae, Mammalia) from the Late Jurassic of Colorado, and a reappraisal of mammaliaform interrelationships. *American Museum Novitates* 3183:1–38.

Rowe, T. 1987. Definition and diagnosis in the phylogenetic system. *Systematic Zoology* 36:208–211.

———. 1988. Definition, diagnosis, and origin of Mammalia. *Journal of Vertebrate Paleontology* 8:241–264.

———. 1993. Phylogenetic systematics and the early history of mammals; pp. 129–145 *in* F. S. Szalay, M. J. Novacek and M. C. McKenna (eds.), *Mammal Phylogeny: Mesozoic Differentiation, Multituberculates, Monotremes, Early Therians, and Marsupials.* New York: Springer-Verlag.

Rowe, T. and J. A. Gauthier. 1992. Ancestry, paleontology, and definition of the name Mammalia. *Systematic Biology* 41:372–378.

Sereno, P. C. 1998. A rationale for phylogenetic definitions, with application to the higher-level taxonomy of Dinosauria. *Neues Jarhbuch für Geologie und Paläontologie Abhandlungen* 210:41–83.

———. 1999. Definitions in phylogenetic taxonomy: critique and rationale. *Systematic Biology* 48:329–351.

———. 2005. The logical basis of phylogenetic taxonomy. *Systematic Biology* 54:595–619.

———. In review. Shoulder girdle and forelimb in *Kryptobaatar dashzevegi* (Mammalia: Multituberculata). *Geodiversitas.*

Sereno, P. C. and M. C. McKenna. 1995. Cretaceous multituberculate skeleton and the early evolution of the mammalian shoulder girdle. *Nature* 377: 144–147.

Simmons, N. B. 1993. Phylogeny of Multituberculata; pp. 146–164 *in* F. S. Szalay, M. J. Novacek and M. C. McKenna (eds.), *Mammal Phylogeny: Mesozoic Differentiation, Multituberculates, Monotremes, Early Therians, and Marsupials.* New York: Springer-Verlag.

Simpson, G. G. 1926. Mesozoic Mammalia. IV. The multituberculates as living
    animals. *American Journal of Science* 2:228–250.

———. 1928a. Further notes on Mongolian Cretaceous mammals. *American
    Museum Novitates* 329:1–14.

———. 1928b. *A Catalogue of the Mesozoic Mammalia in the Geological De-
    partment of the British Museum.* British Museum (Natural History), Lon-
    don: 215 pp.

———. 1959. Mesozoic mammals and the polyphyletic origin of mammals. *Evo-
    lution* 13:405–414.

Sloan, R. E. and L. Van Valen. 1965. Cretaceous mammals from Montana. *Sci-
    ence* 148:220–227.

Sun A.-L. and Y.-H. Li. 1985. The postcranial skeleton of Jurassic tritylodonts
    from Sichuan Province [in Chinese with English abstract]. *Vertebrata PalAsi-
    atica* 23:135–151.

Waddell, P. J., Y. Cao, M. Hasegawa and D. P. Mindell. 1999. Assessing the Cre-
    taceous divergence times within birds and placental mammals by using
    whole mitochondrial protein sequences and an extended statistical frame-
    work. *Sytematic Biology* 48:119–137.

Wible, J. R. 1991. Origin of Mammalia: The craniodental evidence reexamined.
    Journal of Vertebrate Paleontology 11:1–28.

Wible, J. R. and J. A. Hopson. 1993. Basicranial evidence for early mammal
    phylogeny; pp. 45–61 *in* F. S. Szalay, M. J. Novacek and M. C. McKenna
    (eds.), *Mammal Phylogeny: Mesozoic Differentiation, Multituberculates,
    Monotremes, Early Therians, and Marsupials.* New York: Springer-Verlag.

Wilkinson, M. 1995. Coping with abundant missing entries in phylogenetic in-
    ference using parsimony. *Systematic Biology* 44:501–514.

**Tooth Orientation during Occlusion
and the Functional Significance of
Condylar Translation in Primates
and Herbivores**

A. W. Crompton

Daniel E. Lieberman,

and Sally Aboelela

Introduction

The temporomandibular joint (TMJ) in carnivores is a ginglymoid (hinge) joint in which the preglenoid and postglenoid processes limit anteroposterior movements of the mandibular condyle, permitting almost exclusively rotation around a transverse axis (Becht, 1953; Maynard Smith & Savage, 1959; Noble, 1973; Freeman, 1979; Crompton, 1981). In contrast, the TMJ in anthropoid primates and several orders of specialized herbivores (Perissodactyla and Artiodactyla) is a diarthroidal joint that permits the mandibular condyles not only to rotate around a transverse axis but also to slide anteroposteriorly ("translate") relative to the glenoid fossa during jaw closure and opening. In a study on goats, De Vree and Gans (1975) pointed out that lateral shifts of the mandible in animals with a diarthroidal TMJ must occur by the translation of one condyle in its fossa coincident with rotation of the opposite condyle about a vertical axis through the second fossa. There is little precise information on how the active and balancing side condyles move relative to the glenoid fossae during a masticatory cycle. Weijs (1994) claimed that, during closing, medially directed movement of the working side mandible is essentially a pure rotation around the working side condyle, which is braced against the postglenoid flange. Extensive transverse jaw movements obviously require an anteroposteriorly long glenoid. However, it is not obvious why both condyles move anteriorly during opening and posteriorly (at different rates) during closing given that medial and lateral movement of the mandible can be accomplished by rotating the mandible about a stationary working side condyle that is buttressed against the postglenoid flange while translating the balancing side condyle. It is this aspect of condylar translation that we wish to discuss in this paper.

Several hypotheses have been proposed to explain anteroposterior condylar translation in the diarthroidal TMJ. First, Moss (1960, 1975, 1983) suggested that condylar translation is an adaptation to stabilize the axis of rotation of the mandible near the location of the mandibular foramen. Second, following on Moss's observation, Carlson (1977), Hylander (1978, 1992), and Weijs et al. (1989) proposed that condylar translation is an adaptation to maximize the contractile force of the masseter and medial pterygoid muscles during jaw adduction by minimizing their stretching. The position of the masseter and medial pterygoid muscles relative to the TMJ and molar rows constrains not only how far the jaws can open but also the maximum gape at which effective force can be generated by the adductors. Muscle stretching is an important biomechanical consideration because the length-tension curve of sarcomeres becomes less efficient when muscles elongate beyond forty to fifty percent of their resting length (McMahon, 1984). Support for this hypothesis comes from several sources. Smith (1985) found that maximum jaw abduction without translation in humans would stretch the masseter by approximately forty-one percent, approximately twice the twenty-one percent stretching observed with translation. Wall (1995) found that condylar translation during mastication and incision in anthropoid primates moved the mandible's instantaneous center of rotation anteriorly, thereby minimizing stretching and maximizing bite force for the masseter, medial pterygoid, and anterior temporalis over a range of gapes. However, varying the orientation of fibers within different regions of individual muscles can partially compensate for different degrees of stretching and has been shown to characterize the masseter of many mammals with diarthroidal TMJs, including hippopotami, suids, and hyracoids (Herring, 1975, 1980; Herring et al., 1979; Janis, 1979, 1983). In addition, kinematic and electromyogram studies of human mastication indicate that the optimum lengths of the masseter and medial pterygoid occur near resting position at 10–20 mm of gape (Garrett et al., 1964; Manns et al., 1981; Rugh & Drago, 1981). In fact, peak contraction of the masseter in primates, goats, and other mammals occurs just before maximum intercuspation, toward the end of the power stroke when stretching of the adductors would be minimal (De Vree & Gans, 1975; Hylander & Johnson, 1985, 1993; Hylander et al., 1992; Lieberman & Crompton, 2000).

Another hypothesis, proposed by Smith (1985) and others (e.g., Craddock, 1948; DuBrul, 1964; Herring, 1975; Mack, 1984) is that condylar translation is an adaptation to prevent impingement of the esophagus, trachea, carotid sheath, and the larynx by the posterior margins of the mandible, tongue, and suprahyoid muscles during maximum gape. Impingement of these structures is an evident problem in humans in which upright posture, in combination with a descended larynx and a highly flexed cranial base, positions these structures within a few millimeters of the posterior margin of the mandibular ramus at resting position (Smith, 1985). The impingement hypothesis, however, is an unlikely explanation for condylar translation in many quadrupedal mammals with highly extended (flat) cranial bases in which the pharynx is oriented craniocaudally away from the mandibular ramus (Smith, 1985; Hylander, 1992). There is no evidence that the deep mandibular angle and tongue impinge on the pharynx in these species during maximum gape.

Model (A)

It is very likely that condylar translation serves more than one function. We propose here an additional hypothesis: that condylar translation is an adaptation for occlusion. In particular, we suggest that, in nonrodent mammals whose tall mandibular rami position the occlusal plane below the level of the TMJ, the working and balancing side condyles are retracted at slightly different rates from their protracted position at the end of opening to maintain a vertical trajectory of the working side lower molariform teeth relative to the upper molariform teeth during unilateral mastication.

To introduce the above hypothesis, it is useful to review variations in the anatomy of the TMJ in the context of the overall design of the mammalian masticatory system. One major difference between mammals with and without condylar translation is the position and orientation of the jaw adductors relative to the tooth row. In mammals with a ginglymoid (hinge) TMJ, the joint usually lies on or close to the same plane as that of the lower dentition (figs. 11.1A, 11.2A), whereas the diarthroidal TMJ (sliding hinge) usually occurs in association with a vertically tall mandibular ramus that positions the glenoid fossa well above the occlusal plane of the posterior dentition (figs. 11.1D, 11.2B). These differences relate partly to the pattern of molar occlusion. Mammalian carnivores, especially felids and mustelids, have molars designed for crushing and slicing (Mills, 1967; Crompton & Kielan-Jaworowska, 1978). Masticatory movements are primarily orthal, with only a small medially directed component (Mills, 1967; Crompton, 1981; Weijs, 1994). The dentary condyle, which is cylindrically shaped and transversely oriented (fig. 11.1A, B), is tightly held within the temporal bone by postglenoid (also referred to as retroarticular) and preglenoid processes. These processes limit jaw movement to the sagittal plane, permitting only a slight amount of mediolateral shifting of the jaw as a whole. The dominant adductor muscle in mammals with ginglymoid TMJs is the temporalis. The moment arm of the posterior temporalis is a function of the distance of its insertion point on the coronoid process to the TMJ. The combination of a tall coronoid process and extension of the

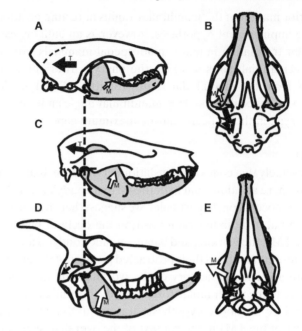

Figure 11.1. Comparison of TMJ and cranial morphology in mammals with ginglymoid and diarthroidal TMJs. (A, B) Lateral and inferior views of a mustelid (*Martes pennanti*) with a ginglymoid TMJ. Note the TMJ lies in the same plane as the postcanine occlusal plane; the dominant adductor is the temporalis, T; the superficial masseter, M, is small, in a parasagittal plane, and anteriorly oriented. The sizes of the arrows indicate schematically the relative size of the temporalis versus the masseter and medial pterygoid complex. (C) Lateral view of opossum (*Didelphis virginianus*) in which the short diarthroidal TMJ lies slightly above the occlusal plane; the temporalis and masseter muscles are roughly equivalent in size; the masseter is more laterally and dorsally oriented than in the carnivore. (D, E) Lateral and inferior views of a goat (*Capra hircus*). Note the tall ramus, which positions the occlusal plane well below the long diarthroidal TMJ; the temporalis is small relative to the size of the masseter. The masseter has a strong dorsal and lateral orientation, whereas the medial pterygoid has a dorsal and medial orientation.

skull posterior to the TMJ permits both a large gape and a powerful bite. The superficial masseter inserts close to the TMJ and therefore adds speed without restricting gape. The postglenoid processes resists the posteriorly directed force of the temporalis and the preglenoid process resists the anteriorly directed force of the superficial masseter. These two muscles acting synchronously on one side form a force couple that tends to move the jaw medially; the flanges prevent this movement.

In herbivores, the masseters and medial pterygoids tend to be the dominant adductor muscles, composing 50–80 percent of total adductor mass, whereas the temporalis composes twenty to thirty percent of adductor mass (Janis, 1979; Weijs, 1994). Herbivores have a tall ascending ramus, which effectively positions the mandibular condyle high above the occlusal plane. A postglenoid flange resists the posteriorly directed force of the temporalis. The orientation of the fibers of the superficial masseter and medial pterygoid is more vertical than in typical mammalian carnivores and a preglenoid flange, which would limit condylar translation, is not present. In some cases (e.g., the pygmy hippopotamus), the mass of the masseter and medial pterygoid muscles is increased by positioning the posteroventral margin of the expanded mandibular angle considerably below the occlusal plane (Herring, 1975). Mammals with a diarthroidal TMJ also tend to have molars primarily designed for grinding (Becht, 1953; Kay & Hiiemae, 1974; Kay, 1978; Lucas, 1982; Janis, 1983; Herring, 1985; Hiiemae & Crompton, 1985; Weijs, 1994). In ungulates and primates, the lower molars and molariform premolars on the working side mandible are drawn transversely across the upper dentition to triturate material. The amount of dorsal movement that accompanies the transverse movement is highly variable: high in most artiodactyls and low in most perissodactyls (Becht, 1953). The flat condyle and long glenoid lacking preglenoid processes (fig. 11.1D) result in a highly mobile TMJ. Differential contraction of the jaw adductors on both sides can generate precisely controlled medially directed movements of the working side mandible (Becht, 1953; Crompton, 1981; Hiiemae & Crompton, 1985; Hylander et al., 1987; Hylander, 1992; Weijs, 1994).

As noted previously (e.g., Osborn, 1987), positioning the TMJ well above the occlusal plane affects the parasagittal trajectory of the molariform teeth during jaw closure (fig. 11.2). In a masticatory system in which the plane of occlusion is in line with the glenoid fossa (fig. 11.2A), the distance between a given tooth and the center of rotation at the TMJ is the same for occluding teeth on the mandible and the maxilla. As a result, the trajectory of the lower tooth relative to the upper tooth in the sagittal plane, $\emptyset$, is a function of X, the distance between the two teeth, and Z, the distance to the TMJ:

$$\emptyset = \arccos(X/2Z)$$

so that $\emptyset$ approaches 90° as X approaches 0. It is important to note, however, that tooth-food-tooth contact—when peak masticatory force is exerted—occurs before maximum intercuspation (when $X = 0$), so that $\emptyset$ is always slightly less than 90°.

In mammals with diarthroidal TMJs (illustrated in fig. 11.2B), the occlusal plane is positioned below the glenoid fossa by a tall ascending

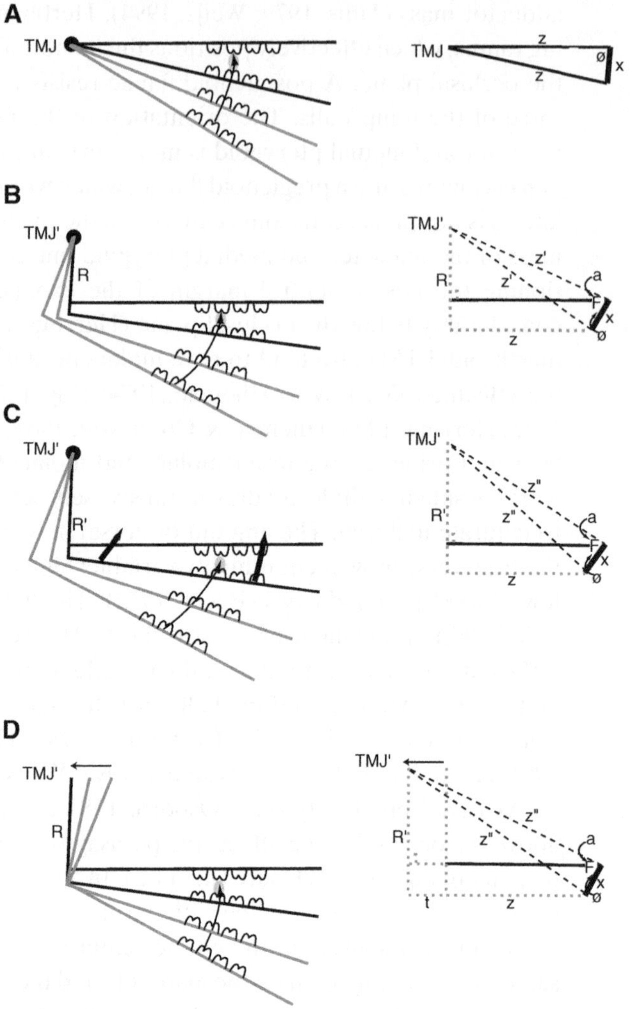

Figure 11.2. Geometric model of TMJ function in relation to occlusal plane position. (A) Ginglymoid TMJ in which the orientation of the mandibular tooth row relative to the maxillary tooth row, ø, at any given gape, *x*, equals arccos(X/2Z). (B) Effect of positioning the occlusal plane below the TMJ by distance *R* so that ø = Φ − α (see text for details), introducing an anterior component to the trajectory of the mandibular teeth relative to the maxillary postcanine occlusal plane. (C) Effect of increasing ramus height to *R'*, which makes ø more acute relative to the postcanine maxillary occlusal plane (see text for details). Note also that ø is more anteriorly oriented (thick arrows) for more posterior teeth that lie closer to the TMJ. (D) Effect of posterior translation on the orientation of the mandibular tooth row relative to the maxillary tooth row, making ø more dorsally oriented.

ramus of length *R*, changing the center of rotation to TMJ'. Assuming the mandibular and maxillary planes of occlusion are perpendicular to the ascending ramus, this configuration increases the distance between occluding teeth and the center of rotation, *Z'*, in proportion to *R*:

$$Z' = (Z^2 + R^2)^{0.5}$$

so that

$$\Phi = \arccos[X/2 \cdot (Z^2 + R^2)^{0.5}].$$

It is important to note, however, that in this configuration Φ does not describe the angle of the sagittal trajectory of a lower tooth relative to the maxillary tooth row, Z. This angle, $\varnothing$, is equal to $\Phi - \alpha$, where

$$\alpha = 90° - [\arctan(Z/R) + (180° - 2\Phi)]$$

As R gets larger, $\varnothing$ becomes smaller for a given X, making the trajectory of a lower tooth relative to an upper tooth in the sagittal plane more acute. In other words, the taller the mandibular ramus, the more acute is the trajectory of the dentition during tooth-food-tooth contact. This effect of ramus length on $\varnothing$ is illustrated in figure 11.2C. Note that the trajectory of $\varnothing$ varies with position along the tooth row and is more acute for more posterior teeth that lie closer to the TMJ (see fig. 11.2C). In other words, the effect of positioning the tooth row below the TMJ on the anterior trajectory of the lower molars is greatest on the posterior teeth, which occlude at wider gapes than more anteriorly positioned teeth. Greaves (1974, 1978, 1980) has claimed that, in herbivores, because of the transverse movement of the jaw during the power stroke, it is important for the upper and lower molariform dentitions to occlude essentially simultaneously as food is ground. This can be true only of herbivores in which the occlusal surface of the molars is horizontal (i.e., at right angles to the sagittal plane). In this case all the teeth can come into occlusion before any transverse movement occurs. However, in most herbivores (especially artiodactyls), the occlusal plane of the molars relative to the sagittal plane is oblique and often as high as 45° (Becht, 1953). In this case, the posterior molars are the first to contact and the more anterior teeth occlude as the jaw moves mediodorsally.

Regardless of the angle of the occlusal plane of the molars relative to the sagittal plane, and the degree to which the jaw moves transversely and/or rotates longitudinally, the important point is that the lower teeth move dorsally and not anterodorsally during occlusion. Many lines of evidence support this point. For example, the pattern of wear of the molars in primates and ungulates indicates that, as the teeth come into occlusion, the lower molariform row moves dorsally and medially but not anteriorly relative to the upper row (fig. 11.3). Weijs (1994) recognized this feature when he pointed out that, at least in the rabbit, the lower tooth row approximates the upper in a vertical path, recognizing that the working side condyle must translate posteriorly as the teeth come into occlusion. It is unlikely, therefore, that the working side condyle is braced against the postglenoid flange before the working side jaw begins to move medially. Posterior translation of the mandibular condyle results in a more vertical trajectory for the

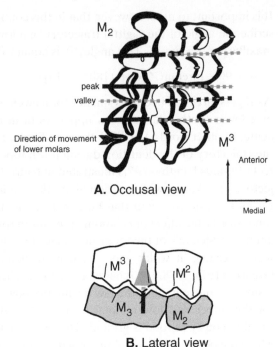

Figure 11.3. (A) Occlusal view of the lower (left) and upper (right) second and third molars of a goat. The arrow indicates movement of the lower molars relative to the uppers; high peaks of the cusps (black squares) move in a medial direction through the valleys (grey squares) on the matching molars. (B) Lateral view of third and fourth molars. Movement of the lowers must be vertical (arrow) and medial relative to the uppers during occlusion.

lower teeth. This effect (illustrated in fig. 11.2D) correctly aligns the lower molar shearing facets relative to those of the occluding upper molars.

We therefore predict that the amount of posterior translation of the working side condyle as the molars approach and enter into occlusion should vary as a function of the height of the TMJ above the postcanine occlusal plane (R, hereafter termed ramus height), independent of mandible length. In particular, the correlation between R and maximum anteroposterior translation distance (translation potential) of the working side condyle is predicted to be strong and independent of mandible length in both ungulate herbivores and primates. In addition, because the effects of positioning the occlusal plane below the TMJ on the orientation of tooth movement in the sagittal plane should be the same for all mammals with diarthroidal TMJs, the slope of the regression between ramus height and translation potential is predicted to be the same for primates, artiodactyls, and perissodactyls. In contrast, the anteroposterior length of the glenoid fossa in mammals with ginglymoid TMJs is expected to be independent of ramus height relative to mandibular length.

In short, condylar translation of the working side condyle during closing minimizes muscle stretching and also aligns the shearing surfaces on occluding molars. We stress that, for the topography of the crowns of the occluding molars to "fit" one another as they come into occlusion, the working side lower jaw must move dorsally and medially but not anteriorly. Even in primitive mammals with tribosphenic molars such as the American opossum, where the lower tooth row is positioned slightly below the TMJ, the working side condyle translates during occlusion. Tribosphenic molars are characterized by embrasure shearing (Crompton & Kielan-Jaworowska, 1978), and, for the lower molars to move into occlusion, they must move vertically relative to the uppers. We suggest that the same constraint on occlusal trajectory was retained in ungulates and anthropoid primates and could have been maintained only if increasing amounts of condylar translation developed in parallel with the elevation of the TMJ above the occlusal surface.

Materials and Methods

Sample

Table 11.1 summarizes the sample studied. Measurements of mandibular and glenoid fossa size were taken on five males and five females from thirty-one mammalian species of various sizes with diarthroidal

Table 11.1. Sample studied (by order)

Primates	Artiodactyla + Perrisodactyla	Carnivora
Allouatta seniculus	*Bison bison* (A)	*Canis latrans*
Ateles paniscus	*Bos taurus* (A)	*Canis lupus*
Cebus olivaceus apliculatus	*Camelus bactrianus* (A)	*Conepatus sonoriensis*
Cercopithecus aethiops	*Capra ibex* (A)	*Eira barbara*
Colobus badius oustaleti	*Madoqua kirki* (A)	*Felis catus*
Gorilla gorilla gorilla	*Equus burchelli boehmi* (P)	*Gulo gulo luscus*
Homo sapiens	*Equus caballus* (P)	*Mustela erminea kanel*
Hylobates syndactylus	*Equus zebra zebra* (P)	*Panthera leo*
Lagothrix lagotricha cana	*Giraffa camelopardalis* (A)	*Panthera tigris*
Macaca fascicularis	*Lama glama* (A)	*Ursus maritimus*
Pan troglodytes schweinfurthii	*Mazama americana* (A)	
Papio hamadryas	*Odocoileus virginianus* (A)	
Pongo pygmaeus	*Ovis ammon poli* (A)	
Saguinus fuseicollis	*Ovis canadensis canadensis* (A)	
Saimiri sciureus oerstedii	*Tapirus terrestris* (P)	

Note: Abbreviations: A, Artiodactyla; P, Perissodactyla.

TMJs from diverse families of primates ($n = 15$), artiodactyls ($n = 11$), and perissodactyls ($n = 4$). In addition, as many of the same measurements as possible were taken on five males and five females from ten carnivoran species of various sizes with ginglymoid TMJs. All individuals were adults with all molars fully erupted and were sampled from the collections of the American Museum of Natural History (New York).

Measurements

Measurements were taken with sliding digital calipers (accurate to 0.01 mm), with the exception of measurements greater than 200 mm, which were taken with plastic measuring tape (accurate to 1 mm). Whenever possible, measurements were taken on both sides of each individual and then averaged. The average standard error of measurement is estimated to be 0.36 mm on the basis of five sets of all measurements (see below) taken on the same individual.

Anteroposterior translation potential was measured in two ways. Maximum anteroposterior translation potential, TP_{max}, was measured using the maximum anteroposterior length in the parasagittal plane between the postglenoid process and the anterior margin of the subchondral surface of the preglenoid plane. Note that TP_{max} may overestimate the actual amount of translation that occurs because it estimates only the maximum amount of space available for anteroposterior translation of the condylar head relative to the articular surface of the glenoid fossa and preglenoid plane. A second estimate of translation, TP_{min}, was therefore measured as the anteroposterior length in the parasagittal plane between the postglenoid process and the ventralmost point on the articular eminence in primates and between the postglenoid process and the midpoint of the articular plane in herbivores. In carnivores, TP_{carn} was measured as the maximum anteroposterior distance between the preglenoid and postglenoid processes. These estimates of anteroposterior translation need to be verified with kinematic studies (e.g., Wall, 1995), but there is some indication that TP_{max} is a reasonably accurate predictor of actual translation potential. Cinefluorographic studies indicate that maximum anteroposterior translation in humans is 18–20 mm (Smith, 1985), which is almost the same as the TP_{max} value reported here (21.2 mm).

The height of the postcanine occlusal plane below the glenoid fossa, R, was measured from the most dorsal point on the condyle perpendicular to the plane of occlusion of the mandibular molars (determined by placing a thin rod along the molar tooth row). Because R correlates strongly across species with mandibular length (Wall, 1995, 1999), the total anteroposterior length of the mandible, ML, was measured in the midline from the gonial angle to the anteriormost point on the symphysis.

Analysis

All measurements were entered into Statview 4.5 (Abacus Concepts, Berkeley) and mean values were determined for all species by sex. Several statistical tests were used to examine the hypothesized relationship between R, the height of the occlusal plane below the TMJ, and estimated anteroposterior translation (TP_{max}, TP_{min}, and TP_{carn}). Most importantly, reduced major axis (RMA) regression of logged measures of R versus TP was used to examine the correlation between and scaling of these variables (Sokal & Rohlf, 1995). Because of the strong correlation between ramus height and mandibular length (Wall, 1995, 1999), partial correlation analysis was used to estimate the strength of the correlation between R and the estimates of anteroposterior translation independent of mandibular length. Significance of Pearson correlation coefficients was determined by Fisher's r-z test; significance of partial correlations was determined by using the significance of partial regression coefficients (Sokal & Rohlf, 1995). Finally, least-squares regressions were also computed for estimated anteroposterior translation versus R standardized by mandibular length as an alternative means of examining the strength of the relationship between translation potential and R independent of mandibular size.

Results

Correlation and partial correlation coefficients for R, TP, and ML are summarized for the pooled sample of primates, perissodactyls, and artiodactyls in table 11.2, for primates in table 11.3, for perissodactyls and artiodactyls in table 11.4, and for carnivorans in table 11.5. Among the pooled sample of primates and ungulates (table 11.2A), there is a strongly correlated and highly significant ($r = 0.927$; $P < 0.001$) negative allometry ($b = 0.594$) between TP_{max} and R (fig. 11.4A). Because of the high correlation between mandibular length and ramus height ($r = 0.927$), only a proportion of this relationship ($r = 0.579$, $P < 0.001$) is attributable to R independent of mandibular length as determined by partial correlation analysis. The relationship between TP_{min}, R, and ML is almost the same as that between TP_{max}, R, and ML (see table 11.2B, and fig. 11.5).

When these predicted relationships are examined for primates and the ungulate herbivores separately, several important patterns are evident. Among primates (table 11.3), there is a strongly correlated and highly significant ($r = 0.974$; $P < 0.001$) negative allometry ($b = 0.74$) between

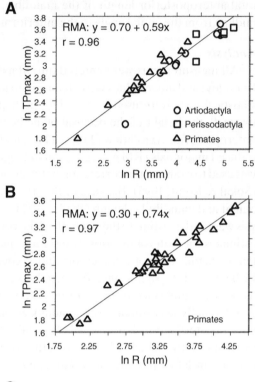

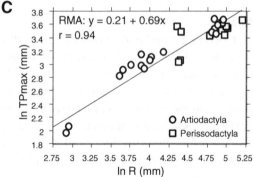

Figure 11.4. Reduced major axis regression of log-transformed R versus TP_{max}. (A) Combined sample of primates and ungulate herbivores; (B) primates; (C) combined herbivore sample (artiodactyls and perissodactyls). The slopes of the primate and herbivore regressions are not significantly different from each other but both are significantly different from isometry ($P < 0.001$).

TP_{max} and R (fig. 11.4B). While the Pearson correlation coefficient between R and TP_{max} is high ($r = 0.963$, $P < 0.001$), mandibular length also correlates strongly with both R ($r = 0.935$, $P < 0.001$) and TP_{max} ($r = 0.972$, $P < 0.001$), raising the possibility that any relationship between R and TP_{max} may be an autocorrelation with mandibular length. Partial correlation analysis, however, indicates that, when the effects of mandibular

Table 11.2. Correlation (A) and partial correlation (B) coefficients for pooled primate, perissodactyl and artiodactyl sample

A Variable	R	TP_{max}	ML
R	—	0.927***	0.927***
TP_{max}	0.579***	—	0.896***
ML	0.580***	0.260 ns	—

B Variable	R	TP_{min}	ML
R	—	0.938***	0.927***
TP_{min}	0.614***	—	0.907***
ML	0.522***	0.291 ns	—

Note: Abbreviations: R, height of the TMJ above the occlusal plane; ML, distance between gonial angle and most anterior point on symphysis; TP_{max}, maximum translation potential of the condyle estimated from the cranium (see Methods); TP_{min}, minimum translation potential of the condyle estimated from the cranium (see Methods). Significance: * = $P < 0.05$; ** = $P < 0.01$; *** = $P < 0.001$; ns = not significant.

Table 11.3. Correlation (A) and partial correlation (B) coefficients for primate sample

A Variable	R	TP_{max}	ML
R	—	0.963***	0.935***
TP_{max}	0.649**	—	0.972***
ML	−0.20 ns	0.752***	—

B Variable	R	TP_{min}	ML
R	—	0.941***	0.935***
TP_{min}	0.612**	—	0.910***
ML	0.564***	0.249 ns	—

Note: Abbreviations as in Table 11.2.

length are held constant, the relationship among primates between R and TP_{max} is moderately strong ($r = 0.649$, $P < 0.01$). The independence of this relationship is corroborated by the results shown in fig. 11.6A, which illustrates the moderately strong correlation ($r = 0.733$, $P < 0.001$) between translation potential and R standardized by mandibular length. Table 11.3B and figures 11.4B and 11.5A confirm that the correlation and

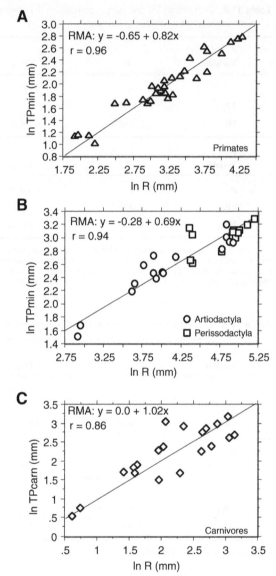

Figure 11.5. Reduced major axis regression of log-transformed R versus TP_{min}. (A) Primates. (B) Combined herbivore sample (artiodactyls and perissodactyls). (C) Carnivores. The slopes of the primate and herbivore regressions but not the carnivores are significantly different from isometry ($P < 0.001$).

strength of the relationship between R and ML with TP_{min} are very similar to those between R, ML and TP_{max}.

As predicted by the above model, the ungulate herbivores exhibit a pattern similar to that in primates in which the coefficients of allometry between TP_{max} and R, and between TP_{min} and R (in both cases, $b = 0.69$), are almost the same as in the primate sample (see table 11.4, figs. 11.4C,

Table 11.4. Correlation (A) and partial correlation (B) coefficients for the combined artiodactyl and perissodactyl sample

A

Variable	R	TP_{max}	ML
R	—	0.903***	0.878***
TP_{max}	0.502**	—	0.923***
ML	0.267 ns	0.635***	—

B

Variable	R	TP_{min}	ML
R	—	0.894***	0.878***
TP_{min}	0.568**	—	0.861***
ML	0.475**	0.354 ns	—

Note: Abbreviations as in Table 11.2.

11.5B). As in primates, Pearson correlation coefficients between TP_{max} and R ($r = 0.903$; $P < 0.001$) and between TP_{max} and ML ($r = 0.878$; $P < 0.001$) are high, but when mandibular length is controlled for using partial correlation analysis, the relationship between R and TP_{max} is moderate and highly significant ($r = 0.502$; $P < 0.01$) (table 11.4A). As figure 11.6 illustrates, this relationship is not as strong ($r = 0.647$; $P < 0.01$) and is significantly steeper than in the primate sample, presumably because total mandibular length relative to the position of the molar tooth row is much greater in most ungulate herbivores than in primates, which have shorter rostra relative to overall cranial size. Table 11.4 indicates that the correlation and partial correlation coefficients of R, ML, and TP_{min} in the ungulate herbivore sample are almost identical to those between R, ML, and TP_{max}.

The relationship between TP_{carn}, R, and ML in the carnivorans with ginglymoid TMJs is quite different in several respects from that observed in primates, artiodactyls, and perissodactyls. First, the coefficient of allometry between R and TP_{carn} is isometric ($b = 1.02$, $r = 0.857$) rather than negative (fig. 11.5C). In addition, while there is a moderately strong correlation between TP_{carn} and R ($r = 0.718$, $P < 0.01$), the partial correlation between these dimensions is close to zero ($r = -0.094$, ns) when the effects of mandibular length are controlled for by partial correlation analysis (table 11.5). When TP_{carn} is plotted against R standardized by mandibular length (fig. 11.6B), the correlation is both low ($r = 0.392$, ns) and negative ($b = -71.17$), in contrast to the moderate and positive relationship

Table 11.5. Correlation coefficients for the carnivoran sample

Variable	R	TP$_{carn}$	ML
R	—	0.718*	0.743***
TP$_{carn}$	−0.094 ns	—	0.983***
ML	0.287 ns	0.964***	—

Note: Abbreviations as in Table 11.2; TP$_{carn}$, maximum anteroposterior distance between pre- and postgle-noid flanges.

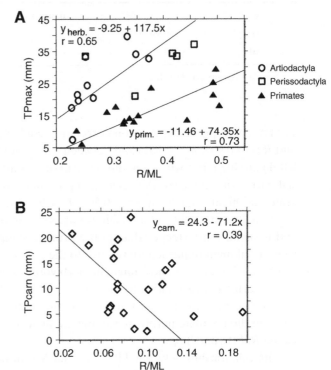

Figure 11.6. Least squares regression plots of R standardized by mandibular length (ML) against (A) TP$_{max}$ in primates, perissodactyls. and artiodactyls; and (B) TP$_{carn}$ in carnivores.

between R and TP/ML in primates, artiodactyls, and perissodactyls with translating TMJs.

Discussion

Condylar translation on the working side during postcanine mastication occurs in taxa whose masticatory systems are dominated by the masseter and medial pterygoid muscles and whose occlusal planes are positioned below the plane of the TMJ by a tall mandibular ramus. Previous studies (e.g., Carlson, 1977; Hylander, 1978; Smith, 1985; Weijs et al., 1989) have focused on two disadvantages of this configuration: it limits gape, and it exacerbates muscle stretching during wide-gape positions. Although both problems are reduced and/or eliminated by condylar translation, there are several reasons to believe that condylar translation may have additional and possibly alternative evolutionary bases.

The above results provide preliminary support for the additional hypothesis, which is not mutually exclusive of those discussed above, that anteroposterior condylar translation is an adaptation to reduce the extent to which the lower postcanine teeth move anteriorly relative to the upper postcanine teeth during jaw closure in mammals whose tooth row lies below the plane of the TMJ. As shown above, in the absence of translation, anteriorly directed movements of the teeth are a direct, unavoidable trigonometric function of the depth of the tooth row below the glenoid fossa: the trajectory of the lower teeth relative to the upper teeth during the power stroke becomes more anteriorly oriented as a function of this distance. In addition, more posterior teeth have more anteriorly oriented trajectories than more anterior teeth. Anteriorly oriented occlusal trajectories are a significant problem, because mammal molars (with the exception of rodents) have morphologies and wear facets that indicate that the jaw moves dorsally, coupled with various amounts of medial movement during occlusion (fig. 11.3). Of course, molar wear can considerably modify the topography of the crowns of recently erupted molars. Given that condylar translation causes a vertical trajectory of the lower molars as they come into occlusion, it could be argued that the orientation of the matching wear facets on occluding molars is simply a by-product of condylar translation, which therefore would not be an adaptation for reorienting tooth movements during occlusion. We reject this view because of the topography of the crowns of unworn molars of primates and ungulates as well as mammals with tribosphenic molars. The unworn molars of these mammals can occlude correctly only if lower jaw movements are vertical and lack an anteriorly directed component. Consequently, changing the instantaneous center of rotation through condylar translation must compensate for the biomechanical advantages of dropping the tooth row below the TMJ.

The hypothesis that translation is an adaptation to reorient the trajectory of the postcanine mandibular teeth relative to maxillary teeth in mammals with occlusal planes below the TMJ specifically predicts a strong correlation between translation potential and R, the height of the postcanine tooth row below the TMJ, independent of mandibular size. As shown above, this prediction is supported by regression analysis and partial correlation analysis for perissodactyls, artiodactyls, and primates, but not for carnivores. Wall (1999) also found a strong relationship between ramus height and the degree of translation in anthropoid primates. In addition, because the trigonometric relationship between R and the sagittal trajectory of the lower teeth relative to the upper teeth must be the same for all mammals, the coefficient of allometry between translation potential and the height of the postcanine tooth row below the TMJ is predicted to be the same in primates and ungulate herbivores. This prediction is also supported by the results presented above for the relationship between R and TP_{max}, which is not significantly different in ungulates and primates but is less strong for the relationship between R and TP_{min}. Negative allometry in this case makes sense because the relationship between the angle of the lower-to-upper molariform teeth ($\varnothing$) is a nonlinear cosine function (see above). Thus, the scaling relationship between antero-posterior linear translation ("t" in fig. 11.2) and R should scale with negative allometry to maintain a constant $\varnothing$ (the slope varying depending on the value of $\varnothing$). In contrast, the length of the glenoid fossa in which any translation can occur in carnivores is isometric with the height of the postcanine tooth row below the TMJ. In other words, glenoid fossa length in carnivores scales with jaw size. In fact, the length of the glenoid fossa in carnivores is essentially equal to the anteroposterior length of the condyle (the slope of the least squares regression of condyle length against TP_{max} is 1.2; $r = 0.98$). In addition, the relationship between translation potential and ramus height in carnivores is not statistically significant and is also negative when one factors out the effects of variation in mandibular size. The slightly different results in primates and ungulates for regressions between R versus TP_{max} and TP_{min} probably reflect differences in how these two measurements estimate translation but need to be verified with *in vivo* kinematic studies.

Given the complexity of condylar translation, better kinematic data are needed to test further the hypothesis that posterior translation of the working side condyle during closing renders the trajectory of the lower molars more vertical than would be the case if the mandible merely rotated around a nontranslating condyle. Except for well-documented evidence (discussed above) that the jaw as a whole is drawn anteriorly during opening and posteriorly during closing, there is little information that precisely correlates the movements of the working and balancing side condyles with the trajectory of the lower molars during jaw closure (Gibbs et al., 1971). Our model assumes that the working side condyle moves posteriorly during most if not all of the closing phase (fast close and the power stroke).

Once again, we stress that the results presented here are not mutually exclusive of hypotheses that condylar translation reduces muscle stretching and/or prevents impingement of the pharynx. In fact, the above predictions and results are reasonably compatible with the impingement hypothesis with one exception. Because many primates, especially humans, tend to be more orthograde than perissodactyls and artiodactyls, one might expect a different allometric relationship between condylar translation and ramus height in these taxa. In particular, impingement should be more problematic, and hence translation potential should be proportionately greater in orthograde mammals in which the pharynx is not directed caudally away from the posterior margin of the ramus. Yet humans, gibbons, chimpanzees, and other highly orthograde primates all fall closely along the same line as more pronograde primates (e.g., macaques, baboons) and ungulate herbivores. The fact that humans exhibit the same relationship between translation potential and R is especially significant because humans have a flexed cranial base that further exacerbates the potential for the posterior margin of the mandible to impinge on crucial pharyngeal structures.

Condylar translation is clearly a complex phenomenon, and it probably evolved because it is adaptive for several reasons. However, any satisfactory understanding of the evolutionary basis for translation probably needs to include some consideration of occlusal function as well as gape and muscle efficiency. Further testing of the hypothesis proposed here as well as previous hypotheses regarding the evolution and functional basis for condylar translation will require comparative, *in vivo* kinematic data on translational movements in a variety of species in conjunction with data on cranial, mandibular, and dental morphology. Kinematic data are also necessary to test the accuracy of any estimates of translation potential based on measurements of the glenoid fossa.

Acknowledgments

We thank Rula Harb for help gathering preliminary data; the Anthropology and Mammalogy Departments at the American Museum of Natural History for access to collections; David Strait, Richard Smith, and Christine Wall for useful discussions; and the referees and editors for helpful, insightful comments.

Literature Cited

Becht, G. 1953. Comparative biological-anatomical research on mastication in some mammals. *Proceedings of the Koninklijke Nederlandse Akademie van Wetenschappen Series C* 56:508–527.

Carlson, D. S. 1977. Condylar translation and the function of the superficial masseter in the rhesus monkey (*M. mulatta*). *American Journal of Physical Anthropology* 47:53–64.

Craddock, F. W. 1948. The muscles of mastication and mandibular movements. *New Zealand Dental Journal* 44:233–239.

Crompton, A. W. 1981. The origin of mammalian occlusion; pp. 3–18 in H. G. Barrer (ed.), *Orthodontics.* Philadelphia: University of Pennsylvania Press.

Crompton, A. W. and Z. Kielan-Jaworowska. 1978. Molar structure and occlusion in Cretaceous mammals; pp. 249–287 *in* P. M. Butler and K. A. Joysey (eds.), *Development, Function and Evolution of Teeth.* New York: Academic Press.

De Vree, F. and C. Gans. 1975. Mastication in pygmy goats *Capra hircus. Annales de la Société Royale Zoologique de Belgique* 105:255–306.

DuBrul, E. L. 1964. Evolution of the temporomandibular joint; pp. 3–27 *in* B. G. Sarnat (ed.), *The Temporomandibular Joint,* 2nd ed. Springfield, IL: Charles Thomas.

Freeman, P. W. 1979. Specialized insectivory: beetle-eating and moth-eating molossid bats. *Journal of Mammalogy* 60:467–479.

Garrett, F. A., L. Angelone, and W. I. Allen. 1964. The effect of bite opening, bite pressure and malocclusion on the electrical response of the masseter muscles. *American Journal of Orthodontics* 50:435–444.

Gibbs, C. H., T. Messerman, J. B. Reswick, and H. J. Derda. 1971. Functional movements of the mandible. *Journal of Prosthetic Dentistry* 26:604–620.

Greaves, W. S. 1974. Functional implications of mammalian joint position. *Forma et Functio* 7:363–376.

———. 1978. The jaw lever system in ungulates: a new model. *Journal of Zoology, London* 184:271–285.

———. 1980. The mammalian jaw mechanism—the high glenoid cavity. *American Naturalist* 184:432–440.

Herring, S. W. 1975. Adaptations for gape in the hippopotamus and its relatives. *Forma et Functio* 8:85–100.

———. 1980. Functional design of cranial muscles: comparative and physiological studies in pigs. *American Zoologist* 20:282–290.

———. 1985. Morphological correlates of masticatory patterns in peccaries and pigs. *Journal of Mammalogy* 66:603–617.

Herring, S. W., A. F. Grimm, and B. R. Grimm. 1979. Functional heterogeneity in a multipennate muscle. *American Journal of Anatomy* 154:563–576.

Hiiemae, K. M. and A. W. Crompton. 1985. Mastication, food transport and swallowing; pp. 262–290 *in* M. E. Hildebrand, D. M. Bramble, K. L. Liem, and B. D. Wake (eds.), *Functional Vertebrate Morphology*. Cambridge, MA: Harvard University Press.

Hylander, W. L. 1978. Incisal bite force direction in humans and the functional significance mammalian mandibular translation. *American Journal of Physical Anthropology* 48:1–8.

———. 1992. Functional anatomy; pp. 60–92 *in* B. G. Sarnat and D. M. Laskin (eds.), *The Temporomandibular Joint: A Biological Basis for Clinical Practice*. Philadelphia: W. B. Saunders.

Hylander, W. L. and K. R. Johnson. 1985. Temporalis and masseter function during incision in macaques and humans. *International Journal of Primatology* 6:289–322.

———. 1993. Modeling relative masseter force from surface electromyograms during mastication in non-human primates. *Archives of Oral Biology* 38: 233–240.

Hylander, W. L., K. R. Johnson, and A. W. Crompton. 1987. Loading patterns and jaw movements during mastication in *Macaca fascicularis:* a bone-strain, electromyographic and cineradiographic analysis. *American Journal of Physical Anthropology* 72: 287–314.

———. 1992. Muscle force recruitment and biomechanical modeling: An analysis of masseter muscle function during mastication in *Macaca fascicularis*. *American Journal of Physical Anthropology* 88: 365–387.

Janis, C. M. 1979. Aspects of the Evolution of Herbivory in Ungulate Mammals. Ph.D. Dissertation, Harvard University, Cambridge, MA.

———. 1983. Muscles of the masticatory apparatus in two genera of hyraxes (*Procavia* and *Heterohyrax*). *Journal of Morphology* 176:61–87.

Kay, R. F. 1978. Molar structure and diet in extant Cercopithecidae; pp. 309–338 *in* P. M. Butler and K. A. Joysey (eds.), *Development, Function and Evolution of Teeth*. New York: Academic Press.

Kay, R. F. and K. M. Hiiemae. 1974. Jaw movement and tooth use in recent and fossil primates. *American Journal of Physical Anthropology* 40:227–256.

Lieberman, D. E. and A. W. Crompton. 2000. Why fuse the mandibular symphysis? *American Journal of Physical Anthropology* 112:517–540.

Lucas, P. W. 1982. Basic principles of tooth design; pp. 154–162 *in* B. Kurtén (ed.), *Teeth: Form, Function and Evolution*. New York: Columbia University Press.

Mack, P. J. 1984. A functional explanation for the morphology of the temporomandibular joint of man. *Journal of Dentistry* 12:225–230.

Manns, A., R. Miralles, and F. Guerrero. 1981. The changes in electrical activity of the postural muscles of the mandible upon varying the vertical dimension. *Journal of Prosthetic Dentistry* 45:438–445.

Maynard Smith, J. and R. J. G. Savage. 1959. The mechanics of mammalian jaws. *School Science Review* 40:289–301.

McMahon, T. 1984. *Muscles, Reflexes and Locomotion*. Princeton, NJ: Princeton University Press.

Mills, J. R. E. 1967. A comparison of lateral jaw movements in some mammals from wear facets on their teeth. *Archives of Oral Biology* 12:645–661.

Moss, M. L. 1960. Functional anatomy of the temporomandibular joint; pp. 73–88 *in* L. Schwartz (ed.)., *Disorders of the Temporomandibular Joint.* Philadelphia: W.B. Saunders.

———. 1975. A functional cranial analysis of centric relation. *Dental Clinics of North America* 19:431–442.

———. 1983. The functional matrix concept and its relationship to temporomandibular joint dysfunction and treatment. *Dental Clinics of North America* 27:445–455.

Noble, H. W. 1973. Comparative functional morphology of the temporomandibular joint. *Oral Science Review* 2:3–28.

Osborn, J. W. 1987. Relationship between the mandibular condyle and the occlusal plane during hominid evolution: some of its effects on jaw mechanics. *American Journal of Physical Anthropology* 73:193–207.

Rugh, J. G. and C. J. Drago. 1981. Vertical dimension: a study of clinical rest position and jaw muscle activity. *Journal of Prosthetic Dentistry* 45:670–675.

Smith, R. J. 1985. Functions of condylar translation in human mandibular movement. *American Journal of Orthodontics* 88:191–202.

Sokal, R. E. and J. Rohlf. 1995. *Biometry,* 2nd ed. San Francisco: W. H. Freeman.

Wall, C. E. 1995. Form and Function of the Temporomandibular Joint in Anthropoid Primates. Ph.D. Dissertation, Department of Anatomical Sciences, Stony Brook University, NY.

———. 1999. A model of temporomandibular joint function in anthropoid primates based on condylar movements during mastication. *American Journal of Physical Anthropology* 109:67–88.

Weijs, W. A. 1994. Evolutionary approach of masticatory motor patterns in mammals. *Advances in Comparative and Environmental Physiology* 18:282–320.

Weijs, W. A., J. A. M. Korfage, and G. J. Langenbach. 1989. The functional significance of the position and center of rotation for jaw opening and closing in the rabbit. *Journal of Anatomy* 162:133–148.

PART FOUR

Ontogeny and Evolution

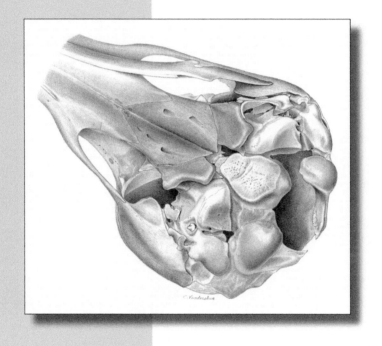

Ontogeny and Evolution

Plate 4. Skull of a juvenile echidna, *Tachyglossus aculeatus*, in right posteroventrolateral view. The basicranium is exposed on the right via removal of the ectotympanic, malleus, and incus. Illustration by Claire Vanderslice, from: J. R. Wible & J. A. Hopson (1995), Homologies of the prootic canal in mammals and non-mammalian cynodonts, *Journal of Vertebrate Paleontology* 14(2): 331–356.

12 Neoteny and the Plesiomorphic Condition of the Plesiosaur Basicranium

F. Robin O'Keefe

Introduction

Historically, the systematics of the Plesiosauria (Reptilia, Sauropterygia) were based largely on postcranial characters (Persson, 1963; Brown, 1981). Several factors account for this bias: plesiosaur skulls tend to be delicate and are often crushed even when preserved, postcranial elements are relatively common and cranial elements are not, lack of knowledge about the relationships of stem-group sauropterygians, and lack of knowledge of plesiosaur cranial anatomy itself. However, recent detailed examinations of plesiosaur cranial anatomy have identified many characters of use in plesiosaur systematics (Brown, 1993; Cruickshank, 1994; Storrs & Taylor, 1996; Storrs, 1997; Carpenter, 1997; Evans, 1999; O'Keefe, 2001, 2004), and the systematics of the group have changed markedly in response (Carpenter, 1997; O'Keefe, 2001, 2004). The work of Rieppel and others has clarified the anatomy and relationships of stem-group sauropterygians (Storrs, 1991; see Rieppel, 2000, for review). This work has laid the anatomic and phylogenetic foundations for a better understanding of plesiosaur cranial anatomy.

The purpose of this paper is to describe the condition of the braincase in stratigraphically early and morphologically primitive plesiosaurs. Information on the braincase of plesiomorphic taxa is important because it establishes the polarity of characters occurring in more derived plesiosaurs. This paper begins with a short review of braincase anatomy in stem-group sauropterygians. Data on braincase morphology of the plesiomorphic plesiosaur genera *Thalassiodracon* and *Eurycleidus* are then presented and interpreted via comparison with other plesiosaurs, stem-group sauropterygians, and stem diapsids (*Araeoscelis*). Early diapsids are relevant because plesiosaur skulls more closely resemble early diapsids than stem-group sauropterygians in several key areas. Plesiosaurs display a broad trend of delayed and reduced ossification compared with stem-group sauropterygians (Storrs, 1991). This trend is linked to the acquisition of a truly pelagic lifestyle in the Plesiosauria (Romer, 1956).

As pointed out by Rieppel (2000, 113), plesiosaurs show several reversals to plesiomorphic character states, such as the reappearance of posterior interpterygoid vacuities. This paper advances the hypothesis that
reversal in this and several other cranial characters is a consequence
of heterochrony, specifically neoteny in the ossification of the pterygoid
(Gould, 1977).

The Braincase in Stem-Group Sauropterygians

One barrier to understanding the plesiosaur braincase is the condition
of the basicranium in stem-group sauropterygians. All "nothosaur"-
grade taxa (i.e., all Sauropterygia exclusive of clade Pistosauria; see
fig. 12.7) lack interpterygoid vacuities in the posterior palate; the pterygoids meet in an unbroken median suture running from the vomers anteriorly and caudad to the occipital condyle and therefore hide the basicranium in ventral view (Rieppel, 1994). In plesiosaurs, the palate is quite
different. Posterior interpterygoid vacuities are present, as in early diapsids, and an anterior interpterygoid vacuity sometimes allows a view of
the relations of the anterior end of the parasphenoid (O'Keefe, 2001).

The braincase in plesiosaurs is an open and poorly ossified structure, and different patterns of ossification are taxonomically informative
(O'Keefe, 2001). The occiput is closed and plate-like in many nothosaur-
grade taxa, further obscuring the more anterior relationships of braincase elements; however, Rieppel used well-preserved material to produce
detailed reconstructions of the braincase in *Simosaurus, Nothosaurus*
(Rieppel, 1994), and *Cymatosaurus* (Rieppel & Werneburg, 1998).
Simosaurus and *Nothosaurus* are nothosauroids possessing a closed occiput, while *Cymatosaurus* is a pistosauroid possessing an open occiput
very similar to that found in all plesiosaurs (Rieppel, 2000). The closed
occiput of nothosauroids is characterized by large and plate-like exoccipitals and opisthotics with a clear suture between the two elements,
whereas more laterally the opisthotic possesses a long dorsolateral suture
with the descending occipital flange of the squamosal (Rieppel, 1994).
The posttemporal fossa in both taxa is reduced to a small foramen. Below
the ventral margin of the exoccipital and opisthotic, the cranioquadrate
passage lies partially open in *Simosaurus,* whereas in *Nothosaurus* the
pterygoid meets the opisthotic in a long transverse suture, closing off the
cranioquadrate passage.

The closed nature of the occiput leaves several identifiable foramina
for the passage of structures into and out of the head; most prominent is
the foramen magnum, on either side of which rest the pillar-like bodies

of the exoccipitals. Just lateral to these pillars and still within the exoc-cipitals are the jugular foramina. Lateral and slightly ventral to this posi-tion are foramina in the pterygoid for the passage of the internal carotid. Lastly, two foramina in the ventromedial aspect of each exoccipital pillar allow passage for the hypoglossal nerve (Rieppel, 1994); plesiosaurs usu-ally have two foramina in the same position for the hypoglossal nerve (Hopson, 1979; Storrs & Taylor, 1996; Carpenter, 1997; Evans, 1999). Ev-ans (1999) illustrates a single foramen in *Muraenosaurus*.

In contrast to the closed occiput in nothosauroids, the occiput is open in the pistosauroids *Corosaurus* and *Cymatosaurus* as well as in all pis-tosaurians including plesiosaurs. The occiput in all plesiosaurs is broadly open, possessing a large posttemporal fenestra bounded dorsally by the squamosal arch and ventrally by the slender paroccipital process (Williston, 1903, 26; Storrs & Taylor, 1996; Carpenter, 1997). The paroc-cipital process is formed by the opisthotic. The exoccipital is reduced to a column forming the boundary of the foramen magnum. The paroccipital process articulates laterally with the median surface of the squamosal in most plesiosaur taxa (O'Keefe, 2001). Below the paroccipital process is a large cranioquadrate passage bordered by the quadrate, squamosal, ptery-goid, basioccipital, and exoccipital/opisthotic (Storrs and Taylor, 1996).

Rieppel described the unusual course of the internal carotid from the occiput to the brain in both *Nothosaurus* (Rieppel, 1994) and *Cymato-saurus* (Rieppel & Werneburg, 1998). In both taxa, the internal carotid travels anteriorly from the occiput within a prominent groove in the dor-sal surface of the quadrate ramus of the pterygoid until reaching the ba-sisphenoid and then courses ventrally between the lateral aspect of the basisphenoid and prootic before dividing into the palatine artery and the cerebral carotid (Rieppel, 1994). This course of the internal carotid—characterized by a groove in the dorsal aspect of the pterygoid—is an apomorphy of the Eusauropterygia, although the course of this vessel is currently unknown in pachypleurosaurs and placodonts (Rieppel, 2000). This character is reversed in most plesiosaurs (see below).

In nothosauroids, the cerebral carotid enters the base of the sella turcica ventrolaterally through two prominent foramina separated by a low ridge of bone (Hopson, 1979; Rieppel, 1994). This ridge of bone runs anteriorly along the median dorsal aspect of the palate and is prob-ably the root of a cartilaginous interorbital septum (Romer 1956, 57) ex-tending back into the hypophyseal fossa and subdividing that structure, as noted by Hopson (1979, 120). An identical condition exists in the pistosauroid *Cymatosaurus* (Rieppel & Werneburg, 1998) and in the plesiosaur *Thalassiodracon* (see below). Last, the basal articulation between ossifications of the palatoquadrate cartilage and the lateral braincase wall is "obliterated" in all stem-group sauropterygians, yielding an akinetic skull (Rieppel, 1994, 2000). The pterygoid encloses the basicranium in these taxa, and a discreet basal articulation is obscured by the extensive contact between pterygoid and basicranium in general. In plesiosaurs, this character is also reversed; a discreet and well-developed basal articulation is present in many taxa (O'Keefe, 2001), although it does not always ossify in later pliosaurids.

The pattern of the semicircular canals is currently unknown in basal sauropterygians. Because of the general reduction of ossification in the endochondral elements of the plesiosaur braincase, however, the paths of the three semicircular canals and the locations of the anterior and posterior ampullae have been identified in several plesiosaur genera (*Peloneustes,* Andrews, 1913; *Cryptoclidus,* Brown, 1981; *Eurycleidus,* Cruickshank, 1994; *Libonectes* and *Dolichorhynchops,* Carpenter, 1997; *Muraenosaurus,* Evans, 1999). All these genera show a similar condition. The posterior ampulla is located in the body of the fused exoccipital/opisthotic, and this bone contains a deep fossa, open anteromedially, for this structure. The horizontal semicircular canal leaves the posterior ampulla laterally and passes directly into the posterior edge of the prootic through the suture between this bone and the exoccipital/opisthotic. The anterior ampulla is located in the prootic, where it resides in a shallow, medially open fossa. The anterior ampulla accepts the horizontal semicircular canal posteriorly, and the canal in the prootic for the anterior portion of this canal is either open or roofed depending on the degree of ossification of the prootic. The posterior vertical semicircular canal ascends through the body of the exoccipital and enters the supraoccipital, where it communicates with the superior utriculus. The anterior vertical semicircular canal leaves the superior utriculus and descends into the prootic, passing through the suture between prootic and supraoccipital. Again, the groove of the anterior vertical semicircular in the prootic can be open or closed depending on the degree of ossification.

In summary, the internal ear in plesiosaurs is not derived and is similar to the condition illustrated in *Captorhinus* by Price (1935). The posterior ampulla resides in the fused exoccipital/opisthotic, and the suture between these two bones is seldom apparent in plesiosaurs. However, Andrews (1913) illustrated a specimen of *Peloneustes* that does preserve this suture. On the medial wall of the foramen magnum, the suture between opisthotic and exoccipital angles from anteroventral to posterodorsal; the jugular foramen is a gap within this suture as is expected given its developmental origin within the metotic fissure (de Beer, 1937). The paroccipital process is formed entirely of opisthotic, and the fossa for the posterior ampulla is contained within the opisthotic, again like *Captorhinus* (Price, 1935). Below I refer to the fused exoccipital/opisthotic as simply the "exoccipital" except in cases in which the suture between the two elements is clear.

Materials and Methods

This study utilizes material from the Permian diapsid *Araeoscelis* as well as the Early Jurassic plesiosaur taxa *Thalassiodracon, Eurycleidus,* and *Plesiopterys.* Material examined is listed by taxon below.

Institutional abbreviations are as follows: FMNH, Field Museum of Natural History, Chicago; OXFUM, Geological Collections, University Museum, Oxford, UK; CAMSM, Sedgwick Museum, Cambridge, UK; SMNS, Staatliches Museum für Naturkunde, Stuttgart, Germany.

Material

Araeoscelis casei. The partial skull figured here (fig. 12.1) is specimen number FMNH UR 2419. The skull is small and may represent a juvenile. At this time members of the genus *Araeoscelis* are assigned to two

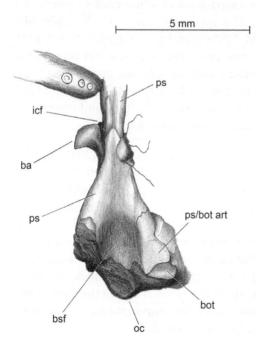

Figure 12.1. Basicranium of a juvenile specimen of *Araeoscelis casei,* FMNH UR 2419. Abbreviations: art, articulation; ba, basal articulation; bot, basioccipital tuber; bsf, basisphenoid fossa; icf, internal carotid foramen; oc, occipital condyle; ps, parasphenoid.

species based mainly on stratigraphic position, although the species are very similar and the genus usually is treated as one form (Vaughn, 1955; Reisz et al., 1984). *Araeoscelis* occurs in three localities in central Texas, two of which are in the Wichita Group; the remaining locality is in the Clear Forks Group (Lower Permian; see Reisz et al., 1984 for details). The skull figured in this paper is poorly preserved except for the basicranium.

Thalassiodracon hawkinsi. This plesiosaur taxon is known from several specimens originating from one locality, the lowermost Lias (Rhaetian/Hettangian) of Street, Somerset, United Kingdom (see Storrs & Taylor, 1996, for a thorough review of this taxon). This genus is one of the stratigraphically earliest plesiosaurs for which adequate material is known and is correspondingly plesiomorphic, sharing several symplesiomorphies with pistosaurids that are lost in later plesiosaurs, including low neural spine height, an angled humerus, presence of nasals, and the presence of a posterior postorbital process (O'Keefe, 2001). The present study deals with the skull figured and described in detail by Storrs and Taylor (1996; CAMSM J.46986). Reference is also made to a skull in the University Museum of Oxford, OXFUM J.10337.

Eurycleidus arcuatus. This taxon was redescribed by Cruickshank (1994) following the extensive preparation of a historical specimen first dealt with by Owen (1840). The specimen resides in the University Museum of Oxford (OXFUM J.28585) and is a fragmentary but well-preserved skull with accompanying postcranial material. The provenance of the material is somewhat unclear, with the only sure information being the source of the fossil, which is "Lower Lias, Lyme Regis." Cruickshank (1994) considered the most likely age to be Hettangian-Lower Sinemurian, although the Pliensbachian is also possible.

Plesiopterys wildi. This taxon was named by O'Keefe (2004) and is a morphologically primitive plesiosaur from the Posidonienschiefer near Holzmaden, Germany. It is of Lower Toarcian age. The holotype of this taxon is a complete skeleton with a crushed and disarticulated skull (SMNS 16812); elements of the braincase are extremely well preserved and are illustrated here. This material was originally referred to *Eurycleidus* by O'Keefe (2001), but further research proved this assignment erroneous (see O'Keefe, 2004, for a discussion of the complex taxonomic issues involved). The basicranium of this taxon is among the most primitive of all known plesiosaurs and is therefore important for documenting the condition from which later taxa are derived.

Description

The taxa *Eurycleidus* (*sensu stricto*) and *Thalassiodracon* are very similar in most aspects of the braincase and are discussed together here; *Plesiopterys* is discussed separately.

Palate and Context. In ventral view, the palate of *Thalassiodracon* is planar, lacking pterygoid flanges or other excrescences projecting ventrally from the plane of the palate, and is similar to the palate in nothosauroids. The palate is dominated by the pterygoids; these bones have a long suture on the midline stretching from the vomers anteriorly to the parasphenoid posteriorly (figs. 12.2, 12.3). This suture is broken by a small anterior interpterygoid vacuity, but behind the vacuity the pterygoids meet again, obscuring the cultriform process of the parasphenoid. This condition is again similar to the nothosauroid configuration and differs from the broadly open palate of *Araeoscelis*. (The genus *Plesiosaurus* is more derived compared with nothosauroids. In this genus, the palate is

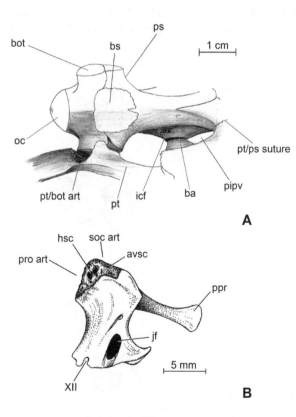

Figure 12.2. (A) Basicranium of *Thalassiodracon hawkinsi*, CAMSM J.46986. (B) Right exoccipital-opisthotic of same in oblique medial view, modified from Storrs and Taylor (1996). Abbreviations as in figure 12.1, and XII, foramen for cranial nerve XII; avsc, anterior vertical semicircular canal; hsc, horizontal semicircular canal; jf, jugular foramen; pipv, posterior interpterygoid vacuity; ppr, paraoccipital process; pro art, prootic articulation; pt, pterygoid; soc art, supraoccipital articulation.

more open and similar to the condition in *Araeoscelis;* Storrs, 1997). Farther posteriorly, the pterygoids separate and form the lateral borders of the posterior interpterygoid vacuities, participate in a broad articulation with the basioccipital tubera, and then continue laterally as the quadrate flanges of the pterygoids (fig. 12.2). Just anterior to the posterior interpterygoid vacuities, a triangular portion of the parasphenoid is exposed on the palate surface, where it sutures with the pterygoids. The posterior interpterygoid vacuities are large fenestrae affording an unobstructed view of the basicranium similar to the condition in *Araeoscelis.* The pterygoids do not meet in a midline suture behind the posterior interpterygoid vacuities in *Thalassiodracon.*

Basicranium. The basicrania of *Thalassiodracon* and *Eurycleidus* are surprisingly plesiomorphic and are best understood by comparison with the basicranium in *Araeoscelis* (fig. 12.1). In *Araeoscelis* (and *Petrolacosaurus;* Reisz, 1981), the parasphenoid is a triangular structure with a narrow cultriform process projecting anteriorly on the midline. Posteriorly, the parasphenoid broadens into a triangular sheet of bone covering the basicranium, with the two lateral cristae ventrolaterales (Reisz, 1981)

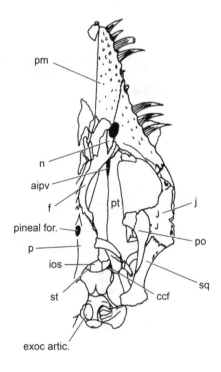

Figure 12.3. Skull of *Thalassiodracon hawkinsi* in dorsal view, OUM J. 10337. Abbreviations as in figure 12.2, and aipv, anterior interpterygoid vacuity; ccf, cerebral carotid foramen; exoc artic, exoccipital articulation; f, frontal; for, foramen; ios, interorbital septum; j, jugal; n, nasal; p, parietal; pm, premaxilla; po, postorbital; sq, squamosal; st, sella turcica. Skull length 18 cm.

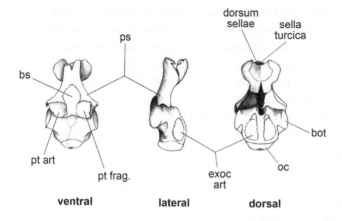

ventral   **lateral**   **dorsal**

Figure 12.4. Braincase of *Eurycleidus arcuatus,* OXFUM J. 28585. Abbreviations as in figures 12.1–12.3, and frag, fragment. Fragment length 4.5 cm.

separated by a deep basisphenoid fossa on the midline. The parasphenoid has prominent articulations with the basioccipital tubera laterally but does not reach to the margin of the occipital condyle on the midline. Two small excavations in the posterior edge of the parasphenoid expose a small portion of the body of the basioccipital near the midline, but the suture between basioccipital and basisphenoid is covered by the parasphenoid. More anteriorly, the basipterygoid processes of the basisphenoid are visible on either side of the cultriform process of the parasphenoid. A groove runs around the anterior aspect of each basipterygoid process and probably carried the internal carotid artery into the floor of the hypophyseal fossa.

In *Thalassiodracon* and *Eurycleidus,* the parasphenoid is also triangular in outline (figs. 12.2, 12.4). The anterior cultriform process is obscured by the midline suture of the pterygoids (but is present in *Plesiosaurus;* O'Keefe, 2001); posteriorly, the bone expands laterally into cristae ventrolaterales that articulate with the basioccipital tubera. The parasphenoid wraps around the basioccipital tubera and participates in the articulation with the pterygoid. The basisphenoid fossa is very shallow, however, and the parasphenoid extends posteriorly for only a short distance on the midline, exposing the suture between basisphenoid and basioccipital. This suture is complex, with a slip of bone reaching posteriorly from the body of the basisphenoid and coursing ventral to the body of the basioccipital. Dorsal to the parasphenoid, the body of the basisphenoid is visible, displaying a prominent basal articulation composed of a shallow basipterygoid process and a deeper process of the pterygoid. A large foramen for the internal carotid artery is present in the body of the basisphenoid posterior to the basal articulation, in contrast to the location of this structure in *Araeoscelis.*

The lateral and dorsal aspects of the basicranium are illustrated best in the isolated braincase of the *Eurycleidus* specimen (fig. 12.4). In lateral view, the basioccipital tubera are prominent structures projecting ventrolaterally from the body of the basioccipital, and their distal ends are wrapped by the posterior margin of the parasphenoid. The articulations for the exoccipitals are clearly delineated, rugose ovoids on the dorsal surface of the basioccipital. The occipital condyle is rather poorly developed. It is a shallow dome with no groove between the condyle and the body of the basioccipital (*Thalassiodracon* and *Plesiosaurus* display a similar condition; O'Keefe, 2001). A notochordal pit is present in *Eurycleidus* and is often present in other plesiosaurs, although its presence varies ontogenetically. The anterior face of the basioccipital carries a deep fossa or notch; its position in the anterior edge of the basioccipital presumably indicates it was overlain by the medulla oblongata (Romer, 1956), although there are no obvious soft tissue structures that might be associated with this notch.

The basisphenoid does not articulate directly with the basioccipital in either *Eurycleidus* (fig. 12.4) or *Thalassiodracon* (fig. 12.2). The only contact between these two bones is formed by the ventroposterior process of the basisphenoid exposed on the palate surface (described above); hence there is a gap in dorsal view between the bodies of the basioccipital and the basisphenoid, floored by the ventrolateral basisphenoid process on the midline and the dorsal surface of the parasphenoid laterally. This condition contrasts that in nothosauroids, where the basisphenoid and basioccipital bodies have a tight, broad connection (Rieppel, 1994). *Thalassiodracon* and *Eurycleidus* also posses a deep notch in the back of the clivus, a trait they share with the pistosauroid *Cymatosaurus*. *Cymatosaurus* also might lack a tight connection between the bodies of the basisphenoid and basioccipital, although the basioccipital is not preserved in the one skull of *Cymatosaurus* that visibly preserves this area (Rieppel & Werneburg, 1998; see also Rieppel, 1997, 2000). If this character can be demonstrated, it will provide another link between the Plesiosauria and pistosauroids; the notch in the clivus mentioned above certainly does link these taxa.

The clivus terminates anteriorly in the dorsum sellae, a poorly developed structure in plesiosaurs consisting of the posterior wall of the sella turcica, without an obvious raised ridge as seen in *Captorhinus* (Price, 1935). The foramina for the cerebral carotid arteries are not preserved in *Eurycleidus* and probably were not ossified (they may be represented by shallow grooves in the lateral edges of the body of the basisphenoid); the foramina are large in *Thalassiodracon* (fig. 12.2). The cerebral carotids entered the floor of the hypophyseal fossa through paired posterolateral foramina (fig. 12.3). The hypophyseal fossa itself is open laterally and anteriorly. A broad, low ridge extends forward from the back of the hypophyseal fossa on the midline and seems to have been confluent with the interorbital septum, similar to the condition seen in nothosauroids (Rieppel, 1994) and *Cymatosaurus* (Rieppel & Werneburg, 1998). The body of the basisphenoid lacks ossified clinoid processes in *Eurycleidus* and *Thalassiodracon,* although the elasmosaur *Libonectes* does have stubby ossifications in this area (Carpenter, 1997). The basipterygoid process is ossified but shallow in *Thalassiodracon,* whereas in *Eurycleidus* no distinct process has ossified.

   Dorsal Braincase Ossifications (*Exoccipital, Supraoccipital, Opisthotic, Prootic*). The exoccipital and supraoccipital in stratigraphically early and morphologically primitive plesiosaurs are illustrated here by disarticulated, well-preserved elements of *Plesiopterys* (fig. 12.5) and by elements from *Thalassiodracon* (fig. 12.2). The form of the exoccipitals is very similar to that described in more derived plesiosaurs (see above), although the elements are more gracile and poorly ossified (for the following discussion see also Storrs & Taylor, 1996). In *Thalassiodracon,* the paroccipital process is narrow medially and flares laterally into an articular process for the squamosal. This process meets the body of the exoccipital near its dorsal margin. The medial surface of the exoccipital anterior to the foramen

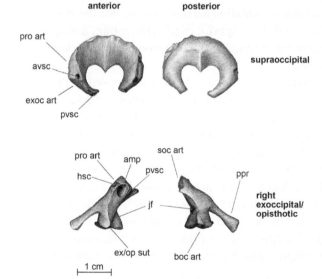

Figure 12.5. Dorsal braincase elements from *Plesiopterys*, SMNS 16812. Abbreviations as in figures 12.1–12.3, and amp, ampulla; bo, basioccipital; ex/op sut, exoccipital/opisthotic suture; pvsc, posterior vertical semicircular canal.

magnum is pierced by a prominent jugular foramen, and passage of the hypoglossal nerve is restricted to a notch at the inferior margin of this bone. The fossa for the posterior ampulla, generally open in plesiosaurs, seems to be closed over except for a fissure above the jugular foramen in *Thalassiodracon*. The suture between exoccipital and opisthotic ossifications is not visible.

The exoccipital of *Plesiopterys* (fig. 12.5) is similar, except that the fossa for the posterior ampulla is open. A foramen for the passage of the horizontal semicircular canal is present on the articular surface for the prootic in both *Plesiopterys* and *Thalassiodracon,* whereas in *Plesiopterys* a grove for the posterior vertical semicircular canal enters into the articular surface for the supraoccipital. Unlike *Thalassiodracon* and like *Peloneustes* (Andrews, 1913), the suture between the opisthotic and exoccipital ossifications is visible in *Plesiopterys,* and this suture contains the jugular foramen. The opisthotic is very similar to that in *Peloneustes,* forming the paroccipital process as well as the anterodorsal portion of the "exoccipital pillar" lateral to the foramen magnum. Also like *Peloneustes* (and *Captorhinus;* Price, 1935), the opisthotic ossification contains a fossa for the posterior ampulla and associated foramina for semicircular canals.

The supraoccipital of *Plesiopterys* is typical of those found in most early Jurassic plesiosaurs (fig. 12.5; O'Keefe, 2001, for *Plesiosaurus*). The bone is shallow anteroposteriorly, with a poorly developed articulation for the parietals on its dorsal margin. Distinct articulations for the exoccipitals and prootics are evident on the anterolateral edge. The prootic articulation is perforated by a foramen for the anterior vertical semicircular canal, and the exoccipital articulation carries a groove for the posterior vertical semicircular canal. The superior edge of the foramen magnum is divided by a prominent, pointed process surmounted by a low ridge on its posterior surface.

The prootic is a dish-shaped, quadrangular bone not exposed in the *Thalassiodracon* material and poorly preserved in *Plesiopterys* (fig. 12.6). In *Plesiopterys,* the prootics are crushed ventrally and anteriorly to their articulations with the body of the basisphenoid; they now reside anterior to the presumed position of the clinoid processes, and their articulations are posterior to this position (fig. 12.6). Both prootics seem to be lying on their medial faces, and little detailed morphology is visible. The ear remains a poorly known region in plesiosaurs. Carpenter (1997) illustrates a large fenestra ovalis bordered by the exoccipital column posteriorly, the prootic anteriorly, and the basisphenoid ventrally in the elasmosaur *Libonectes*. Most plesiosaurs represented by adequate material do have this

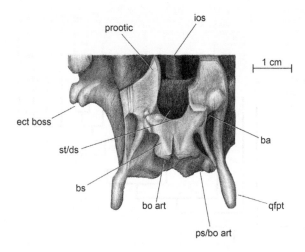

Figure 12.6. Dorsal view of basicranium from *Plesiopterys,* SMNS 16812. Abbreviations as in figures 12.1–12.3, and ds, dorsum sellae; ect, ectopterygoid; qfpt, quadrate flange of the pterygoid.

gap in the lateral braincase wall, although there is some doubt about its homology with the fenestra ovalis. Storrs and Taylor (1996) maintained that the fenestra ovalis is absent in *Thalassiodracon.* The specimen discussed by those authors and in this paper possesses a well-preserved stapes. Storrs and Taylor believed that this element articulated tightly with the body of the exoccipital; the present author could not determine whether the present location of the stapes was in fact an articulation or an artifact of preservation. Considering the development of the otic capsule in squamates (de Beer, 1937), the fenestra ovalis should pierce the lateral wall of the otic capsule in the embryo and persist between the two ossifications derived from the otic capsule in adults (prootic and opisthotic). The opisthotic is reduced in plesiosaurs and has no suture with the prootic; however, the position of the gap in the lateral braincase wall behind the prootic and in front of the opisthotic portion of the exoccipital pillar in both *Plesiopterys* and in *Peloneustes* is consistent with the position of the fenestra ovalis. The gap in the lateral braincase wall of plesiosaurs is probably too large and irregular to represent only the fenestra ovalis; the center of the otic capsule may have failed to ossify completely in plesiosaurs and this may account for some of the difficulty in assigning the relations of the footplate of the stapes.

Discussion

In general, the braincase in plesiosaurs is quite similar to those of more basal diapsids such as *Araeoscelis* and *Petrolacosaurus* and is more similar to those taxa than to the nothosaur-grade sauropterygians from which

Figure 12.7. Cladogram of sauropterygian relationships, representing the combined topologies of Rieppel (2000) and O'Keefe (2001). Character changes discussed in the text occur at the following positions: A, closed occiput → open occiput; B, posterior interpterygoid vacuity closed → open; C, basal articulation obliterated → exposed, anterior interpterygoid vacuity closed → open, grooves in pterygoid for internal carotids present → absent; D, cultriform process obliterated → exposed.

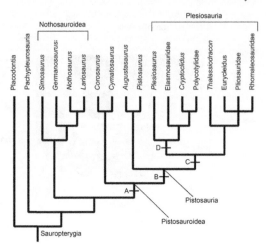

plesiosaurs arose. The possibility exists that plesiosaurs arose independently from basal diapsids; however, there is excellent anatomic evidence supporting the derivation of plesiosaurs from nothosaur-grade sauropterygians, and several parsimony analyses have yielded strong support for this view (Storrs, 1991; O'Keefe, 2001; see above). Symplesiomorphy is therefore a poor explanation for the many characters that show reversals to more primitive states from the position of Nothosauroidea up to plesiosaurs (fig. 12.7). The goals of this discussion are to highlight these reversals, plot them onto the cladogram, and advance an alternative hypothesis linking the derivation of these plesiomorphic states.

The following characters are reversed to more primitive states in plesiosaurs (i.e., states present in *Araeoscelis*) relative to the conditions seen in nothosauroids (also summarized in figure 12.7):

1. The basal articulation between the pterygoids and basisphenoid is well developed in plesiosaurs and is obliterated in more basal sauropterygians.
2. Plesiosaurs do not possess well-developed grooves in the dorsal surface of the pterygoid for the passage of the internal carotid artery, unlike more basal sauropterygians.
3. The most basal plesiosaurs possess both anterior and posterior interpterygoid vacuities, whereas the palate is imperforate in nothosaur-grade sauropterygians.
4. At least some plesiosaurs have an exposed cultriform process of the parasphenoid, unlike nothosaur-grade sauropterygians.

5. The configuration of the basicranium in plesiosaurs is very similar to that in *Araeoscelis* (i.e., presence of basisphenoid fossa, cristae ventrolaterales, articulation of parasphenoid with basioccipital tubera), whereas the configuration of this region is obscured by the pterygoids in more basal sauropterygians.

6. Plesiosaurs and pistosauroids possess what Rieppel (2000) terms an "open occiput," meaning that the opisthotic is restricted to a paroccipital process, leaving the cranioquadrate passage and the posttemporal fenestra widely open. Most, but not all, nothosaur-grade sauropterygians have a closed occiput.

Of the six changes listed above, the first five are associated with the pterygoid, specifically with reduction in the extent of this bone. The sixth change, opening of the occiput, concerns reduction of the pterygoid with respect to the opening of the cranioquadrate passage but also concerns reduction of the opisthotic as well as other bones surrounding the post-temporal fenestra. Any alternative hypothesis meant to explain these six character reversals therefore should account for reduction in the ptery-goid bone.

Delayed and reduced ossification in all bones of the skeleton is a well-known trend in aquatic tetrapods (Romer, 1956), and sauropterygians are no exception. Plesiosaurs are thought to have been more pelagic than more basal sauropterygians (Storrs, 1993) and show a general reduction in ossification of the postcranial skeleton (Storrs, 1991). The reduction in the pterygoid and other skull bones may be another manifestation of this trend. The pterygoid is an intramembranous bone and as such is not preformed in cartilage before ossification, as is the case in endochon-dral elements (Romer, 1956). However, there is still an ontogeny to the ossification of intramembranous bones; ossification begins at one or more ossification centers and spreads from these locations. Knowledge about the ontogeny of ossification of the pterygoid is available for *Lacerta* (Riep-pel, 1992) and will be used as a model here.

The pterygoid is one of the first bones to ossify in the skull of *Lacerta* (and in *Podarcis;* Rieppel, 1987). Ossification begins in the middle portion of the bone close to the processus ascendens of the palatoquadrate cartilage, near the future basal articulation between basisphenoid pro-cess, the processus ascendens ossified as the epipterygoid, and possibly the pterygoid itself (Romer, 1956, 63–64). The pterygoid then ossifies quickly caudad to the quadrate portion of the palatoquadrate cartilage and forms an anterior contact with the ossifying palatine (Rieppel, 1992).

The major relations of the pterygoid therefore ossify very early in
Lacerta—namely, the contact with the basal articulation in the center of
the bone, with the quadrate via the quadrate flange and anterior and laterally with the palate surface. These contacts are correspondingly plesiomorphic and present in a wide range of reptiles and are thought to be
the primitive conditions for amniotes (Romer, 1956).

Unfortunately, no information exists on skull ossification patterns in
nothosaur-grade sauropterygians. However, working from the supposi-
tion that ossification of the pterygoid proceeded in the same way in these
taxa as in other amniotes, the plesiomorphic contacts and relations of the
pterygoid would be expected to ossify first, followed later by the conditions of the hypertrophied pterygoid that characterize nothosaur-grade
sauropterygians: the closing of anterior and posterior interpterygoid
vacuities and resulting obscuration of cultriform process and basicranium, obliteration of the basal articulation, and closing of the cranioquadrate passage. The closing of the cranioquadrate passage requires the
presence of a foramen in the pterygoid for the passage of the internal
carotid artery (present in *Simosaurus* and *Nothosaurus;* Rieppel, 1994),
and the presence of bone between the occiput and basisphenoid requires
a groove for the passage of the same artery. All these features might be
expected to ossify later in ontogeny in accordance with von Baer's first
and second laws of development (see discussion by Gould, 1977, 56). The
ossification of the pterygoid around the braincase in sauropterygians may
be compared with water filling a bathtub; the first water in the tub runs to
the drain, the lowest point in the tub and representing the plesiomorphic
contacts of the pterygoid in the analogy. As water continues to fill the tub,
fenestrae in the skull are closed and structures around the braincase are
enclosed: basal articulation, cultriform process, and internal carotid artery. A reduction of ossification in the plesiosaur pterygoid linked to an
aquatic lifestyle would simply reverse the above process. The more derived conditions of the pterygoid would fail to ossify, leaving only the
more plesiomorphic conditions in the adult. The pterygoid of adult plesiosaurs therefore would be neotenous relative to more basal sauropterygians (Gould, 1977).

The above interpretation is strongly supported by a phylogenetic
argument. The reversals in the characters discussed above are plotted
onto a cladogram of sauropterygian relationships in figure 12.7. If plesiosaurs had an independent derivation from basal diapsids, all these characters should reverse on the branch directly below the Plesiosauria.
However, the changes are spread out over four branches, establishing a
pterygoid transformation series between plesiosaurs and nothosaur-grade

sauropterygians. [Transformation series of this type also exist for several aspects of the postcranial skeleton (reviewed by O'Keefe, 2001)]. The assignment of these reversals to symplesiomorphy rather than homoplasy is therefore unlikely, and the conclusion that plesiosaurs are derived from nothosaur-grade sauropterygians seems firm. There is no clear pattern in the sequence of reversals and no ontogenetic series available for comparison with the phylogeny.

The hypothesis that the basicranium in plesiosaurs reverts to a more plesiomorphic condition due to neoteny in the ossification of the pterygoid is testable in two ways. The first is the identification and study of ontogenetic series in stem-group sauropterygians in an attempt to gather data on the ossification history of the pterygoid. Various pachypleurosaur taxa are represented by good growth series (Rieppel, 1989; Sander, 1989; O'Keefe et al., 1999), including at least one embryo. Given the early ossification of the pterygoid in *Lacerta,* however, the available material may be too late in ontogeny to offer much help. A second avenue of research would be to thin section or computerized axial tomography (CAT) scan a well-preserved nothosauroid skull. The presence and configuration of features such as the basal articulation and parasphenoid might be discernible in this way. Last, the six character state changes listed above can be accounted for by one hypothesis—reduction in ossification of the pterygoid—and so might bear consideration as a single character.

Acknowledgments

Thanks are due to R. Wild, R. Schoch, D. Norman, and P. Manning for access to specimens in their care and institutional assistance of all kinds. M. Maisch, O. Rieppel, and an anonymous reviewer read and improved an earlier version of this manuscript. This study was funded in part by the University of Chicago Hinds Fund. This paper is dedicated to Jim Hopson, teacher, mentor, and friend.

Literature Cited

Andrews, C. W. 1913. *A Descriptive Catalogue of the Marine Reptiles of the Oxford Clay, Part II.* London: British Museum (Natural History).

Brown, D. S. 1981. The English Upper Jurassic Plesiosauroidea (Reptilia) and a review of the phylogeny and classification of the Plesiosauria. *Bulletin of the British Museum of Natural History (Geology)* 35(4):253–347.

Brown, D. S. 1993. A taxonomic reappraisal of the families Elasmosauridae and Cryptoclididae (Reptilia: Plesiosauroidea). *Revue de Paléobiologie, volume spéciale* 7:9–16.

Carpenter, K. 1997. Comparative cranial anatomy of two North American Cretaceous plesiosaurs; pp. 191–216 *in* J. M. Callaway and E. L. Nicholls (eds.), *Ancient Marine Reptiles.* San Diego: Academic Press.

Cruickshank, A. R. I. 1994. A juvenile plesiosaur (Plesiosauria: Reptilia) from the lower Lias (Hettangian: Lower Jurassic) of Lyme Regis, England: a pliosauroid-plesiosauroid intermediate? *Zoological Journal of the Linnean Society* 112:151–178.

de Beer, G. R. 1937. *The Development of the Vertebrate Skull.* Oxford: Oxford University Press.

Evans, M. 1999. A new reconstruction of the skull of the Callovian elasmosaurid plesiosaur *Muraenosaurus leedsii* Seeley. *Mercian Geologist* 14(4): 191–196.

Gould, S. J. 1977. *Ontogeny and Phylogeny.* Cambridge, MA: Harvard University Press.

Hopson, J. A. 1979. Paleoneurology; pp. 39–146 *in* C. Gans, R. G. Northcutt, and P. Ulinski (eds.), *Biology of the Reptilia. Vol. 9. Neurology.* London: Academic Press.

O'Keefe, F. R. 2001. A cladistic analysis and taxonomic revision of the Plesiosauria (Reptilia: Sauropterygia). *Acta Zoologica Fennica* 213:1–63.

———. 2004. Preliminary description and phylogenetic position of a new plesiosaur (Reptilia: Sauropterygia) from the Toarcian of Holzmaden, Germany. *Journal of Paleontology* 78(5):973–988.

O'Keefe, F. R., O. Rieppel, and P. M. Sander. 1999. Shape disassociation and inferred heterochrony in a clade of pachypleurosaurs (Reptilia, Sauropterygia). *Paleobiology* 25(4):504–517.

Owen, R. 1840. *Report on British Fossil Reptiles, Part 1.* Report of the ninth meeting of the British Association for the Advancement of Science, Birmingham, 1839, London; pp. 42–126.

Persson, P. O. 1963. A revision of the classification of the Plesiosauria with a synopsis of the stratigraphical and geological distribution of the group. *Lunds Universites Årsskrift N. F. Ard. 2,* 59(1):1–57.

Price, L. I. 1935. Notes on the braincase of *Captorhinus. Proceedings of the Boston Society of Natural History* 40(7):377–386.

Reisz, R. R. 1981. A diapsid reptile from the Pennsylvanian of Kansas. *Special Publication of the Museum of Natural History, University of Kansas* 7:1–74.

Reisz, R. R., D. S. Berman, and D. Scott. 1984. The anatomy and relationships of the lower Permian reptile *Araeoscelis. Journal of Vertebrate Paleontology* 4(1):57–67.

Rieppel, O. 1987. The development of the trigeminal jaw adductor musculature and associated skull elements in the lizard *Podarcis sicula* (Rafinesque). *Journal of Zoology, London* 212:131–150.

———. 1989. A new pachypleurosaur (Reptilia: Sauropterygia) from the Middle Triassic of Monte San Giorgio, Switzerland. *Philosophical Transactions of the Royal Society of London B* 323:1–73.

———. 1992. Studies on skeleton formation in reptiles, III. Patterns of ossification in the skeleton of *Lacerta vivipara* Jacquin (Reptilia, Squamata). *Fieldiana (Zoology) new series* 68:1–25.

———. 1994. The braincases of *Simosaurus* and *Nothosaurus:* monophyly of the Nothosauridae (Reptilia: Sauropterygia). *Journal of Vertebrate Paleontology* 14(1):9–23.

———. 1997. Revision of the sauropterygian reptile genus *Cymatosaurus* v. Fritsch, 1894, and the relationships of *Germanosaurus* Nopcsa 1928, from the Middle Triassic of Europe. *Fieldiana (Geology), new series* 36:1–38.

———. 2000. *Sauropterygia 1. Encyclopedia of Paleoherpetology, part 12A.* Munich: Verlag Dr. Friedrich, Pfeil.

Rieppel, O. and R. Werneburg. 1998. A new species of the sauropterygian *Cymatosaurus* from the lower Muschelkalk of Thuringia, Germany. *Palaeontology* 41(4):575–589.

Romer, A. S. 1956. *Osteology of the Reptiles.* Chicago: University of Chicago Press.

Sander, P. M. 1989. The pachypleurosaurids (Reptilia: Nothosauria) from the Middle Triassic of Monte San Giorgio (Switzerland), with the description of a new species. *Philosophical Transactions of the Royal Society of London B* 325:561–670.

Storrs, G. W. 1991. Anatomy and relationships of *Corosaurus alcovensis* (Diapsida: Sauropterygia) and the Triassic Alcova Limestone of Wyoming. *Bulletin of the Peabody Museum of Natural History* 44:1–151.

———. 1993. Function and phylogeny in sauropterygian (Diapsida) evolution. *American Journal of Science* 293A:63–90.

———. 1997. Clarification of the genus *Plesiosaurus;* pp. 145–190 *in* J. M. Callaway and E. L. Nicholls (eds.), *Ancient Marine Reptiles.* San Diego: Academic Press.

Storrs, G. W. and M. A. Taylor. 1996. Cranial anatomy of a new plesiosaur genus from the lowermost Lias (Rhaetian/Hettangian) of Street, Somerset, England. *Journal of Vertebrate Paleontology* 16(3):403–420.

Vaughn, P. P. 1955. The Permian reptile *Araeoscelis* restudied. *Bulletin of the Museum of Comparative Zoology* 113(5):305–467.

Williston, S. W. 1903. North American plesiosaurs, Part 1. *Field Columbian Museum Publication (Geology)* 73(2):1–77.

Richard W. Blob

13 Scaling of the Hind Limb Skeleton in Cynognathian Cynodonts: Implications for Ontogeny and the Evolution of Mammalian Endothermy

Introduction

Much of the fascination of paleontology stems from the glimpses that fossils provide into the life of the past and into the evolution of living organisms through time. Fossils often can reveal the steps through which major transitions in morphology or function unfolded within phylogenetic lineages. One of the best-documented transitions in the vertebrate fossil record is the "reptile-to-mammal" transition that occurred within the synapsid lineage (Hopson, 1987, 1991, 1994). The evolution of mammals from "mammal-like reptile" ancestors was the result of a gradual accumulation of many changes among an extensive array of intermediate forms (Hopson, 1994; Sidor & Hopson, 1998; Hopson & Kitching, 2001; fig. 13.1). Modern mammals can be clearly distinguished from other clades of extant tetrapods by numerous morphological and functional features. However, examination of the gradual tempo of morphological character acquisition in the fossil record of synapsids (e.g., Sidor & Hopson, 1998) suggests that many of the functional features commonly viewed as characterizing modern mammals actually may have originated among ancestral mammal-like reptiles. Evaluating when in the evolutionary history of synapsids that various "mammalian" aspects of function first appeared represents a major challenge, and opportunity, for functional studies of fossil synapsids.

One primary focus among functional analyses of fossil synapsids has been the origin of "mammalian" endothermy. A wide range of studies, using several different lines of evidence, have yielded a general consensus that internal heat production and high metabolic rates first evolved in synapsid taxa prior to the origin of mammals (Brink, 1956; Geist, 1972; Hopson, 1973; McNab, 1978; Bennett & Ruben, 1986; Hillenius, 1992, 1994; Ruben, 1995). However, many of the analyses applied to evaluate the metabolic status of therapsids have been found to have serious shortcomings. For instance, Bakker (1980) cited two types of paleoecological data as evidence of endothermy among therapsids. First, the paleogeographic distribution of fossil therapsids includes regions such as southern Africa, which, in the Late Permian, would have been located at high southern latitudes and, perhaps, subject to cold climates. Thus, endothermy might have been required of the dominant members of the fauna in this region. Second, a decrease in the ratio of predator biomass to prey biomass between assemblages of basal "pelycosaur" synapsids and assemblages of more derived therapsids, from a level similar to that of modern communities of ectothermic reptiles to a level closer to that of modern communities of endothermic mammals, might suggest that predatory therapsids required a standing crop of prey similar to that needed by mammals because both were endothermic and had similar energy requirements. But objections have been raised to both of these ecological arguments. For example, paleoclimatic data actually indicate a warm, equable climate in southern Africa during the Late Permian (Parrish et al., 1986; Bennett & Ruben, 1986). Furthermore, several studies have cast doubt on the reliability of biomass and community structure estimates for assemblages of fossil vertebrates (Farlow, 1980; Hotton, 1980).

   Other studies attempting to evaluate the metabolic status of extinct synapsids have focused on the identification of anatomical features in fossil taxa that are thought to correlate with endothermy among living animals. However, complications also have arisen among many proposed anatomical indicators of endothermy. For example, the identification of a fibrolamellar histology in the limb bones of a variety of therapsid taxa has been cited as evidence that these animals were endothermic (de Ricqlès, 1976). Fibrolamellar bone is a well-vascularized tissue that is deposited by numerous primary osteons, producing a matrix of compact bone that

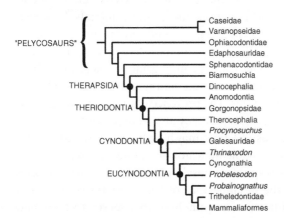

Figure 13.1. Cladogram illustrating phylogenetic relationships among synapsid lineages, based on the analyses of Sidor and Hopson (1998) and Hopson and Kitching (2001), with the clade including Caseidae and Varanopseidae based on analyses by Sidor (1996). Dinocephalia equals (Anteosauridae + Estemmenosuchidae) of Sidor and Hopson (1998).

appears woven or fibrous with few lines of arrested growth (de Ricqlès et al., 1991; Ruben, 1995). Fibrolamellar bone is generally regarded as an indicator of rapid, sustained growth (de Ricqlès, 1976; Bennett & Ruben, 1986; de Ricqlès et al., 1991; Chinsamy & Rubidge, 1993), which, it has been argued, is possible only among taxa with high, endothermic metabolic rates (de Ricqlès, 1976). In contrast, the presence of poorly vascularized lamellar-zonal bone, which is deposited by a small number of primary osteons and possesses a layered appearance in which lines of arrested growth are common, has been thought to indicate slower growth rates more typical of ectotherms (de Ricqlès, 1976; Bennett & Ruben, 1986; de Ricqlès et al., 1991; Ruben, 1995). However, the correlation between metabolic rate and bone histology is not straightforward because living endotherms and ectotherms do not consistently exhibit distinct histological patterns. Fibrolamellar bone is absent in many small, rapidly growing endotherms (Bennett & Ruben, 1986; Ruben, 1995) but present in some species of ectothermic crocodilians (Reid, 1984, 1997; Chinsamy, 1995) as well as some Paleozoic amphibians (Enlow, 1969). As a result, the utility of bone histology as an indicator of the metabolic rates of fossil taxa remains uncertain.

In light of the difficulties encountered among many of the proposed indicators of metabolic rates in extinct synapsids, Bennett and Ruben (1986) suggested three criteria for the identification of a valid indicator of endothermy in fossil taxa. First, the feature should have a clear and logical association with high metabolic rates. Second, the feature should not be explainable through factors unrelated to high metabolic rates. Finally, the feature should not be observed in ectotherms. On the basis of these three criteria, Bennett and Ruben (1986) concluded that many traits previously viewed as suggesting endothermy among fossil therapsids were inconclusive or untenable indicators of metabolic rate in these animals. However, two indicators still stood out as viable signs of endothermy in extinct therapsids. First, monotremes and therian mammals share a large number of thermoregulatory and metabolic traits, including hair, sweat glands, a high capacity of the blood for carrying oxygen, and metabolic rates substantially greater than those of reptiles. The presence of these features in both monotremes and therians suggests that they also were present in the common ancestor of both lineages and, thus, may have been inherited from derived therapsids. The second viable indicator of endothermy among therapsids was the presence of respiratory nasal turbinals in several fossil taxa. Respiratory turbinals in the nasal cavities of living mammals are bones or cartilages covered with a moist epithelium that serves to filter and humidify air during inspiration and reduce water loss during expiration, functions that are especially important during the rapid ventilation typical of endothermic taxa (Bennett & Ruben, 1986; Hillenius, 1992, 1994). The presence of respiratory nasal turbinals in later cynodont therapsids such as *Thrinaxodon* and *Massetognathus* suggests that endothermy likely had been attained among more derived cynodonts (Bennett & Ruben, 1986); moreover the identification of respiratory turbinals in the therocephalian therapsid *Glanosuchus* potentially indicates a more basal origin for synapsid endothermy (Hillenius, 1994).

Subsequent to Bennett and Ruben's (1986) analysis, Carrier and Leon (1990) proposed an additional anatomical feature that might provide insight into the metabolic rates of extinct taxa. In many vertebrates, some form of locomotor activity begins very soon after hatching or birth. However, juvenile skeletal tissues are often intrinsically weaker than those of adults, with lower mineral contents, material stiffnesses, and resistance to stress (Bonfield & Clark, 1973; Currey & Butler, 1975; Torzilli et al., 1982). Among rapidly growing endothermic taxa such as jackrabbits (Carrier, 1983) and gulls (Carrier & Leon, 1990), changes in limb bone shape appear to compensate for the weakness of juvenile bone tissue, allowing the limbs to function effectively in locomotion throughout ontogeny. Specifically, in rapidly growing endothermic taxa, juvenile limb bones used in locomotion are relatively more robust than those of adults. This contrast is manifested in ontogenetic patterns of limb bone scaling. Two indicators of limb bone robusticity are midshaft diameter and second moment of area, a measure of the distribution of material about a bone cross section that indicates the resistance of the bone to bending in a particular direction (Wainwright et al., 1976). Through ontogeny in endotherms, diameters and second moments of area of weight-bearing limb bones tend to exhibit negatively allometric scaling with respect to body mass (fig. 13.2). In other words, as fast-growing endotherms increase in size, growth in bone diameter and second moment of area are slower than growth in mass, producing relatively thinner limb bones as the animals age. In contrast, among typically slower-growing ectotherms such as alligators (Dodson, 1975) and lizards (Peterson & Zernicke, 1986, 1987), the mechanical properties of bone change little during ontogeny, and limb bone thickness scales isometrically or with positive allometry relative to body mass (fig. 13.2). Therefore, examining the ontogenetic allometry of limb bone robusticity in an extinct species could indicate whether its growth rate corresponded more closely to modern endothermic or ectothermic species, thereby suggesting the likely metabolic status of the fossil taxon (Carrier & Leon, 1990).

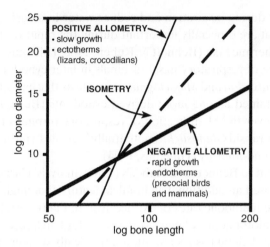

Figure 13.2. Schematic graphical depiction of contrasting general patterns of ontogenetic limb bone scaling between ectotherms and endotherms, as proposed by Carrier and Leon (1990). The dashed line depicts isometric growth of bone diameter relative to bone length, whereas the thin solid line depicts faster growth of bone diameter relative to bone length (positive allometry of diameter relative to length) and the thick solid line depicts slower growth of bone diameter relative to bone length (negative allometry of diameter relative to length). Carrier and Leon (1990) suggested that positively allometric increases in bone robusticity during growth would be expected to characterize slowly growing ectothermic taxa, but negatively allometric increases in bone robusticity during growth would be expected to characterize quickly growing endothermic taxa.

In the context of Carrier and Leon's (1990) proposal, ontogenetic scaling data from the limb bones of fossil mammal-like reptiles could provide several insights into the functional evolution of synapsids. First, such data would facilitate evaluations of the evolution of tetrapod limb bone growth patterns, addressing the following question: if mammals exhibit different patterns of limb bone growth than reptiles, when in the ancestry of mammals did these differences first emerge? In addition, ontogenetic scaling data from the limb bones of mammal-like reptiles could also be useful for evaluating the correlation between scaling patterns and metabolic rate. If the correlation between ontogenetic scaling patterns and metabolic rate is accepted, then an examination of ontogenetic limb bone scaling among fossil synapsids would provide an additional avenue through which the possibility of endothermy could be evaluated in these taxa. However, if therapsids display reptilian ontogenetic scaling patterns, but other evidence suggests that these animals were endothermic, then the utility of scaling patterns as an indicator of metabolic status might be called into question.

In this study, measurements were collected from a broad size range of limb bones from fossil cynognathian cynodont therapsids, a lineage of basal eucynodonts (*sensu* Sidor & Hopson, 1998; fig. 13.1) that lived in southern Africa during the Early and Middle Triassic (Scythian to Early Anisian: Kitching, 1995). Data were collected from these specimens for analyses of ontogenetic limb bone scaling with two primary goals: to help evaluate the evolution of limb bone growth patterns in tetrapods, and to examine the utility of ontogenetic scaling data as an indicator of metabolic status.

Material and Methods

Specimens

Specimens of cynognathian femora and tibiae from two collections were examined and measured. The first collection, from the Bernard Price Institute (Johannesburg, South Africa), included four femora and five tibiae among a wide range of postcranial specimens that were excavated from a single bone bed and collectively assigned the catalogue designation BP/1/1675. The second collection, from the National Museum (Bloemfontein, South Africa), included twenty-seven femora catalogued as part of NMB.QR 1206 and thirteen tibiae (as well as an additional, twenty-eighth femur) catalogued as NMB.QR 1208. Tibia specimens spanned a nearly twofold range of lengths, and femur specimens spanned more than a threefold range of lengths (fig. 13.3; appendix 13.1). In both collections, specimens were identified as *Cynognathus/Diademodon,* presumably because the bones were disarticulated when they were collected and diagnostic cranial material could not be associated with individual elements. *Cynognathus* and *Diademodon* both belong to the cynodont clade Cynognathia (*sensu* Hopson & Kitching, 2001) and, as Jenkins (1971) has noted, the hind limb skeletons of these taxa are morphologically indistinguishable. The results of analyses performed in this study, therefore, reflect a general cynognathian pattern of limb bone scaling during growth, rather than a species-specific pattern. After my examination of these specimens, Botha and Chinsamy (2000) identified histologic features that can be used to distinguish the limb elements of *Cynognathus* and *Diademodon.* Future application of their method to the specimens examined in this analysis could allow more specific taxonomic identifications and refine insights into the biology of these animals. Nonetheless, the present study provides a starting point for the evaluation of changes in limb bone geometry during growth among cynodonts and, thereby,

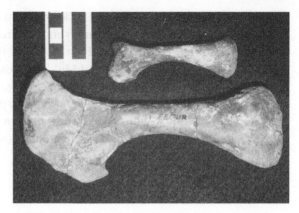

Figure 13.3. Photograph of small and large femora from cynognathian cynodonts (*Cynognathus/Diademodon*, BP/1/1675) in dorsal view. Proximal is to the left (femoral head), distal is to the right (knee), anterior is toward the top of the page, and posterior is toward the bottom of the page. Increments on scale equal 1 cm.

provides a starting point for evaluating the evolutionary transition from reptilian to mammalian patterns of postnatal limb bone growth.

Measurements

Many of the specimens in these collections had experienced postfossilization breaks transverse to the long axis of the bone and required reassembly. Three measurements were taken from each reassembled specimen using digital calipers (sensitive to 0.01 mm): bone length (L), midshaft diameter in the direction of knee flexion and extension (FED), and midshaft diameter perpendicular to the direction of flexion and extension (PD). However, in many of the bones (seventeen femora, four tibiae), one of the transverse breaks had occurred within ± 10 percent of the midshaft, exposing the cross section of the specimen and allowing the measurement of additional geometric parameters. For these specimens, camera lucida tracings of the cross sections were made either from photographs of the specimens (fig. 13.4) or from enlargements of tracings made directly from the specimens (Blob, 2001). Direct tracings were made by placing transparent adhesive tape on the cross section and tracing the endosteal and periosteal edges on the tape with needle-pointed steel forceps. The etched tape then was lifted from the specimen and placed in a cardstock notebook with the correct anatomical orientation noted. A digitizing tablet was used to enter the geometric data from these outline tracings of the bone cross-sections into a computer. Custom programs written in QuickBasic (provided by A. Biewener) or Matlab (provided by W. Corning) were then used to calculate the following four variables for each midshaft cross section: cross-sectional area (A), second moment of area in the direction of knee flexion and extension (IFE), second moment

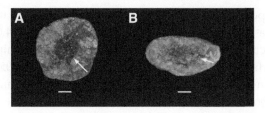

Figure 13.4. Midshaft cross sections of (A) cynodont femur and (B) cynodont tibia, illustrating internal geometry. The dark area in the center of each cross section is the medullary cavity; the margin of this dark central region (indicated by white arrow) represents the endosteal surface. This margin was traced and digitized, along with the periosteal surface, to calculate cross-sectional geometric variables for fossil specimens. Scale = 2 mm.

of area in the direction perpendicular to knee flexion and extension (*IP*), and polar moment of area (*J*).

Elements from both the left and right sides were measured to provide adequate sample sizes for the analyses. Thus, it is possible that some individual animals may be represented by two elements in the study. However, the differences in length between most left and right elements suggest that this is unlikely for most of the specimens. Even by the most conservative estimate (i.e., counting elements from only one side), femora from twenty-three different individuals and tibiae from thirteen different individuals are represented in these analyses.

Analyses

All data were logarithmically transformed (base 10) before analysis. Scaling relationships for limb bone dimensions were estimated by Model II reduced major axis (RMA) regression. RMA is the most appropriate method of regression for evaluating structural relationships between variables when both are subject to error (McArdle, 1988; LaBarbera, 1989; Sokal & Rohlf, 1995). RMA slopes equal ordinary least-squares (OLS) regression slopes divided by R, the correlation coefficient. Regressions were calculated for each diameter (*FED* and *PD*) of each bone (femur and tibia) on the length of the appropriate bone. In addition, for the femur, regressions were calculated for each of the four cross-sectional geometric variables *A, IFE, IP,* and *J* on bone length, for a total of eight linear regressions based on the original log-transformed measurements. Regressions for cross-sectional measurements on tibia length were not calculated because of the small sample of tibia midshaft cross sections available. Evaluation of allometry in these scaling relationships was based on whether 95 percent confidence intervals (CIs) for the RMA slopes included slope values expected for isometric scaling (1.0 for length-diameter relationships, 2.0 for length-area relationships, 4.0 for length-second moment of area and length-polar moment relationships). If 95 percent CIs overlapped values predicted for isometry by >0.005, then slopes were judged not to deviate significantly from geometric similarity (Blob, 2000). CIs about RMA slopes were calculated as recommended by Jolicoeur and Mosimann (1968) using custom computer routines (written by N. Espinoza and M. LaBarbera) and account for the asymmetric distribution of the RMA estimator of the slope (McArdle, 1988).

It should be noted that the negative ontogenetic allometry of limb bone diameters and second moments of area observed by Carrier (1983) and Carrier and Leon (1990) in endothermic taxa was relative to body mass rather than bone length. Therefore, to compare scaling patterns between these taxa and extinct fossil synapsids, two options are available. The first option is to estimate body masses for the fossil specimens and calculate mass-diameter and mass-second moment of area regressions that could be directly compared with Carrier's analyses (Carrier, 1983; Carrier & Leon, 1990). However, this approach is impractical for two reasons. First, available methods of estimating body masses for nonmammalian synapsids rely on correlations between mass and anatomical measurements of features other than the limbs (Blob, 2001). Use of these methods would require much more complete, associated fossil specimens than are available over the broad size range required for an analysis of ontogenetic allometry. Second, although body masses for a wide range of fossil taxa are frequently estimated from limb bone dimensions (Anderson et al., 1985; Damuth & MacFadden, 1990) such approaches would not be appropriate for a study of limb bone scaling as they would introduce analytical circularity—limb bone dimensions would be scaled against masses calculated from predictive regressions for variables closely correlated with the measurements of interest (Blob, 1998).

In light of these difficulties, I used an alternative option to compare the ontogenetic scaling patterns of cynognathian cynodonts with the predictions of Carrier and Leon's (1990) model: I retained the scaling relationships of measurements relative to bone length calculated for the fossil taxa and, instead, recalculated the predictions of Carrier and Leon's model for regressions of variables on bone length, rather than body mass, in extant taxa. The strategy I used to recalculate scaling relationships relative to bone length for extant taxa depended on the data available. For gulls, D. Carrier kindly supplied original data from the Carrier and Leon (1990) study, which I then used to directly calculate RMA slopes and 95 percent CIs for relationships of bone diameter and second moment of area on length in this species (following methods outlined for fossil taxa). However, complete original data from Carrier's (1983) jackrabbit study were not available at the time of the present analysis. Instead, parameters from the OLS scaling equations published for jackrabbits by Carrier (1983) were used to evaluate RMA scaling relationships (with 95 percent CIs) for measurements relative to bone length (see Appendix 13.2 for details of these calculations). For these two endothermic taxa, four of five recalculated regressions of second moment of area on bone length, as well as four of six regressions of diameter on length, show negative allometry (table 13.1). Three recalculated regressions for gulls show isometry; however, two of these cases also exhibited isometry in the original analyses from which the data were derived (diameter relative to body mass for the tibiotarsus and tarsometatarsus; Carrier and Leon, 1990, 380), and the third showed isometry only marginally (95 percent CI overlapped isometric slope by 0.008). None of the measurements for either gulls or jackrabbits exhibits positive allometry with respect to bone length. Thus, with a slight clarification, Carrier and Leon's (1990) predictions for ontogenetic scaling differences between endotherms and ectotherms remain the same when recast in terms of bone length. Although both endotherms and ectotherms may exhibit isometry of some measures of limb bone robusticity, negative allometry of diameter and second moment of area relative to length typify endotherms, whereas positive allometry of diameter and second moment of area relative to length typify ectotherms. The analysis of limb bone scaling patterns in fossil cynognathian cynodonts, and their implications for the evolution of limb bone growth patterns and thermoregulation, therefore can proceed in the same context outlined by Carrier and Leon's (1990) original proposed correlations between metabolic function and limb bone ontogeny.

Results

Measurements and calculations of geometric variables for the hind limb bones of cynognathian cynodonts are reported in appendix 13.1. The results of each RMA scaling regression are presented in table 13.2, and representative plots of scaling relationships are illustrated in figure 13.5. Correlation coefficients for the regressions are generally high, with all R^2 values > 0.86. All eight scaling relationships exhibit either isometry or positive allometry, based on whether CIs for the slope overlap values predicted for geometric similarity. Three of the four length-diameter relationships (both for the tibia and *PD* for the femur) show positive

Table 13.1. Exponents of scaling relationships for rabbit (*Lepus californicus*: Carrier, 1983) and gull (*Larus californicus*: Carrier & Leon, 1990) limb bone dimensions relative to bone length, recalculated from data and analyses in original studies of the scaling of bone dimensions relative to body mass (see text for details of these calculations)

Species	Bone	Recalculated regression parameters						Allometry	
		D on *L*			*I* on *L*				
		N	R^2	RMA slope (95% CI)	*N*	R^2	RMA slope (95% CI)	*D* on *L*	*I* on *L*
Rabbit	Femur	49	0.945	0.767 (0.716–0.822)	*	*	2.600 (2.277–2.977)	–	–
	Tibia	51	0.939	0.797 (0.742–0.855)	NR	NR	NR	–	NR
	MT	20	0.925	0.735 (0.642–0.842)	*	*	2.888 (2.371–3.516)	–	–
Gull	Femur	23	0.888	0.742 (0.638–0.863)	22	0.933	3.553 (3.150–4.008)	–	0
	Tibiotarsus	23	0.790	0.878 (0.714–1.079)	22	0.901	3.196 (2.761–3.700)	0	–
	TMT	22	0.838	0.926 (0.769–1.116)	21	0.729	2.851 (2.226–3.651)	0	–

Note: Exponents are reported as RMA regression slopes, with asymmetric 95 percent CIs evaluated following recommendations of Jolicoeur and Mosimann (1968). Negative allometry is indicated by a "–" in the column titled "Allometry"; isometry is indicated by a "0" in this column. Four of five relationships between second moment of area and bone length, as well as four of six relationships between diameter and length, show negative allometry (95% CIs exclude slope values predicted for isometry by ≤0.005, see text). No case exhibits positive allometry, and only the relationships of diameter relative to length in the gull tibiotarsus and tarsometatarsus exhibit clear isometry; however, these bones also exhibited isometry in the original study of diameter relative to body mass from which data were derived (Carrier & Leon, 1990). Abbreviations: *L*, bone length; *D*, bone diameter; *I*, bone second moment of area; MT, metatarsal; TMT, tarsometatarsus; NR, not reported. * Relationship calculated from slopes of regressions with different sample sizes; see Appendix 2 for explanation.

Table 13.2. Scaling exponents (with asymmetric 95 percent CIs), evaluated as RMA regression slopes, for limb bone dimensions relative to bone length in fossil cynognathian cynodonts

Bone	X	Y	N	R^2	RMA slope (95% CI)	Allometry
Femur	Length	FED	34	0.925	0.986 (0.893 – 1.088)	0
	Length	PD	34	0.890	1.250 (1.109 – 1.408)	+
	Length	A	17	0.865	2.250 (1.841 – 2.751)	0
	Length	IFE	17	0.887	4.105 (3.415 – 4.934)	0
	Length	IP	17	0.888	4.826 (4.019 – 5.796)	+
	Length	J	17	0.890	4.465 (3.724 – 5.353)	0
Tibia	Length	FED	18	0.935	1.342 (1.173 – 1.536)	+
	Length	PD	18	0.917	1.341 (1.152 – 1.561)	+

Note: Positive allometry is indicated by a "+" in the column titled "Allometry"; isometry is indicated by a "0" in this column. Abbreviations: *X*, regression abscissa; *Y*, regression ordinate; *N*, sample size; *FED*, bone midshaft diameter in the direction of knee flexion and extension; *PD*, bone midshaft diameter perpendicular to *FED*; *A*, bone midshaft cross-sectional area; *IFE*, bone midshaft second moment of area in the direction of knee flexion and extension; *IP*, bone midshaft second moment of area in the direction perpendicular to *IFE*; *J*, bone midshaft polar moment of area.

allometry of bone length on diameter. Only one of the four cross-sectional measurements, the second moment of area perpendicular to knee flexion and extension (*IP*), exhibits definitive positive allometry. However, although the slope CIs overlap the isometric slope for the other three cross-sectional variables, the actual calculated slope is greater than the isometric slope for each of these variables. No scaling relationship displays negative allometry.

Discussion

Implications for the Evolution of Ontogenetic Shape Change in Tetrapod Limb Bones

The cynognathian cynodonts examined in this study exhibit isometry or positive allometry of limb bone robusticity relative to length during postnatal growth. The prevalence of positive allometry in these relationships contrasts with patterns thought to characterize mammals, in which negative allometry of limb bone robusticity has been identified during growth (Carrier, 1983; Carrier & Leon, 1990). Thus, despite the considerable increases in "mammalness" that had occurred among synapsids by the evolution of cynognathian cynodonts (Sidor & Hopson, 1998), including changes in several features of the locomotor skeleton (Jenkins,

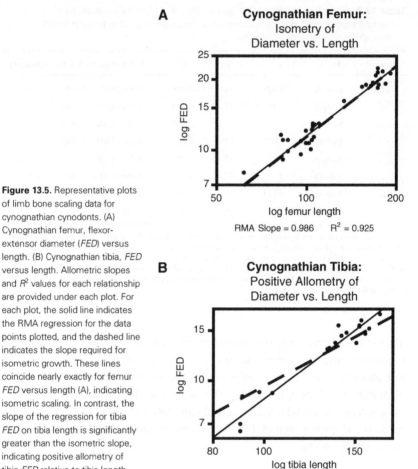

Figure 13.5. Representative plots of limb bone scaling data for cynognathian cynodonts. (A) Cynognathian femur, flexor-extensor diameter (*FED*) versus length. (B) Cynognathian tibia, *FED* versus length. Allometric slopes and *R²* values for each relationship are provided under each plot. For each plot, the solid line indicates the RMA regression for the data points plotted, and the dashed line indicates the slope required for isometric growth. These lines coincide nearly exactly for femur *FED* versus length (A), indicating isometric scaling. In contrast, the slope of the regression for tibia *FED* on tibia length is significantly greater than the isometric slope, indicating positive allometry of tibia *FED* relative to tibia length among cynognathian cynodonts.

A

Cynognathian Femur:
Isometry of
Diameter vs. Length

RMA Slope = 0.986 R^2 = 0.925

B

Cynognathian Tibia:
Positive Allometry of
Diameter vs. Length

RMA Slope = 1.342 R^2 = 0.935

1971; Kemp, 1980, 1985), these animals appear to display the same patterns of limb bone growth and shape change found among reptiles.

Ontogenetic limb bone shape change has been analyzed in only a few species. Thus, it may be premature to describe negative allometry of limb bone robusticity as a mammalian pattern, because the potential for positive allometry of limb bone robusticity among mammalian lineages has not been fully explored. Characterization of positive allometry in limb bone robusticity as reptilian also may be premature for similar reasons. However, mounting evidence from recent studies suggests that positive and negative allometry of limb bone robusticity are distributed among tetrapod taxa much as Carrier and Leon (1990) predicted. Preliminary RMA analyses show significant positive allometry of midshaft diameter on length for the femur and tibia in growth series of the lizards *Iguana iguana* and *Varanus bengalensis* (R. W. Blob, unpublished data). In contrast, RMA analyses of data (provided by R. German) from the study of Maunz and German (1997) on the marsupial *Didelphis virginiana* show significant negative allometry of bone diameter relative to length. Among seven *D. virginiana* in which the limb bones were measured on more than sixty separate days between the age of forty days and attaining adult size (>3 kg, >300 days of age), average RMA coefficients (and 95 percent CIs) for diameter relative to length were 0.912 (0.884–0.941) for the femur and 0.955 (0.924–0.987) for the tibia. Moreover, although Lammers and German (2002) report isometric growth of bone diameter relative to length in the femur and tibia of another marsupial, *Monodelphis domestica,* they found negative allometry of length on diameter in these bones among three other mammals they examined (the rabbit *Oryctolagus cuniculus* and the rodents *Chinchilla laniger* and *Rattus norvegicus*). Thus, no reptile examined to date has exhibited negative ontogenetic allometry of hind limb bone robusticity, and no mammal examined to date has exhibited positive ontogenetic allometry of hind limb bone robusticity. The presence of positive ontogenetic allometry of limb bone robusticity among cynognathian eucynodonts thus suggests that, despite the close evolutionary relationship of these animals to mammals, mammalian patterns of limb bone growth had not yet appeared at this point in the synapsid lineage.

Further exploration of the phylogenetic distribution of ontogenetic scaling patterns is needed before their evolutionary history can be traced with confidence. However, in some respects, the persistence of a reptilian pattern of bone shape change during the growth of nonmammalian cynodonts would not be surprising. For example, Botha and Chinsamy (2000) report histological evidence that *Diademodon* exhibited cyclical cessations of limb bone growth similar to those seen among modern reptiles. If the tempo of bone growth were similar in cynognathians and reptiles, similar patterns of shape change in these groups would not be unexpected.

Implications for Evaluating the Evolution of Metabolic Rates

As reviewed in the introduction, a wide range of studies using several different lines of evidence have concluded that endothermy likely evolved within the synapsid lineage at least by the appearance of cynodont therapsids, if not earlier. Although the merits of some specific indicators of endothermy have been challenged, a few (such as nasal respiratory turbinates) have survived repeated scrutiny and remain convincing evidence that cynodonts, and perhaps some earlier synapsids, were endothermic. However, the results of ontogenetic analyses of limb bone scaling among cynognathian cynodonts appear to stand at odds with this conclusion. Carrier and Leon (1990) proposed that endothermic species should exhibit negative allometry of limb bone robusticity during ontogeny, but cynognathian cynodonts display positively allometric scaling of several measurements reflecting limb bone robusticity. Two potential conclusions might be drawn from these conflicting results: the scaling data suggest that either cynognathian cynodonts were ectothermic, or that the correlation between ontogenetic scaling pattern and metabolic rate is not as straightforward as previously was believed. It is difficult to be certain about which of these two conclusions is more likely to be correct. There are reasons to be reluctant to overturn earlier conclusions in favor of metabolic evaluations based on ontogenetic scaling data. For example, the small sample of species from which the association between scaling patterns and metabolic rates is derived leaves open the possibility that the association is not as clear as currently available data indicate. Carrier and Leon (1990) originally proposed this association on the basis of data from two mammalian, two avian, and three reptilian species, and subsequent analyses (cited earlier in this Discussion) have tested this pattern with only two additional reptilian and five additional mammalian species (Maunz & German, 1997; Lammers & German, 2002; Blob, unpublished data). In contrast, the correspondence between nasal respiratory turbinates and endothermy is based on a much wider sampling of taxa (Hillenius, 1992, 1994). However, in an evolutionary scenario suggested by an anonymous reviewer of this manuscript, the presence of nasal turbinates might be reconciled with ectothermy in cynodont therapsids, if turbinates originally arose among these animals in relation to an increase in *olfactory* epithelium, and only later acquired a thermoregulatory function. This suggestion is intriguing, although it is not yet clear how this hypothesis could be tested.

If cynognathian cynodonts were endothermic, then the evolutionary change in metabolic physiology that occurred in the synapsid lineage might have preceded any evolutionary change in growth strategy. This conclusion must be regarded as preliminary, as it currently hinges on scaling data from only a single lineage of nonmammalian synapsids. Nonetheless, this sequence of character changes would raise the possibility that endothermy may not have been a correlate but, rather, a prerequisite for a growth strategy in which relatively robust limb bone shapes in juveniles allow locomotor integrity to be maintained despite weak juvenile tissue. This possibility will require considerable further study before it can be evaluated, but it seems likely that data from the fossil record will play a key role in establishing the evolutionary sequence of these functional changes.

Acknowledgments

My work on fossil synapsids has been inspired by Jim Hopson, who has always encouraged me to keep asking new questions. My sincere thanks to Bruce Rubidge (Bernard Price Institute) and Johann Welman (National Museum, Bloemfontein) for access to specimens and for their hospitality during my visit to the collections in their care. Thanks also to Dave Carrier and Rebecca German, who generously provided original data from their previously published papers on ontogenetic limb bone scaling, and to Andy Lammers, who kindly supplied an advance copy of results from ontogenetic limb bone scaling analyses reported in a manuscript in press. This paper benefited substantially from comments by two anonymous reviewers and from discussions about scaling with Nora Espinoza. Nora Espinoza, Mike LaBarbera, Andy Biewener, and Will Corning provided custom software for several analyses; Mike LaBarbera provided camera lucida equipment; and Laura Panko provided a backup camera and spare brain during travel to museum collections. Sally Brock helped tremendously with the preparation of figure 13.4. Travel to South Africa was supported by the National Science Foundation (IBN-9520719). Support during initial analyses was provided by a Society of Vertebrate Paleontology Predoctoral Fellowship. Support during final analyses and manuscript preparation was provided by the Clemson University Department of Biological Sciences.

Literature Cited

Anderson, J. F., A. Hall-Martin, and D. A. Russell. 1985. Long-bone circumference and weight in mammals, birds, and dinosaurs. *Journal of Zoology (London) (A)* 207:53–61.

Bakker, R. T. 1980. Dinosaur heresy-dinosaur renaissance; pp. 351–462 *in* R. D. K. Thomas and E. C. Olson (eds.), *A Cold Look at the Warm-Blooded Dinosaurs.* AAAS Selected Symposium 28. Boulder, CO: Westview Press.

Bennett, A. F. and J. A. Ruben. 1986. The metabolic and thermoregulatory status of therapsids; pp. 207–218 *in* N. Hotton III, P. D. MacLean, J. J. Roth, and E. C. Roth (eds.), *The Ecology and Biology of Mammal-Like Reptiles.* Washington, DC: Smithsonian Institution Press.

Blob, R. W. 1998. Evaluation of vent position from lizard skeletons for estimation of snout-vent length and body mass. *Copeia* 1998:792–801.

————. 2000. Interspecific scaling of the hind limb skeleton in lizards, crocodilians, felids and canids: does limb bone shape correlate with limb posture? *Journal of Zoology (London)* 250:507–531.

————. 2001. Evolution of hindlimb posture in nonmammalian therapsids: biomechanical tests of paleontological hypotheses. *Paleobiology* 27:14–38.

Bonfield, W. and E. A. Clark. 1973. Elastic deformation of compact bone. *Journal of Material Science* 8:1590–1594.

Botha, J. and A. Chinsamy. 2000. Growth patterns deduced from the bone histology of the cynodonts *Diademodon* and *Cynognathus*. *Journal of Vertebrate Paleontology* 20:705–711.

Brink, A. S. 1956. Speculations on some advanced mammalian characteristics in the higher mammal-like reptiles. *Palaeontologia Africana* 4:77–95.

Carrier, D. R. 1983. Postnatal ontogeny of the musculo-skeletal system in the black-tailed jack rabbit (*Lepus californicus*). *Journal of Zoology (London)* 201:27–55.

Carrier, D. R. and L. R. Leon. 1990. Skeletal growth and function in the California gull (*Larus californicus*). *Journal of Zoology (London)* 222:375–389.

Chinsamy, A. 1995. Ontogenetic changes in the bone histology of the Late Jurassic ornithopod *Dryosaurus lettowvorbecki*. *Journal of Vertebrate Paleontology* 15:96–104

Chinsamy, A. and B. S. Rubidge. 1993. Dicynodont (Therapsida) bone histology: phylogenetic and physiological implications. *Palaeontologia Africana* 30:97–102.

Currey, J. D. and G. Butler. 1975. The mechanical properties of bone tissue in children. *Journal of Bone and Joint Surgery* 57:810–814.

Damuth, J. and B. J. MacFadden. 1990. *Body Size in Mammalian Paleobiology: Estimation and Biological Implications.* Cambridge: Cambridge University Press.

de Ricqlès, A. J. 1976. On bone histology of fossil and living reptiles, with comments on its functional and evolutionary significance; pp. 123–150 *in* A. d'A. Bellairs and C. B. Cox (eds.), *Morphology and Biology of Reptiles.* Linnean Society Symposium Series, Number 3.

de Ricqlès, A. J., F. J. Meunier, J. Castanet, and H. Francillon-Vieillot. 1991. Comparative microstructure of bone; pp. 1–78 *in* B. K. Hall (ed.), *Bone, Volume 3.* Boca Raton, FL: CRC Press.

Dodson, P. 1975. Functional and ecological significance of relative growth in *Alligator. Journal of Zoology (London)* 175:315–355.

Enlow, D. H. 1969. The bone of reptiles; pp. 45–80 *in* C. Gans, A. d'A. Bellairs, and T. S. Parsons (eds.), *Biology of the Reptilia, Vol. 1 (Morphology A).* New York: Academic Press.

Farlow, J. O. 1980. Predator/prey biomass ratios, community food webs, and dinosaur physiology; pp. 55–83 *in* R. D. K. Thomas and E. C. Olson (eds.),

A Cold Look at the Warm-Blooded Dinosaurs. AAAS Selected Symposium 28. Boulder, CO:Westview Press.

Geist, V. 1972. An ecological and behavioral explanation of mammalian characteristics, and their implications to therapsid evolution. *Zeitschrift für Säugetierekunde* 37:1–15.

Hillenius, W. J. 1992. The evolution of nasal turbinates and mammalian endothermy. *Paleobiology* 18:17–29.

———. 1994. Turbinates in therapsids: evidence for Late Permian origins of mammalian endothermy. *Evolution* 48:207–229.

Hopson, J. A. 1973. Endothermy, small size, and the origin of mammalian reproduction. *American Naturalist* 107:446–452.

———. 1987. The mammal-like reptiles: a study of transitional fossils. *American Biology Teacher* 49:16–26.

———. 1991. Systematics of the nonmammalian Synapsida and implications for patterns of evolution in synapsids; pp. 635–693 *in* H.-P. Schultze and L. Trueb (eds.), *Origins of the Higher Groups of Tetrapods: Controversy and Consensus*. Ithaca, NY: Comstock.

———. 1994. Synapsid evolution and the radiation of non-eutherian mammals; pp. 190–217 *in* D. R. Prothero and R. M. Schoch (eds.), *Major Features of Vertebrate Evolution. Short Courses in Paleontology 7.* Knoxville, TN: Paleontological Society.

Hopson, J. A. and J. W. Kitching. 2001. A probainognathian cynodont from South Africa and the phylogeny of nonmammalian cynodonts. *Bulletin of the Museum of Comparative Zoology* 156:5–35.

Hotton, N., III. 1980. An alternative to dinosaur endothermy; pp. 311–350 *in* R. D. K. Thomas and E. C. Olson (eds.), *A Cold Look at the Warm-Blooded Dinosaurs. AAAS Selected Symposium 28.* Boulder, CO: Westview Press.

Jenkins, F. A., Jr. 1971. The postcranial skeleton of African cynodonts. *Bulletin of the Peabody Museum of Natural History* 36:1–216.

Jolicoeur, P. and J. E. Mosimann. 1968. Intervalles de confiance pour la pente de l'axe majeur d'une distribution normale bidimensionelle. *Biometrie-Praximetrie* 9:121–140.

Kemp, T. S. 1980. Aspects of the structure and functional anatomy of the Middle Triassic cynodont *Luangwa. Journal of Zoology (London)* 191:193–239.

———. 1985. A functional interpretation of the transition from primitive tetrapod to mammalian locomotion; pp. 181–191 *in* J. Reiß and E. Frey (eds.), *Principles of Construction in Fossil and Recent Reptiles.* Stuttgart: Universität Stuttgart/Universität Tübingen.

Kitching, J. W. 1995. Biostratigraphy of the *Cynognathus* Assemblage Zone; pp. 40–45 *in* B. S. Rubidge (ed.), *Biostratigraphy of the Beaufort Group (Karoo Supergroup). Series 1.* Pretoria: South African Committee for Stratigraphy.

LaBarbera, M. 1989. Analyzing body size as a factor in ecology and evolution. *Annual Review of Ecology and Systematics* 20:97–117.

Lammers, A. and R. Z. German. 2002. Ontogenetic allometry in the locomotor skeleton of specialized half-bounding mammals. *Journal of Zoology* 258: 485–495.

Maunz, M. and R. Z. German. 1997. Ontogeny and limb bone scaling in two New World marsupials, *Monodelphis domestica* and *Didelphis virginiana*. *Journal of Morphology* 231:117–130.

McArdle, B. H. 1988. The structural relationship: regression in biology. *Canadian Journal of Zoology* 66: 2329–2339.

McNab, B. K. 1978. The evolution of endothermy in the phylogeny of mammals. *American Naturalist* 112:1–21.

Parrish, J. M., J. T. Parrish, and A. M. Ziegler. 1986. Permian-Triassic paleogeography and paleoclimatology and implications for therapsid distribution; pp. 109–131 *in* N. Hotton III, P. D. MacLean, J. J. Roth, and E. C. Roth (eds.), *The Ecology and Biology of Mammal-Like Reptiles.* Washington, DC: Smithsonian Institution Press.

Peterson, J. A. and R. F. Zernicke. 1986. The mechanical properties of limb bones in the lizard *Anolis equestris. American Zoologist* 26:133A.

———. 1987. The mechanical properties of limb bones in the cursorial lizard *Dipsosaurus dorsalis. American Zoologist* 27:64A.

Reid, R. E. H. 1984. Primary bone and dinosaurian physiology. *Geological Magazine* 121:580–598.

———. 1997. How dinosaurs grew; pp. 403–413 *in* J. O. Farlow and M. K. Brett-Surman (eds.), *The Complete Dinosaur.* Bloomington: Indiana University Press.

Ruben, J. A. 1995. The evolution of endothermy in mammals and birds: from physiology to fossils. *Annual Review of Physiology* 57:69–95.

Sidor, C. A. 1996. Early synapsid evolution, with special reference to the Caseasauria. *Journal of Vertebrate Paleontology* 16:65A–66A.

Sidor, C. A. and J. A. Hopson. 1998. Ghost lineages and "mammalness": assessing the temporal pattern of character acquisition in the Synapsida. *Paleobiology* 24:254–273.

Sokal, R. R. and F. J. Rohlf. 1995. *Biometry* (3rd ed.). New York: W. H. Freeman.

Torzilli, P. A., K. Takebe, A. H. Burstein, M. Zika, and K. G. Heiple. 1982. The material properties of immature bone. *ASME Journal of Biomechanical Engineering* 104:12–20.

Wainwright, S. A., W. D. Biggs, J. D. Currey, and J. M. Gosline. 1976. *Mechanical Design in Organisms.* Princeton: Princeton University Press.

Appendix 13.1

This appendix reports limb bone measurements used to calculate scaling relationships for the femur and tibia of fossil cynognathian cynodonts (table A13.1). In the specimen column of table A13.1, extensions in parentheses after the catalog number were assigned unofficially by me and include "L" and "R" designations to refer to bones from the left and right sides for NMB specimens, and "F" or "T" designations to refer to the femur or tibia for BP/1 specimens. Abbreviations: *L*, bone length; *FED*, bone midshaft diameter in the direction of knee flexion and extension; *PD*, bone midshaft diameter perpendicular to *FED; A*, bone midshaft cross-sectional area; *IFE*, bone midshaft second moment of area in the direction of knee flexion and extension; *IP*, bone midshaft second moment of area in the direction perpendicular to *IFE; J*, bone midshaft polar moment of area; NMB.QR, National Museum (Bloemfontein); BP/1, Bernard Price Institute, Johannesburg.

Table A13.1. Limb bone measurements from fossil specimens of cynognathian cynodants

Specimen	Bone	*L* (mm)	*FED* (mm)	*PD* (mm)	*A* (mm^2)	*IFE* (mm^4)	*IP* (mm^4)	*J* (mm^4)
NMB.QR.1206(1R)	Femur	158.31	18.78	18.58	—	—	—	—
NMB.QR.1206(2R)	Femur	174.39	18.17	20.14	231.32	5454.72	6615.01	12069.73
NMB.QR.1206(3R)	Femur	185.41	18.93	21.52	—	—	—	—
NMB.QR.1206(4R)	Femur	173.43	21.21	23.09	—	—	—	—
NMB.QR.1206(5R)	Femur	171.11	20.55	21.58	286.91	8180.96	9720.02	17900.98
NMB.QR.1206(6R)	Femur	173.84	21.95	23.16	209.46	9552.69	10859.79	20412.48
NMB.QR.1206(7R)	Femur	154.20	18.33	20.73	311.25	8880.33	11086.88	19967.21
NMB.QR.1206(8R)	Femur	83.59	9.04	8.88	53.48	441.81	389.07	830.88
NMB.QR.1206(9R)	Femur	107.56	12.62	10.55	—	—	—	—
NMB.QR.1206(10R)	Femur	104.57	10.86	10.12	—	—	—	—
NMB.QR.1206(11R)	Femur	62.30	7.90	7.67	—	—	—	—
NMB.QR.1206(12R)	Femur	96.11	10.85	9.90	—	—	—	—
NMB.QR.1206(13R)	Femur	102.53	10.55	8.97	—	—	—	—
NMB.QR.1206(14R)	Femur	96.39	9.44	8.60	58.10	630.26	514.55	1144.81
NMB.QR.1206(15R)	Femur	106.72	11.20	9.73	—	—	—	—
NMB.QR.1206(16R)	Femur	82.84	11.51	13.03	—	—	—	—
NMB.QR.1206(17R)	Femur	86.92	9.77	9.10	95.95	1182.40	1151.10	2333.50
NMB.QR.1206(18R)	Femur	104.06	9.94	9.61	67.18	573.57	421.10	994.67
NMB.QR.1208(R)	Femur	164.27	18.50	20.20	—	—	—	—
NMB.QR.1206(1L)	Femur	180.00	21.35	22.79	338.24	11988.99	16585.20	28574.18
NMB.QR.1206(2L)	Femur	184.60	19.00	22.50	218.36	6322.41	9225.45	15547.86
NMB.QR.1206(3L)	Femur	134.37	15.94	15.63	—	—	—	—

(continued)

Table A13.1. *(continued)*

Specimen	Bone	L (mm)	FED (mm)	PD (mm)	A (mm²)	IFE (mm⁴)	IP (mm⁴)	J (mm⁴)
NMB.QR.1206(4L)	Femur	110.22	12.86	11.56	—	—	—	—
NMB.QR.1206(5L)	Femur	105.76	12.51	9.77	69.04	600.27	463.21	1063.48
NMB.QR.1206(6L)	Femur	100.03	10.60	9.63	89.55	1021.72	785.12	1806.83
NMB.QR.1206(7L)	Femur	100.13	11.99	12.82	74.70	658.69	658.25	1316.94
NMB.QR.1206(8L)	Femur	84.02	10.87	9.63	64.72	632.95	499.14	1132.09
NMB.QR.1206(9L)	Femur	104.98	12.12	10.82	70.62	959.61	754.65	1714.26
NMB.QR.1206(10L)	Femur	105.16	12.41	11.00	73.95	860.46	687.12	1547.58
NMB.QR.1206(11L)	Femur	105.64	12.84	10.87	85.60	1341.91	946.56	2288.47
NMB.QR.1208(1R)	Tibia	90.22	6.70	10.42	28.85	79.61	184.86	264.47
NMB.QR.1208(2R)	Tibia	89.99	7.09	12.03	48.86	186.03	554.95	740.98
NMB.QR.1208(3R)	Tibia	141.42	14.35	21.90	—	—	—	—
NMB.QR.1208(4R)	Tibia	139.91	15.57	25.12	—	—	—	—
NMB.QR.1208(5R)	Tibia	154.59	16.19	24.38	—	—	—	—
NMB.QR.1208(6R)	Tibia	153.49	15.48	22.78	—	—	—	—
NMB.QR.1208(7R)	Tibia	153.37	15.64	21.82	—	—	—	—
NMB.QR.1208(8R)	Tibia	144.81	14.73	23.77	—	—	—	—
NMB.QR.1208(9R)	Tibia	97.87	9.14	13.84	—	—	—	—
NMB.QR.1208(1L)	Tibia	151.52	13.51	23.50	—	—	—	—
NMB.QR.1208(2L)	Tibia	168.24	17.08	28.57	187.33	4010.21	10639.13	14649.34
NMB.QR.1208(3L)	Tibia	157.57	14.55	23.05	171.60	1822.85	8074.52	9897.36
NMB.QR.1208(4L)	Tibia	133.15	12.95	21.81	—	—	—	—
BP/1/1675(F.1)	Femur	191.28	20.95	24.67	—	—	—	—
BP/1/1675(F.2)	Femur	174.68	18.73	19.57	—	—	—	—
BP/1/1675(F.3)	Femur	167.78	19.08	20.42	—	—	—	—
BP/1/1675(F.4)	Femur	86.80	10.95	9.96	—	—	—	—
BP/1/1675(T.1)	Tibia	160.48	15.27	25.52	—	—	—	—
BP/1/1675(T.2)	Tibia	137.55	13.57	22.37	—	—	—	—
BP/1/1675(T.3)	Tibia	103.92	9.08	13.96	—	—	—	—
BP/1/1675(T.4)	Tibia	134.20	13.05	21.93	—	—	—	—
BP/1/1675(T.5)	Tibia	90.91	8.85	14.61	—	—	—	—

Appendix 13.2

This appendix details the calculation of RMA scaling relationships for limb bone measurements relative to bone length for Carrier's (1983) jackrabbit data.

I generated RMA diameter-on-length relationships for jackrabbit hind limb bones in two steps. First, I divided Carrier's (1983) published OLS slopes for the scaling of length on diameter by their correlation coefficients to produce RMA slopes for length on diameter. Second, I took the reciprocal of each resulting length-on-diameter RMA slope to generate an RMA relationship for diameter on length. This procedure is valid for RMA slopes because, unlike OLS equations, RMA equations minimize squared deviations from the calculated line for both variables being compared. I then calculated CIs for these new RMA slopes using the same methods that I applied in analyses of fossil taxa.

In contrast to the bone diameter data, no equations in Carrier's (1983) study directly relate second moment of area to bone length. However, his study does report scaling exponents for OLS regressions of length on mass and second moment of area on mass for the jackrabbit femur and third metatarsal. To obtain exponents for RMA regressions of second moment of area on length for each bone, I calculated RMA slopes for the length-mass and second moment of area-mass relationships and then divided the exponent for the regression of second moment of area on mass by the exponent for the regression of length on mass. Because the original relationships of length on mass and second moment of area on mass for each bone are based on different sample sizes, RMA slopes for second moment of area on length that are calculated in this fashion provide an estimate of the scaling relationship for which CIs cannot be calculated by standard RMA methods (e.g., Jolicoeur and Mosimann, 1968). Therefore, CIs for second moment of area on length relationships in jackrabbits were calculated as follows. After calculating RMA slopes for jackrabbit length-mass and second moment of area-mass relationships, I calculated RMA ninety-five percent CIs for each of these relationships, producing a ninety-five percent confidence range for both the numerator (second moment of area on mass slope) and denominator (length on mass slope) that would be used to calculate the slope of the second moment of area on length relationship. The lowest value of the ninety-five percent CI for the second moment of area on length slope was calculated by dividing the smallest possible value for its numerator by the largest possible value for its denominator; similarly, the highest value of the ninety-five percent CI for the second moment of area on length slope was calculated by dividing the largest possible value for its numerator by the smallest possible value for its denominator.

Cranial Variability, Ontogeny, and Taxonomy of *Lystrosaurus* from the Karoo Basin of South Africa

Frederick E. Grine

Catherine A. Forster

Michael A. Cluver

and Justin A. Georgi

Introduction

The herbivorous dicynodonts comprise a widespread and diverse therapsid radiation that flourished in the Permian and persisted into the Triassic. The taxonomic diversity of this group reached its acme in the Late Permian, when dicynodonts constitute the most abundant component of the terrestrial vertebrate fauna (King, 1988; Rubidge & Sidor, 2001). At least fifteen genera are recognized from the Permian strata of the South African Karoo Basin alone (King, 1993), where they constitute the vast bulk of vertebrate fossils (Hotton, 1986). This group is comparatively depauperate in the Early Triassic, with only the genera *Myosaurus* and *Lystrosaurus* recognized from the lower Tarkastad Subgroup rocks in South Africa. *Myosaurus* comprises a single, small, and exceedingly rare species, *M. gracilis,* that is known also from the Fremouw Formation of Antarctica (Cluver, 1974; Kitching, 1977; Hammer & Cosgriff, 1981).

Lystrosaurus, on the other hand, is represented by abundant specimens from numerous localities in South Africa (Kitching, 1977; Keyser & Smith, 1978) and elsewhere. Indeed, the earliest Triassic fauna of the Karoo Basin is overwhelmingly dominated by *Lystrosaurus* (Smith, 1995). Its abundance in Katberg Sandstone Formation was recognized by Broom (1909) in his proposed biostratigraphic zonation of the Beaufort Group strata. Keyser and Smith (1978) have observed that, although the *Lystrosaurus* Assemblage Zone is largely confined to the Katberg Sandstone Formation of the Tarkastad Subgroup, *Lystrosaurus* fossils occur both above and below it. In particular, they are known from the uppermost part of the underlying Palingkloof Member of the Balfour Formation (Keyser & Smith, 1978; Smith & Ward, 2001).

In this regard, Broom (1940) had suggested that *Lystrosaurus rubidgei* derived from the underlying *Cistecephalus* Zone (= *Dicynodon* Assemblage Zone) strata at Lootsburg Pass, and Hotton (1967) subsequently recorded a section about 60 m thick there in which *Lystrosaurus* was observed to co-occur with components of the *Dicynodon* Assemblage

Zone. Although Kitching (1977) doubted this association, being of the opinion that these *Lystrosaurus* specimens were deposited in sediments that had filled erosion channels in the older rocks, he noted that specimens of *Daptocephalus* [= *Dicynodon* (Cluver & Hotton, 1981)] have been recovered elsewhere from strata containing components of the *Lystrosaurus* Assemblage Zone fauna. Smith (1995) documented a narrow (about 25 m) overlap in the ranges of *Dicynodon* and *Lystrosaurus* within the Palingkloof Member of the Balfour Formation, and recent work (Smith & Ward, 2001) has extended this to some 40 m. Smith (1995) has argued that the Permian-Triassic boundary should be placed at the last occurrence of *Dicynodon,* which thus would position this horizon within the lower part of the *Lystrosaurus* Assemblage Zone.

In addition to being abundantly represented in the South African Karoo Basin, *Lystrosaurus* specimens have been recognized also from the Madumabisa Mudstones, Zambia (King & Jenkins, 1997), the Fremouw Formation, Antarctica (Elliot et al., 1970; Kitching et al., 1972; Colbert, 1974; Cosgriff et al., 1982), the Panchet Formation, India (Das Gupta, 1922; Tripathi, 1962; Tripathi & Satsangi, 1963), the Noyan Somon Formation, Mongolia (Gubin & Sinitza, 1993), the Guodikeng and Jiucaiyuan Formations of the Tunghungshan Series and the 3rd fossiliferous stratum of Taoshuyuanzi, China (Yuan & Young, 1934; Young, 1935, 1939; Sun, 1964, 1973; Olson, 1989; Cheng, 1993), the Northampton District of western Australia (Thulborn, 1983, 1990), the Pou Say sandstones, Laos (Repelin, 1923), and the Vokhmian Gorizonot Formation, Russia* (Kalandadze, 1975; Lozovskiy, 1983; Battail & Surkov, 2000).

The uppermost Guodikeng Formation preserves a small species of *Lystrosaurus* together with the Permian anomodonts *Dicynodon* and *Jimusauria,* whereas the overlying Jiucaiyuan Formation contains only specimens of a large species of *Lystrosaurus* (Cheng, 1993; Lucas et al., 1994). In the Madumabisa Mudstones, *Lystrosaurus* occurs together with typical Permian anomodonts such as *Dicynodon, Diictodon,* and *Oudenodon* (King & Jenkins, 1997). Thus, the Chinese and Zambian records provide support for the evidence from Lootsberg Pass and the Palingkloof Member of the Balfour Formation in South Africa for the contemporaneity of *Lystrosaurus* with Late Permian taxa (Lucas et al., 1994;

---

*Efremov (1938, 1940) attributed material from the Donguz Suite sediments on the banks of the Donguz River to *Lystrosaurus klimovi.* Subsequently, however, he (1951) transferred this species to *Rhadiodromus,* which is recognized as a sinokannemeyeriine by King (1988).

Smith, 1995; King & Jenkins, 1997). These occurrences serve to place *Ly-strosaurus* securely within the Late Permian.

There is sufficient morphological (King & Cluver, 1991) and circumstantial (Groenewald, 1991; Smith, 1995) evidence to suggest, moreover, that *Lystrosaurus* was a capable excavator and a probable burrower. In particular, Smith (1995) and King and Jenkins (1997) have argued that the feeding/digging adaptations of *Lystrosaurus,* which would have enabled it to deal with resistant vegetation and to grub for rhizomes, permitted it to track the expanding dry floodplain environments of the end-Permian (Rayner & Coventry, 1985; Rayner, 1992, 1995) at the expense of other dicynodonts that preferred wetter environs. Smith (1995) has suggested that the burrowing habits of *Lystrosaurus* may be partly responsible for its comparative abundance in the fossil record.

Thus, because *Lystrosaurus* was one of very few terrestrial genera to survive the end-Permian extinction event, this genus is significant for any assessment of the faunal transition across the Permo-Triassic boundary. At the same time, however, because any evaluation of the faunal transition across this critical geological boundary depends on an understanding of taxonomic diversity, it is essential to establish reasonable species-level taxonomies.

While there is certainly a decline in dicynodont generic diversity from the Late Permian to the Early Triassic, the level of turnover has been questioned by King (1993). In part, the problem reflects the fact that an unnecessarily profligate number of taxa have been proposed for Permo-Triassic anomodont fossils from the Karoo. Thus, for example, Tollman et al. (1980) listed fourteen species regarded by them as junior synonyms of *Aulacephalodon baini,* and King (1988) listed thirty-five species as junior synonyms of *Oudenodon bainii.* King (1988) also suggested that fewer than half of the 111 species of *Dicynodon* listed by Haughton and Brink (1954) should be recognized as valid. Although King (1993) has pointed out that numerous taxa have been "over-split," especially in the latest Permian *Dicynodon* Assemblage Zone, she accepts the presence of at least seven dicynodont genera comprising twenty-four species as indicated by Kitching (1977) and Keyser and Smith (1978) as stemming from that biozone. If *Lystrosaurus* is added to this *Dicynodon* Assemblage Zone fauna, the level of taxonomic diversity becomes even more difficult to assess because of the number of species that have been proposed for this genus. Thus, it is necessary to address the question of species diversity in *Lystrosaurus* to better assess the level of taxonomic diversity on either side of the Permo-Triassic boundary. It is, moreover, necessary to first establish species diversity of *Lystrosaurus* in the Permian and Trias-sic before erecting adaptive hypotheses that purport to explain a mainte-nance or increase in diversity. Finally, but of no less significance, it is nec-essary to determine species diversity in *Lystrosaurus* as a prerequisite for any attempt to address the question of possible adaptive/ecologic re-placements among the dicynodonts comprising the *Dicynodon* and *Ly-strosaurus* Assemblage Zone fauna.

The Species Taxonomy of *Lystrosaurus*

The taxonomy of *Lystrosaurus* has been confounded by a proplifera-tion of species names that have been applied to incomplete, poorly pre-served fossils, and erected on questionable diagnostic criteria. This situ-ation has been exacerbated by the sheer abundance of fossils that have been collected from the Karoo Basin. These specimens not only span a considerable size range, but they also evince considerable morphologic variation. Although *Lystrosaurus* skulls are quite distinct from those of other dicynodont genera (Cluver, 1971), morphologic variation of "un-certain significance" (Cosgriff et al., 1982), together with the consider-able size range that is spanned by these fossils, makes species determina-tion a difficult task.

To date, thirty-seven species names have been applied to *Lystrosaurus* fossils (table 14.1). Of these, twenty-seven have been erected on skulls and/or skull fragments from South Africa alone; the remaining ten have been applied to specimens from Russia, China, Laos, and India. Only one of these purported species (*L. weidenreichi* Young, 1939) has been diag-nosed on the basis of postcranial remains.

Huxley (1859) described the first specimen of *Lystrosaurus*, attribut-ing it to a novel species of the genus *Dicynodon, D. murrayi.* The follow-ing year, Owen (1860) recognized the distinctive nature of South African dicynodont skulls with short, deep snouts, such as had been described by Huxley, and proposed the genus *Ptychognathus* to accommodate *D. mur-rayi* as well as three new species, *P. declivis, P. latirostris,* and *P. verticalis.* Owen (1862, 1876) subsequently described three additional species of *Ptychognathus, P. alfredi, P. boopis,* and *P. depressus.* In 1870, Cope in-troduced the genus name *Lystrosaurus* for the species *L. frontosus.*

Lydekker (1890) observed that the genus name *Ptychognathus* had been preoccupied by Stimpson in 1858 for a crustacean and proposed the name *Ptychosagium* to accommodate the species that Owen had attrib-uted to *Ptychognathus.* Seeley (1898), however, argued that, because Cope's *L. frontosus* was clearly attributable to the genus *Ptychognathus,*

Table 14.1. Species names attributable to *Lystrosaurus*

Taxa based on fossils from South Africa

Dicynodon murrayi	Huxley, 1859		
Ptychognathus verticalis	Owen, 1860	= *Ptychosagium*	Lydekker, 1890
Ptychognathus declivis	Owen, 1860	= *Ptychosagium*	Lydekker, 1890
Ptychognathus latirostris	Owen, 1860	= *Ptychosagium*	Lydekker, 1890
Ptychognathus alfredi	Owen, 1862	= *Ptychosagium*	Lydekker, 1890
Lystrosaurus frontosus	Cope, 1870		
Dicynodon curvatus	Owen, 1876		
Ptychognathus boopis	Owen, 1876	= *Ptychosagium*	Lydekker, 1890
Ptychognathus depressus	Owen, 1876	= *Ptychosagium*	Lydekker, 1890
Dicynodon copei	Seeley, 1889		
Mochlorhinus platyceps	Seeley, 1898		
Rhabdotocephalus mccaigi	Seeley, 1898		
Lystrosaurus andersoni	Broom, 1907		
Dicynodon strigops	Broom, 1913		
Lystrosaurus oviceps	Haughton, 1915		
Lystrosaurus breyeri	Van Hoepen, 1915		
Lystrosaurus putterilli	Van Hoepen, 1915		
Lystrosaurus jeppei	Van Hoepen, 1916		
Lystrosaurus wageri	Van Hoepen, 1916		
Lystrosaurus wagneri	Van Hoepen, 1916		
Lystrosaurus jorisseni	Van Hoepen, 1916		
Lystrosaurus theileri	Van Hoepen, 1916		
Prolystrosaurus natalensis	Haughton, 1917		
Lystrosaurus rubidgei	Broom, 1940		
Lystrosaurus bothai	Broom, 1941		
Lystrosaurus amphibius	Brink, 1951		
Lystrosaurus primitivus	Toerien, 1954		

Taxa based on fossils from India, Laos, China, and Russia

Dicynodon orientalis	Huxley, 1865
Dicynodon incisivum	Repelin, 1923
Lystrosaurus broomi	Yuan & Young, 1934
Lystrosaurus hedini	Young, 1935
Lystrosaurus weidenreichi	Young, 1939
Lystrosaurus rajurkari	Tripathi & Satsangi, 1963
Lystrosaurus youngi	Sun, 1964
Lystrosaurus latifrons	Sun, 1973b
Lystrosaurus robustus	Sun, 1973b
Lystrosaurus georgi	Kalandadze, 1975

the latter's preoccupation would result in *Lystrosaurus* Cope, 1870, having precedence over *Ptychosagium.* In the same paper, Seeley (1898) proposed two subgenera for *Lystrosaurus* (= *Ptychognathus*)—namely *Rhabdotocephalus,* type species *R. mccaigi,* and *Mochlorhinus,* type species *M. platyceps.* Broom (1907, 1940, 1941), Haughton (1915), van Hoepen (1915, 1916), Brink (1951), and Toerien (1954) added a further thirteen species from the Karoo Basin to this genus. Haughton (1917) introduced the genus *Prolystrosaurus,* attributing to it a form described several years earlier by Broom (1913) as *Dicynodon strigops,* and a new species *Prolystrosaurus natalensis.*

As aptly noted by Broom, "*Lystrosaurus* is by far the most difficult genus of South African fossil reptile with which we have to deal" (1932, 242). In particular, Broom observed that "the determination of the species of this genus is a matter of the greatest difficulty." This perspicacious pronouncement has lost none of its veracity in the intervening seven decades.

There have been numerous revisions of the taxonomy of *Lystrosaurus,* including those by Broom (1932), Brink (1951, 1986), Kitching (1968, 1977), Cluver (1971), Colbert (1974), Cosgriff et al. (1982), and King (1988). In the first major revision of this genus, Broom (1932) recognized five species, proposing that seventeen species names be regarded as junior synonyms of *L. murrayi* (table 14.2). Brink (1951) recognized thirteen or fourteen species, and Kitching (1968) recognized nine. Cluver (1971) also recognized nine species, although only six of these were also recognized by Kitching. Colbert (1974) erroneously claimed that Kitching (1968) had recognized six species and proposed to accept these because they were agreed to be valid by both Kitching and Cluver (table 14.2). Only in 1977 did Kitching publish his opinion concerning the validity of these six species. However, in that paper, Kitching (1977, 24) stated that "Kitching (unpublished notes, 1961) recognized six species [of *Lystrosaurus*]," even though he lists nine as being valid in his 1968 article. It would seem, therefore, that Colbert (1974) must have been citing a personal communication from Kitching rather than the 1968 paper to which he refers. Nevertheless, Cluver (1971), Colbert (1974), and Kitching (1977) were in agreement that at least six species of *Lystrosaurus* (*L. murrayi, L. declivis, L. mccaigi, L. platyceps, L. oviceps,* and *L. curvatus*) are represented in the Karoo assemblage.

Cosgriff et al. (1982) also provisionally recognized a maximum of six species of *Lystrosaurus* from South Africa, although they were of the opinion that this number might be reduced to a minimum of three (table 14.2). They noted that *L. murrayi* and L. *declivis* "are obviously very closely allied if, indeed, they are separate species," and followed Cluver

Table 14.2. Species and synonyms recognized in the principal revisions of *Lystrosaurus* from the Karoo Basin of South Africa

Valid species	Proposed synonyms				
			Broom, 1932		
L. murrayi	*P. declivis,*	*P. depressus,*	*D. strigops,*	*Pr. natalensis,*	*L. jorisseni,*
	P. latirostris,	*P. verticalis,*	*L. theileri,*	*L. oviceps,*	
	P. alfredi,	*P. boopis,*	*L. wageri,*	*L. putterilli,*	
	L. frontosus,	*D. copei,*	*L. wagneri,*	*L. breyeri*	
L. curvatus					
L. platyceps					
L. mccaigi					
L. andersoni					
			Brink, 1951		
L. murrayi	*L. frontosus,*	*L. theileri,*	*L. wagneri,*	*L. jorisseni,*	*L. wageri,*
	L. jeppei				
L. declivis	*L. alfredi,*	*L. boopis,*	*L. latirostris*		
L. oviceps	*L. breyeri,*	*L. depressus*			
L. andersoni					
L. bothai					
L. curvatus					
L. mccaigi					
L. outterilli					
L. platyceps					
L. rubidgei					
L. verticalis					
L. amphibius					
L. strigops	*?L. natalensis*				
			Kitching, 1968		
L. murrayi	*L. bothai,*	*L. putterilli,*	*?L. rubidgei,*	*?L. verticalis*	
L. platyceps					
L. declivis					
L. mccaigi					
L. curvatus					
L. oviceps					
L. amphibius					
L. andersoni					
L. primitivus					

Table 14.2. *(continued)*

Valid species	Proposed synonyms				
			Cluver, 1971		
L. murrayi	*L. frontosus,*	*L. jeppei,*	*P. boopis,*	*D. strigops,*	*P. verticalis,*
	Pr. natalensis				
L. curvatus	*L. wageri,*	*L. jorisseni,*	*L. theileri*		
L. platyceps	*L. andersoni*				
L. oviceps	*L. breyeri*				
L. declivis	*P. latirostris,*	*P. alfredi,*	*P. depressus,*	*L. wagneri,*	*L. primitivus*
L. mccaigi	*L. amphibius*				
L. bothai					
L. putterilli					
L. rubidgei					
			Colbert, 1974		
L. murrayi	*P. verticalis,*	*L. frontosus,*	*P. boopis,*	*D. strigops,*	*L. jeppei,*
	Pr. natalensis,	*L. rubidgei,*	*L. bothai*		
L. mccaigi	*L. putterilli,*	*L. amphibius*			
L. declivis	*L. latirostris,*	*L. alfredi,*	*L. depressus,*	*L. wagneri,*	*L. primitivus*
L. curvatus	*L. wagneri,*	*L. jorisseni,*	*L. theileri*		
L. platyceps	*L. andersoni*				
L. oviceps	*L. breyeri*				
			Kitching, 1977		
L. murrayi	*L. putterilli,*	*L. rubidgei*			
L. curvatus	*L. bothai*				
L. mccaigi	*L. amphibius*				
L. declivis					
L. oviceps					
L. platyceps					
			Cosgriff et al., 1982		
L. murrayi	*? L. declivus*				
L. curvatus	*? L. platyceps*				
L. oviceps					
L. mccaigi					
			Brink, 1986		
L. murrayi	*P. verticalis,*	*L. frontosus,*	*P. boopis,*	*D. strigops,*	*L. jeppei,*
	Pr. natalensis,	*L. wageri*			

(continued)

Table 14.2. (continued)

Valid species	Proposed synonyms				
L. platyceps	R. mccaigi,	L. amphibius,	L. andersoni,	L. putterelli	
L. declivis	P. latirostris,	P. alfredi,	P. depressus,	L. primitivus	
L. curvatus	L. bothai,	L. breyeri,	L. jorisseni,	L. oviceps,	L. theileri
		King, 1988			
L. murrayi	P. verticalis,	L. frontosus,	P. boopis,	D. strigops,	L. jeppei,
	Pr. natalensis,	L. rubidgei,	L. bothai,	D. copei	
L. curvatus	L. wageri,	L. jorisseni,	L. theileri		
L. declivis	P. latirostris,	P. alfredi,	P. depressus,	L. wagneri,	L. primitivus
L. mccaigi	L. putterilli,	L. amphibius			
L. oviceps	L. breyeri				
L. platyceps	L. andersoni				

(1971) in regarding *L. curvatus* and *L. platyceps* as forming a primitive species complex that included *L. oviceps*. These same six species were accepted as valid by King (1988), whereas Brink (1986) recognized only four. The principal difference between these later taxonomic schemes concerns the recognition of *L. mccaigi*. Thus, whereas Cluver (1971), Colbert (1974), Cosgriff et al. (1982), and King (1988) accepted *L. mccaigi* as a separate species, and whereas Cluver (1971), Colbert (1974), and Cosgriff et al. (1982) regarded it as a member of the more derived species group, Brink (1986) considered *L. mccaigi* (together with *L. amphibius* and *L. putterilli*) as a junior synonym of *L. platyceps*.

The apparent uncertainties concerning the species taxonomy of *Lystrosaurus* are reflected in a comparison of the junior synonyms that have been proposed by different authorities for those species they regard as valid (table 14.2). Thus, for example, *L. bothai* was considered a valid species by Cluver (1971); it was regarded as junior synonym of *L. murrayi* by Colbert (1974) and King (1988) but as a junior synonym of *L. curvatus* by Kitching (1977) and Brink (1986). This is of some interest considering the basic differences in cranial morphology that have been posited to separate the more primitive and derived species groups that include *L. curvatus* and *L. murrayi*, respectively. In this regard, Broom appears to have regarded *L. platyceps* and *L. mccaigi* as being relatively closely related inasmuch as he noted that the type specimen of *L. andersoni* is "badly preserved but manifestly belongs to the same group as *L. platyceps* and *L. mccaigi*" (1932, 258).

With reference to these taxonomic revisions, few workers have attempted to provide differential diagnoses for the species they consider valid. Notable exceptions have been Broom (1932), Cluver (1971)*, and Brink (1986). Broom (1932) indicated the five species recognized by him could be distinguished on the basis of ten characters, although only three features (orbit size, prefrontal boss development, and the length of the snout) were used to distinguish among more than two taxa (table 14.3). With reference to possible ontogenetic variation, Broom (1932) noted that in *L. murrayi* the septomaxilla contacted the lacrimal in "younger" (i.e., smaller) specimens but that these elements were excluded by a contact between the nasal and maxilla in "older" individuals; in *L. curvatus,* on the other hand, the nasal contacted the maxilla in crania of all sizes. Several of Broom's characters (orbit size, relative breadth of the skull roof, size of the snout, and size of the preparietal) are readily amenable to measurement, although they have not been subjected to any kind of quantitative analysis.

The development of the snout and prefrontal bosses noted by Broom (1932) were elaborated upon by Cluver (1971) as part of a character matrix that he identified as distinguishing among nine species of *Lystrosaurus* (table 14.4). These features also included reference to Broom's observations on interorbital width and the longitudinal ridges on the snout, although Cluver (1971) recognized quite different distributions of their states. In particular, he (1971) noted that interorbital width served to distinguish *L. curvatus* (and *L. bothai*) from both *L. platyceps* and *L. mccaigi* but that it did not differentiate *L. platyceps* and *L. mccaigi* from one another (cf. tables 14.3, 14.4).

Cluver (1971) recorded that the form of the nasofrontal suture is variable in *L. curvatus* and *L. platyceps,* and Cosgriff et al. (1982) argued that the position of this suture is correlated with snout development and the angle at which the facial plane meets the skull roof (i.e., two of the characters in table 14.4). Cosgriff et al. (1982) also dismissed the size of the canine tusk as a viable character on the basis that it "could not be evaluated objectively" and that this feature simply reflects "individual variation and/or sexual dimorphism." However, the objective assessment of this

---

*Abridged versions of Cluver's (1971) specific diagnoses have been reiterated by both Colbert (1974) and Cosgriff et al. (1982), but neither study provided additional species characterization.

Table 14.3. Distribution of diagnostic characters of *Lystrosaurus* species according to Broom (1932)

Character	L. murrayi	L. curvatus	L. platyceps	L. mccaigi	L. andersoni
Orbit size	moderate	larger than L. murrayi	fairly large	large	smaller than L. platyceps and L. mccaigi
Prefrontal boss	moderate horn-like processes	distinct antorbital development	well-developed preorbital ridge	very marked & thick orbital margin	
Dorsal border of Skull roof	prefrontal divided into frontal & facial planes		face curves smoothly round from frontal plane		
Contact of bones of the snout	septomax-lacrimal or nasal-maxilla (larger)	nasal-maxilla			
Snout	elongate		moderately elongate	upper facial height greater than L. murrayi & L. platyceps	
Longitudinal ridges on snout	not marked		absent		
Interorbital diameter			only slightly greater than intertemporal	much greater than intertemporal	
Frontal development	differs greatly from L. mccaigi			differs greatly from L. murrayi	
Nostril depression		relatively larger than in L. murrayi			
Preparietal size			very large		

Table 14.4. Distribution of diagnostic characters of *Lystrosaurus* species according to Cluver (1971)

Character	L. curvatus	L. bothai	L. platyceps	L. oviceps	L. declivis	L. mccaigi	L. putterilli	L. murrayi	L. rubidgei
Snout	shallow	moderately deep	deep	deep	deep	deep	deep	moderately deep	moderately deep
Longitudinal ridges on snout	absent	absent	absent	median (strong)	occasional	present		median	
Fronto-nasal ridge	absent	absent	absent	weak	strong	variable (far posterior)	present	present	absent
Frontal bosses and/or ridges	absent	absent	absent	tuberosities	prominent tuberosities	bosses and ridges	present	present	
Prefrontal (orbital) boss	absent	absent	absent	absent	absent	present (protrude forward)	present	absent	absent
Interorbital width	relatively narrow	relatively narrow	broad	broad	broad	broad	broad	broad	broad
Dorsal border of skull roof	smooth arc	smooth arc	two planes	smooth arc	three planes	two planes	two planes	three planes	
Angle of facial plane to skull roof	curved at angle	curved at angle	flat at angle	curved at angle	flat at angle	flat perpendicular	flat perpendicular	flat perpendicular	
Nasals in plane of premaxilla	only partly	only partly	entirely	only partly	partly but variable with size	entirely	no	variable with size	entirely
Flare of ventral ramus of squamosal	posterior and lateral	lateral	lateral	lateral	lateral	lateral	lateral	lateral	lateral
Frontonasal suture	interdigitated or straight transverse	interdigitated	interdigitated or straight transverse	straight transverse	straight transverse				interdigitated
Canine size	weak	weak	weak		large	large	large	large	

Table 14.5. Distribution of diagnostic characters of *Lystrosaurus* species according to Brink (1986)

Character	L. murrayi	L. declivis	L. curvatus	L. platyceps
Skull size	small			very large
Transverse contour of snout	smoothly convex	very angular	smoothly convex	sharp angles and/or ridges
Orbit size			very large	
Orbital bosses	absent	slight (frontal)	absent	very prominent (pre- and postfrontal)
Orbit position	not protruding	not protruding	slight dorsal protrusion	prominent dorsal protrusion
Snout development	shorter and more vertical			
Dorsal border of skull roof			smoothly curved	
Angle of facial plane to skull roof	perpendicular	flat at angle	curved at angle	
Fronto-nasal ridge	absent	present		

feature seems fairly straightforward inasmuch as the cross-sectional diameter and/or length of the canine are easily measured. Furthermore, although canine size may well reflect taxonomic differences and/or sexual dimorphism in *Lystrosaurus,* this has yet to be demonstrated empirically. Among the characters Cluver (1971) cited, the orientation of the ventral ramus of the squamosal seems especially prone to the effects of diagenetic distortion. Indeed, although this process was claimed to be flared laterally in all species except *L. curvatus* (table 14.4), it conceals the occipital condyles in lateral view in several of the taxa illustrated by him [1971, fig. 68 (*L. platyceps*), fig. 74 (*L. mccaigi*), figs. 75, 77 (*L. murrayi*)]. With reference to probable ontogenetic variation, Cluver (1971) noted that the degree to which the nasal bones conformed to the plane formed by the premaxillae is variable in *L. murrayi* and *L. declivis* and that the variation is related to skull size in both species. He observed that in smaller specimens the nasals meet the premaxillae at an angle, while in larger crania approximately two-thirds of the length of the nasal lies in the plane formed by the premaxillae.

Brink (1986) cited a number of diagnostic features for the four South African species that he recognized (table 14.5), but in some cases it is unclear how the character states are distributed among them, and in other instances his characterization of a species contrasts with that by other workers. Thus, for example, whereas Brink (1986) stated that *L. murrayi* lacks a frontonasal ridge, Cluver (1971) pointed to its presence in this species. Cluver (1971) had earlier dismissed the relative height of the dorsal border of the orbit as an unreliable character because it is easily affected by postmortem deformation. Brink (1986) regarded *L. murrayi* as a small species and *L. platyceps* as a very large species, but it is not clear where *L. declivis* and *L. curvatus* fall along this gradient, unless he regarded them as being intermediate between these extremes.

Cluver (1971) recognized that two basic cranial morphotypes could be distinguished on the basis of seven characters: snout development, the angle at which the facial and frontal planes intersect, canine tusk size, frontonasal ridge development, prefrontal boss development, frontal tuberosity development, and the presence of longitudinal ridges on the snout. He stressed the basic similarity of the crania assignable to the two groups, noting that they differ by modification of common structures rather than alteration of fundamental skull form. The more primitive species group, comprising *L. curvatus* and *L. platyceps,* retained the ancestral condition of a smoothly rounded skull roof ("Type B"), while skulls of the more derived ("Type A") species group (*L. murrayi, L. declivis,* and *L. mccaigi*) had more distinctly delineated facial and dorsal planes, variably developed longitudinal and frontonasal ridges, and variably developed frontal and prefrontal bosses (Cluver, 1971).

Cluver (1971) was of the opinion that *L. putterilli* was closely related to *L. mccaigi,* noting that the type (and only) specimen of *L. putterilli* differs in "the relationships of the nasals, which curve back from the premaxillary facial surface" such that "the prefrontal bosses do not protrude forward of the facial plane as is the case in *L. mccaigi.*" Similarly, he considered that *L. rubidgei* was closely allied to *L. platyceps* and only provisionally recognized the validity of the former, noting that the type (and only) skull is "considerably crushed." At the same time, while Cluver (1971) observed that *L. oviceps* shared a closer "morphological link" with the species comprising the more primitive group, he noted that the sort of variation exhibited by both *L. oviceps* and *L. bothai* muddied these typological distinctions.

Cluver's (1971) recognition of two basic cranial morphs was presaged in some measure by both Seeley (1898) and Broom (1932). Seeley (1898) observed that in some *Lystrosaurus* specimens the snout and dorsal surface of the skull are separated by an angular ridge (e.g., "*Ptychognathus*" *declivis*), whereas in others (e.g., "*Mochlorhinus*" *platyceps*) the two surfaces curve smoothly into one another. Broom (1932, 250) noted that "in any collection of *Lystrosaurus* skulls it is possible to divide the larger ones into two groups—those with comparatively smooth frontal and prefrontal regions and those with much ridged skulls and with large prefrontal growths at the upper and anterior margins of the orbits." He regarded the comparatively smooth skulls as probable females and the more highly ornamented as probable males, although he noted further that "young skulls are more difficult to separate as the ridging is only slightly developed as [sic] young males." It is not clear, however, whether Broom therefore regarded the specimens assigned by him to *L. murrayi* and *L. mccaigi* as being males, and the skulls of *L. curvatus* and *L. platyceps* as females, for while he used the development of ridges and bosses in taxonomic differentiation, he recognized that this variation had the potential to be influenced by sex as well.

Ontogeny, Sexual Dimorphism, and Species Taxonomy of *Lystrosaurus*

The question of the degree to which ontogenetic changes and/or sexual dimorphism in skull size and shape may influence the characters that have been used to differentiate the various species of *Lystrosaurus* has been addressed by a number of workers concerned about the taxonomy of this genus. As noted above, Broom (1932) was of the opinion that male and female skulls could be identified on the basis of variation in certain nonmetrical features but that the crania of both sexes possessed canine tusks, a position that Tripathi and Satsangi (1963) supported.

Broom also observed that "individuals of *Lystrosaurus* seem to have grown almost indefinitely like tortoises" (1932, 244), and he attributed at least some of the taxonomic confusion surrounding *Lystrosaurus* to the effect of prolonged growth on cranial morphology. Thus, for example, he regarded the type specimens of *Prolystrosaurus natalensis* and *L. latirostris* as "young" and "well-grown" females of *L. murrayi,* respectively. Earlier, however, he (1907) had proposed that the species of *Lystrosaurus* could be divided into two evolutionary branches, one comprising the smaller taxa, such as *L.murrayi* and *L. verticalis,* and the other composed by larger taxa, such as *L. mccaigi.* Brink (1951), too, subscribed to the presence of small and large species lineages. Kitching (1968) also recognized the problem of ontogenetic alteration of skull morphology, noting that the type of *L. rubidgei* could be regarded alternatively as either a "young" specimen of *L. mccaigi* (=*L. amphibius*) or an "adult male" of *L. murrayi.* He proposed a *L. murrayi* growth series comprising eight small crania, ranging from 42 to 111 mm in basal length. He noted that in the smallest specimens

(42 to < 50 mm), the canines had yet to erupt, that they were just emergent in specimens between 50 and 58 mm, and that they were fully erupted only in an individual with a length of 111 mm. These same eight specimens were attributed to *L. curvatus* by Cluver (1971).

Cluver (1971) discussed the evidence for sexual differences in skull morphology in *Lystrosaurus* and concluded that none of the species was dimorphic. With reference to the effects of ontogenetic age on cranial morphology, he presented evidence that certain diagnostic features, such as frontonasal ridges and frontal bosses, only begin to make their appearance in skulls with a length between 73 and 90 mm. He noted, too, that skulls with smoothly curved roofs may be well over twice the size of those with frontal bosses and a distinct frontonasal ridge, indicating that these features are not simply correlates of increased size. Cluver (1971, 241) concluded that "it seems unlikely that age is a factor in the determination of the major differences in skull proportions within the genus." Cosgriff et al. (1982) concurred with Cluver's (1971) conclusion regarding the extent of sexual dimorphism in *Lystrosaurus* but were of the opinion that the influence of growth allometry on cranial variation was highly uncertain. Nevertheless, they concluded that *L. curvatus, L. declivis, L. murrayi,* and *L. mccaigi* each attained large adult size and that each maintained its definitive characteristics through a range of increasing size. They noted, however, that most of the crania attributed to *L. mccaigi* are from large individuals.

Thackeray et al. (1998) undertook a multivariate study of crania attributed to *L. murrayi* and *L. declivis* using least-squares linear regression analyses. They compared measurements of a reference specimen assigned to *L. murrayi* with those obtained from a sample of skulls provisionally assigned to *L. murrayi* and *L. declivis.* They concluded that all the crania in their series represented younger and older individuals of a single, sexually dimorphic species, in which adult males and females differed in the development of supraorbital bosses.

Studies of possible growth-related changes in cranial measurements of Permian and Triassic therapsids (Grine et al., 1978; Bradu & Grine, 1979; Tollman et al., 1980; Abdala & Giannini, 2000) and Cretaceous protoceratopsian dinosaurs (Dodson, 1976) have demonstrated that allometric relationships of these variables may account for many of the differences observed between specimens that comprise hypothesized ontogenetic series. Thus, Grine et al. (1978; Bradu & Grine, 1979) demonstrated that a suite of morphometric variables in differently sized diademodontine crania are allometrically related according to a single mathematical model,

which led to the conclusion that these specimens represent an ontogenetic series of a single species. This, in turn, led to the proposal that
some twenty-four species names be recognized as junior synonyms of
Diademodon tetragonus. Qualitative differences observed among the
skulls suggested the presence of two cranial morphs, but neither bivariate
nor multivariate analyses revealed any pattern of clustering that corresponded to these morphotypes. Grine et al. (1978; Bradu & Grine, 1979)
thus concluded that differences in these qualitative features might be
attributable to sexual dimorphism in this species. Abdala and Giannini
(2000) similarly used the demonstrably allometric (bivariate) relationships of a number of cranial measurements to propose that three species
names be recognized as junior synonyms of the Middle Triassic cynodont
Massetognathus pascuali.

Tollman et al. (1980) undertook a bivariate morphometric analysis of
the skull of the Permian dicynodont, *Aulacephalodon baini,* and concluded that their sample, which included specimens attributed to at least
six different species, represented a morphometrically homogeneous ontogenetic series. They (1980) also examined nasal boss development in
these crania and concluded that this feature probably reflected sexual dimorphism. In particular, these bony excrescences began to assume different configurations when the skull attained a snout length of some
80–95 mm in a series that ranged from about 50 to about 220 mm for this
measurement.

Despite the attention that has been paid to the potential influences of
ontogeny and sexual dimorphism on the species taxonomy of *Lystrosaurus,* these factors have yet to be evaluated in a systematic manner for
more than the two species included by Thackeray et al. (1998). This is
rather surprising in view of the comparative abundance of *Lystrosaurus*
skulls in museum collections. In light of the analyses that have been conducted on other Permo-Triassic therapsids (Grine et al., 1978; Bradu &
Grine, 1979; Tollman et al., 1980; Abdala & Giannini, 2000) and the
preliminary assessment of *L. murrayi* and *L. declivis* by Thackeray et al.
(1998), we undertook a quantitative and qualitative assessment of cranial
variation exhibited by *Lystrosaurus* skulls of various sizes from the
Karoo Basin in an attempt to throw light on the question of species
diversity in this genus.

Material and Methods

This study represents an analysis of quantitative variables and nonmetric characters of the skull of *Lystrosaurus,* with special emphasis on
those features that have been regarded as being diagnostic of different

species and/or species groups. Measurements and qualitative observations were recorded by two of us (F.E.G. and M.A.C.) working together or, in some instances, by one of us (F.E.G.) working alone. Variation in the nonmetric features was assessed by rating the development of a given character against a series of stereophotographic standards compiled by two of us (F.E.G. and M.A.C.) to ensure a degree of objective consistency in scoring.

To evaluate the validity of the diagnostic characterizations of purported species, the skulls were also provisionally attributed a priori to one of six taxa according to Cluver's (1971) definitions for *L. murrayi, L. declivis, L. curvatus, L. platyceps, L. oviceps,* and *L. mccaigi,* or to *Lystrosaurus* sp. indet. when sufficient diagnostic criteria were lacking (table 14.6). These attributions were made a priori to evaluate patterns of variability in the distribution of both quantitative and qualitative parameters in relation to purported species hypodigms.

The null hypothesis under investigation is that morphologic and morphometric variability in the skull can be related to an ontogenetic series of a single species. Thus, it is expected that the features that have been related to species differences will covary and that their expression will depend on cranial size, suggesting the presence of a single, morphologically variable species of *Lystrosaurus* in the Karoo Basin.

Material

The present sample comprised 112 variably preserved skulls from localities in South Africa attributable to *Lystrosaurus* (table 14.6). This series is not exhaustive, as it does not sample every specimen for which measurements and nonmetric observations could be recorded. At least thirty-four other such specimens have been listed in various publications (e.g., Kitching, 1968; Cluver, 1971; Cosgriff et al., 1982; Brink, 1986). In addition, several specimens for which cranial dimensions have been published were not included here because they could not be located in the relevant museum collections at the time they were studied by us. Thus, for example, Kitching (1968) recorded a series of measurements for four small skulls in the collection of the Bernard Price Institute for Palaeontological Research (BP/1/267, BPI MN410, 411, 413), and Cosgriff et al. (1982) cite basal skull length for an additional four specimens from that collection (BPI 464, 1379, 1754, 3976) that are absent from our sample. Cluver (1971) and Cosgriff et al. (1982) cite basal skull length for three specimens (SAM-PK-K 1362; TM 19; TM 196) not included here. Thackeray et al. (1998) included eleven specimens housed by the Transvaal Museum and thirteen specimens from the collection of the Council for

Table 14.6. *Lystrosaurus* skulls included in the present study

Spec.	Museum no.	Taxonomic designation	Var.	CL	This study
1	NMQR C 282	*L. murrayi* (Cluver, 1971)	37	111	*L. murrayi*
2	NMQR C 608		34	105	*L. murrayi*
3	NMQR C 211	*L. murrayi* (Cluver, 1971)	36	86	*L. murrayi*
4	SAM-PK-K 1397		27	95	*L. murrayi*
5	NMQR C 270		25	106	*L. murrayi*
6	SAM-PK 11167		21	–	*L. murrayi*
7	SAM-PK-K 1495	*L. murrayi* (Cluver, 1971)	43	115	*L. murrayi*
8	SAM-PK 11181		25	146	*L. murrayi*
9	BMNH 36253	*L. boopis* (type)	21	83	*L. murrayi* (Cluver, 1971)
10	BMNH 36224	*L. verticalis* (type)	25	100	*L. murrayi* (Cluver, 1971)
11	BMNH 1291	*L. murrayi* (type)	15	159	*L. murrayi*
12	UMZC T 783		10	46	*L.* sp. indet.
13	UMZC T 784		10	44	*L.* sp. indet.
14	UMZC T 763		44	103	*L. murrayi*
15	SAM-PK-K 1165		31	105	*L. murrayi*
16	BMNH R.5710		23	101	*L. murrayi*
17	SAM-PK 11171		15	168	*L.* sp. indet.
18	BMNH 47324	*L. alfredi* (type)	22	126	*L. declivis* (Cluver, 1971)
19	BMNH 36222	*L. latirostris* (type)	31	143	*L. declivis* (Cluver, 1971)
20	BMNH 36221	*L. declivis* (type)	30	140	*L. declivis*
21	UMZC T. 778		9	–	*L. declivis*
22	SAM-PK-K 4800		40	124	*L. declivis*
23	NMQR C 391	*L. declivis* (Cluver, 1971)	39	113	*L. declivis*
24	SAM-PK-K 1398		44	135	*L. declivis*
25	SAM-PK 3596		36	121	*L. declivis*
26	SAM-PK 11184		34	124	*L. declivis*
27	SAM-PK 3593		22	203	*L. declivis*
28	SAM-PK 3455		29	134	*L. declivis*
29	SAM-PK 1362	*L. curvatus* (Cluver, 1971)	39	80	*L. curvatus*
30	SAM-PK-K 1381		25	–	*L. curvatus*
31	UMZC T 788		22	146	*L. curvatus*
32	NMQR C 154		44	117	*L. curvatus*
33	SAM-PK 3531		22	51	*L.* sp. indet.
34	SAM-PK-K 3		32	134	*L. curvatus*
35	SAM-PK-K 15		21	–	*L. curvatus*
36	BMNH 3597		40	197	*L. curvatus*
37	SAM-PK 3715	*Pr. natalensis* (type)	22	94	*L.* sp. indet.

Table 14.6. (continued)

Spec.	Museum no.	Taxonomic designation	Var.	CL	This study
38	SAM-PK-K 16		26	—	*L. curvatus*
39	SAM-PK 641	*L. oviceps* (type)	25	—	*L. oviceps*
40	SAM-PK 11186		27	95	*L. oviceps*
41	SAM 706	*L. platyceps* (Cluver, 1971)	30	201	*L. platyceps*
42	SAM 7843		41	170	*L. platyceps*
43	SAM-PK-K 116	*L. mccaigi* (Cluver, 1971)	29	238	*L. mccaigi*
44	SAM-PK-K 9		14	—	*L. mccaigi*
45	BMNH 3749/3754		11	—	*L. mccaigi*
46	BMNH 2211		25	105	*L.* sp. indet.
47	BMNH 5711		19	115	*L.* sp. indet.
48	BMNH 5709		20	132	*L.* sp. indet.
49	BMNH 6760		29	93	*L.* sp. indet.
50	UMZC T 780		14	119	*L.* sp. indet.
51	BMNH R.7915		18	133	*L.* sp. indet.
52	UMZC T 1057		35	103	*L.* sp. indet.
53	BPI MN 109		27	141	*L. declivis*
54	BPI/1 3980		31	94	*L. declivis*
55	TM 139		29	113	*L. declivis*
56	TM 21		45	166	*L. declivis*
57	TM 3580		32	185	*L. declivis*
58	TM 4113		23	96	*L. declivis*
59	TM 4465		36	144	*L. declivis*
60	TM 4049		37	105	*L. declivis*
61	TM 20	*L. wagneri* (type)	29	136	*L. declivis* (Cluver, 1971)
62	TM 117		25	101	*L. declivis*
63	BPI/1 1750		28	110	*L. declivis*
64	BPI/1 4275		27	97	*L. declivis*
65	BPI/1 1754	*L. mccaigi* (Cosgriff et al., 1982)	27	230	*L. declivis*
66	BPI/1 3900		17	96	*L. murrayi*
67	BPI MN 106	*L. murrayi* (Brink, 1986)	31	110	*L. murrayi*
68	BPI/1 4798		41	100	*L. murrayi*
69	BPI/1 4314		15	—	*L. murrayi*
70	TM 69		31	106	*L. murrayi*
71	TM 115		18	87	*L. murrayi*
72	TM 71		38	116	*L. murrayi*
73	TM 27	*L. platyceps* (Cluver, 1971)	32	93	*L.* sp. indet

(*continued*)

Table 14.6. (continued)

Spec.	Museum no.	Taxonomic designation	Var.	CL	This study
74	TM 33		27	127	*L. murrayi*
75	TM 2881		32	144	*L. murrayi*
76	TM 4051		29	134	*L.* sp. indet.
77	TM 3		27	96	*L.* sp. indet.
78	TM 67		28	102	*L. murrayi*
79	TM 164		29	100	*L. murrayi*
80	TM 4050		45	99	*L.* sp. indet.
81	TM 4466		29	114	*L.* sp. indet.
82	TM 4429		33	157	*L.* sp. indet.
83	TM 34		15	105	*L.* sp. indet.
84	TM 165		29	97	*L.* sp. indet.
85	TM 215		21	108	*L.* sp. indet.
86	BPI/1 4053		7	—	*L. mccaigi*
87	BPI/1 2241		10	—	*L. mccaigi*
88	BPI/1 4052		12	—	*L. mccaigi*
89	BPI MN 380	*L. mccaigi* (Cluver, 1971)	41	270	*L. mccaigi*
90	BPI MN 136	*L. amphibius* (type)	14	216	*L. mccaigi* (Cluver, 1971)
91	TM 155		34	254	*L.* sp. indet.
92	TM 196 (R.44)	*L. putterilli* (type)	14	225	*L. mccaigi* (cf. Cluver, 1971)
93	BPI/1 1746		18	119	*L. curvatus*
94	BPI MN 330	*L. primitivus* (type)	3	—	*L.* sp. indet.
95	BPI/1 4039		26	106	*L. curvatus*
96	BPI MN 413	*L. murrayi* (Kitching, 1968)	30	81	*L. curvatus* (Cluver, 1971)
97	BPI/1 1269		26	102	*L. curvatus*
98	TM 213		17	98	*L. curvatus*
99	TM 156		27	128	*L. curvatus*
100	TM 16	*L. theileri* (type)	21	130	*L. curvatus* (Cluver, 1971)
101	TM 191	*L. curvatus* (Brink, 1986)	15	115	*L. curvatus*
102	BPI/1 3976		30	193	*L. platyceps*
103	BPI MN 110	*L. curvatus* (Brink, 1986)	30	180	*L. curvatus*
104	TM 17	*L. jorisseni* (type)	29	142	*L. curvatus* (Cluver, 1971)
105	TM 2039		33	227	*L. platyceps*
106	BPI MN 407	*L. murrayi/L. curvatus**	24	41	*L.* sp. indet.
107	BPI MN 409	*L. murrayi/L. curvatus**	22	40	*L.* sp. indet.
108	BPI MN 412	*L. murrayi/L. curvatus**	27	52	*L.* sp. indet.
109	BPI MN 408	*L. murrayi/L. curvatus**	24	42	*L.* sp. indet.

Table 14.6. *(continued)*

Spec.	Museum no.	Taxonomic designation	Var.	CL	This study
110	TM 18	*L. wageri* (type)	24	92	*L.* sp. indet.
111	TM 112		36	105	*L.* sp. indet.
112	TM 70		23	–	*L.* sp. indet.

Note: Institutional abbreviations: NMB, National Museum, Bloemfontein; SAM, South African Museum, Cape Town; BMNH, British Museum (Natural History); CUM, Cambridge University Museum; BPI, Bernard Price Institute for Palaeontological Research, University of the Witwatersrand, Johannesburg; TM, Transvaal Museum, Pretoria. Taxonomic designation refers to the first species designation published for a particular specimen. Var. = number of measurements recorded. CL = basal cranial length (mm). This study = species designation accorded crania in this study applying the diagnostic criteria of Cluver (1971); where indicated, a specimen initially identified by Cluver (1971). * = these four specimens were attributed to *L. murrayi* by Kitching (1968) and to *L. curvatus* by Cluver (1971) but are assigned to an unknown species category here together with all other specimens with CL < 80 mm.

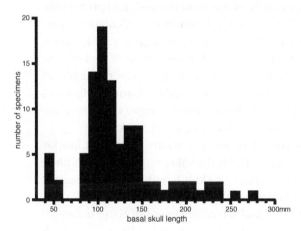

Figure 14.1. Size distribution of *Lystrosaurus* crania comprising the present sample. *N* = 98. Basal skull length ranges between 40 and 270 mm. Each specimen is represented by a short box covering a 10-mm span (e.g., there is one specimen only with a basal skull length between 270 and 280 mm).

Geoscience (Geological Survey of South Africa), Pretoria, that are absent from the present sample.

Nevertheless, the sample used here is reasonably large, most of the specimens are reasonably complete, and it includes crania of the species that have been regarded as valid by the more recent assessments (Cluver, 1971; Colbert, 1974; Kitching, 1977; Cosgriff et al., 1982; Brink, 1986; King, 1988). The sample includes fifteen type specimens. The crania included here span a considerable size range and represent some of the smallest and largest that are known: a nearly sevenfold difference in basal skull length is represented (40–270 mm). The size distribution of the sample (fig. 14.1) is kurtotic toward smaller specimens, with most crania between 90 and

150 mm in length. This may reflect the comparative rarity of large individuals across the paleolandscape, or it may simply be a taphonomic artifact, whereby very large skulls are less likely to be preserved intact.

Measurements

A possible forty-six possible measurements was recorded for each skull (fig. 14.2). Quantitative variables were not recorded for the mandible because comparatively few specimens preserve it intact. All measurements were recorded to the nearest millimeter. Approximately half the variables (measurements 2, 5–21, 36–40) pertain to the snout, being the region upon which almost all species diagnoses have concentrated. Most of these facial measurements could be recorded for the majority of specimens comprising the sample.

Measurements 5 and 6, 7 and 8, and 9 and 10 (fig. 14.2) reflect the arc and chord dimensions, respectively, of the frontal, nasal, and premaxilla. The arc measurements (5, 7, 9) were used only in the calculation of indices (chord/arc × 100) that express the degree of sagittal profile curvature for these bones; they were not treated as independent variables in the bivariate or multivariate analyses. Chord-arc indices were calculated to provide quantitative expression to purportedly diagnostic features such as the sagittal profile of the skull roof and the angle(s) at which the various facial planes intersect one another (tables 14.3–14.5).

In several instances, the cranial measurements are redundant insofar as the combination of two or more yields the value recorded for another. Thus, for example, variables 2 (anterior skull length) and 3 (posterior skull length) equal variable 1 (basal skull length). Similarly, variables 41–43 combine to yield variable 44, and the combination of these measurements together with variable 40 is equivalent to variable 1 (fig. 14.2). Such redundancy was eliminated from the multivariate analyses through the elimination of variables 1, 5, 7, 9, and 44.

Statistical Analyses

The metric variables were subjected to bivariate and multivariate analyses. In the first instance, the allometric relationship of each variable against basal skull length was determined by fitting both measurements to the simple biparametric power function $y = \alpha x \beta$, where y is the variable in question, x is basal skull length, α is a numerical constant, and β is the ratio of the specific growth rates of the two variables (Gould, 1966; Alexander, 1985). Following the procedures adopted in other studies (e.g., Grine et al., 1978; Tollman et al., 1980; Emerson & Bramble 1993), basal skull length was chosen as the measure by which to evaluate

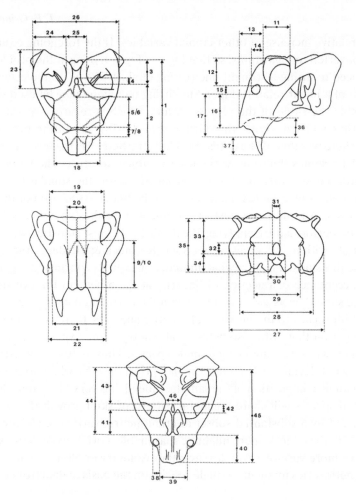

Figure 14.2. Measurements recorded for *Lystrosaurus* skulls. 1, basal skull length, from tip of snout to back of occipital condyles; 2, anterior skull length, from tip of snout to anterior margin of pineal foramen; 3, posterior skull length, from anterior margin of pineal foramen to back of occipital condyles; 4, length of the pineal foramen; 5, arc length of frontal (in midline from preperietal—frontal suture to frontonasal suture); 6, chord length of frontal; 7, arc length of nasal (in midline from frontonasal suture to nasopremaxillary suture); 8, chord length of nasal; 9, arc length of premaxilla (in midline from nasopremaxillary suture to tip of snout); 10, chord length of premaxilla; 11, orbit length; 12, orbit height; 13, anteroposterior (AP) length of snout from anterior margin of orbit; 14, AP distance between anterior margin of orbit and anterior margin of nares; 15, dorsoventral (DV) distance between inferior margin of orbit and inferior margin of nares; 16, DV height of snout from inferior margin of nares to tip of premaxilla; 17, DV height of caniniform process of maxilla from inferior margin of nares; 18, minimum interorbital distance; 19, maximum distance between the lateral margins of the prefrontals; 20, minimum distance between nares; 21, maximum distance between lateral margins of caniniform processes of maxillae; 22, maximum distance between anterior quadrate foramina; 23, maximum AP length of temporal fossa; 24, maximum breadth of temporal fossa from the lateral margin of the postorbital; 25, minimum breadth of the braincase between the temporal fossae; 26, maximum distance between the lateral margins of the temporal

the relative increase of other cranial variables. If the foregoing equation is converted to logarithms, so that $\log y = \log a + \beta \log x$, the problem is reduced to fitting a straight line (Grine et al., 1978). With the establishment of a linear relationship, a line-fitting technique may be used to determine the slope of the regression line, which is regarded as the allometric coefficient. Following Grine et al. (1978), the fitting procedure used here was that of Bartlett's "best fit" (Bartlett 1949) as this regression model assumes that both variables are subject to error. The 95 percent confidence intervals for β were calculated, as was the standard correlation coefficient, r, which is a measure of the linear association of the two variables. An additional test of the linearity of the functional relation can be performed by applying Bartlett's (1949) test statistic, which is a t-variate with $n-3$ degrees of freedom (Kermack & Haldane, 1950). Because this statistic is somewhat limited, however, its conclusion regarding the acceptance or rejection of linearity at the 5 percent level of significance should be assessed together with the correlation coefficient r.

With regard to the bivariate allometric analyses, there is no universal assumption that either isometry or allometry is the expected condition for a growth series involving a single species. Thus, neither isometry nor significant deviations from it form an a priori basis by which to infer the taxonomic composition of the sample. Indeed, it is very likely that closely related species will share a general pattern of skull growth that will be reflected by a substantial subset of craniometric variables. However, in particular instances, there might be reasonable a priori expectations that one or more variables will display either isometry or allometry in an ontogenetic series involving a single species on the basis of data from extant

---

Figure 14.2. (*continued*)

arches; 27, maximum width of the back of the skull across the lateral margins of the squamosals; 28, maximum distance between the lateral margins of the quadrates; 29, minimum distance between the posttemporal foraminae; 30, breadth across the occipital condyles; 31, maximum breadth of the foramen magnum; 32, height of the foramen magnum; 33, DV distance between the inferior margin of the foramen magnum and the skull roof; 34, DV distance between the inferior margin of the foramen magnum and the basioccipital tuber; 35, DV distance between the basiocciptal tuber and the skull roof; 36, depth of the palate to the inferior margin of caniniform process of the maxilla; 37, height of the canine tusk; 38, maximum mediolateral (ML) diameter of the canine tusk; 39, minimum distance between the canine tusks; 40, AP length of the palate; 41, maximum AP length of the internal choanae; 42, AP length of the interpterygoid fossa; 43, AP distance between anterior margin of interpterygoid fossa and back of occipital condyles; 44, AP distance between the posterior margin of the palate and the back of occipital condyles; 45, AP distance between the tip of the snout and the posterior margin of the quadrate; 46, minimum ML breadth across the pterygoids.

species. For these variables, deviations from the expected slope of the regression line may suggest species heterogeneity in the sample. An example of such a variable might be orbit size in diurnal terrestrial vertebrates (Gould 1966; Emerson & Bramble 1993). At the same time, a low correlation coefficient (r) of the slope also might indicate that more than one allometric coefficient can be fitted to the data. In such instances, this might well imply taxonomic heterogeneity in the sample, where different species have dissimilar ontogenetic trajectories.

In the second instance, the data were assessed by applying multivariate statistical models. Because these models require a complete data matrix,* and because many measurements could not be recorded for all specimens, two separate data sets were distilled from the original matrix (table 14.7). Multivariate data set 1 maximized the number of specimens at the expense of variables (sixty-eight specimens, twelve variables). Multivariate data set 2 maximized the number of variables, resulting in fewer crania (twenty-nine specimens, twenty-one variables). Both multivariate data sets were subjected to multivariate analysis of allometry and principal components analyses. Measurements were log-transformed before analysis, and all analyses were performed with the statistical software package Matlab.

Shea (1985) has discussed the implementation of multivariate allometry and its relationship to bivariate methods of analysis. Briefly, multivariate approaches to allometry employ analyses of variable loadings on the first axis of a principal components analysis (PCA). Principal components analyses calculate eigenvectors and corresponding eigenvalues from multivariate data matrices. On the assumption that the first calculated eigenvalue accounts for the overwhelming majority of variance, and on the assumption that the variable loadings (i.e., eigenvector coefficients) of the first principal axis are all positive, the variable loadings can be interpreted as scaling factors, or coefficients of allometry. In addition, the null hypothesis of isometry can be evaluated as one over the square root of the number of variables measured.

An important distinction between bivariate and multivariate approaches to allometry relates to the size-standard against which changes in other variables are evaluated. The bivariate analyses of allometry used here followed the standard practice of employing a single variable (basal

*It is, of course, possible to calculate values for missing data using a variety of techniques (e.g., Holt & Benfer, 2000) but, ultimately, such values contain a high degree of circularity. Because of this, we have eschewed such procedures in the present study.

Table 14.7. Composition of the multivariate data sets

Multivariate Data Set 1

Specimens (*n* = 68)

1, 2, 3, 4, 5, 7, 8, 9, *14*, 16, 18, 19, *22*, 23, *24*, *25*, 26, 28, *29*, *32*, 34, *36*, 40, 41, 42, 43, 46, *52*, 53, *54*, 55, 56, 58, *59*, *60*, 61, 62, 63, 64, 65, *67*, 68, *70*, *72*, *73*, *74*, 75, *76*, *77*, *78*, *79*, *80*, *81*, *82*, 85, 89, *91*, 95, 96, 97, 99, 102, 104, 105, 107, 108, 109, *111*

Variables (*n* = 12)

2, 3, 4, 6, 8, 10, 13, 14, 16, 17, 18, 25

Multivariate Data Set 2

Specimens (*n* = 29)

1, 2, 3, 14, 22, 24, 25, 29, 32, 36, 52, 54, 56, 59, 60, 67, 70, 72, 73, 74, 76, 77, 78, 79, 80, 81, 82, 91, 111

Variables (*n* = 21)

2, 3, 4, 6, 8, 10, 11, 12, *13, 14*, 15, *16, 17, 18*, 19, 20, 21, *25*, 32, 33, 38

Note: Specimen numbers in bold italics in Multivariate Data Set 1 compose multivariate data set 2. Variable numbers in bold italics in Multivariate Data Set 2 compose multivariate data set 1.

skull length) as a size comparator with respect to all other morphometric variables. This approach implicitly assumes the isometry of the variable upon which size comparisons are based. In contrast, multivariate analyses of allometry implicitly define the size comparator as the geometric mean, which is the *n*th root of the product of the *n* variables measured for each sample (O'Keefe et al., 1999). In this regard, isometry need not be assumed for any given variable.

Nevertheless, multivariate analyses did support the assumption of isometry in basal skull length that was used in the bivariate analyses. That is, a variation of Multivariate Data Set 2, run with variable 1 rather than variables 2 and 3 (see above), revealed that basal skull length has a scaled isometry value that is very weakly negative (0.9092), but the 95 percent confidence interval for this variable includes isometry. Given the small sample size, and the lack of strength of this signal, there is every reason to consider basal skull length as an isometric size comparator in the bivariate allometry analyses.

Principal components analysis is designed to analyze sets of correlated variables. Because it does not assume a priori grouping of specimens, it allows for the discovery of groups. Therefore, in addition to using the first principal components analyses to generate multivariate analyses of allometry, we also used PCA to identify possible grouping of specimens by evaluating the overlap in specimen distributions on the first three axes generated from PCA.

Nonmetric Features

Morphologic variants of six cranial features that are not readily amenable to measurement were scored. These characters have featured prominently in the species diagnoses provided by Broom (1932), Cluver (1971; see also Colbert, 1974 and Cosgriff et al., 1982), and Brink (1986).

Sagittal facial profile. This character was noted by Broom (1932) as distinguishing *L. murrayi* from *L. platyceps,* and by Brink (1986) as a feature that presumably distinguished *L. curvatus* from all others. Following Cluver (1971), three states were recognized: (1) the premaxillae, nasals, and frontals form a generally smooth arc in profile without a distinct angular break between any of these elements; (2) the premaxillae, nasals, and frontals form a series of distinct, variably angled breaks so that the profile consists of three individual planes; and (3) the premaxillae and nasals form a single plane, while the frontals are distinctly angled to this plane so that the profile consists of two planes.

Prefrontal-nasal crest. Broom (1932) recognized the prefrontal-nasal crest as a diagnostic feature, and Cluver (1971) observed that the "frontonasal" ridge together with the presence of a prefrontal boss along the orbital margin served to discriminate among species. However, because the presence of prefrontal bosses along the superoanterior orbital margin is directly related to the development of the prefrontal-nasal ridge, which terminates laterally at this point, the prefrontal bosses were not considered here separately from the prefrontal-nasal crest. Four states of prefrontal-nasal crest development were recognized, ranging from absence (state 1) to marked development of a transverse ridge that crosses the prefrontal bones and the upper part of the nasals (state 4). The specimens employed as standards of comparison were NMQR-C 154 (absent), NMQR-C 211 (slight), NMQR-C 608 (moderate), and SAM-PK-K 4800 (marked).

Postorbital boss. Brink's (1986) recognition of a single character—orbital bosses—is perhaps misleading, because bony bosses may be present on the prefrontal and/or postorbital bones, and they appear to vary independently in their development. Thus, these features were recorded here as separate entities. Two states of postorbital boss development were recognized: (1) absent, and (2) present. The specimens employed as standards of comparison were SAM-PK-K 4800 (absent) and BP/1/879 (present).

Frontal rugosities. Broom (1932) stated that "frontal development" differs greatly between *L. murrayi* and *L. mccaigi,* but it is unclear in what manner he regarded these taxa as differing. It is possible that he was referring to the development of bony ridges and/or tuberosities, although Cluver (1971) cites these as being present in both taxa. We here regard frontal rugosities to include bosses as well as the sagittal and radiating ridges that may be variably developed on the frontal bone (Cluver, 1971). Four states were recognized, ranging from absence (state 1) to marked development of bony excrescences (state 4). The specimens used as comparative standards were NMQR-C 154 (absent), NMQR-C 211 (slight), SAM-PK-K 4800 (moderate), and NMQR-C 282 (marked).

Premaxillary sagittal ridge. Broom (1932) mentioned that the presence of longitudinal ridges on the snout distinguishes *L. murrayi* from *L. platyceps,* although it is not clear whether he was referring to the median premaxillary ridge, parasagittal premaxillary ridges, or both. Cluver (1971) considered parasagittal ridges together with the median premaxillary sagittal ridge as comprising a single feature, although he noted that they do not necessarily covary (table 14.4). As a result, we regard the median sagittal premaxillary ridge as being independent of the parasagittal premaxillary ridges, which are related to the degree to which the premaxilla is squared-off (see below). Four states of median premaxillary sagittal ridge development were recognized, ranging from absence (state 1) to a marked, sharp crest (state 4). The specimens employed as comparative standards were SAM-PK-K1381 (absent), SAM-PK641 (slight), NMQR-C 282 (moderate), and SAM-PK-K 4800 (marked).

Transverse contour of premaxilla. The premaxilla was described by Brink (1986) as either having a smoothly rounded profile or as being sharply angled with or without corresponding parasagittal ridges. Here we relate parasagittal ridge development to the degree to which the premaxilla is squared-off. Two states of premaxillary contour were recognized: (1) the premaxilla forms an evenly and gently rounded contour, and (2) there is an angular demarcation between anterior and lateral surfaces of the premaxilla accompanied by a crest of variable expression. The specimens employed as comparative standards were NMQR-C 154 (rounded) and SAM-PK-K 1381 (angled).

Results

Bivariate Allometric Analysis

The range of basal skull length in the current sample (40–270 mm) can be accommodated comfortably within the ontogenetic size range exhibited by the skulls of large-bodied extant sauropsids such as *Caiman crocodilus* (Gans, 1980), *Alligator mississippiensis* (Mook, 1921; Dodson, 1975), and *Crocodylus porosus* (Webb & Messel, 1978). The coefficients of allometry as well as the correlation coefficients and results of the statistical tests for linearity recorded for the forty-two morphometric variables against basal skull length are presented in table 14.8.

For most measurements, correlation with basal skull length is rather high (i.e., >0.85), and the hypothesis of bivariate linearity cannot be rejected. Thus, most variables reflect a comparatively high degree of morphometric homogeneity within this sample. However, there are six measurements for which the correlation coefficients are comparatively low (0.59–0.81), and another six that do not necessarily conform to a quadratic relationship with basal skull length; all but one of these is associated with positive allometry (table 14.8). Selected bivariate plots of the cranial measurements discussed below are presented in figure 14.3.

Significantly, five of the variables that exhibit a low *r* values with basal skull length (variables 8, 13, 14, 15, and 37) are related in some way to characters cited in species diagnoses by Broom (1932), Cluver (1971), and/or Brink (1986). Thus, the low *r* value associated with nasal bone length (variable 8) may be related to variation in the degree of nasofrontal ridge development. Alternatively, it could be influenced by the angle between the nasal bones and the facial plane of the premaxilla or by the location of the nasofrontal suture. These features have been cited as being diagnostic by Cluver (1971), but they also may exhibit individual (i.e., intraspecific) variation (Cluver, 1971; Cosgriff et al., 1982). Variables thirteen and fourteen, which are also associated with low *r* values, reflect the anteroposterior (AP) elongation of the snout. Their variability may be related to differences in snout development among various species noted by Broom (1932), Cluver (1971), and Brink (1986), although these authors clearly viewed this character more in relation to dorsoventral (DV) height than AP length. On the other hand, measurement 15, which reflects the height of the superior portion of the snout, is certainly related to snout development as envisioned by them. In particular, Broom (1932) claimed that upper facial height differentiates *L. mccaigi* from *L. murrayi* and *L. platyceps* (table 14.3). Measurement fifteen is associated with the lowest

Table 14.8. Summary of data concerning relative growth and variability in cranial dimensions of *Lystrosaurus*

Y	n	α	β		β-CI	r	t	df	RL
2	94	−0.65	1.05	0.97	1.13	0.96	−1.39	91	no
3	94	−0.69	0.91	0.76	1.07	0.85	1.45	91	no
4	90	−1.73	0.78	0.99	1.21	0.94	−0.60	84	no
6	87	−1.52	1.06	0.96	1.17	0.94	−0.08	84	no
8	86	−1.46	0.91	0.71	1.11	0.80	0.41	83	no
10	86	−1.53	1.19	1.11	1.28	0.97	0.98	83	no
11	82	−0.89	0.95	0.87	1.03	0.96	1.16	79	no
12	89	−1.12	0.98	0.87	1.09	0.92	0.85	86	no
13	90	−1.61	1.06	0.85	1.27	0.81	−0.97	87	no
14	84	−1.34	0.90	0.70	1.11	0.76	0.09	81	no
15	74	−4.52	1.42	0.96	1.93	0.59	1.68	71	no
16	81	−2.45	1.27	1.13	1.41	0.94	−1.05	78	no
17	77	−2.11	1.25	1.11	1.38	0.94	−2.50	74	yes
18	96	−0.79	0.98	0.85	1.10	0.90	−1.33	93	no
19	90	−1.24	1.14	1.04	1.23	0.95	−1.45	87	no
20	92	−1.09	0.91	0.84	0.98	0.96	−0.92	89	no
21	84	−0.89	1.05	0.97	1.13	0.97	−1.58	81	no
22	21	−0.81	1.09	0.90	1.30	0.94	−2.26	18	no
23	53	−0.79	1.00	0.88	1.12	0.95	0.16	50	no
24	51	−1.88	1.18	1.03	1.32	0.95	−2.24	48	yes
25	96	−1.42	0.98	0.88	1.08	0.93	0.67	93	no
26	41	−0.82	1.16	1.02	1.31	0.96	−2.01	38	no
27	23	−0.77	1.16	1.02	1.33	0.98	0.56	20	no
28	22	−0.22	0.98	0.83	1.14	0.96	−0.91	19	no
29	30	−1.38	1.13	0.94	1.31	0.96	1.34	27	no
30	53	−1.26	0.90	0.80	1.00	0.94	1.60	50	no
31	44	−1.19	0.77	0.61	0.92	0.89	0.11	41	no
32	55	−1.79	0.95	0.77	1.13	0.88	1.86	52	no
33	64	−1.75	1.17	1.05	1.29	0.94	2.45	61	yes
34	40	−3.33	1.35	1.14	1.55	0.93	1.74	37	no
35	38	−1.80	1.26	1.14	1.38	0.97	2.37	35	yes
36	12	−2.50	1.17	0.93	1.39	0.98	0.65	9	no
37	26	−5.23	1.70	1.01	2.57	0.74	−0.05	23	no
38	80	−3.10	1.10	0.92	1.29	0.86	1.37	77	no
39	72	−0.83	0.93	0.83	1.02	0.95	−0.44	69	no

Table 14.8. *(continued)*

Y	n	α	β	β-CI		r	t	df	RL
40	14	−2.20	1.21	1.03	1.41	0.97	2.47	11	yes
41	11	−0.66	0.87	0.43	1.26	0.87	−0.93	9	no
42	14	−2.55	1.03	0.60	1.62	0.80	−2.08	11	no
43	15	−1.10	1.02	0.86	1.19	0.97	0.46	12	no
44	12	−0.26	0.97	0.86	1.09	0.99	0.27	9	no
45	19	0.27	0.93	0.82	1.03	0.99	−0.36	16	no
46	25	−1.81	1.01	0.90	1.12	0.97	3.90	22	yes

Note: Y = comparative variable number; n = number of specimens; α = log a; β = slope of line; β-CI = 95% confidence interval for β; r = correlation coefficient, t = Bartlett's test statistic; df = degrees of freedom; RL = rejection of linearity in favor of another (e.g., parabolic) relationship.

correlation coefficient among all measurements (r = 0.59), and it exhibits positive allometric growth (β = 1.42), although the lower 95 percent confidence limit for β ranges to just below isometry (0.96). Finally, canine tooth height (variable 37) exhibits a low correlation with basal skull length and is also positively allometric, exhibiting the highest β value of any measurement (table 14.8). This may reflect several phenomena that are not necessarily mutually exclusive. In the first instance, it may reflect a common albeit sexually dimorphic pattern of canine eruption in *Lystrosaurus*. By this possibility, the tusks, which have not yet emerged in the smallest skulls (Kitching, 1968), subsequently emerge at a rate that outstrips the increase in overall cranial size in both sexes, but their emergence (i.e., length) is influenced by the sex of the individual. Males may have erupted their tusks at a faster rate, resulting in longer canines throughout the period of sexual maturity. Another possibility is that c anine length was moderated by its use, so that apical wear might have positively influenced the rate of eruption. In either case, the prolonged eruption of the tusk would affect its length to a greater degree than its cross-sectional diameter (variable 38) if the degree of longitudinal tapering exhibited by the canine decreases toward the cervix. That is, unless the tusk is constructed as a cone whose height is equivalent to its basal radius, its continued eruption will increase its length to a greater degree than its basal diameter. Finally, Cluver (1971) has suggested that there is a taxonomic difference in tusk size, with *L. curvatus* and *L. platyceps* possessing weaker canines than *L. murrayi*, *L. declivis*, and *L. mccaigi*.

Thirteen of forty-one measurements are positively allometric. As noted above, variable 15, which measures upper facial height, is positive,

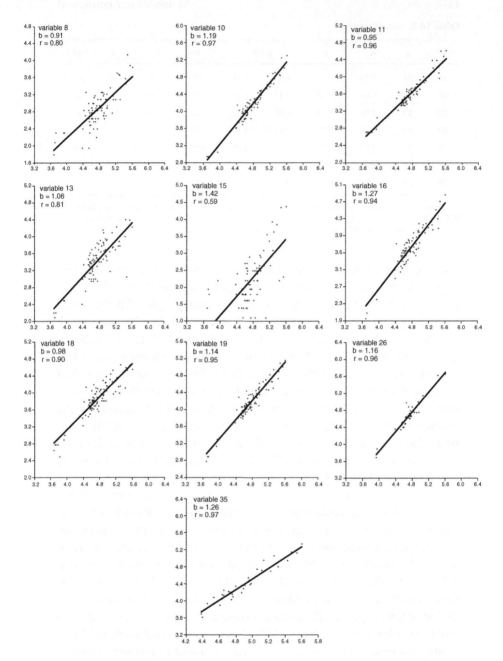

Figure 14.3. Selected bivariate plots related to measurements of the skull of *Lystrosaurus*. The slopes represent the calculated line of Bartlett's "best fit" and are plotted as logarithmic against both axes, but the measurements along both axes have not been log converted from millimeters. In all plots, the *x* axis is basal skull length (variable 1).

as are variables 10, 16, and 17, which relate to the height of the lower portion of the snout. These four variables are clearly related to the differential snout development cited by Broom (1932), Cluver (1971), and Brink (1986) in their species diagnoses (tables 14.3–14.5). Similarly, measurement 19, which would be directly influenced by the development of prefrontal protuberances (a character sometimes cited as diagnostic; tables 14.3–14.5) also exhibits positive allometry. By way of comparison, the width between the nasal bosses, which is also positively allometric in the anomodont *Aulacephalodon,* is regarded as probably having been sexually dimorphic in that taxon (Tollman et al., 1980). Not only are snout height and the antorbital width across the skull roof positively allometric to basal skull length, so too is the width of the temporal fossa (reflected in variables 24 and 26) and the breadth (variable 27) and height (variables 33–35) of the back of the cranium. The latter three parameters may be related to relative enlargement in the size of the jaw adductor musculature in larger crania.

Overall, those measurements that pertain to the height of the snout, the height of the back of the skull, and the overall breadth of the skull are all positively allometric. This translates to the observation that the skull of *Lystrosaurus* becomes relatively broader and deeper as it increases in length. The situation in *Lystrosaurus* contrasts rather noticeably with that recorded for *Aulacephalodon,* where only three of thirty-one cranial measurements displayed positive allometry (Tollman et al., 1980). The cranium of *Aulacephalodon,* like that of *Lystrosaurus,* exhibited positive allometric increase in palatal length and total skull width.

Only three cranial measurements are associated with negative allometry in *Lystrosaurus* (table 14.8). Variable 20, which records the breadth of the premaxillae anterior to the nares, exhibits negative allometry, whereas premaxilla length (variable 10) is positively allometric. This results in an accentuation of the appearance of an elongate, narrow beak in the larger *Lystrosaurus* skulls. Variable thirty (breadth of the occipital condyles) exhibits negative allometry, and this parameter is related to the size of the foramen magnum (variable 31), which is also negatively allometric. Of potential interest is the observation that orbit size, as determined by either length (variable 11) or height (variable 12), is essentially isometric. The β values for these measurements are only marginally negative (0.95 and 0.98, respectively), and the upper 95 percent confidence limits of both slopes include positive allometry (1.03 and 1.09, respectively). This is of potential biological significance because orbital size is characteristically negatively allometric in the ontogenetic growth series of living higher vertebrates, as smaller, juvenile specimens in any given species have relatively large orbits. Both Broom (1932) and Brink (1986) cited orbit size as differentiating among *Lystrosaurus* species. Although they differ in their views on the distribution of orbit size, it is perhaps noteworthy that Broom (1932) in particular cited the largest species (*L. platyceps* and *L. mccaigi*) as possessing the relatively largest orbits. This observation is largely in accord with our findings (see fig. 14.8).

Post-Hoc Taxonomic Assessment of Quantitative Variables

A number of the features that have been said to differ among the purported species of *Lystrosaurus* are amenable to measurement, and these have been discussed in the preceding section. Although some appear to be related simply to allometric patterns of cranial growth that characterize the genus, others appear to reflect species level differences. Here we examine the bivariate arrangement of specimens that have been attributed a priori for this purpose to one of six purported species (table 14.6).

In the first instance, it is necessary to establish the size ranges of the crania attributed to these species. Basal skull length could be measured for twenty-three skulls attributed here to *L. murrayi*, twenty-three attributed to *L. declivis*, fifteen attributed to *L. curvatus*, four attributed to *L. mccaigi*, one of the two attributed to *L. oviceps*, and all four attributed to *L. platyceps*. This measurement could also be recorded for twenty-eight crania assigned a priori to *Lystrosaurus* sp. indet. A series of very small crania in the collection of the Bernard Price Institute for Palaeontological Research (University of the Witwatersrand) were attributed by Kitching (1968) to *L. murrayi* and by Cluver (1971) to *L. curvatus,* but we have refrained here from attributing these or any other specimen with a basal skull length of less than 80 mm to a particular species.

The ranges of cranial length in the current samples of *L. murrayi, L. declivis,* and *L. curvatus* exhibit considerable overlap, with most of specimens attributed to these three species being between about 80 and 150 mm (table 14.9, fig. 14.4). The larger specimens attributed to *L. declivis* and *L. curvatus* overlap the range of *L. platyceps,* and the largest skulls of *L. declivis* and *L. platyceps* fall within the observed *L. mccaigi* range. However, all specimens of *L. platyceps* and *L. mccaigi* are larger than any specimen of *L. murrayi,* and all specimens of *L. mccaigi* are larger also than those attributed to *L. curvatus* (fig. 14.4).

Brink (1986), who considered *L. platyceps* and *L. mccaigi* to be synonymous, distinguished "very large" *L. platyceps* from "small" *L. murrayi* (table 14.5). Brink's (1986) observation is supported insofar as all eight

Variability in *Lystrosaurus*

Table 14.9. Range of basal skull length (mm) recorded for specimens of purported species of *Lystrosaurus* comprising the present sample

Species	*n*	Min.	Median	Max.	Range %
L. murrayi	23	83	121	159	48
L. declivis	23	94	162	230	59
L. curvatus	15	80	139	197	54
L. oviceps	1	—	—	—	—
L. platyceps	4	170	199	227	25
L. mccaigi	4	216	243	270	20
L. sp.	28	40	—	254	—

Note: Basal skull length was measurable for only one specimen attributed to *L. oviceps* (95 mm). Range % is calculated as (max. − min.)/max. × 100. Note that the minimum was established a priori at 80 mm for any purported species hypodigm.

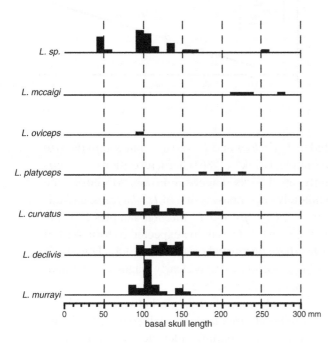

Figure 14.4. Size distribution of *Lystrosaurus* crania composing the present putative species samples. Each specimen is represented by a short box covering a 10-mm span (e.g., there is one specimen attributed to *L. murrayi* with a basal skull length between 150 and 160 mm).

specimens assigned here to *L. platyceps* and *L. mccaigi* have a basal length of ≥170, and all twenty-three crania assigned here to *L. murrayi* are less than 160 mm in length. Certainly there is a substantial gap in size between the largest *L. murrayi* skull (159 mm) and the smallest *L. mccaigi* specimen (216 mm) in the present sample. However, because almost all

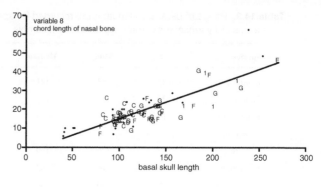

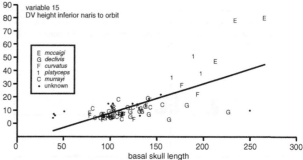

Figure 14.5. Distribution of *Lystrosaurus* crania attributed a priori to one of six putative species for the chord length of the nasal bones (variable 8) and for superior facial height (variable 15) versus basal skull length.

the specimens attributed to *L. declivis* and *L. curvatus* fall within the size range of the *L. murrayi* sample, Brink's (1986) characterization of *L. murrayi* as being particularly small lacks diagnostic power. Moreover, the range of *L. declivis* completely encompasses that of *L. platyceps,* and it is evident that the present *L. platyceps* and *L. mccaigi* samples lack the smaller (i.e., younger) specimens that would be expected to form part of the ontogenetic ranges for these purported species (fig. 14.4). Thus, their characterization as very large (Brink, 1986) is clearly diagnostically inadequate because not all of them could have been.

As noted above, the relatively weak correlation between nasal bone length (variable 8) and overall skull size may be explained by several phenomena that are not necessarily of taxonomic relevance. This appears to be borne out by the lack of separation among crania that have been attributed to different putative species (fig. 14.5). Thus, specimens of *L. curvatus, L. murrayi,* and *L. declivis* are randomly distributed around the regression line in the lower part of the cranial size range (<150 mm), and crania of *L. declivis* and *L. platyceps* are so distributed in the upper part of that range (>150 mm). On the other hand, the chord/arc index of the nasal bones serves to distinguish most specimens attributed to

Table 14.10. Chord/arc index values recorded for the nasal bone in crania of purported
species of *Lystrosaurus* comprising the present sample

Species	n	x	Min.	Max.
L. murrayi	23	94.0	85.0	100.0
L. declivis	23	92.0	82.6	96.3
L. curvatus	15	97.2	91.7	100.0
L. oviceps	2	97.1	94.1	100.0
L. platyceps	4	100.0	100.0	100.0
L. mccaigi	2	99.2	98.4	100.0

Note: Some species samples include specimens for which basal skull length could not be measured.

L. curvatus, L. oviceps, L. platyceps, and *L. mccaigi* from those assigned
to *L. murrayi* and *L. declivis,* irrespective of cranial length (table 14.10).
Indeed, there is no overlap of the *L. declivis* sample range with those of
L. platyceps and *L. mccaigi.* This index reveals the nasals to be more
sharply curved or angled in *L. murrayi* and *L. declivis* than in the other
four purported species.

The three smallest skulls for which the nasal chord/arc index could be
determined (lengths of 40–42 mm) were attributed by Kitching (1968) to
L. murrayi and by Cluver (1971) to *L. curvatus.* These specimens have index values (100.0) typical of *L. curvatus.* However, the next smallest specimen (SAM-PK 3531), with a basal skull length of 51 mm, displays the
more highly curved nasal profile (index = 90.0) typical of *L. murrayi* and
L. declivis, whereas the similarly sized cranium BPI/1/4190 (basal skull
length = 52 mm) displays the flat nasal profile typical of *L. curvatus.* This
suggests that variability in nasal profile was expressed early in ontogeny
in *Lystrosaurus* (i.e., at a size in which the skull has increased in length to
less than five percent of the largest known skull). Moreover, it suggests
that taxonomic differences in this feature were also expressed at an ontogenetically early stage if the present species samples are valid.

The height of the superior portion of the snout (variable 15), which
displays the lowest correlation with and is positively allometric to skull
length, was seen by Broom (1932) as distinguishing *L. mccaigi* from *L.
murrayi* and *L. platyceps* (table 14.3). However, the distribution of specimens reveals that most crania attributed to *L. murrayi* fall below the common regression line, and all six skulls of *L. platyceps* and *L. mccaigi* fall
above it; so too do several much smaller *L. murrayi* specimens. A pattern
of random taxonomic dispersal is observed when variables 15 and 16 are
combined to represent total facial height (fig. 14.6). This indicates that
the depth of the snout, which has been regarded as diagnostic by all

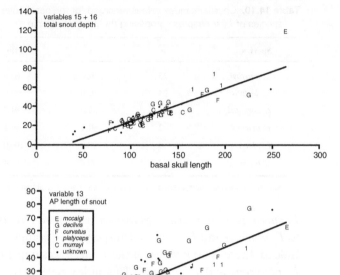

Figure 14.6. Distribution of *Lystrosaurus* crania attributed a priori to one of six putative species for total facial height (variables 15 + 16) and for the AP length of the snout (variable 13) versus basal skull length.

previous workers (Broom, 1932; Cluver, 1971; Colbert, 1974; Cosgriff et al. 1982; Brink, 1986), simply increases in a positively allometric manner in all purported species. It does not distinguish among them in a manner that is independent of overall cranial size.

At the same time, however, snout development also may be considered in relation to its length, where the facial plane of the premaxilla may be aligned either perpendicular to or at an angle to the plane of the skull roof (tables 14.4, 14.5). In this regard, the distribution of specimens for variable 13 (AP length of the snout) against basal skull length reveals that those attributed to *L. curvatus* are randomly scattered around the regression line (fig. 14.6). Crania of *L. murrayi*, *L. platyceps*, and *L. mccaigi* tend to fall below it, whereas *L. declivis* skulls tend to fall above it (fig. 14.6). Thus, the snout projects relatively farther forward in *L. declivis* than in these other purported species. This observation may be in keeping with the differences cited by Cluver (1971) regarding the angle at which the snout is inclined to the roof of the skull in *L. declivis* (angled) and *L. murrayi* (perpendicular). However, a similar relationship between snout elongation and inclination is not readily apparent for skulls of

Table 14.11. Chord/arc index values recorded for the snout in crania of purported species of *Lystrosaurus* comprising the present sample

Species	*n*	*x*	SD	Min.	Max.
L. murrayi	22	84.3	3.6	76.0	89.9
L. declivis	22	89.2	3.7	79.4	93.1
L. curvatus	16	88.6	3.2	82.9	97.5
L. oviceps	2	89.0	—	86.0	91.9
L. platyceps	4	89.6	1.5	87.4	90.9
L. mccaigi	2	81.8	—	79.4	84.1

Note: This index, calculated as (vars 6 + 8 + 10)/(vars 5 + 7 + 9) × 100, reflects the overall degree of curvature of the frontal, nasals, and premaxillae in the sagittal plane. Some samples include specimens for which basal skull length could not be measured.

L. platyceps, in which the snout is angled forward, but it is relatively short by comparison with crania attributed to *L. declivis* (cf. table 14.4, fig. 14.6). The snout of *L. mccaigi,* like that of *L. murrayi,* has a perpendicular facial plane and is relatively short AP. The total chord/arc index of the snout [variables $(6 + 8 + 10)/(5 + 7 + 9)$] is not sensitive to overall cranial size, and the values for crania attributed to *L. curvatus* encompass those of all other purported species (table 14.11). Nevertheless, in general, the snouts of *L. murrayi* skulls are somewhat more sharply curved than those of *L. declivis,* which tend to resemble more closely the crania of *L. curvatus, L. oviceps,* and *L. platyceps* in this particular index (table 14.11). The overall degree of curvature of the snout is the same in specimens attributed to *L. mccaigi* as in those of *L. murrayi* (table 14.11).

Broom (1932) suggested that the skull roof between the orbits was relatively broader (compared with the intertemporal diameter) in *L. mccaigi* than in *L. platyceps.* According to Cluver (1971), interorbital width serves to distinguish the relatively narrow skulls of *L. curvatus* (and *L. bothai*) from those of all other species. In this regard, the distribution of specimens for interorbital diameter (variable 18) reveals that most of those attributed to *L. curvatus* and *L. platyceps* fall below the common regression line, whereas *L. murrayi* and *L. declivis* crania tend to fall above it (fig. 14.7). The single specimen of *L. oviceps* and the two skulls of *L. mccaigi* for which both measurements could be recorded fall on the common regression line ($r = 0.72$). This is strongly suggestive of a distinction in the relative breadth of the skull roof between *L. curvatus* and *L. platyceps,* on the one hand, and *L. murrayi* and *L. declivis,* on the other. Indeed, regression lines fitted to the specimens composing these

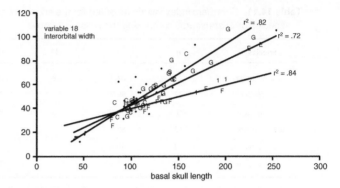

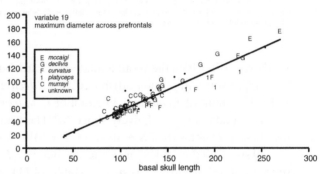

Figure 14.7. Distribution of *Lystrosaurus* crania attributed a priori to one of six putative species for interorbital diameter (variable 18) and for prefrontal breadth (variable 19) versus basal skull length. With regard to variable 18, the r^2 value = 0.72; specimens attributed to *L. murrayi* and *L. declivis* align along a steeper slope with a higher r^2 value of 0.82, and specimens attributed to *L. curvatus* and *L. platyceps* align along a lower slope with a higher r^2 value of 0.84.

two groups have very different slopes and higher r values (0.84 and 0.82, respectively) than the common regression line (fig. 14.7). These data corroborate Cluver's (1971) observations regarding the difference of *L. curvatus* from *L. murrayi*, *L. declivis*, and perhaps *L. mccaigi*, but they differ from his characterization of *L. platyceps* as possessing a broad skull roof. These data also corroborate Broom's (1932) observation that the skull roof of *L. platyceps* tends to be relatively narrower than that of *L. mccaigi*. The different slopes obtained here imply different ontogenetic allometric trajectories in *L. curvatus* and *L. platyceps* than were realized by *L. murrayi* and *L. declivis*: in the latter, the skull roof grew in width at a relatively faster rate.

As noted above, the distance between the lateral margins of the prefrontals (variable 19), which is variable and positively allometric, is influenced by the development of bosses on these bones. The distribution of specimens for variable 19 against basal skull length reveals that most attributed to *L. curvatus* and *L. platyceps* fall below the regression line, whereas the majority referred to *L. murrayi*, *L. declivis*, and *L. mccaigi* fall above it (fig. 14.7). This indicates a distinction in the relative breadth

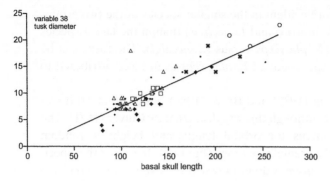

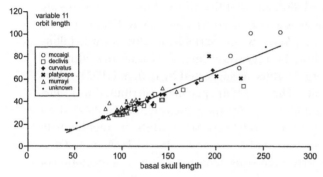

Figure 14.8. Distribution of *Lystrosaurus* crania attributed a priori to one of six putative species for the cross-sectional diameter of the canine tusk (variable 38) and for orbit size (variable 11) versus basal skull length.

of the skull roof across the prefrontals between *L. curvatus* and *L. platyceps,* on the one hand, and *L. murrayi* and *L. declivis* on the other. At this point, however, it is not clear whether this distinction simply reflects a generally broader skull roof in *L. murrayi* and *L. declivis* (as reflected by variable 18), or the differential development of bony protuberances along the prefrontal margin of the orbit in these specimens.

The size (especially length) of the canine tusk was noted above as being positively allometric and poorly correlated with skull size. Species group differences have been proposed as a potential explanation for this, with *L. curvatus* and *L. platyceps* supposedly possessing weaker canines than *L. murrayi, L. declivis,* and *L. mccaigi* (Cluver, 1971). A scatterplot of those crania attributed a priori to purported species involves comparatively few specimens, although the five included *L. curvatus* crania fall below *L. murrayi* and *L. declivis* specimens of similar size. A scatterplot for the cross-sectional diameter of the tusk increases the available sample but suggests little in the way of a taxonomic distinction, although here, too, most *L. murrayi* and *L. declivis* specimens have relatively larger tusk diameters than similarly sized *L. curvatus* specimens (fig. 14.8). This suggests that if there is any taxonomic difference in this feature (*sensu*

Cluver, 1971), it is more evident in the smaller species of the two groups (*L. curvatus* versus *L. murrayi* and *L. declivis*) than in the larger species of the same groups (*L. platyceps* versus *L. mccaigi*). The tusks can be relatively as large in specimens of *L. platyceps* as in those attributed to *L. mccaigi.*

Finally, both Broom (1932) and Brink (1986) posited that orbit size differs among species, although this was dismissed by Cluver (1971). The distribution of specimens for orbital length and height is random throughout most of the cranial size range (fig. 14.8). It is clear that specimens attributed to *L. curvatus* do not possess relatively "very large" orbits, contrary to Brink (1986), and that the orbits of these specimens are no larger than those assigned to *L. murrayi*, contrary to Broom (1932). Crania attributed to *L. platyceps* are variable in this regard, falling equally above and below the regression line, indicating that although some possess "fairly large" orbits, as suggested by Broom (1932), others have "fairly small" orbits. Three of four specimens attributed to *L. mccaigi* fall above the regression lines for these measurements; these absolutely large crania tend to have relatively large orbits, in keeping with Broom's (1932) characterization of *L. mccaigi* (table 14.3). As noted previously, this is a particularly interesting departure from the expectation of negative allometric growth in orbit size and suggests that *L. mccaigi* is rather distinctive in this regard.

Multivariate Analysis of Quantitative Variables
In the multivariate analysis of isometry, the first eigenvalue of the variance-covariance matrix accounts for 83.2% of the variance of Data Set 1 and 73.3% of the variance in Data Set 2, indicating that the eigenvectors in both analyses are good indicators of the scaling factors. The isometry expectations are 0.2887 (Data Set 1) and 0.2182 (Data Set 2) for the variables in the analyses, and the eigenvectors are normalized to this expectation as scaled isometry values (table 14.12). The bootstrap analysis produced eigenvectors consistent with those previously calculated and permitted the determination of 95 percent confidence intervals around each vector. The confidence limits range from ±0.0810 (variable 6) to ±0.2618 (variable 13) in subset 1 and from ±0.1141 (variable 10) to ±1.1388 (variable 15) in subset 2.

In the first data set (which maximizes the number of specimens), four variables have scaled isometry values that lie outside the 95 percent confidence interval: 4 (length of the pineal foramen), 10 (chord of the premaxilla), 16 (height of snout from anterior premaxilla to ventral naris),

Table 14.12. Multivariate allometric analyses of multivariate data sets 1 and 2

Variable	Data Set 1 (12 variables/68 specimens)			Data Set 2 (21 variables/29 specimens)		
	Loading on PC1	Differential	Scaled Isometry Value	Loading on PC1	Differential	Scaled Isometry Value
isometry =	*0.2887*			*0.2182*		
2	0.2796	−0.0091	0.9685	0.2274	+0.0092	1.0421
3	0.2455	−0.0432	0.8504	0.1514	−0.0668	**0.6939**
4	0.2284	−0.0603	**0.7911**	0.1889	−0.0293	0.8657
6	0.2770	−0.0117	0.9595	0.2319	+0.0137	1.0628
8	0.2825	−0.0062	0.9785	0.2732	+0.0550	1.2521
10	0.3332	+0.0445	**1.1540**	0.2317	+0.0135	1.0612
11				0.2147	−0.0035	0.9840
12				0.2325	+0.0143	1.0655
13	0.2810	−0.0077	0.9733	0.2824	+0.0642	1.2942
14	0.2299	−0.0588	0.7963	0.2411	+0.0229	1.1049
15				0.1886	−0.0296	0.8643
16	0.3610	+0.0723	**1.2504**	0.2407	+0.0225	1.1031
17	0.3592	+0.0705	**1.2442**	0.2378	+0.0196	1.0898
18	0.2595	−0.0292	0.8989	0.1858	−0.0324	0.8515
19				0.2338	+0.0156	1.0715
20				0.1709	−0.0473	**0.7832**
21				0.2071	−0.0111	0.9491
25	0.2887	0.0000	1.0000	0.2135	−0.0047	0.9785
32				0.1634	−0.0548	**0.7489**
33				0.2095	−0.0087	0.9601
38				0.2046	0.0136	0.9377

Note: Scaled isometry values in bold lie outside the scaled 95% confidence interval.

17 (height of snout from the maxillary caniniform process to ventral naris). Whereas variable number 4 scales negatively, the other three, which are related to the height of the premaxilla, scale positively with size. These four variables were also included in subset 2, where they were found to scale in identical directions although not to the degree seen in subset 1.

In the second data set (which maximizes the number of measurements), three variables have scaled isometry values that lie outside the

95 percent confidence interval, and all scale in the negative direction. These three variables include number 3, which measures the length of the skull posterior to the anterior margin of the pineal foramen and thus includes the length of the pineal foramen itself. However, the scaled isometry value of variable 3 (0.6939) is lower than that of variable 4 in either analysis (0.7911 in subset 1, and 0.8657 in subset 2), suggesting that pineal length alone does not account for the negative scaling of the rear of the skull in *Lystrosaurus*. Other variables related to the rear of the skull (25, 33) also scale negatively, albeit to a lesser degree, suggesting a decrease in the size of the rear of the skull relative to the snout. Variable 3 is included in subset 1, where it similarly scaled in the negative direction.

Further comparison of scaled isometry values between the two data sets show that variables common to both analyses do not necessarily show consistent patterns of allometry. Variables 2, 6, 8, 13, and 14, although within the ninety-five percent confidence interval, scale in the negative direction in the first data set but in the positive direction in data set 2. Similarly, variable 25 is isometric in data set 1 but slightly negatively allometric in the second data set. Variable 18 scales slightly negatively in both analyses. These small discrepancies likely are due to differences in the specimens and thus the putative taxa included in each analysis. For example, the biplot of variable 13 (AP length of the snout) reveals a clear separation with increasing size between specimens attributed a priori to *L. curvatus* and *L. platyceps,* on the one hand, and those attributed to *L. declivis* on the other. Although Data Set 1 includes examples of all taxa, data set 2 includes no *L. platyceps* and only one large *L. curvatus* specimen. As a result, the absence in data set 2 of large specimens with relatively low values for variable 13 artificially skews the allometric analysis in the positive direction. When these specimens are included in the analysis, as in Data Set 2, variable 13 becomes slightly negatively allometric.

Although individual results do not necessarily agree with each other, or with the information obtained from the bivariate analysis, there is concordance in that the first data set as well as the bivariate analysis indicate that the height of the snout (variables 16 and 17) enlarges in a positively allometric manner (tables 14.8, 14.12). At the same time, both the second data set and the bivariate analysis suggest that the width of the snout between the nares (variable 20) is negatively allometric, decreasing proportionately with age.

Principal components analysis of the first multivariate data set (fig. 14.9) reveals no clear separation among crania along the first three major axes with the exception that the three smallest skulls included in the

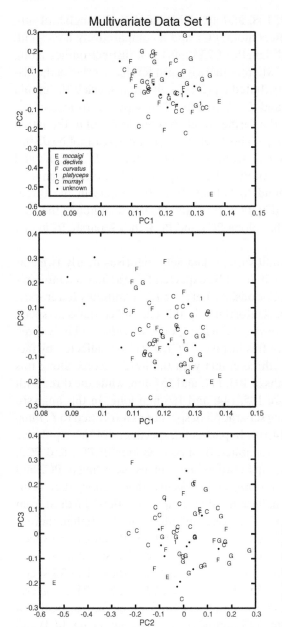

Figure 14.9. Principal components analysis of multivariate Data Set 1, which maximizes specimens (sixty-eight) at the expense of variables (twelve). Specimens attributed a priori to *L. murrayi* and *L. declivis* tend to separate along PC2.

analysis (specimens 107, 108, and 109 with basal skull lengths of 40–52 mm) are distinct outliers along PC1 and a comparatively large skull (specimen 43; basal skull length of 238 mm) is a distinct outlier along PC2. In general, however, whereas there is considerable overlap in the distribution of specimens attributed a priori to different species, the crania assigned to *L. murrayi* and *L. declivis* tend to separate from one another along PC2, where the former tend to score lower than the latter. The *L. curvatus, L. platyceps,* and *L. oviceps* specimens cannot be distinguished from one another or from those attributed to *L. murrayi* and *L. declivis* along PC1, PC2, or PC3. The two putative *L. mccaigi* crania included in this analysis tend to fall out from the others in the three-dimensional space formed by PC1, PC2, and PC3, and, although they are quite similar to one another along PC1 and PC3, they separate from one another along PC2.

Because the second multivariate data set comprises nearly twice as many variables as the first, it might be expected to be somewhat more reliable with regard to the relationships between the admittedly fewer specimens to which it pertains. Unfortunately, this data set contains no specimens attributed to *L. platyceps, L. oviceps,* or *L. mccaigi,* and only three assigned to *L. curvatus.* In this instance, PC1 is strongly affected by factors other than size. The three crania with the lowest scores along this axis have basal skull lengths of 103, 111, and 157 mm, while the three with the highest PC1 scores are 105, 144, and 166 mm long. In this analysis, there is a reasonably strong separation of specimens attributed to *L. murrayi* and *L. declivis* (fig. 14.10). In particular, specimens of these putative species are almost totally separated by a combination of PC1 and PC3, and there is very little overlap between their ranges according to PC2 and PC3. The specimens of *L. murrayi* and *L. declivis* that lie along their common border are not the smallest in either sample, and those furthest from that common border are not the largest in either sample. Rather, there is a seemingly random distribution of basal skull length among the specimens comprising each cluster. Thus, it is evident that size does not account for the morphometric differences revealed by these data for specimens attributed here a priori to *L. murrayi* and *L. declivis.* This suggests that the shape differences between these crania are already evident at a comparatively small size in both samples (105–110 mm; see fig. 14.4) and that they are not affected by increasing size.

The distinction that is apparent here between specimens attributed to *L. murrayi* and *L. declivis* is not size dependent. This strongly suggests that these two groups of crania do not represent sexual dimorphs of a

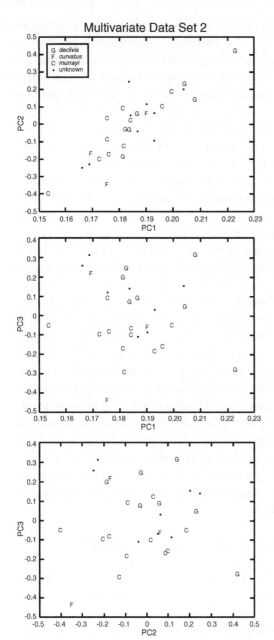

Figure 14.10. Principal components analysis of multivariate Data Set 2, which maximizes variables (twenty-one) at the expense of specimens (twenty-nine). Specimens attributed a priori to *L. murrayi* and *L. declivis* separate along PC1 and PC3. Note that size alone does not account for this separation. See text for further explanation.

single species. In this regard, our results contradict the conclusions reached by Thackeray et al. (1998). However, their analysis was based on only eight measurements of the skull (two of which pertain to the size of the orbit). In addition, the technique employed by them is strongly biased by the relative dimensions of the variables, whereby equivalent numbers of very small and very large variables should be used, but their data set included only one comparatively small variable ("maximum tooth diameter"). Furthermore, although their analysis included a number of specimens not measured by us, their specific attributions of a number of the specimens are questionable. Thus, for example, whereas they employed specimen TM 20 as the reference skull of *L. murrayi* (to which the other eighteen crania of *L. murrayi* and twenty-five skulls of *L. declivis* were compared), Cluver (1971) specifically referred this skull to *L. declivis*. Similarly, two of the specimens in their *L. murrayi* sample (TM 17, TM 213) display overall cranial morphology consistent with their attribution to *L. curvatus* (e.g., see Cluver, 1971, with regard to TM 17, the type of *Lystrosaurus jorisseni* Van Hoepen, 1916). It appears, therefore, that the study by Thackeray et al. (1998) simply demonstrates that the attribution of some specimens in the catalogues of the two collection they sampled are erroneous. On the other hand, because their sample included skulls of a very different overall configuration (e.g., those attributable to *L. curvatus* versus those attributable to *L. murrayi* or *L. declivis*), it would appear that the technique employed by Thackeray et al. (1998) is incapable of distinguishing between closely related species. Indeed, this appears to be the case. In the application of this technique to a sample of fossil hominin primates, Thackeray et al. (1997) were unable to distinguish among crania that are widely accepted as representing different species, if not different genera (e.g., *Australopithecus, Paranthropus,* and *Homo*) (Wood & Richmond, 2000).

Distribution of Qualitative Cranial Features in Lystrosaurus

Table 14.13 records the distribution of the six qualitative features according to cranial size. Three of these features—the prefrontal-nasal crest, frontal rugosity, and premaxillary sagittal ridge—display some relationship to cranial size, insofar as they are absent from any specimen whose length is less than about 85 mm (i.e., in the lower fifth of the total range). However, all degrees of expression of the frontal rugosity and premaxillary sagittal ridge may be manifest by skulls between c. 85 and 100 mm. Thus, any increase in cranial length beyond about 100 mm does not appear to result in a greater degree of development of these two features.

Table 14.13. Expression of nonmetric cranial features in *Lystrosaurus* according to cranial size

Character	State	Basal skull length (mm)
Sagittal facial profile	1 (rounded)	40–197
	2 (three planes)	52–254
	3 (two planes)	83–270
Prefrontal-nasal crest	1 (absent)	40–254
	2 (slight)	86–270
	3 (moderate)	83–216
	4 (marked)	121–230
Frontal rugosity	1 (absent)	40–227
	2 (slight)	83–185
	3 (moderate)	93–203
	4 (marked)	87–270
Premaxillary sagittal ridge	1 (absent)	40–254
	2 (slight)	83–270
	3 (moderate)	86–227
	4 (marked)	97–230
Postorbital boss	1 (absent)	40–230
	2 (present)	216–270
Transverse contour of premaxilla	1 (angled)	40–230
	2 (rounded)	44–270

The prefrontal-nasal crest attains its maximum expression in individuals of somewhat larger size (about 120 mm), and postorbital bosses are developed only in specimens within the uppermost quartile (215–270 mm) of the observed total size range (table 14.13). Significantly, skulls that lack any expression of these four features span the size range of those that do (83–254 mm). In this regard, the sagittal facial profile may present a gently rounded contour or be composed of three obtusely angled planes in crania smaller than about 85 mm, but a profile in which the nasals occupy the plane formed by the premaxilla is exhibited only by crania that exceed 85 mm in length. The premaxilla may have a rounded or angled transverse contour in specimens throughout the entire size range (table 14.13).

It appears, therefore, that, in *Lystrosaurus,* a variety of bony features did not become manifest in small individuals. Thus, crania that are less than about 85 mm long do not exhibit any form of prefrontal-nasal crest,

Table 14.14. Co-occurrence of nonmetric cranial features in *Lystrosaurus*

		Facial profile	Prefrontal-nasal crest	Frontal rugosity	Premaxillary sagittal ridge	Postorbital boss	Transverse contour
Sagittal facial profile	1	—	1 2 3	1 2 3	1 2 3	1	1 2
	2	—	1 2 3 4	1 2 3 4	1 2 3 4	1 2	1 2
	3	—	1 2 3 4	1 2 3 4	1 2 3 4	1 2	1 2
Prefrontal-nasal crest	1		—	1 2 3 4	1 2 3	1 2	1 2
	2		—	1 2 3 4	1 2 3	1 2	1 2
	3		—	1 2 3 4	2 3 4	1 2	1 2
	4		—	2 3 4	3 4	1	1 2
Frontal rugosity	1			—	1 2 3 4	1	1 2
	2			—	1 2 3 4	1	1 2
	3			—	2 3 4	1	1 2
	4			—	1 2 3 4	1 2	1 2
Premaxillary sagittal ridge	1				—	1 2	1 2
	2				—	1 2	1 2
	3				—	1	1 2
	4				—	1	1 2
Postorbital boss	1					—	1 2
	2					—	2

Note: Character state assignations (numbers) are as described in the text, pp. 459–460.

frontal rugosity, premaxillary sagittal ridge, postorbital boss, or a snout in which the nasals occupy the facial plane formed by the premaxilla. Most of these features, however, gain expression in skulls that are between about 85 and 100 mm long, and any further enlargement in size beyond about 100 mm does not affect the degree to which these features may be developed. The two exceptions to this are the prefrontal-nasal crest, which attains its maximum expression only in crania of somewhat larger size (about 120 mm), and the postorbital boss, which becomes manifest only in much larger individuals (about 215 mm).

The co-occurrence of the six nonmetric characters is recorded in table 14.14. There is no relationship between the transverse contour of the snout and any other feature save for the observation that the snout is "squared-off" in all crania with postorbital bosses. Skulls with a rounded sagittal facial profile may lack a prefrontal-nasal crest, frontal rugosities, and a premaxillary sagittal crest, or they may exhibit slight to moderate development of these features. However, the prefrontal-nasal crest, frontal rugosities, and premaxillary crest are well developed only in specimens in which the facial profile comprises two or three distinctly angled planes. A well-developed prefrontal-nasal crest does not occur in any specimen that lacks frontal rugosities. There is a reasonably strong relationship between the development of the prefrontal-nasal crest and the premaxillary sagittal ridge; the latter is not strongly expressed in any specimen that does not also show at least moderate development of the prefrontal-nasal crest. Finally, a postorbital boss is exhibited only by skulls that also display angled sagittal and transverse facial profiles and that have a well-developed frontal rugosity but in which the premaxillary sagittal ridge is either lacking or poorly defined. Apart from the co-occurrences enumerated above, there is no relationship between the degrees of development of any two characters.

With regard to the post hoc taxonomic distribution of the nonmetric features (table 14.15), all crania attributed here to *L. murrayi* exhibit some expression of the prefrontal-nasal crest and premaxillary sagittal ridge and lack a postorbital boss. Within the *L. murrayi* sample, there is no relationship between cranial size and the degree to which the prefrontal-nasal crest and premaxillary sagittal ridge are developed. These specimens exhibit all states of the frontal rugosity, from its absence to its extreme development, and a snout that is either rounded or squared-off transversely. In only one specimen attributed to *L. murrayi* (SAM-PK-K 1495; basal skull length = 115 mm; Cluver 1971) does the skull exhibit a rounded facial contour; in all others, this comprises three planes (fifteen specimens ranging between 86 and 159 mm), or two planes (ten specimens ranging between 83 and 146 mm).

All specimens attributed to *L. declivis* also exhibit some expression of the prefrontal-nasal crest and premaxillary sagittal ridge and lack a postorbital boss. However, unlike the *L. murrayi* sample, there is a distinct trend among specimens attributed to *L. declivis* for the prefrontal-nasal crest and premaxillary sagittal ridge to become better developed with increasing cranial size. Crania of *L. declivis*, like those attributed to *L. murrayi*, exhibit all states of the frontal rugosity, from its absence to its extreme development, and either a rounded or an angled transverse facial contour. In all specimens of *L. declivis*, the facial contour is formed either by three planes (four specimens ranging between 126 and 185 mm) or, much more commonly, by two distinctly angled planes (eighteen specimens ranging between 94 and 230 mm).

All crania attributed to *L. curvatus* have a gently curved facial profile. In them, the prefrontal-nasal crest, frontal rugosity, and interpremaxillary sagittal ridge are either absent or only weakly developed, and there is no postorbital boss. Specimens of *L. curvatus* may evince either a rounded or

Table 14.15. Distribution of nonmetric cranial features by skull size among purported species of *Lystrosaurus*

Character	State	*L. murrayi* (83–159)	*L. declivis* (94–230)	*L. mccaigi* (216–270)	*L. platyceps* (170–227)	*L. oviceps* (95)	*L. curvatus* (80–197)
Sagittal facial profile	1	1 (115)				2 (95)	14 (80–197)
	2	15 (86–159)	4 (126–185)	1 (225)			
	3	9 (83–146)	18 (94–230)	8 (216–270)	3 (170–227)		10 (80–197)
Prefrontal-nasal crest	1			4 (200–270)	4 (170–227)		
	2	8 (86–115)	3 (96–110)	4 (225–270)		2 (95)	5 (98–133)
	3	16 (83–159)	8 (94–166)	1 (216)			
	4		12 (121–230)				
Frontal rugosity	1	2 (115–127)	2 (110–136)		4 (170–227)		11 (80–197)
	2	8 (86–105)	7 (94–185)			2 (95)	3 (98–119)
	3	8 (96–159)	11 (105–203)				
	4	2 (87–111)	3 (121–230)	9 (216–270)			
Premaxillary sagittal ridge	1			2 (254)	2 (170–201)	2 (95)	7 (80–133)
	2	9 (82–146)	3 (97–113)	5 (225–270)	2 (193–227)		6 (102–197)
	3	9 (86–127)	8 (101–203)	1 (216)			
	4	1 (–98)	9 (121–230)				
Postorbital boss	1	20 (83–159)	21 (94–230)	9 (216–270)	4 (170–227)	2 (95)	15 (80–197)
	2			9 (216–270)			
Transverse contour	1	13 (99–159)	14 (94–203)	9 (216–270)	1 (170)	2 (95)	7 (81–128)
	2	6 (83–106)	5 (113–230)	9 (216–270)	7 (170–227)		9 (80–197)

Note: Character state assignations are as described in the text, pp. 459–460. Numbers in parentheses refer to basal skull length.

a squared-off transverse facial contour. In the expression of these features, some 25 percent of the *L. curvatus* specimens resemble the two crania attributed here to *L. oviceps.*

The sagittal facial profile of all crania attributed here to *L. platyceps* comprises two distinctly angled planes. Despite the fact that the *L. platyceps* specimens are generally larger than most attributed to *L. curvatus,* the skulls attributed to *L. platyceps* resemble those of *L. curvatus* in that they lack any development of the prefrontal-nasal crest, frontal rugosity, or postorbital boss. Indeed, the only potential difference between the *L. platyceps* and *L. curvatus* specimens other than their sagittal facial profile concerns the development of the premaxillary sagittal ridge, which may attain moderate expression in the former but only slight expression in the latter. However, this feature may also be absent in specimens attributed to *L. platyceps,* and its stronger development in some may be related to their comparatively large size.

All nine specimens attributed to *L. mccaigi* exhibit postorbital bossing, well-developed frontal rugosities, and a squared-off transverse facial contour. In the *L. mccaigi* specimens, like those of *L. declivis,* the facial contour comprises either three planes (one specimen with a basal skull length of 225 mm) or two planes (eight specimens ranging between 216 to 270 mm). However, in the *L. mccaigi* sample, unlike the *L. declivis* sample, the prefrontal-nasal crest and premaxillary sagittal ridge tend to be absent or only weakly developed, and there is no tendency for these features to exhibit stronger development with increasing cranial size.

Thus, certain of the nonmetric features that have been employed in the differential diagnoses of *Lystrosaurus* species appear to covary insofar as their extreme manifestation occurs only under certain circumstances. For example, the prefrontal-nasal crest, frontal rugosities, and interpremaxillary crest are well developed only in specimens in which the facial profile is distinctly angled. A well-developed prefrontal-nasal crest, moreover, occurs only in the presence of frontal rugosities. The expression of some of the qualitative characters also appears to be influenced by overall cranial (i.e., individual) size insofar as they are manifest only in skulls within a certain size range. For example, crania that are less than about 85 mm long lack any prefrontal-nasal crest, frontal rugosity, premaxillary sagittal ridge, or postorbital bossing.

At the same time, however, these qualitative features do not uniformly occur together, which suggests that they may be largely independent of one another. Also, most of these characters gain expression in skulls that are between about 85 and 100 mm long; any further enlargement in cranial size does not necessarily affect their degree of development. It is also significant that the various bony protuberances may be altogether lacking in comparatively large crania (i.e., those that attain a length of almost 200 mm). These observations, together with the distribution of the various nonmetric characters among skulls attributed here a priori to different species, suggest that these features may exhibit greater variability than previously indicated. Nevertheless, it is evident that these features vary according to three or perhaps four different patterns among the *Lystrosaurus* crania examined here, suggesting the existence of a corresponding number of species within this sample.

Discussion

The foregoing analyses effectively refute the null hypothesis that variability in morphometric and qualitative characters of the skull of *Lystrosaurus* can be attributed to ontogenetic growth changes in single, sexually dimorphic species. Although the overall range in basal skull length (40–270 mm) of the present sample can be accommodated within the ontogenetic ranges of extant large-bodied sauropsids (e.g., *Caiman crocodilus, Alligator mississippiensis, Crocodylus porosus*), a variety of morphometric parameters suggest that these *Lystrosaurus* crania sample individuals of different size from more than a single species.

In general, cranial ontogeny in *Lystrosaurus* involved a relatively greater increase in both width and depth (especially the depth of the snout) compared with increase in skull length, whereas the posterior part of the cranium grew at a negatively allometric rate. This observation appears to hold for all species of the genus because these variables are associated with high correlation coefficients, which suggests little (if any) differentiation within the genus.

Orbit size, which is characteristically negatively allometric in ontogenetic series of living diurnal vertebrate species, is effectively isometric for the present *Lystrosaurus* sample. Broom (1932), who cited orbit size as differing among *Lystrosaurus* species, maintained that the largest species possessed the relatively largest orbits. His observation is borne out by the present sample, in which the largest crania (e.g., those attributed to *L. mccaigi*) evince unexpectedly large orbits. In this context, it is interesting to note that the recently described lystrosaurid, *Kwazulusaurus shakai*, is characterized by having very large orbits, which even exceed the size of the temporal fenestra (Maisch, 2002). The type and only specimen of *K. shakai* has a basal skull length of approximately 160 mm (determined from Maisch, 2002, fig. 1), which is about seventy-five percent of the size of the smallest skull attributed to *L. mccaigi* in our sample.

The exceedingly large orbits of the skull of *K. shakai* suggest that it may represent a juvenile of *L. mccaigi,* although other morphologic features discussed below appear to mitigate against this interpretation.

A number of measurements exhibit comparatively low degrees of correlation with basal skull length in this *Lystrosaurus* sample, and most of these are related directly to features that have been variously cited as differing among species (Broom, 1932; Cluver, 1971; Brink, 1986). Finally, several qualitative characters of the cranium in *Lystrosaurus* appear to be related to skull size insofar as they become manifest only when the skull has attained a given size (e.g., prefrontal-nasal crest, frontal rugosities, premaxillary sagittal ridge, and coplanar nasals and premaxillae appear only in crania with a basal length of 100 mm, whereas postorbital bosses are developed only in specimens with a basal length of more than 215 mm). However, even these features appear to be largely independent of size after they first appear (i.e., throughout the upper eight percent of the size range exhibited by *Lystrosaurus*) insofar as they do not become better developed with increasing size.

Bivariate and multivariate analyses of quantitative variables according to the a priori assignment of crania to individual species reveal rather unambiguous patterns of differentiation that are largely consistent with such assignments. Thus, PCA serves to differentiate skulls attributed to *L. murrayi* from those of *L. declivis,* and the distinction is unrelated to size. The few specimens attributed to *L. mccaigi* also tend to be differentiated on the basis of a principal components analysis. On the other hand, no such distinction can be made among crania attributed to *L. curvatus,* *L. platyceps,* and *L. oviceps* on the basis of either multivariate (principal components) or bivariate analyses. A number of quantitative variables, including both raw measurements and indices, serve to differentiate *Lystrosaurus* specimens in keeping with their a priori species assignments. Thus, for example, the skull tends to be broader both across the prefrontals and between the orbits in specimens attributed to *L. declivis* and *L. murrayi* than those attributed to *L. curvatus* and *L. platyceps.* The chord/arc index that expresses nasal bone curvature indicates that these bones are flatter in specimens attributed to *L. curvatus, L. platyceps,* and *L. mccaigi* than in those assigned to *L. murrayi* and *L. declivis.* On the other hand, the chord/arc index that expresses the overall sagittal facial profile shows the snout to be more highly angled (or curved) in specimens attributed to *L. murrayi* and *L. mccaigi* than in those assigned a priori to other purported species.

The distribution of both metric features and nonmetric characters considered here suggests the presence of four groups of crania within the present sample of *Lystrosaurus* fossils. These groups correspond largely to the individual species that were recognized a priori, with the exception that the two specimens attributed here a priori to *L. oviceps* cannot be distinguished adequately from similarly sized crania assigned to *L. curvatus,* and the skulls attributed to *L. curvatus* and *L. platyceps* appear to conform to a single group, whose members differ principally in features that appear to reflect the generally larger size of the crania attributed to the latter. Thus, with the sole exception of the sagittal facial profile, these various characters do not serve to differentiate among specimens attributed a priori to *L. curvatus* and *L. platyceps.* Inasmuch as the crania assigned here to *L. platyceps* tend to be larger than those assigned to *L. curvatus,* this feature may be size (age and/or sex) related. In this instance, the larger individuals tend to have a facial profile composed of two planes rather than an evenly curved profile.

The metric and qualitative characters examined here suggest that, in addition to *L. curvatus,* recognition should be accorded *L. murrayi, L. declivis,* and *L. mccaigi* as valid, diagnosable species. With regard to individual quantitative parameters, the nasal bones of *L. murrayi* and *L. declivis* are curved or angled, compared with their flat condition in *L. curvatus* and *L. mccaigi.* Similarly, the skull roof is relatively broader across the prefrontals and between the orbital margins in *L. murrayi* and *L. declivis* than in *L. curvatus,* whereas *L. mccaigi* appears to be intermediate with regard to the latter dimension. *Lystrosaurus murrayi* and *L. declivis* appear to differ from *L. curvatus* in the size of the tusk in larger individuals, which may relate to a relatively faster rate of canine eruption in *L. curvatus.* In *L. murrayi,* the snout is both shorter (AP) and overall more strongly curved or angled than in *L. declivis.* In its overall form, the snout is more strongly curved or angled in *L. murrayi* and *L. mccaigi* than in *L. declivis* and *L. curvatus.* Finally, quantitative variables indicate that the orbits of *L. mccaigi* are relatively and unexpectedly large in individuals of large cranial size.

In *L. murrayi,* the snout comprises three distinct planes, and a prefrontal-nasal crest and premaxillary sagittal ridge are present in crania over 80 mm in length. The transverse contour of the snout is variable in *L. murrayi,* being either squared-off or rounded, and the development of frontal rugosities in this species is variable; they may be lacking or well-developed. Postorbital bosses are not present in *L. murrayi.* In *L. declivis,* as with *L. murrayi,* the prefrontal-nasal crest and premaxillary sagittal ridge are present, the frontal rugosity and transverse contour of the snout are variable in morphology, and postorbital bosses are absent. The facial plane in *L. declivis* is variable; most commonly, it comprises only two separate planes, but in some individuals it is formed of three planes.

The series of morphometric and nonmetric characters considered here suggest that taxonomic differences appear in crania with a basal length less than 80 mm (the value employed by us in the a priori assignment of specimens). In particular, the facial profiles and nasal chord/arc index values observed for two skulls with a length of about 50 mm are consistent with their recognition as small juveniles of *L. murrayi* (specimen 33; SAM-PK 3531, length = 51 mm) and *L. curvatus* (specimen 108; BP/1/4190, length = 52 mm). The latter was so identified by Cluver (1971). However, the smallest specimens in the present sample (40–46 mm) remain unattributable to individual species on the basis of the characters recognized here. At the same time, these characters suggest that one previously indeterminate cranium of large size can be attributed to *L. declivis* (specimen 91; TM 155, length = 254 mm). This specimen extends the size range for *L. declivis* in the present sample by some 10 percent, from 230 to 254 mm, and extends it into the upper third of the observed range for the overall largest species, *L. mccaigi*.

The current sample of specimens attributed a priori to *L. mccaigi* range from 216 to 270 mm in basal skull length. The question that arises in this context pertains to the presence of smaller, juvenile specimens of this species in this sample. Our expectation from the foregoing is that specimens of *L. mccaigi* with a basal skull length of less than 216 mm (the smallest in the present sample) should possess relatively large orbits (although this might be impossible to discern as a distinguishing feature in small individuals), a frontonasal angle with either state 2 or 3, an absent or weakly formed prefrontal-nasal crest (states 1 or 2), some expression of the frontal rugosities (states 2 or 3), an absent or weakly developed premaxillary ridge (states 1 or 2), and an angled transverse contour of the premaxilla (state 2). Although all the specimens of *L. mccaigi* assigned here a priori display postorbital bosses, it is easy to imagine that the development of this feature is strongly influenced by ontogenetic size and possible sex. These criteria are met by very few specimens in the present sample with a basal skull length less than 216 mm. These specimens are TM 112 (length = 105 mm) and UMZC-T 780 (length = 119 mm). Two other crania included here (TM 3, length = 96 mm; TM 4051, length = 134 mm) also meet these criteria except that the prefrontal nasal crest on each is somewhat better developed than would be expected. Finally, as noted above, the type and only specimen of *Kwazulusaurus shakai* (BPI/1/2792) is characterized by exceptionally large orbits, which would be expected in a juvenile of

L. mccaigi. It also possesses slight parasagittal ridges that demarcate an angled premaxillary contour below the external nares. However, several features mitigate against this interpretation, including its exceedingly narrow skull roof, rather foreshortened snout, and a sagittal facial profile that resembles more closely that of *L. curvatus.* In addition, this cranium also has very thin postorbital bars and appears to lack any expression of bony excrescences such as a prefrontal-nasal crest or frontal rugosities. The presence of a separate postfrontal bone and the peculiar posterior midline extension of the premaxilla between the nasals may also mitigate against BPI/1/2792 being interpreted as a juvenile of *L. mccaigi.* If, however, any of these four or five specimens is, indeed, attributable to *L. mccaigi,* it would serve to bring the lower limit of the observed size range for this species to that recorded here for specimens attributable to *L. declivis, L. curvatus,* and *L. murrayi.*

The specific attributions proposed here extend the ranges for each of the four species so that all are known to range between 100 and 160 mm. However, in the present series, no cranium larger than about 160 mm is certainly attributable to *L. murrayi.* The other species range upward to about 230 mm (*L. curvatus*), 250 mm (*L. declivis*), or 270 mm (*L. mccaigi*). It appears, therefore, that Brink's (1986) diagnostic criterion of overall size may be applicable in modified form if the present sample is, indeed, exhaustive of the range for each of these species. In other words, if the maxima of the observed ranges established by the present specimens are close to being accurate, *L. murrayi* could be distinguished from the other taxa by having a maximum skull length of about 160 mm. However, because we doubt that the specimens in our sample cover the entire range for all *Lystrosaurus* species, we hesitate to suggest that overall size be included in the diagnosis of any. For the moment, it is perhaps prudent to suggest only that *L. murrayi* did not achieve the same overall cranial size as the other three species.

It is argued here that four species of *Lystrosaurus—L. murrayi, L. declivis, L. curvatus,* and *L. mccaigi*—can be recognized from the Beaufort strata of the South African Karoo Basin. *Lystrosaurus mccaigi* appears to be confined to the lower horizons of the *Lystrosaurus* Assemblage Zone (Kitching, 1977), and R. Smith (personal communication) has identified this species at the base of an approximately 40-m overlap of the *Lystrosaurus* and *Dicynodon* Assemblage Zones in the Palingkloof Member of the Balfour Formation. Thus, both *L. curvatus* and *L. mccaigi* have been identified in the Late Permian of southern Africa (King & Jenkins, 1997; Smith, personal communication). With the exception of *L. mccaigi,* which may be confined to the Permian* (Smith, personal communication), it does not appear that these species are separated stratigraphically. Current records indicate that *L. murrayi*, *L. declivis,* and *L. curvatus* occur throughout the thickness of the *Lystrosaurus* Assemblage Zone sediments in the Karoo Basin. However, the possibility of geochronological differences in these species ranges warrants further field work.

The phylogenetic relationships among these and other putative *Lystrosaurus* species from localities beyond the Karoo Basin is beyond the scope of this study but should be studied in view of the paleobiogeographic implications that these relationships may have on the dispersal of this genus throughout Gondwana. We hope the alpha taxonomy provided here for the forms from the Karoo Basin might serve as a basis for such an analysis. Similarly, some of the morphologies that have been defined here as differentiating among taxa (e.g., snout length and angle, bony ridges, etc.) may be of functional (i.e., biomechanically adaptive) significance in reconstruction of the feeding habits or other behaviors of these Permo-Triassic dicynodonts.

We provide here differential diagnoses for the four species we recognize from the Karoo Basin of South Africa.

Species Paleontology of Lystrosaurus in the Karoo Basin of Southern Africa

Order Therapsida Broom, 1905
Suborder Anomodontia Owen, 1859
Infraorder Dicynodontia Owen, 1859
Superfamily Pristerodontoidea Cluver and King, 1983
Family Dicynodontidae Cluver and King, 1983
Subfamily Kannemeyeriinae von Huene, 1948
Tribe Lystrosaurini Broom, 1903
Lystrosaurus Cope, 1870a

Synonyms. Ptychognathus Owen, 1860 [preoccupied] = *Ptychosagium* Lydekker, 1890; *Mochlorhinus* Seeley, 1898; *Rhabdotocephalus* Seeley, 1889; *Prolystrosaurus* Haughton, 1917.

Type species. Lystrosaurus murrayi (Huxley, 1859) Broom, 1932 (= *Dicynodon murrayi* Huxley, 1859)

Diagnosis. Dicynodont therapsid lacking teeth but with maxillary tusks present in both sexes. Orbits high, with nares placed immediately

---

* In this context, it is interesting to note that the type of *Kwazulusaurus shakai* also derives from the Late Permian *Dicynodon* Assemblage Zone (Maisch, 2002).

anterior to them. Basicranial axis shortened with inferiorly projecting snout, which is formed by elongated maxillae and premaxilla.

Lystrosaurus murrayi (Huxley, 1859) Broom, 1932

Synonyms. Dicynodon murrayi Huxley, 1859; *Ptychognathus verticalis* Owen, 1860; *Lystrosaurus frontosus* Cope, 1870b; *Ptychognathus boopis* Owen, 1876; *Dicynodon copei* Seeley, 1889 [*Ptychognathus copei* (Seeley, 1889) Lydekker, 1890]; *Dicynodon strigops* Broom, 1913 [*Prolystrosaurus strigops* (Broom, 1913) Haughton, 1917]; *Lystrosaurus jeppei* Van Hoepen, 1916; *Prolystrosaurus natalensis* Haughton, 1917; *Lystrosaurus rubidgei* Broom, 1940; *Lystrosaurus bothai* Broom, 1941.

Type Specimen. BMNH R.1291

Type Locality. Colesburg District, Cape Province, South Africa.

Diagnosis. Skull (fig. 14.11) with a snout that is relatively short AP; skull roof relatively broad across the frontals and prefrontals; snout highly curved in sagittal profile and often composed of three distinct planes separating frontals, nasals, and premaxilla; premaxillary sagittal ridge present; prefrontal–nasal crest present and of variable development; skull lacks postorbital bosses; frontal rugosity variably present and developed.

Referred Specimens. BMNH R.5710, 36253, 36224; UMZC-T 763; NMQR-C 211, C 270, C 282; SAM-PK-K 1165, K 1397, K 1459, 3531, 11167, 11181; BPI/1/267, 1/314, 1/3900, 1/4798, 1/267; TM 33, 67, 69, 71, 115, 164, 2881.

Lystrosaurus declivis (Owen, 1860) Brink, 1951

Synonyms. Ptychognathus declivis Owen, 1860; *Ptychognathus latirostris* Owen, 1860; *Ptychognathus alfredi* Owen, 1862; *Ptychognathus depressus* Owen, 1876; *Lystrosaurus wagneri* Van Hoepen, 1916; *Lystrosaurus primitivus* Toerien, 1954.

Type Specimen. BMNH 36221.

Type Locality. Rhenosterburg, South Africa.

Diagnosis. Skull (fig. 14.12) with a snout that is relatively long AP associated with the facial plane of the premaxilla and nasals inclined at an

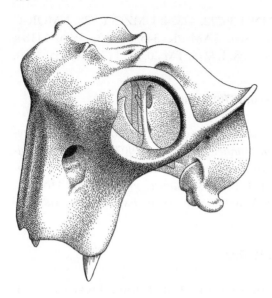

Figure 14.11. Cranium of *Lystrosaurus murrayi* in left anterodorsolateral view.

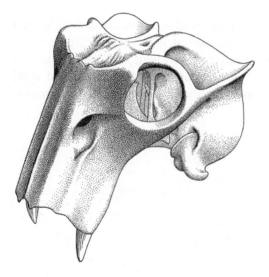

Figure 14.12. Cranium of *Lystrosaurus declivis* in left anterodorsolateral view.

obtuse angle to the skull roof; skull roof relatively broad across the frontals and prefrontals; snout composed most commonly of two distinct planes separating the premaxillae and nasals from the frontals; premaxillary sagittal ridge present; prefrontal–nasal crest present and of variable development; skull lacks postorbital bosses; frontal rugosity variably present and developed.

Referred Specimens. BMNH 36222, 47324; UMZC-T 778; NMQR-C 391; SAM-PK-K 1398, K 4800, SAM-PK 3455, 3593, 3596, 11184; BPI/1/477, 1/1750, 1/1754, 1/3908, 1/4275; TM 20, 21, 117, 139, 155, 3580, 4049, 4113, 4465.

Lystrosaurus curvatus (Owen, 1876) Broom, 1932

Synonyms. Dicynodon curvatus Owen, 1876; *Mochlorhinus platyceps* Seeley, 1898; *Lystrosaurus andersoni* Broom, 1907; *Lystrosaurus oviceps* Haughton, 1915; *Lystrosaurus breyeri* Van Hoepen, 1916; *Lystrosaurus wageri* Van Hoepen, 1916; *Lystrosaurus jorisseni* Van Hoepen, 1916; *Lystrosaurus theileri* Van Hoepen, 1916; *Lystrosaurus rubidgei* Broom, 1940.

Type Specimen. BMNH R.3792.

Type Locality. Elandburg, Cradock District, Cape Province, South Africa.

Diagnosis. Skull (fig. 14.13) with a snout that is moderately elongated AP associated with either a gently curved sagittal facial profile or, in some larger individuals (basal skull length > about 160), a facial plane in which the premaxilla and nasals form an obtuse angle with the skull roof; skull roof relatively narrow across the frontals and prefrontals; premaxil-

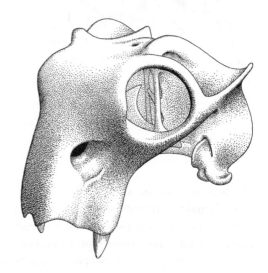

Figure 14.13. Cranium of *Lystrosaurus curvatus* in left anterodorsolateral view.

lary sagittal ridge most commonly absent but may be weakly defined; prefrontal–nasal crest most commonly absent but may be weakly developed; skull lacks postorbital bosses; frontal rugosity absent or only weakly developed.

Referred Specimens. BMNH 3597; UMZC-T 788; NMQR-C 154; SAM-PK-K 3, K 15, K 16, K 1381, SAM-PK 641, 1362, 11186; BPI/1/464, 1/1269, 1/1746, 1/4039, 1/4190, BPI MN 413; TM 16, 17, 18, 156, 191, 213.

Lystrosaurus mccaigi (Seeley, 1898) Broom, 1903

Synonyms. Rhabdotocephalus mccaigi Seeley, 1898; *Lystrosaurus puterilli* Van Hoepen, 1915; *Lystrosaurus amphibius* Brink, 1951.

Type Specimen. AMNH 404.

Type Locality. Elandburg, Cradock District, Cape Province, South Africa.

Diagnosis. Skull (fig. 14.14) with a snout that is moderately elongated AP associated with a highly angled sagittal profile composed of two or

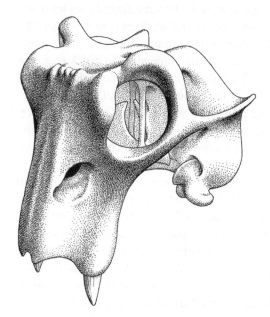

Figure 14.14. Cranium of *Lystrosaurus mccaigi* in left anterodorsolateral view.

three distinct planes separating frontals, nasals, and premaxilla; prefrontal–nasal crest absent or only weakly developed; premaxillary sagittal ridge absent or only weakly developed; frontal rugosities present and well developed (at least in larger individuals); postorbital bosses present (at least in larger individuals); premaxilla has an angular transverse contour; orbits relatively large in larger individuals.

Referred Specimens. BMNH 3749/3754; ?CUM T 780; SAM-PK-K 9, K 116; BPI/1/879, 1/2241,1/3972, 1/4052, 1/4053; ?TM 3, ?TM 112, ?TM 4051.

Summary and Conclusions

The dicynodont genus *Lystrosaurus* is important in the assessment of the faunal transition across the Permo-Triassic boundary because it is one of very few terrestrial taxa to survive the end-Permian extinction event. Twenty-seven species of *Lystrosaurus* have been named for South African fossils; recent reviews suggest that there are fewer, but the number that are diagnostically valid has eluded consensus largely because of the potential influences of ontogeny and sexual dimorphism. This study addresses the species taxonomy of *Lystrosaurus* using both morphometric and nonmetric characters of the skull. A total of 112 variably preserved skulls ranging in length from 40 to 270 mm were examined, and a possible forty-six craniometric variables and six nonmetric characters were recorded for each. The fossils comprising this sample had been referred to at least seventeen different species. Bivariate allometric analyses reveal low correlation coefficients and/or differential distribution of specimens for a number of variables that have been cited as taxonomically diagnostic (e.g., low r values for anteroposterior elongation of the snout and dorsoventral height of the superior portion of the snout; differential distribution of specimens for interorbital and prefrontal widths). There is also an interesting departure from the expectation of negative allometry in orbit size, which has also been cited as a taxonomically relevant character by previous workers. These results suggest the existence of three species groups: (1) *L. curvatus* and *L. platyceps,* (2) *L. mccaigi,* and (3) *L. murrayi* and *L. declivis.* Specimens attributable to the latter group, however, may be differentiated from one another on the basis of relative elongation of the snout, and multivariate allometric (principal components) analyses suggest that size does not account for the morphometric differences between specimens attributed to *L. murrayi* and *L. declivis.* The few specimens attributed to *L. mccaigi* also tend to be differentiated in these analyses, whereas no such distinction can be afforded crania assigned to *L. curvatus* and *L. platyceps.* The nonmetric features tend also to vary along these same lines, indicating the presence of four definable taxa in the present sample: *L. curvatus* (= *L. platyceps*), *L. murrayi, L. declivis,* and *L. mccaigi.* It appears that some taxonomic differences are detectable in crania as small as 50 mm in length. Most bony ridges and excrescences appear only at about 80 mm length, and they attain their maximum expression by 120 mm. Postorbital bosses, however, appear only in substantially larger crania (215 mm) attributable to *L. mccaigi.* Crania of each of the four recognized species span the size range between about 100 and 160 mm. In the present sample, no specimen attributable to *L. murrayi* exceeds 160 mm, whereas the other species maxima are considerably larger, at 230 mm for *L. curvatus,* 250 mm for *L. declivis,* and 270 mm for *L. mccaigi. Lystrosaurus curvatus* is the only species recognized from the Late Permian strata of southern Africa, and *L. mccaigi* appears to be confined to the lower horizons of the *Lystrosaurus* Assemblage Zone. Otherwise, the species of *Lystrosaurus* do not appear to be separated stratigraphically, as *L. curvatus, L. murrayi,* and *L. declivis* have been documented throughout the thickness of the Lower Triassic *Lystrosaurus* Assemblage Zone sediments of the South African Karoo Basin.

Acknowledgments

    The authors thank the curators at the following institutions for permitting us to study specimens in their care: The Bernard Price Institute for Palaeontological Research, University of the Witwatersrand, Johannesburg; The Natural History Museum, London; Cambridge University Museum, Cambridge; The National Museum, Bloemfontein; The Transvaal Museum, Pretoria. The late James W. Kitching provided many enlightening discussions about the geographic and temporal distribution of therapsids in the Karoo Basin. We thank M. Carrano, R. Blob, C. Sidor, L. Panko, and C. Sullivan for their insightful comments and helpful suggestions about the manuscript. We are grateful to Luci Betti-Nash and Cedric Hunter for their skilled execution of the artwork. We thank Matthew Carrano and Rick Blob for their invitation to contribute to this volume in honor of the official retirement of Professor James A. Hopson from The University of Chicago. Our sincere thanks to Jim for his friendship, wise counsel, and constant encouragement.

Literature Cited

Abdala, F. and N. P. Giannini. 2000. Gomphodont cynodonts of the Chañares Formation: the analysis of an ontogenetic sequence. *Journal of Vertebrate Paleontology* 20:501–506.

Alexander, R. McN. 1985. Body support, scaling and allometry; pp. 27–37 *in* M. Hildebrand and D. B. Wake (eds), *Functional Vertebrate Morphology*. Cambridge, MA: Harvard University Press.

Bartlett, M. S. 1949. Fitting a straight line when both variables are subject to error. *Biometrika* 5:207–212.

Battail, B. and M. V. Surkov. 2000. Mammal-like reptiles from Russia; pp. 86–119 *in* M. J. Benton, M. A. Shiushkin, D. M. Unwin and E. N. Kurochkin (eds.), *The Age of Dinosaurs in Russia and Mongolia*. Cambridge: Cambridge University Press.

Bradu, D. and F. E. Grine. 1979. Multivariate analysis of diademodontine crania from South Africa and Zambia. *South African Journal of Science* 75:441–448.

Brink, A.S. 1951. On the genus *Lystrosaurus* Cope. *Transactions of the Royal Society of South Africa* 33(1):107–120.

———. 1986. *Illustrated Bibliographical Catalogue of the Synapsida. Handbook 10, Part 1*. Pretoria: Geological Survey of South Africa.

Broom, R. 1903. On the classification of the theriodonts and their allies. *Reports of the South African Association for the Advancement of Science* 1:286–295.

———. 1905. On the use of the term Anomodontia. *Records of the Albany Museum* 1:266–69.

———. 1907. Reptilian fossil remains from Natal. Part 1. On reptilian remains from the supposed Beaufort Beds of the Umkomanzan River in western Natal. *Reports of the Geological Survey of Natal* 3:93–95.

———. 1909. An attempt to determine the horizons of the fossil vertebrates of the Karroo. *Annals of the South African Museum* 7:285–289.

———. 1913. On four new fossil reptiles from the Beaufort Series, South Africa. *Records of the Albany Museum* 2:397–401.

———. 1932. *The Mammal-Like Reptiles of South Africa and the Origin of Mammals*. London: H. F. and G. Witherby.

———. 1940. On some new genera and species of fossil reptiles from the Karroo Beds of Graaff-Reinet. *Annals of the Transvaal Museum* 20:157–192.

———. 1941. Some new Karroo reptiles with notes on a few others. *Annals of the Transvaal Museum* 20:194–213.

Cheng, Z.-W. 1993. On the discovery and significance of the nonmarine Permo-Triassic Transition Zone at Dalongkopu in Jimusar, Xinjiang, China; pp. 65–67 *in* S. G. Lucas and M. Morales (eds.), *The Nonmarine Triassic. Bulletin of the New Mexico Museum of Natural History and Science 3*, Albuquerque.

Cluver, M. A. 1971. The cranial morphology of the dicynodont genus *Lystrosaurus*. *Annals of the South African Museum* 56:155–274.

———. 1974. The cranial morphology of the Lower Triassic dicynodont *Myosaurus gracilis*. *Annals of the South African Museum* 66:35–54.

Cluver, M. A. and Hotton, N. 1981. The genera *Dicynodon* and *Diictodon* and their bearing on the classification of the Dicynodontia. *Annals of the South African Museum* 83:99–146.

Cluver, M. A. and King, G. M. 1983. A reassessment of the relationships of Permian Dicynodontia (Reptilia, Therapsida) and a new classification of dicynodonts. *Annals of the South African Museum* 91:195–273.

Colbert, E. H. 1974. *Lystrosaurus* from Antarctica. *American Museum Novitates* 2535:1–44.

Cope, E. D. 1870a. On the homologies of some of the cranial bones of the Reptilia, and on the systematic arrangement of the class. *Proceedings of the American Association for the Advancement of Science* 19:194–247.

———. 1870b. On the skull of dicynodont Reptilia. *Lystrosaurus frontosus* from Cape Colony. *Proceedings of the American Philosophical Society* 11:419.

Cosgriff, J. W., W. R. Hammer, and W. J. Ryan. 1982. The Pangean reptile, *Lystrosaurus maccaigi* in the Lower Triassic of Antarctica. *Journal of Paleontology* 56:767–784.

Das Gupta, H. C. 1922. Note on the Panchet reptile. *Sir Asutosh Mukherjee Silver Jubilee Volumes* 2:237–241.

Dodson, P. 1975. Functional and ecological significance of relative growth in *Alligator. Journal of Zoology* 175:315–355.

———. 1976. Quantitative aspects of relative growth and sexual dimorphism in *Protoceratops. Journal of Paleontology* 50:929–940.

Efremov, J. A. 1938. The recovery of a Triassic anomodont in the Orenburg Province. *Doklady Akademii Nauk SSSR* 20:227–229.

———. 1940. Preliminary description of the new Permian and Triassic tetrapoda from USSR. *Trudy Paleontologicheskogo Instituta Akademiy Nauk SSSR* 10:1–139.

———. 1951. On the structure of the knee joint in the higher Dicynodontidae. *Doklady Akademii Nauk SSSR* 77:483–485.

Elliot, D. H., E. H. Colbert, W. J. Breed, J. A. Jensen and J. S. Powell. 1970. Triassic tetrapods from Antarctica: evidence for continental drift. *Science* 169: 1197–1201.

Emerson, S. B. and D. M. Bramble. 1993. Scaling, allometry and skull design; pp. 384–416 *in* J. Hanken and B. K. Hall (eds), *The Skull. Volume 3.* Chicago: University of Chicago Press.

Gans, C. 1980. Allometric changes in the skull and brain of *Caiman crocodilus. Journal of Herpetology* 14:297–301.

Gould, S. J. 1966. Allometry and size in ontogeny and phylogeny. *Biological Review* 41:587–670.

Grine, F. E., B. D. Hahn, and C. E. Gow. 1978. Aspects of relative growth and variability in *Diademodon* (Reptilia: Therapsida). *South African Journal of Science* 74:50–58.

Groenewald, G. H. 1991. Burrow casts from the *Lystrosaurus-Procolophon* Assemblage Zone, Karoo Sequence, South Africa. *Koedoe* 34:13–22.

Gubin, Y. M. and S. M. Sinitza. 1993. Triassic terrestrial tetrapods of Mongolia and the geological structure of the Sain-Sar-Bulak locality; pp. 169–170 *in*

S. G. Lucas and M. Morales (eds.), The Nonmarine Triassic. *Bulletin of the New Mexico Museum of Natural History and Science 3,* Albuquerque.

Hammer, W. R. and J. W. Cosgriff. 1981. *Myosaurus gracilis,* an anomodont reptile from the Lower Triassic of Antarctica and South Africa. *Journal of Paleontology* 55:410–424.

Haughton, S. H. 1915. On some new anomodonts. *Annals of the South African Museum* 12:61–63.

———. 1917. Descriptive catalogue of the Anomodontia, with special reference to the examples in the South African Museum. *Annals of the South African Museum* 12:167–172.

Haughton, S. H. and Brink, A. S. 1954. A bibliographic list of the Reptilia from the Karroo Beds of Africa. *Palaeontologia Africana* 2:1–187.

Holt, B. and R. A. Benfer. 2000. Estimating missing data: an iterative regression approach. *Journal of Human Evolution* 39:289–296.

Hotton, N. 1967. Stratigraphy and sedimentation in the Beaufort Series (Permian-Triassic), South Africa; pp. 380–428 *in* C. Teichert and E. L. Yochelson (eds.), *Essays in Paleontology and Stratigraphy. Raymond C. Moore Commemorative Volume, University of Kansas Special Publications 2.* Lawrence: University of Kansas Press.

———. 1986. Dicynodonts and their role as primary consumers; pp. 71–82 *in* N. Hotton, P. D. MacLean, J. J. Roth, and E. C. Roth (eds.), *The Ecology and Biology of Mammal-Like Reptiles.* Washington, DC: Smithsonian Institution Press.

Huene, F. von. 1948. Review of the lower tetrapods; pp. 65–106 *in* A. L. DuToit (ed.), *Robert Broom Commemorative Volume.* Cape Town: Special Publications of the Royal Society of South Africa.

Huxley, T. H. 1859. On a new species of dicynodont (*D. murrayi*) from near Colesburg, South Africa; and on the structure of the skull in dicynodonts. *Quarterly Journal of the Geological Society of London* 15:649–659.

———. 1865. On a collection of vertebrate fossils from the Panchet rocks, Raniganj Coalfield. *Memoirs of the Geological Survey of India, Palaeontologia Indica* 1:2–24.

Kalandadze, N. N. 1975. The first *Lystrosaurus* find from the territory of the European part of the USSR. *Palaeontologicheskiy Zhurnal* 4:140–142.

Kermack, K. A. and J. B. S. Haldane. 1950. Organic correlation and allometry. *Biometrika* 37:30–41.

Keyser, A. W. and R. M. H. Smith. 1978. Vertebrate biozonation of the Beaufort Group with special reference to the western Karoo basin. *Annals of the Geological Survey of South Africa* 12:1–35.

King, G. M. 1988. Anomodontia; pp. 1–174 *in* P. Wellnhofer (ed.), *Encyclopedia of Paleoherpetology.* New York: Gustav Fischer.

———. 1993. Species longevity and generic diversity in dicynodont mammal-like reptiles. *Palaeogeography, Palaeoclimatology, Palaeoecology* 102: 321–332.

King, G. M. and M. A. Cluver. 1991. The aquatic *Lystrosaurus:* an alternative lifestyle. *Historical Biology* 4:323–341.

King, G. M. and I. Jenkins. 1997. The dicynodont *Lystrosaurus* from the Upper Permian of Zambia: evolutionary and stratigraphical implications. *Palaeontology* 40:149–156.

Kitching, J. W. 1968. On the *Lystrosaurus* Zone and its fauna with special reference to some immature Lystrosauridae. *Palaeontologia Africana* 11:61–76.

———. 1977. The distribution of the Karroo vertebrate fauna. *Memoirs of the Bernard Price Institute for Palaeontological Research* 1:1–131.

Kitching, J. W., J. W. Collinson, D. H. Elliot and E. H. Colbert. 1972. *Lystrosaurus* Zone (Triassic) fauna from Antarctica. *Science* 175:524–527.

Lozovskiy, V. R. 1983. The age of *Lystrosaurus* Beds in Moscow Syneclise. *Doklady Akademii Nauk SSSR* 272:1433–1437.

Lucas, S. G., Z. Cheng, and R. M. H. Smith. 1994. The *Dicynodon-Lystrosaurus* overlap zone and the Permian-Triassic boundary in nonmarine strata. *Journal of Vertebrate Paleontology* 14:34A.

Lydekker, R. 1890. *Catalogue of the Fossil Reptilia and Amphibia in the British Museum of Natural History. Part IV;* pp. 33–44. London: British Museum.

Maisch, M. W. 2002. A new basal lystrosaurid dicynodont from the Upper Permian of South Africa. *Palaeontology* 45:343–359.

Mook, C. C. 1921. Individual and age variations in the skulls of recent Crocodilia. *Bulletin of the American Museum of Natural History* 44:51–66.

O'Keefe, F. R., O. Rieppel, and P. M. Sander. 1999. Shape disassociation and inferred heterochrony in a clade of pachypleurosaurs (Reptilia, Sauropterygia). *Paleobiology* 25:504–517.

Olson, E. C. 1989. Problems of Permo-Triassic terrestrial vertebrate extinctions. *Historical Biology* 4:323–341.

Owen, R. 1859. On the orders of fossil and recent Reptilia and their distribution in time. *Reports of the British Association for the Advancement of Science* 1859:153–166.

———. 1860. On some reptilian fossils from South Africa. *Quarterly Journal of the Geological Society of London* 16:49–54.

———. 1862. On the dicynodont Reptilia, with a description of some fossil remains brought by H. R. H. Prince Alfred from South Africa, November, 1860. *Philosophical Transactions of the Royal Society (B)* 152:455–462.

———. 1876. *Descriptive and Illustrated catalogue of the Fossil Reptilia of South Africa in the Collection of the British Museum of Natural History.* London: British Museum.

Rayner, R. J. 1992. Phyllotheca: the pastures of the Late Permian. *Palaeogeography, Palaeoclimatology,Palaeoecology* 92:31–40.

———. 1995. The palaeoclimate of the Karoo: evidence from plant fossils. *Palaeogeography, Palaeoclimatology, Palaeoecology* 119:385–394.

Rayner, R. J. and M. K. Coventry. 1985. A *Glossopteris* flora from the Permian of South Africa. *South African Journal of Science* 81:21–32.

Repelin, J. 1923. Sur un fragment de crâne de *Dicyndon,* recueilli par H. Cuonil-
lon dans les environs de Luang-Prabang (Haut-Laos). *Bulletin du Service
Géologique Indochine* 12:1–7.

Rubidge, B. S. and C. A. Sidor. 2001. Evolutionary patterns among Permo-
Triassic therapsids. *Annual Review of Ecology and Systematics* 32:449–480.

Seeley, H. G. 1889. Researches on the structure, organization, and classification
of the fossil Reptilia. Part VI. On the anomodont Reptilia and their allies.
Philosophical Transactions of the Royal Society (B) 180:215–296.

———. 1898. On the skull of *Mochlorhinus platyceps* from Bethulie, Orange
Free State, preserved in the Albany Museum, Grahamstown. *Annals and
Magazine of Natural History* 1:164–176.

Shea, B. T. 1985. Bivariate and multivariate growth allometry: statistical and
biological considerations. *Journal of Zoology, London* 206:367–390.

Smith, R. M. H. 1995. Changing fluvial environments across the Permian-Triassic
boundary in the Karoo Basin, South Africa and possible causes of tetrapod
extinctions. *Palaeogeography, Palaeoclimatology, Palaeoecology* 117:81–104.

Smith, R. M. H. and P. D. Ward. 2001. Pattern of vertebrate extinctions across
an event bed at the Permian-Triassic boundary in the Karoo Basin of South
Africa. *Geology* 29:1147–1150.

Sun, A. L. 1964. Preliminary report of a new species of *Lystrosaurus* of Sinkiang.
Vertebrata Palasiatica 4:67–81.

———. 1973. Permo-Triassic dicynodonts from Turfan, Sinkiang. *Memoirs of
the Institute of Vertebrate Palaeontology and Paleoanthropology, Academia
Sinica* 10:53–68.

Thackeray, J. F., C. L. Bellamy, D. Bellars, G. Bronner, L. Bronner, C. Chi-
mimba, H. Fourie, A. Kemp, M. Krüger, I. Plug, S. Prinsloo, R. Toms,
A. S. van Zyl, and M. J. Whiting. 1997. Probabilities of conspecificity: application of a morphometric technique to modern taxa and fossil specimens
attributed to *Australopithecus* and *Homo. South African Journal of Science*
93:195–196.

Thackeray, J. F., J. F. Durand, and L. Meyer. 1998. Morphometric analysis of
South African dicynodonts attributed to *Lystrosaurus murrayi* (Huxley,
1859) and *L. declivis* (Owen, 1860): probabilities of conspecificity. *Annals
of the Transvaal Museum* 36:413–420.

Thulborn, R. A. 1983. A mammal-like reptile from Australia. *Nature* 303:
330–331.

———. 1990. Mammal-like reptiles of Australia. *Memoirs of the Queensland
Museum* 28:1–169.

Toerien, M. J. 1954. *Lystrosaurus primitivus,* sp. nov. and the origin of the genus
Lystrosaurus. Annals and Magazine of Natural History 7:407–426.

Tollman, S. M., F. E. Grine, and B. D. Hahn. 1980. Ontogeny and sexual dimorphism in *Aulacephalodon* (Reptilia, Anomodontia). *Annals of the South
African Museum* 81:159–186.

Tripathi, C. 1962. On the *Lystrosaurus* remains from the Panchets of the Rani-ganj coalfield. *Records of the Geological Survey of India* 89:407–426.

Tripathi, C. and Satsangi, P. P. 1963. *Lystrosaurus* fauna of the Panchet series of the Raniganj coalfield. *Palaeontologia Indica* 37:1–65.

Van Hoepen, E. C. N. 1915. Contributions to the knowledge of the reptiles of the Karroo Formation. 3. The skull and other remains of *Lystrosaurus put-terilli*, n. sp. *Annals of the Transvaal Museum* 5:70–82.

———. 1916. Preliminary description of some new lystrosauri. *Annals of the Transvaal Museum* 5:214–216.

Webb, J. W. and H. Messel. 1978. Morphometric analysis of *Crocodylus porosus* from the North Coast of Arnhem Land, Northern Australia. *Australian Jour-nal of Zoology* 26:1–27.

Wood, B. A. and B. G. Richmond. 2000. Human evolution: taxonomy and paleo-biology. *Journal of Anatomy* 196:19–60.

Young, C.-C. 1935. On two skeletons of Dicynodontia from Sinkiang. *Bulletin of the Geological Survey of China* 14:483–517.

———. 1939. Additional Dicynodontia remains from Sinkiang. *Bulletin of the Geological Survey of China* 19:111–146.

Yuan, P. L. and C.-C. Young. 1934. On the occurrence of *Lystrosaurus* in Sinkiang. *Bulletin of the Geological Survey of China* 13:575–580.

Reflections on
James Allen Hopson

Plate 5. Jim Hopson's great-grandfather, William F. Hopson. An artist and engraver from New Haven, CT, William often worked for Yale paleontologist O. C. Marsh.

15   James Allen Hopson: A Biography

A. W. Crompton
Farish A. Jenkins Jr.
Susan Hopson
Timothy J. Gaudin, *and*
Matthew T. Carrano

Student and Professor

James Allen Hopson was born to Frank R. and Henrietta Learnard Hopson in 1935 in New Haven, Connecticut. He attended New Haven Public schools and, when Yale University offered scholarships to local students, Jim applied and was accepted, graduating in 1957 with a degree in Geology.

While an undergraduate student, Jim requested a bursary assignment at the Peabody Museum of Natural History. This work excited his interest in paleontology, an interest nourished by Joe Gregory, Curator of Vertebrate Paleontology in the museum at that time. Gregory offered Jim a student assistantship to work with the collections and later introduced him to field work.

Jim helped to organize the Peabody's vertebrate paleontology collections and, while doing so, discovered a piece of family history that linked him personally to the field of vertebrate paleontology. While paging through a biography of O. C. Marsh during his coffee break, he was heard to exclaim, "Hey—that's my great-grandfather!" It turned out that William F. Hopson, one of the best known American book-plate designers and engravers of his era, had been an engraver and illustrator for O. C. Marsh. William Hopson's magnificent woodcuts are included in Marsh's monographs, *Odontornithes: A Monograph of Extinct Birds of North America* and *The Gigantic Mammals of the Order Dinocerata.*

Although he majored in geology, Jim was always interested in opportunities to expand his understanding of biology. As a consequence, after he graduated from Yale in 1957, he opted to pursue graduate studies at the University of Chicago because of its interdisciplinary program in Paleozoology. Chaired by Jim's mentor, Everett C. Olson, the program freed students from the rigid requirements of a single department and facilitated their broad exploration of the biology of living organisms as they endeavored to interpret the physiology, ecology, or functional morphology of extinct forms. The program is still productive today (although now

Figure 15.1. Jim Hopson (left), Kenneth Kermack (center), and George Gaylord Simpson (right) listening to a scientific presentation at the "Early Mammals" symposium in London, 1970.

known as the Committee on Evolutionary Biology) and has trained and employed an impressive array of paleobiologists.

In 1963, before completing the requirements for his Ph.D. degree, Jim accepted an NSF-funded appointment as Curatorial Assistant/Associate at Yale's Peabody Museum, a position assigned to cataloguing and up-grading the vertebrate paleontology collections. Always curious about the oddities he found in the collections, Jim published his first paper on pseudo-toothed birds (Hopson, 1964). While he was putting great effort into his curatorial responsibilities, Jim allocated sufficient time to com-plete his Ph.D. thesis and simultaneously completed several papers that set the course of his future research.

His study of the braincase of *Bienotherium* addressed the complexity of problems underlying the evolutionary emergence of mammals (Hopson, 1964). His analysis of tooth replacement in mammal-like reptiles revealed how growth patterns and details of dental anatomy can open insights into the biology of extinct organisms (Hopson, 1964, 1971). At Yale, Jim en-countered many colleagues with whom he would share both a professional interest and friendship for the rest of his life—Fuzz Crompton, Farish Jenkins, John Ostrom, Dale Russell, and Keith Thomson, among many others.

In 1967, Jim again left the Peabody Museum to take up an appoint-ment as Assistant Professor in the Department of Anatomy at the Uni-versity of Chicago (currently the Department of Organismal Biology and Anatomy). Ronald Singer, who then chaired the Department, wished to broaden its scope by introducing an evolutionary perspective to anatomy, a foresight that promoted an intellectually vibrant venture. Len Radinsky and Leigh Van Valen were also hired at that time. Future appointments would broaden this scope even more, and the Department continues today as one of the strongest centers of organismic and evolutionary biology in the country.

The University of Chicago turned out to be a harmonious, productive home for Jim. The close working relationships between members of the Department of Anatomy, the Department of Geophysical Sciences, and the Field Museum provided a stimulating environment for students, postdoctoral fellows, faculty, and visitors. Collegial and personal relationships emerged as Jim shared ideas, information, occasional (and frequent) arguments, and always-spirited discussions with Edgar Allin, Andy Biewener, John Bolt, John Flynn, Dave Jablonski, Mike LaBarbara, Eric Lombard, Charles Oxnard, Len Radinsky, Dave Raup, Tom Schopf, Jack Sepkoski, Paul Sereno, Olivier Rieppel, Dave Wake, and, more recently, Mike Coates and Neil Shubin. Jim Clark, Desui Miao, Guillermo Rougier, Juri van den Heever, and John Wible worked closely with Jim in Chicago as postdoctoral fellows. He also developed a close working and publishing relationship with José Bonaparte (Argentina) and James Kitching and

Figure 15.2. Jim Hopson (center) talking with Zofia Kielan-Jaworowska (left) and George Gaylord Simpson (right), at the "Early Mammals" symposium in London, 1970.

Bruce Rubidge (South Africa). Both Herb Barghusen (University of Illinois at Chicago) and Armand de Ricqlès (Université de Paris) worked with Jim as visiting professors in Chicago.

Teacher and Mentor

Jim taught an undergraduate course in Chordate Biology every year that he was in residence at Chicago: first with Dave Wake, then with Eric Lombard, and, since 1990, on his own. He was an enthusiastic teacher in the classroom, but small groups of students and one-on-one reading courses were his preference. A popular lecturer, he was awarded the Quantrell Award for Excellence in Undergraduate Teaching in 1996. Starting in 1969, Jim taught in the graduate Vertebrate Paleobiology course at the University of Chicago. For most of those years, he was responsible for teaching the section of the course covering nonmammalian vertebrates, with Len Radinsky and Leigh Van Valen teaching the mammals section of the course. Jim also taught numerous smaller graduate seminar courses.

Graduate students always assumed an important role in Jim's life in Chicago. He appears as junior author on some of their papers (notably those of Jim Clark, Tim Gaudin, and Chris Sidor), but as a rule he encouraged them to publish solely under their own names. Indeed, the diversity of research topics pursued by his students and postdocs exceeded even his own considerable diversity of interests, serving as one of the hallmarks of Jim's career as a mentor. Jim did not limit his students' inquiry to any one particular taxonomic group or to any particular research

Figure 15.3. Jim Hopson (rear) and his University of Chicago students in a classroom at the Field Museum of Natural History. Clockwise from lower left: Sharon Swartz, James Clark, Scott Schaefer, two unidentified students, and John Clay Bruner. Note the photograph in the background, showing Hopson's advisor Everett Olson teaching his students in the same classroom.

methodology but encouraged his students to ask interesting questions and seek answers to those questions in a rigorous fashion.

The diversity of topics included in this volume and the quality of the contributions stand as a testament to his skills as a mentor. The majority of chapters (by Blob, Carrano, Munter & Clark, Gaudin & Wible, O'Keefe, Parrish, Rougier & Wible, Sidor & Rubidge) were authored or coauthored by former students or postdocs. In addition, a number of former students and postdocs who were unable to participate in this volume have gone on to productive careers in museum curation and research (James Mead, National Museum of Natural History, Smithsonian Institution; Desui Miao, Museum of Natural History, University of Kansas), university research (Arthur Busbey, Texas Christian University; Juri van den Heever, University of Stellenbosch), teaching and research at biomedical institutions (Gaylord Throckmorton, University of Texas Medical Center; Steven Zehren, University of Alabama–Birmingham Medical Center; Robert Fisk, West Virginia College of Osteopathic Medicine), and undergraduate

Figure 15.4. Fieldwork in Mexico during the early 1980s, with (from left to right) Jim Hopson, Paul Sereno, James Clark, and Rene Hernández.

teaching (William Stevens, Georgetown College; Laura Panko, North-western University).

Researcher and Colleague

Jim Hopson's contributions to our understanding of the systematics, structure, distribution, and biology of mammal-like reptiles (nonmammalian synapsids) and their early Mesozoic mammalian descendents represent an unmatched record of scientific achievement. As a graduate student with Everett ("Shorty" or "Ole") Olson in the late 1950s, Jim was introduced to synapsid evolution and in particular to Olson's view that mammalian characters had evolved independently in various groups of mammal-like reptiles. This hypothesis was to remain a dominant theme in Jim's future research as he explored the morphological evolution of synapsids with a level of detail never before undertaken.

In numerous analyses of the phylogeny of nonmammalian synapsids, from the classic paper coauthored by Herb Barghusen in 1986 (Hopson & Barghusen, 1986) to his recent paper with James Kitching (Hopson & Kitching, 2001), Jim amassed overwhelming evidence that supported Olson's views on the pervasiveness of parallel appearance of mammalian features in synapsid evolution. However, Jim did not support the conclusion of Olson and many of his contemporaries that the various groups of Mesozoic mammals had a polyphyletic origin. Rather, in several papers (Hopson & Crompton, 1969; Hopson, 1970) Jim argued strongly that the known groups of Mesozoic mammals had a common ancestor defined by unique mammalian features such as a tricuspid dentition with a large central cusp and a diphyodont pattern of tooth replacement. The monophyly of Mammalia (or of Mammaliaformes, as many would now label the group including the common ancestor of *Morganucodon* and extant mammals) has been widely accepted since the publication of these landmark works. Without Jim's careful and extensive analyses, our appreciation of the relationships of synapsids and early mammals and the selective forces governing the acquisition of mammalian characters would have remained on far less secure grounds.

Never content to rely on the published literature, Jim traveled widely throughout his career to study classic museum collections firsthand. He likewise participated in studies of abundant, newly discovered material of mammal-like reptiles and early mammals that were made during the last half-century. Since 1971, his active research program has involved him in projects in South Africa, Europe (including Russia), Mexico, Argentina, and Australia, and his studies have spanned the breadth of synapsid anatomy, systematics, and evolution.

In one of his very early papers in 1966, Jim was the first to offer a functional explanation for the evolutionary transition of the mammalian middle ear. Ever since the homologies of the mammalian middle ear bones were recognized in the mid-19th century, paleontologists and comparative anatomists had been intrigued by the problem of how, why, and when the postdentary bones shifted from the lower jaw to the skull to form part of the middle ear. Jim suggested that in advanced "cynodonts" and early mammals, the reflected lamina of the angular supported a tympanic membrane that had migrated forward from a "reptilian" postquadrate position. He convincingly demonstrated how the quadrate and articular, while still forming the jaw joint, could conduct sound via the stapes to the inner ear.

However, Edgar Allin, a close colleague of Jim, proposed an alternative hypothesis. He agreed with Jim that the reflected lamina of advanced cynodonts and early mammals supported a tympanic membrane but claimed that it was not the homologue of the postquadrate membrane of typical reptiles. To resolve these two competing hypotheses, Jim and Edgar jointly explored details of the cranial anatomy of fossil and extant synapsids. Their collaborative paper on the evolution of the auditory system in Synapsida (Allin & Hopson, 1992) provides a magnificent account of the structure and evolution of the middle and internal ear from the Permian pelycosaurs to modern mammals.

Another of Jim's notable collaborations included Guillermo Rougier, John Wible, and José Bonaparte. Jointly and separately, they described various aspects of the skull of *Vincelestes,* the oldest known and most completely preserved mammal with nearly tribosphenic molars (e.g., Rougier et al., 1992; Hopson & Rougier, 1993). Their analyses of *Vincelestes* represent a major contribution toward an understanding of the phylogenetic history of mammals. It also provided Jim and Guillermo with the information they needed to explain the origin of the fundamentally different configuration of the lateral wall of the braincase in therian and nontherian (monotreme) mammals (Hopson & Rougier, 1993).

Jim's research has not been restricted to synapsid evolution alone; he has also made several important contributions to archosaurian biology. His paper on the cranial crests of hadrosaurian dinosaurs (Hopson, 1975) interpreted these structures as functioning primarily in intraspecific display. This study was among the first to suggest that the evolution of various peculiar dinosaurian features could be explained as a result of sexual selection. Jim brought a reasoned and measured approach to the heated controversies that raged around speculations on dinosaurian endothermy during the 1970s. Stressing the oversimplification inherent in the view that dinosaur physiology was uniform ("one size does not fit all," he stated), he related dinosaurian brain sizes to hypothesized activity levels and concluded that only coelurosaurs were as active as birds and mammals. All other dinosaurs, he believed, had metabolic levels that ranged between those of living reptiles and living endotherms (e.g., Hopson, 1977, 1980).

Actively promoting the highest standards in paleontological research, Jim served as a coeditor (with Tom Schopf) of the journal *Paleobiology* for four years (1977–1980) and as sole editor for three subsequent years (1981–1983). He was also an Associate Editor of the *Journal of Vertebrate Paleontology* from 1984 to 1988. Always an active member of the Society of Vertebrate Paleontology, Jim served as its Vice-President in 1983 and its President in 1984.

Husband, Father, and Friend

Jim has been married to Susan Hopson since 1961. They have two sons, Andrew (born in 1964, now a freelance sound designer) and Peter (born in 1966, a general manager for a resort in Australia) and two grandchildren (Katie, born in 1994, and Callum, born in 1998). Jim and Sue's closeness and quiet pride in their family have always been readily apparent to colleagues and students alike.

But Jim and Sue have done more than raise their own fine family. Over the years, they have opened their Hyde Park home to countless graduate students, postdocs, and visiting scientists, collaborators and colleagues, for periods of time ranging from days to months, occasionally even years. The "Hotel Hopson," as it is sometimes affectionately known, provided a hospitable respite for scientists at a variety of levels who, for a variety of

Figure 15.5. Jim Hopson in retirement at his Ludington, Michigan, home in 2002.

reasons, found their way to the Department of Anatomy at the University of Chicago.

Science is ultimately a human enterprise, one in which interpersonal relationships matter. Jim and Sue have played an exemplary role in the scientific community in this respect, in the gracious manner in which they have treated their students and colleagues, and in the comfortable, family atmosphere they have provided to nurture both science and scientists.

All those who have collaborated with Jim, or studied under his direction, roundly and freely attest to his enthusiasm, kindness, generosity, and wise counsel. In retirement, he continues to give to paleontology the wisdom of his insights, and to colleagues and former students, his faithful interest and friendship.

We hope this book will serve as an enduring appreciation of Jim Hopson and his many contributions to the field of vertebrate paleontology and its students. The emergence of this volume will surprise few in vertebrate paleontology. Nor will many overlook the fact that it has taken the efforts of many authors to approach the breadth of its eponym—such is the legacy of the great scientist and the great teacher.

It is our great pleasure to honor both.

Jim, thank you.

A P P E N D I X
James Allen Hopson: A Bibliography
(1964–2005)

Matthew T. Carrano

and

Timothy J. Gaudin

1964

Hopson, J. A. 1964. *Pseudodontornis* and other large marine birds from the Miocene of South Carolina. *Postilla* 83:1–19.

Hopson, J. A. 1964. Tooth replacement in cynodont, dicynodont and therocephalian reptiles. *Proceedings of the Zoological Society of London* 142:625–654.

Hopson, J. A. 1964. The braincase of the advanced mammal-like reptile *Bienotherium. Postilla* 87:1–30.

1965

Hopson, J. A. 1965. The braincase of the advanced mammal-like reptile *Bienotherium.* Ph.D. Dissertation, University of Chicago, Chicago.

1966

Hopson, J. A. 1966. The origin of the mammalian middle ear. *American Zoologist* 6(3): 437–450.

1967

Hopson, J. A. 1967. Comments on the competitive inferiority of the multituberculates. *Systematic Zoology* 16(4): 352–355.

Hopson, J. A. 1967. Mammal-like reptiles and the origin of mammals. *Discovery, Peabody Museum of Natural History, Yale University* 2(2): 25–33.

1969

Hopson, J. A. 1969. Origin and adaptive radiation of mammal-like reptiles and nontherian mammals. *Annals of the New York Academy of Sciences* 167(A1): 199.

Hopson, J. A. and A. W. Crompton. 1969. Origin of mammals; pp. 15–72 *in* T. Dobzhansky, M. K. Hecht, and W. C. Steere (eds.), *Evolutionary Biology,* volume 3. New York: Appleton-Century-Crofts.

1970

Barghusen, H. R. and J. A. Hopson. 1970. Dentary-squamosal joint and the origin of mammals. *Science* 168(3931): 573–575.

Hopson, J. A. 1970. The classification of nontherian mammals. *Journal of Mammalogy* 51(1): 1–9.

Hopson, J. A. 1970. Postcanine tooth replacement in the gomphodont cynodont *Diademodon. Biological Journal of the Linnean Society* 2:317–318.

1971

Hopson, J. A. 1971. Postcanine replacement in the gomphodont cynodont *Diademodon. Zoological Journal of the Linnean Society* 50(50): 1–21.

1972

Hopson, J. A. 1972. Book Review: *The Postcranial Skeleton of African Cynodonts,* Bulletin of the Peabody Museum of Natural History, Yale University, by F. A. Jenkins, Jr. *Quarterly Review of Biology* 47:209–210.

Hopson, J. A. 1972. Endothermy, small body size, and the origin of mammalian reproduction. Society of Vertebrate Paleontology 32nd Annual Meeting abstracts:5.

Hopson, J. A. and J. W. Kitching. 1972. A revised classification of cynodonts (Reptilia: Therapsida). *Palaeontologia Africana* 14: 71–85.

1973

Hopson, J. A. 1973. Endothermy, small size, and the origin of mammalian reproduction. *American Naturalist* 107:446–452.

1974

Hopson, J. A. 1974. The functional significance of the hypocercal tail and lateral fin fold of anaspid ostracoderms. *Fieldiana: Geology (New Series)* 33(5): 83–93.

1975

Hopson, J. A. 1975. The evolution of cranial display structures in hadrosaurian dinosaurs. *Paleobiology* 1:21–43.

Hopson, J. A. 1975. On the generic separation of the ornithischian dinosaurs *Lycorhinus* and *Heterodontosaurus* from the Stormberg Series (Upper Triassic) of South Africa. *South African Journal of Science* 71:302–305.

1976

Hopson, J. A. 1976. Hot-, cold-, or lukewarm-blooded dinosaurs? Review of *The Hot-Blooded Dinosaurs: A Revolution in Paleontology,* by Adrian J. Desmond. *Paleobiology* 2(3): 271–275.

1977

Hopson, J. A. 1977. Relative brain size and behavior in archosaurian reptiles. *Annual Review of Ecology and Systematics* 8:429–448.

Hopson, J. A. 1977. Brain size and behavior of dinosaurs. *Journal of Paleontology* 51(2, suppl., part III): 15.

1978

Hopson, J. A. 1978. Waren die Dinosaurier Warmblüter? *Manheimer Forum* 78/79:123–184.

1979

Barghusen, H. R. and J. A. Hopson. 1979. The endoskeleton: The comparative anatomy of the skull and the visceral skeleton; pp. 265–326 in M. H. Wake (ed.), *Hyman's Comparative Vertebrate Anatomy,* 3rd edition. Chicago: University of Chicago Press.

Hopson, J. A. 1979. Paleoneurology; pp. 39–146 in R. G. Northcutt and P. Ulinksi (eds.), *The Biology of the Reptilia. Volume 9. Neurology A.* London: Academic Press.

1980

Hopson, J. A. 1980. Relative brain size in dinosaurs: implications for dinosaurian endothermy; pp. 287–310 in E. C. Olson and R. D. K. Thomas (eds.), *A Cold Look at the Warm-Blooded Dinosaurs.* Boulder, CO: Westview Press.

Hopson, J. A. 1980. Tooth function and replacement in early Mesozoic ornithischian dinosaurs: implications for aestivation. *Lethaia* 13:93–105.

Hopson, J. A. and L. B. Radinsky. 1980. Status of paleontology—1980. Vertebrate paleontology—new approaches and new insights. *Paleobiology* 6(3): 250–270.

1981

Hopson, J. A. 1981. Tooth function and replacement in heterodontosaurid ornithischians of the Stormberg Series—implications for aestivation. *Palaeontologia Africana* 24:9.

Hopson, J. A. and W.-E. Reif. 1981. The status of *Archaeodon reuningi* von Huene, a supposed late Triassic mammal from Southern Africa.

Neues Jährbuch für Geologie und Paläontologie, Monatshefte 5:307–310.

McKenna, M. C., J. A. Hopson, and H.-P. Schultze. 1981. Vertebrate paleontology. *Geotimes* 26:56–57.

Throckmorton, G. S., J. A. Hopson, and P. Parks. 1981. A redescription of *Toxolophosaurus cloudi* Olson, a Lower Cretaceous herbivorous sphenodontid reptile. *Journal of Paleontology* 55:586–597.

1982

Boucot, A., N. Eldredge, and J. A. Hopson. 1982. Sources and comments on scientific creationism. *Journal of Paleontology* 56:1320–1321.

van den Heever, J. A. and J. A. Hopson. 1982. The systematic position of "therocephalian B" (Reptilia: Therapsida). *South African Journal of Science* 78(10): 424–425.

1983

Hopson, J. A. 1983. Synapsids and evolution: Review of *Mammal-like Reptiles and the Origin of Mammals,* by T. S. Kemp. *Science* 219: 49–50.

1984

Hopson, J. A. 1984. Late Triassic traversodont cynodonts from Nova Scotia and southern Africa. *Palaeontologia Africana* 25:181–201.

1985

Clark, J. M. and J. A. Hopson. 1985. Distinctive mammal-like reptile from Mexico and its bearing on the phylogeny of the Tritylodontidae. *Nature* 315:398–400.

Hopson, J. A. 1985. Morphology and relationships of *Gomphodontosuchus brasiliensis* von Huene (Synapsida, Cynodontia, Tritylodontoidea) from the Triassic of Brazil. *Neues Jährbuch für Geologie und Paläontologie, Monatshefte* 1985(5): 285–299.

1986

Hopson, J. A. 1986. Book Review: *The Structure, Development, and Evolution of Reptiles,* edited by M. J. Ferguson. *Quarterly Review of Biology* 61:118–119.

Hopson, J. A. 1986. Leonard B. Radinsky, 1937–1985. *Society of Vertebrate Paleontology News Bulletin* 138:63–65.

Hopson, J. A. 1986. Book Review: *The Evolution of Mammalian Characters,* by D. M. Kermack and K. A. Kermack. *Geological Journal* 21(1): 89–91.

Hopson, J. A. and H. R. Barghusen. 1986. An analysis of therapsid relationships; pp. 83–106 *in* N. Hotton, III, P. D. MacLean, J. J. Roth, and E. C. Roth (eds.), *The Ecology and Biology of Mammal-like Reptiles.* Washington, DC: Smithsonian Institution Press.

1987

Fastovsky, D. E., J. M. Clark, and J. A. Hopson. 1987. Preliminary report of a vertebrate fauna from an unusual paleoenvironmental setting, Huizachal Group, Early or Mid-Jurassic, Tamaulipas, Mexico. *Occasional Papers of the Tyrrell Museum of Palaeontology* 3:82–87.

Hopson, J. A. 1987. Synapsid phylogeny and the origin of mammalian endothermy. *Journal of Vertebrate Paleontology* 7(3, suppl.): 18.

Hopson, J. A. 1987. The mammal-like reptiles: a study of transitional fossils. *The American Biology Teacher* 49(1): 16–26.

1988

Fastovsky, D. E., O. D. Hermes, J. M. Clark, and J. A. Hopson. 1988. Volcanoes, debris flows, and Mesozoic mammals: Huizachal Group (Early or Middle Jurassic), Tamaulipas, Mexico. *Geological Society of America Abstracts-with-Programs* A317–A318.

Hopson, J. A. and J. W. Kitching. 1988. A *Chiniquodon*-like cynodont from the Early Triassic of South Africa and the phylogeny of advanced cynodonts. *Journal of Vertebrate Paleontology* 8(3, suppl.): 18A.

1989

Hopson, J. A. 1989. Leonard Burton Radinsky (1937–1985); pp. 2–12 *in* D. R. Prothero and R. M. Schoch (eds.), *The Evolution of Perissodactyls. Oxford Monographs on Geology and Geophysics 15.* Oxford: Oxford University Press.

Hopson, J. A., Z. Kielan-Jaworowska and E. Allin. 1989. The cryptic jugal of multituberculates. Journal of Vertebrate Paleontology 9(2): 201–209.

Hopson, J. A., J. F. Bonaparte, and G. W. Rougier. 1989. Braincase structure of a non-tribosphenic therian mammal from the Early Cretaceous of Argentina. *Journal of Vertebrate Paleontology* 9(3, suppl.): 25A.

Wible, J. R. and J. A. Hopson. 1989. Book Review: *The Phylogeny and Classification of Tetrapods. Vol. 2: Mammals,* edited by M. J. Benton. *Journal of Vertebrate Paleontology* 9:237–238.

1990

Hopson, J. A. 1990. Cladistic analysis of therapsid relationships. *Journal of Vertebrate Paleontology* 10(3, suppl.): 28A.

Rubidge, B. S. and J. A. Hopson. 1990. A new anomodont therapsid from South Africa and its bearing on the ancestry of Dicynodontia. *South African Journal of Science/Suid-Afrikaanse Tydskrif vir Wetenskap* 86:43–45.

Wible, J. R. and J. A. Hopson. 1990. Interrelationships of Recent mammals and Mesozoic relatives; the basicranial evidence. *Journal of Vertebrate Paleontology* 10(3, suppl.): 48A.

Wible, J. R., D. Miao, and J. A. Hopson. 1990. The septomaxilla of fossil and recent synapsids and the problem of the septomaxilla of monotremes and armadillos. *Zoological Journal of the Linnean Society* 98:203–228.

1991

Clark, J. M., M. Montellano, J. A. Hopson, and R. Hernandez. 1991. Mammals and other tetrapods from the Early Jurassic La Boca Formation, northeastern Mexico. *Journal of Vertebrate Paleontology* 11(3, suppl.): 42A.

Hopson, J. A. 1991. Convergence in mammals, trithelodonts and tritylodonts. *Journal of Vertebrate Paleontology* 11(3, suppl.): 36A.

Hopson, J. A. 1991. Robert Thomas Zanon, 1957–1990. *Society of Vertebrate Paleontology News Bulletin* 153:48–49.

Rougier, G. W., J. R. Wible, and J. A. Hopson. 1991. The cranial vascular system in the Early Cretaceous mammal *Vincelestes neuquenianus*. *Journal of Vertebrate Paleontology* 12(3, suppl.): 53A.

Rougier, G. W., J. R. Wible and J. A. Hopson. 1991. Reconstruccion arterial y venosa del craneo de *Vincelestes neuquenianus* (Mammalia, Theria); implicancias en la evolución vascular del craneo mamaliano. *Ameghiniana* 28(3–4): 412.

1992

Allin, E. F. and J. A. Hopson. 1992. Evolution of the auditory system in Synapsida ("mammal-like reptiles" and primitive mammals) as seen in the fossil record; pp. 587–614 *in* D. B. Webster, R. R. Fay and A. N. Popper (eds.), *The Evolutionary Biology of Hearing.* New York: Springer-Verlag.

Hopson, J. A. 1992. Convergent evolution in the manus and pes of therapsids. *Journal of Vertebrate Paleontology* 12(3, suppl.): 33A–34A.

Hopson, J. A. 1992. Systematics of the nonmammalian Synapsida and implications for patterns of evolution in synapsids; pp. 635–693 *in* H.-P. Schultze and L. Trueb (eds.), *Origins of the Higher Groups of Tetrapods: Controversy and Consensus.* Ithaca, NY: Comstock Publishing Associates.

Hopson, J. A. and E. F. Allin. 1992. Evolutionary origin of the mammalian auditory system as seen in the fossil record. Abstracts of 15th Midwinter Research Meeting, Association for Research in Otolaryngology.

Rougier, G. W., J. R. Wible, and J. A. Hopson. 1992. Reconstruction of the cranial vessels in the Early Cretaceous mammal *Vincelestes neuquenianus:* implications for the evolution of the mammalian cranial vascular system. *Journal of Vertebrate Paleontology* 12(2): 188–216.

Sues, H.-D., J. A. Hopson, and N. H. Shubin. 1992. Affinities of ?*Scalenodontoides plemmyridon* Hopson, 1984 (Synapsida: Cynodontia) from the Upper Triassic of Nova Scotia. *Journal of Vertebrate Paleontology* 12:168–171.

1993

Hopson, J. A. and G. W. Rougier. 1993. Braincase structure in the oldest known skull of a therian mammal: implications for mammalian systematics and cranial evolution. *American Journal of Science* 293-A:268–299.

Hopson, J. A. and J. R. Wible. 1993. Cranial vascular patterns in Multituberculata; the postglenoid foramen that wasn't there. *Journal of Vertebrate Paleontology* 13(3, suppl.): 42A.

Wible, J. R. and J. A. Hopson. 1993. Basicranial evidence for early mammal phylogeny; pp. 45–62 *in* F. S. Szalay, M. J. Novacek, and M. C. McKenna (eds.), *Mammal Phylogeny. Mesozoic Differentiation, Multituberculates, Monotremes, Early Therians and Marsupials.* New York: Springer-Verlag.

1994

Clark, J. M., D. E. Fastovsky, M. Montellano, J. A. Hopson, and R. Hernandez. 1994. Additions to the Middle Jurassic vertebrate fauna of Huizachal Canyon, Tamaulipas, Mexico. *Journal of Vertebrate Paleontology* 14(3, suppl.): 21A.

Clark, J. M., M. Montellano, J. A. Hopson, R. Hernandez, and D. E. Fastovsky. 1994. An Early or Middle Jurassic tetrapod assemblage from the La Boca Formation, northeastern Mexico;

pp. 295–302 *in* N. C. Fraser and H.-D. Sues (eds.), *In the Shadow of the Dinosaurs: Early Mesozoic Tetrapods.* Cambridge: Cambridge University Press.

Hopson, J. A. 1994. Synapsid evolution and the radiation of non-eutherian mammals; pp. 190–219 *in* D. R. Prothero and R. M. Schoch (eds.), *Major Features of Vertebrate Evolution.* Knoxville: The University of Tennessee, The Paleontological Society Short Course.

Wible, J. R. and J. A. Hopson. 1994. Homologies of the prootic canal in mammals and non-mammalian cynodonts. *Journal of Vertebrate Paleontology* 14(3, suppl.): 52A.

1995

Fastovsky, D. E., J. M. Clark, N. H. Strater, M. Montellano, R. Hernandez, and J. A. Hopson. 1995. Depositional environments of a Middle Jurassic terrestrial vertebrate assemblage, Huizachal Canyon, Mexico. *Journal of Vertebrate Paleontology* 15(3): 561–575.

Gaudin, T. J., J. R. Wible, J. A. Hopson, and W. D. Turnbull. 1995. Cohort Epitheria (Mammalia, Eutheria); the morphological evidence reexamined. *Journal of Vertebrate Paleontology* 15(3, suppl.): 31A.

Hopson, J. A. 1995. The Jurassic mammal *Shuotherium dongi;* "pseudotribosphenic therian," docodontid, or neither? *Journal of Vertebrate Paleontology* 15(3, suppl.): 36A.

Hopson, J. A. 1995. Patterns of evolution in the manus and pes of non-mammalian therapsids. *Journal of Vertebrate Paleontology* 15(3): 615–639.

Montellano, M., J. A. Hopson, J. M. Clark, D. E. Fastovsky, and R. Hernandez. 1995. Mammals from the Middle Jurassic of Huizachal Canyon, Tamaulipas, Mexico. *Journal of Vertebrate Paleontology* 15(3, suppl.): 45A.

Sidor, C. A. and J. A. Hopson. 1995. The taxonomic status of the Upper Permian eotheriodont therapsids of the San Angelo Formation (Guadalupian), Texas. *Journal of Vertebrate Paleontology* 15(3, suppl.): 53A.

Wible, J. R. and J. A. Hopson. 1995. Homologies of the prootic canal in mammals and non-mammalian cynodonts. *Journal of Vertebrate Paleontology* 15(2): 331–356.

1996

Clark, J. M., J. A. Hopson, and R. Hernandez. 1996. An uncrushed "rhamphorhynchoid" pterosaur from the Middle Jurassic of Mexico. *Journal of Vertebrate Paleontology* 16(3, suppl.): 28A.

Gaudin, T. J., J. R. Wible, J. A. Hopson, and W. D. Turnbull. 1996.
Reexamination of the morphological evidence for the cohort Epitheria (Mammalia, Eutheria). *Journal of Mammalian Evolution* 3(1):
31–79.

Hopson, J. A. 1996. Cranial morphology of *Ptilodus montanus*
(Ptilodontoidea, Multituberculata) revisited. *Journal of Vertebrate
Paleontology* 16(3, suppl.): 42A.

Montellano-Ballesteros, M., R. Hernandez-Rivera, J. M. Clark, D. E.
Fastovsky, V. H. Reynoso-Rosales, N. H. Strater, and J. A. Hopson.
1996. Avances en el estudio de la fauna de vertebrados Jurasicos del
Canon del Huizachal, Tamaulipas, Mexico. *Boletin de la Sociedad
Geologica Mexicana* 52(3-4): 11–20.

Rougier, G., J. R. Wible, and J. A. Hopson. 1996. Basicranial anatomy of
Priacodon fruitaensis (Triconodontidae, Mammalia) from the Late
Jurassic of Colorado, and a reappraisal of mammaliaform interrela-
tionships. *American Museum Novitates* 3183:1–38.

Rubidge, B. S. and J. A. Hopson. 1996. A primitive anomodont therapsid from the base of the Beaufort Group (Upper Permian) of South
Africa. *Zoological Journal of the Linnean Society* 117:115–139.

1997

Clark, J. M., J. A. Hopson, R. Hernandez, D. E. Fastovsky, and M. Montellano. 1997. A primitive pterosaur from the Jurassic of Mexico and
its implications for pterosaur foot posture. *Journal of Vertebrate Paleontology* 17(3, suppl.): 38–39A.

Hopson, J. A. 1997. Is cusp C of the upper molars of *Kuehneotherium*
homologous with the metacone of *Peramus* and tribosphenic mammals? *Journal of Vertebrate Paleontology* 17(3, suppl.): 53A.

Sidor, C. A. and J. A. Hopson. 1997. Patterns of increasing "mammalness"; comparing ghost lineage duration and measures of morphological change in synapsid evolution. *Journal of Vertebrate Paleontology* 17(3, suppl.): 76A.

1998

Clark, J. M., J. A. Hopson, R. Hernández R., D. E. Fastovsky, and
M. Montellano. 1998. Foot posture in a primitive pterosaur. *Nature*
391:886–888.

Hopson, J. A. and L. M. Chiappe. 1998. Pedal proportions of living and
fossil birds indicate arboreal or terrestrial specialization. *Journal of
Vertebrate Paleontology* 18(3, suppl.): 52A.

Sidor, C. A. and J. A. Hopson. 1998. Ghost lineages and "mammal-ness": assessing the temporal pattern of character acquisition in the Synapsida. *Paleobiology* 24(2): 254–273.

1999

Foote, M. and J. A. Hopson. 1999. J. John Sepkoski Jr. (1948–1999)—in memoriam. *Acta Palaeontologica Polonica* 44(2): 235–236.

Hopson, J. A. 1999. Mammals, Mesozoic and non-therian; pp. 691–701 *in* R. Singer (ed.), *Encyclopedia of Paleontology*. Chicago: Fitzroy Dearborn Publishing.

Hopson, J. A. 1999. Synapsids; pp. 1193–1195 *in* R. Singer (ed.), *Encyclopedia of Paleontology*. Chicago: Fitzroy Dearborn Publishing.

Hopson, J. A. 1999. Therapsids; pp. 1256–1266 *in* R. Singer (ed.), *Encyclopedia of Paleontology*. Chicago: Fitzroy Dearborn Publishing.

Hopson, J. A., G. F. Engelmann, G. W. Rougier, and J. R. Wible. 1999. Skull of a new paurodontid mammal (Holotheria, Dryolestoidea) from the Late Jurassic of Colorado. *Journal of Vertebrate Paleontology* 19(3, suppl.): 52–53A.

2000

Beck, A. M., R. W. Blob, and J. A. Hopson. 2000. Interpreting limb posture in fossil tetrapods: morphological indicators of sprawling and non-sprawling locomotion. *Journal of Vertebrate Paleontology* 19(3, suppl.): 29A.

Beck, A. M., R. W. Blob, and J. A. Hopson. 2000. Morphological indicators of sprawling and non-sprawling limb posture in tetrapods. *American Zoologist* 40(6): 939.

2001

Barbarena, M. C., J. F. Bonaparte, A. W. Crompton, J. A. Hopson, C. Schultz, and R. Rupert. 2001. A new fauna of very mammal-like cynodonts from the Late Triassic of Brazil. *PaleoBios* 21(2): 28.

Hopson, J. A. 2001. Ecomorphology of avian and nonavian theropod phalangeal proportions: implications for the arboreal versus terrestrial origin of bird flight; pp. 211–235 *in* J. Gauthier and L. F. Gall (eds.), *New Perspectives on the Origin and Evolution of Birds: Proceedings of the International Symposium in Honor of John H. Ostrom*. New Haven, CT: Peabody Museum of Natural History, Yale University Press.

Hopson, J. A. 2001. Origin of mammals; pp. 88–94 *in* D. E. G. Briggs and P. R. Crowther (eds.), *Paleobiology II*. Oxford: Blackwell Scientific Publications.

Hopson, J. A. and J. W. Kitching. 2001. A probainognathian cynodont from South Africa and the phylogeny of nonmammalian cynodonts. *Bulletin of the Museum of Comparative Zoology* 156(1): 5–35.

2002

Barbarena, M. C., J. F. Bonaparte, C. Schultz, R. Rupert, J. A. Hopson, and A. W. Crompton. 2002. Advanced non-mammalian cynodonts from the Late Triassic of Brazil: implications for the origin of mammals. *Eighth International Symposium on Mesozoic Terrestrial Ecosystems,* Cape Town, South Africa.

2003

Hopson, J. A. 2003. The transition from nonmammalian cynodonts to mammals. *Journal of Vertebrate Paleontology* 23(3, suppl.): 63A

2004

Sidor, C. A., J. A. Hopson, and A. W. Keyser. 2004. A new burnetiamorph therapsid from the Teekloof Formation, Permian, of South Africa. *Journal of Vertebrate Paleontology* 24(4): 938–950.

2005

Rich, T. H., J. A. Hopson, A. M. Musser, T. F. Flannery, and P. Vickers-Rich. 2005. Independent origins of middle ear bones in monotremes and therians. *Science* 307:910–914.

Rich, T. H., J. A. Hopson, A. M. Musser, T. F. Flannery, and P. Vickers-Rich. 2005. Response to comments on "Independent origins of middle ear bones in monotremes and therians." *Science* 309:1492c.

Contributors

SALLY W. ABOELELA, Department of Physiology and Pharmacology, Oregon Health & Science University, Portland, OR

RICHARD W. BLOB, Department of Biological Sciences, Long Hall, Clemson University, Clemson, SC

JOHN R. BOLT, Department of Geology, Field Museum of Natural History, Chicago, IL

MATTHEW T. CARRANO, Department of Paleobiology, Smithsonian Institution, Washington, D.C.

JAMES M. CLARK, Department of Biological Sciences, George Washington University, Washington, D.C.

MICHAEL A. CLUVER, South African Museum, Cape Town, South Africa

A. W. CROMPTON, Museum of Comparative Zoology, Harvard University, Cambridge, MA

CATHERINE A. FORSTER, Department of Anatomical Sciences, Stony Brook University, Stony Brook, NY

TIMOTHY J. GAUDIN, Department of Biological & Environmental Sciences, University of Tennessee at Chattanooga, Chattanooga, TN

JUSTIN GEORGI, Department of Anatomical Sciences, Stony Brook University, Stony Brook, NY

FREDERICK A. GRINE, Department of Anthropology, Stony Brook University, Stony Brook, NY

FARISH A. JENKINS JR., Museum of Comparative Zoology, Harvard University, Cambridge, MA

DANIEL E. LIEBERMAN, Department of Biological Anthropology, Harvard University, Cambridge, MA

R. ERIC LOMBARD, Department of Organismal Biology and Anatomy, University of Chicago, Chicago, IL

REBECCA C. MUNTER, Department of Biological Sciences, George Washington University, Washington, D.C.

F. ROBIN O'KEEFE, Department of Anatomy, New York College of Osteopathic Medicine, Old Westbury, NY

MICHAEL J. PARRISH, Department of Biological Sciences, Northern Illinois University, DeKalb, IL

GUILLERMO W. ROUGIER, Department of Anatomical Science & Neurobiology, University of Louisville, Louisville, KY

BRUCE S. RUBIDGE, Bernard Price Institute, University of the Witwatersand, Johannesburg, South Africa

PAUL C. SERENO, Department of Organismal Biology and Anatomy, University of Chicago, Chicago, IL

CHRISTIAN A. SIDOR, Department of Biology, University of Washington, Seattle, WA

HANS-DIETER SUES, National Museum of Natural History, Smithsonian Institution, Washington, D.C.

JOHN R. WIBLE, Section of Mammals, Carnegie Museum of Natural History, Pittsburgh, PA

SUBJECT INDEX

Page numbers in italics refer to figures and tables.

acetabulum, 136–38. *See also* foramina, acetabular

Africa, 411, 415

afrotherians, 321

alligators, 413. See also *Alligator* (Taxonomic Index)

allometry. *See* scaling

amniotes, 1–3, 8–9, 406. *See also* Amniota (Taxonomic Index)

amphibians, bone growth in, 412

amphilestids, 53, 271

analyses and tests, ancestor-descendant, 233, 237, 238–39; bivariate, 454–57, 461–74; cineradiographic, 316, 349; decay, 346; independent contrasts, 230–31; kinematic, 368, 384–85; least-squares regression, 377, *382;* minimum, 233, 236–37; modified minimum, 234, 239–40; partial correlation, 377–78, 381; patristic distance, 232, 235; phylogenetic, 157–98; 339, 345–47, 358; principal components (PCA), 457–59, 475–80, 487; reduced major axis regression (RMA), 377, *378, 380,* 417–19, 420–23, 429; Spearman-rank correlation, 232; squared-change parsimony, 231–32; subclade, 233, 237

Anisian, 415

anomodonts, 98, 99, 433, 465. *See also* Anomodontia; Dicynodontia (all in Taxonomic Index)

Antarctica, 432–33

anthracosaurs, 2, 21

archosaurs, 202–3, 207, 208. *See also* Archosauria (Taxonomic Index)

Argentina, 114

Arizona, 116

armadillos, 5, 153; euphractine, 154, 156, 174; eutatine, 154, 156, 174, 175; phylogeny, 157–98, *160, 169. See also* Dasypodidae; Euphracta; *Peltephilus; Stegotherium* (all in Taxonomic Index)

arteries: cerebral carotid, 393; diploëtica magna, 293, 296, 297; external carotid, 293; internal carotid, 293, 294, 295, 393, 399, 404; occipital, 160, 170, 195; palatine, 393; ramus inferior, 276, 285, 295; ramus infraorbitalis, 295; ramus mandibularis, 295; ramus superior, 295, 297; ramus supraorbitalis, 295; ramus temporalis, 191; stapedial, 293, 295, 296; vertebral, 118, 293

artiodactyls, 371, 374, 376–85. *See also* Artiodactyla (Taxonomic Index)

asioryctitheres, 286. *See also* Asioryctitheridae (Taxonomic Index)

Australia, 433

australosphenidans, 279. *See also* Australosphenida (Taxonomic Index)

baboons, 385

Bajocian, 71

baphetids, 51. See also *Megalocephalus* (Taxonomic Index)

bears, 144

Beaufort Group, 76, 77, 94, 95, 96, 97, 432

biarmosuchians, 3–4, 76–79, 99, 100, 102, 104. *See also* Biarmosuchia (Taxonomic Index)

bipedalism, bipedality, 203, 210, 216–17

birds, 318, *414,* 424; limb bone growth in, 413, 418–20, 424; reproduction, 250. *See also* Aves (Taxonomic Index); gulls; Neornithes (Taxonomic Index)

TAXONOMIC INDEX

Page numbers in italics refer to figures and tables.

Printed and bound by CPI Group (UK) Ltd, Croydon, CR0 4YY

27/10/2024

14580397-0005